Handbook of
Chromatography

Handbook of Chromatography

Edited by
Carol Evans

WILLFORD PRESS

www.willfordpress.com

Published by Willford Press,
118-35 Queens Blvd., Suite 400,
Forest Hills, NY 11375, USA

ISBN: 978-1-68285-574-4

Cataloging-in-Publication Data

Handbook of chromatography / edited by Carol Evans.
    p. cm.
Includes bibliographical references and index.
ISBN 978-1-68285-574-4
1. Chromatographic analysis. 2. Chemistry, Analytic. I. Evans, Carol.
QD79.C4 H36 2019
543.8--dc21

For information on all Willford Press publications
visit our website at www.willfordpress.com

WILLFORD PRESS

Contents

Preface

In my initial years as a student, I used to run to the library at every possible instance to grab a book and learn something new. Books were my primary source of knowledge and I would not have come such a long way without all that I learnt from them. Thus, when I was approached to edit this book; I became understandably nostalgic. It was an absolute honor to be considered worthy of guiding the current generation as well as those to come. I put all my knowledge and hard work into making this book most beneficial for its readers.

This book is compiled in such a manner, that it will provide in-depth knowledge about the theory and practice of chromatography used to separate mixtures into their components. Chromatography has two branches- preparative chromatography and analytical chromatography. This book offers extensive information about the key theories and concepts related to chromatography. It is a valuable compilation of topics, ranging from the basic to the most complex advancements in this field. The various sub-fields of chromatography along with technological progress that have future implications are also glanced at in this book. It will prove to be immensely beneficial to students and researchers in this field. It is a complete source of knowledge on the present status of this important field.

I wish to thank my publisher for supporting me at every step. I would also like to thank all the authors who have contributed their researches in this book. I hope this book will be a valuable contribution to the progress of the field.

Editor

Active Ingredient Estimation of Clopyralid Formulation by Reversed Phase HPLC

Singh A, Tandon S* and Sand NK

Department of Chemistry (Division of Agricultural Chemicals), College of Basic Sciences and Humanities,
G. B. Pant University of Agriculture & Technology, Pantnagar 263145, U.S. Nagar, Uttarakhand, India

Abstract

A simple, accurate, and reliable high-performance liquid chromatography (HPLC) method was developed for determination, separation and estimation of clopyralid active ingredient in its formulation. The active ingredient in formulation was estimated by two methods and cleanup was performed using C18 SPE. Estimation was done by HPLC using 5 µm, C18 column, and mobile phase 0.02 % acetic acid in methanol: acetonitrile (90:10 v/v) and detection at 229 nm. The efficiency of clean up for standard clopyralid was found to be 98.02-98.66% at different concentrations. Overall recoveries for different methods ranged from 98.3-100.1 %. The LOD and LOQ were 0.1 and 0.3 ng respectively. The detector response was linear within concentrations range 0.25-10.0 ng at RSD 1.87%.

Keywords: Clopyralid; Soluble liquid; Solid phase extraction; Formulation; HPLC

Introduction

Pesticides are indispensable for sustained agricultural production. The successful employment of any pesticide depends upon stabilities of its formulation. Every formulation contains a toxicants mixed with an inert diluents or carrier. The active ingredient in a formulation sometime does not contain the reported concentration for various reasons. Indian farmers spend about US $2.6 billion on unregistered or counterfeit chemicals. The untested and unregulated products in turn cause about $1.3 billion in crop damage, according to a new report by the Agrochemicals Policy Group (APG) [1]. In order to check passage of spurious/sub-standard formulation to the consumer to save him from avoidable losses from pests, it is necessary to develop a convenient and reliable method of estimation of active ingredient in a formulation.

Clopyralid (3, 6-dichloropicolinic acid) is a selective, auxins type herbicide of pyridine carboxylic acid group used to control broad leaf weeds of the family polygonaceae, compositae, leguminosae and umbelliferae in sugar beet, fodder beet, oilseed rape, brassicas, onions, strawberries, flax and grassland (lawn and turf also) and commercially available as Soluble liquid (SL), Emulsifiable concentrate (EC) and Water Dispersible Granules (WDG) formulations [2]. About 8,91,662 lbs active ingredient of clopyralid was used in the US for controlling weeds in various crops [3] .

Persistence/dissipation studies of clopyralid in soil/water/crops has been reported by various workers [4-9]. As per our literature search, no method has been reported for the estimation of clopyralid in its formulation using SPE method and determination by RP-HPLC. The registration of clopyralid is proposed for use is under consideration by Central Insecticide Board, India. Thus, the present investigation was undertaken to optimized method that would be rapid, efficient, precise, accurate, specific, economic and green method for estimation of active ingredient (a.i.) in formulation (10 SL) by reversed phase HPLC.

Experimental

Chemicals and glasswares

The formulation clopyralid, 10 SL and technical (95% pure) were obtained from M/S Willowood Ltd, Hong Kong. The technical sample was recrystallized prior to use. All the glasswares used were of Corning or Borosil make. All the solvents used were of analytical (AR) or HPLC grade. Triple distilled water was prepared in the laboratory by double distillation of single metal distilled water in all quartz double distillation assembly.

Instruments

Beckman model 322 Gradient HPLC systems equipped with 100 A pump 420 gradient microprocessor controller, 7725i rheodyne injector, 160 selectable wavelengths UV detector, 5 µL loop and HP 3395 series integrator recorder. Systronics P.C. Controlled Double Beam (PMT detector) UV-Visible Spectrophotometer Model 2101. Buchii rotavapour, Analytichem International SPE VAC ELUT Model AI 6000, SPE miniplastic C-18 columns (6.0 × 0.75 cm id) packed with 500 mg C-18 packing material (Varian/Supelco, USA) were used for estimation.

Calibration curve

Different concentrations i.e. 0.05, 0.1, 0.2, 0.5 and 1.0 ppm concentration of clopyralid were used for preparation of calibration curve. Five µL of each concentration in duplicate was injected in HPLC by full loop injection method. The average detector response in terms of peak area of each concentration was used for plotting the graph.

HPLC optimization

The Beckman Gradient HPLC system was used for estimation. The operating parameters were Discovery C-18, 5 µm, (150 × 4.6 mm) column, mobile phase 0.02 % acetic acid in 9:1 v/v methanol:acetonitrile with isocratic mode at a flow rate of 1 mL/min and UV detection at 229 nm, aufs 0.02. Five microliter volume of sample was injected each

*Corresponding author: Tandon S, Department of Chemistry (Division of Agricultural Chemicals), College of Basic Sciences and Humanities, G. B. Pant University of Agriculture & Technology, Pantnagar 263145, U.S. Nagar, Uttarakhand, India, E-mail: shishir_tandon2000@yahoo.co.in

time in HPLC by full loop injection method and chromatograms were recorded by HP (model 3395) integrator.

Formulation analysis

Liquid-liquid extraction (Indirect method): Formulation of clopyralid (0.1 mL) was mixed with water (50 mL) and liquid-liquid extraction was repeated three times with ethyl acetate (30 + 20 + 10 mL). Organic layer was collected, dried and dissolved in 2 mL mobile phase and one part was used for analysis by HPLC while other was subjected to cleanup.

Clean-up of sample was done using SPE C-18 cartridge. The SPE C-18 cartridge was prewashed and conditioned with 3 mL each of acetonitrile and methanol. One mL of sample was loaded on cartridge and washed with 2 mL water followed by 1 mL methanol and washings was discarded then finally eluted with 2 mL methanol. Eluent was collected, dried under the stream of nitrogen and re-dissolved in 2 mL mobile phase and used for HPLC analysis. The samples were filtered through 0.22 μm PTFE disc filter prior to HPLC.

Solid phase extraction (Direct methods): For formulation analysis two methods were used. First method involved preparation of 100 ppm stock solution by dissolving 0.1 mL of formulation (clopyralid 10% SL) in 100 mL of distilled water. From this solution serial dilution of 10, 5, 2, 1, and 0.5 ppm were made. In another method methanol was used instead of water. Lower concentrations (2, 1 and 0.5 ppm) were taken for cleanup and analysis. The concentration of formulation reported by the company i.e. 10% was assumed to be correct. The percent a.i. obtained from formulation was compared with pure standard compound of clopyralid and percentage of a.i. in formulation was calculated. All the experiment was done in triplicate. Since pure formulation contains lot of impurities which may cause damage to the column hence, diluted formulation was not injected and cleanup was done prior to analysis.

Cleanup: Clean-up of formulation is necessary before analysis as it may have non-polar impurities which may interfere with the object peak of interest. Clean-up of formulation of both methods were done using SPE C-18 cartridges. The SPE C-18 cartridge was prewashed and conditioned with 3 mL each of acetonitrile and methanol. Diluted solution (2, 1 and 0.5 ppm) of formulation was loaded (0.5 mL) on cartridge and washed with 2 mL water followed by 1 mL methanol and washings was discarded then finally eluted with 2 mL methanol. Eluent was collected, dried under the stream of nitrogen and re-dissolved in 2 mL mobile phase and used for HPLC analysis. The samples were filtered through 0.22 μm PTFE disc filter prior to HPLC.

Recovery: For recovery of clopyralid from its formulation between two batches of 1ml each were taken in graduated stopper test tubes. In one batch 1 ml of 1 ppm standard solution in methanol was added while in other only methanol was added which was treated as control. Extraction and cleanup procedure were followed as described above. Difference between peak areas of both batches was taken for recovery calculation. All samples were taken in triplicate.

Validation: The method was validated by evaluating it in terms of specificity, linearity, precision, accuracy, Limit of Detection (LOD) and Limit of Quantification (LOQ).

Results and Discussion

The chromatograms obtained using mobile phase, 0.02 % acetic acid in methanol: acetonitrile (90:10 v/v) at a flow rate of 1.0 mL min⁻¹, showed retention time 9.8 minute for clopyralid. The Indirect method of formulation analysis which was without cleanup showed many interfering peaks at retention time of clopyralid which were removed after cleanup but overall recovery was poor (around 80%) (Figure 1). The efficiency (recovery) of cleanup by SPE for standard clopyralid was 98.02-98.66% at different concentrations Table 1. Hence direct method was adapted using methanol and water as diluting solvent. The recoveries using water and methanol as diluents were almost same and ranged from 98.3-100.1%. The chromatogram revealed that there was no interference or co elution of any peak at the time of clopyralid elution peak. Among all methods for the analysis of active ingredient in formulation the method using water seems best in terms of recovery and cost effectiveness. The average active ingredient in the formulation using methnol/water as diluting agent (direct method) was found to be about 9.91%.

Method validation

Specificity: Specificity of the assay was demonstrated by obtaining chromatograms for blank and observing the lack of interfering peaks at the retention time for the clopyralid. Since there was no interfering matrix peak showing good specificity of the method

Linearity: Linearity of the HPLC assay was evaluated in triplicate at five concentration levels consisting of 0.05, 0.1, 0.2, 0.5 and 1.0 μg mL-1 clopyralid. Average detector response in terms of integrator counts was used for preparation of calibration curve. Detector response was linear to the concentration as the value of coefficient of determination (R^2) 0.9995.

Precision and accuracy: The intra-day and inter-day precision of the HPLC analytical method were evaluated at different concentration and estimated variation was between 4 percent. The accuracy of an analytical method is defined as the closeness of measured values to their known nominal values. Accuracy of the method was deemed acceptable if the value of percent RSD value is lower than 2. In our study the value percent RSD was 1.87 which is acceptable.

LOD and LOQ: LOD and LOQ were the concentrations of a compound at which its signal-to-noise ratios (S/N) were detected as 3:1 and 10:1 respectively. They were obtained by diluting the reference standard in a stepwise manner using the established LC conditions. The LOD and LOQ of clopyralid were 0.1 and 0.3 ng respectively.

Conclusion

The proposed method was developed and validated and shows method is simple, rapid, accurate, precise, sensitive, and eco-friendly. The detection of clopyralid by the HPLC method can be performed rapidly as there is noteworthy less retention time with less or no interference of matrix. Therefore, the method would be useful for both

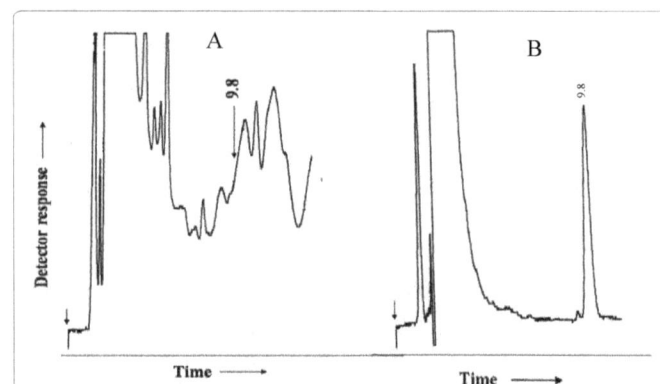

Figure 1: Chromatogram of formulation before cleanup (A) and after cleanup (B)

Method used	Concentration (ppm)	SPE Efficiency	% Recovery	Average Recovery (%)	% a.i. in formulation	Average a.i. in formulation
Standard pure compound	2	98.66 ± 0.03	-	-	-	-
	1	98.30 ± 0.2				
	0.5	98.02 ± 0.09				
Ethyl acetate*	2	-	86.37 ± 2.13	80.04	9.67 ± 1.43	9.353 ± 1.43
	1		79.24 ± 4.33		9.36 ± 1.05	
	0.5		74.53 ± 4.56		9.03 ± 0.91	
Methanol#	2	-	96.98 ± 0.21	94.03	10.01 ± 0.03	9.966 ± 0.072
	1		94.10 ± 0.01		9.98 ± 0.15	
	0.5		91.00 ± 0.12		9.83 ± 0.19	
Water#	2	-	96.56 ± 0.34	93.05	9.95 ± 0.2	9.883 ± 0.094
	1		93.51 ± 0.30		9.87 ± 0.31	
	0.5		89.08 ± 0.46		9.82 ± 0.22	

* Indirect method (LLE)

Direct method (SPE)

Table 1: Recovery of clopyralid from standard and different methods

qualitative and quantitative analysis of clopyralid

Application of the proposed method

Green, rapid and sensitive method developed may be used for determination of active ingredient in the formulation of clopyralid for the purpose of quality control of the commercial product and routine analysis by Agrochemical and research laboratories.

Acknowledgement

The authors are thankful to M/s Willowood Ltd., Honk Kong for providing the technical material of clopyralid and its formulation 10 SL for the research purpose.

References

1. Farm Chemical International (2010) Indian Farmers Face Counterfeit Pesticides.

2. Tomlin CDS (2000) Pesticide manual, a world compendium 12th (Edn). British crop protection council, Berkshire, UK 1276.

3. Gianessi LP, Marcelli MB (2000) Pesticide use in U.S. crop production: 1997. National Summary Report, National Center for Food and Agricultural Policy (NCFAP), Washington, DC 51

4. Galoux MP, Bernes AC, Van Damme JC (1985) Gas chromatographic determination of 3,6-dichloropicolinic acid residues in soils and its application to the residue dissipation in a soil. J Agric Food Chem 33: 965-968.

5. Lauren DR, Taylor HJ, Rahman A (1988) Analysis of the herbicides dicamba, clopyralid and bromacil in asparagus by high-performance liquid chromatography. J Chromatog 439: 470-475.

6. Akerblom M, Thoren L, Staffas A (1990) Estimation of pesticides in drinking water. Var Foda 42: 236-243.

7. Schaner A, Konecny J, Luckey L, Hickes H (2007) Determination of chlorinated acid herbicides in vegetation and soil by liquid chromatography /electrospray-tandem mass spectrometry. J AOAC Int 90:1402-1410

8. Ahmad R, Rahman A, Holland PT, McNaughton DE (2003) Improved analytical procedure for determination of clopyralid in soil using gas chromatography. Bull Environ Contam Toxicol 71: 414–421.

9. Singh A, Tandon S, Sand, NK (2009) HPLC determination of herbicide clopyralid in soil and water. Pestic Res J 21: 187-190.

2

Determining Odor-Active Compounds in a Commercial Sample of *Cinnamomum cassia* Essential Oil Using GC-MS and GC-O

Valentina Bongiovanni[1]*, **Maria Laura Colombo[2]**, **Andrea Cavallero and Daniela Talarico[1]***

[1]*Kerry Ingredients and Flavors, Via Capitani di Mozzo 12/16, 24030, Mozzo, BG, Italy*
[2]*Department of Science and Pharmaceutical Technology, University of Turin, Via P. Jury 9, 10125 Turin, Italy*

Abstract

The volatiles of a commercial sample of *Cinnamomum cassia* (Nees and T. Nees) J. Presl. essential oil were analyzed by gas chromatography-mass spectrometry (GC-MS). The identification of the components was confirmed by Kovats retention index and their quantities were established using internal standard. These analyses had led to the identification of 72 chemicals and quantification of 41 of them. The majority of volatiles identified belongs to oxygenated compounds (e.g., aldehydes) while non oxygenated terpenes represent about 18% of the oil. The odor quality of cassia essential oil was assessed by Gas Chromatography-Olfactometry (GC-O). Among the 26 components identified with GC-O, AEDA (Aroma Extract Dilution Analysis) has allowed to establish a number of components with high dilution factor (strongly odorous) such as cinnamaldehyde, 3-phenylpropanal, guaiacol and 2-phenylethanol.

Keywords: *Cinnamon cassia*; GC-O; AEDA; Essential oil

Abbreviations: GC: Gas Chromatography; MS: Mass spectrometry; O: Olfactometry; AEDA: Aroma Extract Dilution Analysis; EO: Essential Oil; F.I.D./FID: Flame Ionization Detector; FD: Flavor Dilution (factor); VCF: Volatile Compounds in Food; FEMA: Flavor and Extract Manufacturers Association (ingredients generally recognized as safe under conditions of intended use as flavors from the FEMA expert panel are listed); RI: Retention Index.

Introduction

Cinnamomum cassia (Nees and T. Nees) J. Presl, Lauracee family, is one of the oldest known spices. The volatile oil from leaf and bark and the oleoresin from bark are used in soaps, perfumes, spice essences, food and beverages. In traditional medicine it is widely used to treat dyspepsia, gastritis, blood circulation disturbance and inflammatory diseases [1]. The essential oil (EO) obtained from cassia is widely used in cosmetics and foods especially for its antioxidant, antifungal and antibacterial properties [2,3]. Beside its use in fragrances, cassia is used as a flavoring agent in foods for its spicy, sweet and warm aromatic notes both in savory (ham and meat) and sweet food (e.g., beverages such as cola), although with some limits provided by law (Regulation EC N°1334/2008).

The need of analytical methods for quality assessment is directly related to the economic importance of cassia and his derivatives as raw materials in the food industry. This point is crucial if cassia end-use is to make flavors.

In this study a commercial-grade cassia essential oil was analyzed with GC-MS and GC-O was used to investigate the olfactory profile [4]. In literature, several sample of different origin and grade of Cassia EO, from laboratory scale to commercial, have been screened with significant variability on reported results [5]. Moreover, quantitative analyses of constituents in cassia oil are not always available as GC percentages are more often reported.

Literature papers tend agree on cassia oil major component being cinnamic aldehyde, ranging from 50% to 93%, o-methoxy cinnamaldehyde (0.1-25.4%), cinnamyl acetate (0.3-7.6%), benzaldehyde (0.3-2.9%) [5,6]. In literature, the number of compounds identified with GC-olfactometric analysis in food product usually varies with the nature of the raw material.

Literature data [7] for spices such as black pepper shows that important odorants are about 14, mainly terpenes and oxygenated compounds. For some spices, the odor is well represented by that of the main component of the volatile fraction. An example is cinnamon (*Cinnamomum zeylanicum, C. aromaticum, C. burmanii*) whose odor is mainly characterized by the high content in cinnamic aldehyde [7].

Generally the concentration of cinnamic aldehyde determines the flavor quality of cassia too, low levels being known to represent material of low quality [8,9] for these reasons, other aromatic compounds in the oil have not been examined in depth so there is a lack of odor profile characterization of cassia oil in literature.

Despite the considerable amount of cinnamic aldehyde, cassia oil have a more complex odor profile with spicy, warm and sweet notes with a woody and earthy background.

Gas chromatography-olfactometry (GC-O) can help achieve this goal since GC-O allows to determinate the contribution of single constituents to the overall flavor of a product. GC-O enables the assessment of odor-active components in complex mixtures, through the specific correlation with the chromatographic peaks of interest; this is possible because two detectors, one of them being the human olfactory system, perceive the eluted substances simultaneously.

Different GC-O methods are available such as dilution, time-

*Corresponding author: Valentina Bongiovanni, Kerry Ingredients and Flavors, Via Capitani di Mozzo 12/16, 24030, Mozzo, BG, Italy
E-mail: valentina.bongiovanni@kerry.com

intensity, detection frequency and posterior intensity [10,11]. Dilutions methods are the most used methods and they are based on successive dilutions of an aroma extract. AEDA-Aroma Extraction Dilution Analysis-is one of the most used GC-O methods because it permits to identify the most important aroma compounds [12,13].

In AEDA the odorants are separated by gas chromatography on a capillary column (Figure 1). To determine the retention times of the aroma substances, the carrier gas stream, after leaving the capillary column, goes to sniffing port for detection by panelist (GC-O). The sensory assessment of a single GC run is not very meaningful because the perception of aroma substances in the carrier gas stream depends on limited quantities, e.g., the degree of concentration of the volatile fraction, and the amount of sample separated by gas chromatography [7]. These limitations are overcame by the stepwise dilution of the volatile fraction with solvent (Figure 2), followed by the gas chromatographic/olfactometric analysis of each dilution. The dilution process is repeated until no more aroma substance can be detected by the panelist.

Our GC-O equipment splits the flux into sniffing port (so human nose as detector) and analytical detector F.I.D. In order to get a precise identification of the peaks, a GC-MS analysis of the EO was run as well. GC-MS has been proven to be a powerful and suitable tool for the determination of volatile compounds because of its high separation efficiency and sensitive detection [14]. In order to avoid errors due to peaks overlapping, an equipment with double column was used.

Materials and Methods

Essential oil and reagents

A commercial grade cassia (*Cinnamomum cassia* (Nees and T. Nees) J. Presl.) essential oil obtained by steam distillation from cassia bark, leaves and twigs was used for all tests. Ethyl alcohol (96°, food grade) was used in the experiments.

Analysis of volatile composition

The identification of volatile aroma compound was performed on a gas-chromatograph Thermo Trace coupled with a Thermo ISQ mass-spectrometer. The GC-MS system was equipped with a DB1 (30 m, 0.25 mm i.d., 0.25 μm film thickness) capillary column (Agilent JandW). The starting temperature of the column was 50°C, which was held for 3 min, then increased 5°C/min to 280°C, where was held for 10 min. The constant column flow was 1,5 ml/min, using helium as carrier gas, the injector was in split mode at 280°C. Mass spectrometer parameters were as follow: ionization mode EI at 70 eV, source temperature 250°C, scan range m/z 33 to 350, scan mass 1 s. The components were identified by comparing their spectra with those present in the Wiley and NIST spectra collection and in an authentic chemicals spectra library.

The mass spectra identifications were confirmed using Kovats Indices (n-alkanes) on DB1 and DB1701 columns (Figure 3). To calculate the Kovats Indices a GC analysis was performed using an Agilent 5890 gas-chromatograph equipped with an auto sampler. The two capillary columns (Agilent JandW) DB1 and DB1701 (30 m, 0.25 mm i.d., 0.25 μm film thickness) were assembled on the same injector. Helium (flow 1.5 ml/min) was used as carrier gas, the injector was in split mode (ratio 1:50) and two F.I.D. detector were used. The injector and detectors temperatures were 280°C and 285°C respectively. The column temperature was initially maintained at 50°C for 3 min before increasing to 280°C at a rate of 5°C/min and held for 10 min.

The Kovats Indices were compared to those of authentic chemicals elute in the same conditions. For quantitative analysis, the response factor relative to the internal standard (benzyl benzoate) was previously

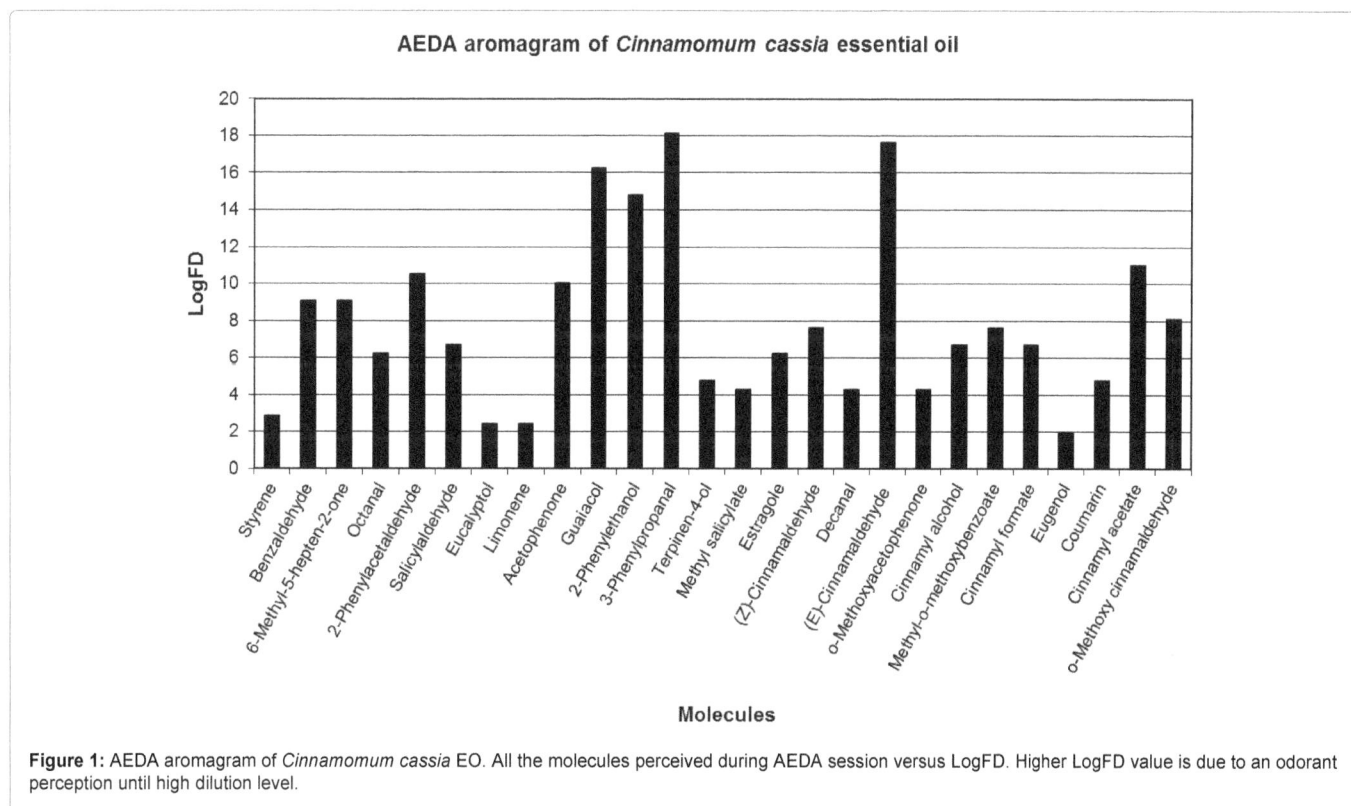

Figure 1: AEDA aromagram of *Cinnamomum cassia* EO. All the molecules perceived during AEDA session versus LogFD. Higher LogFD value is due to an odorant perception until high dilution level.

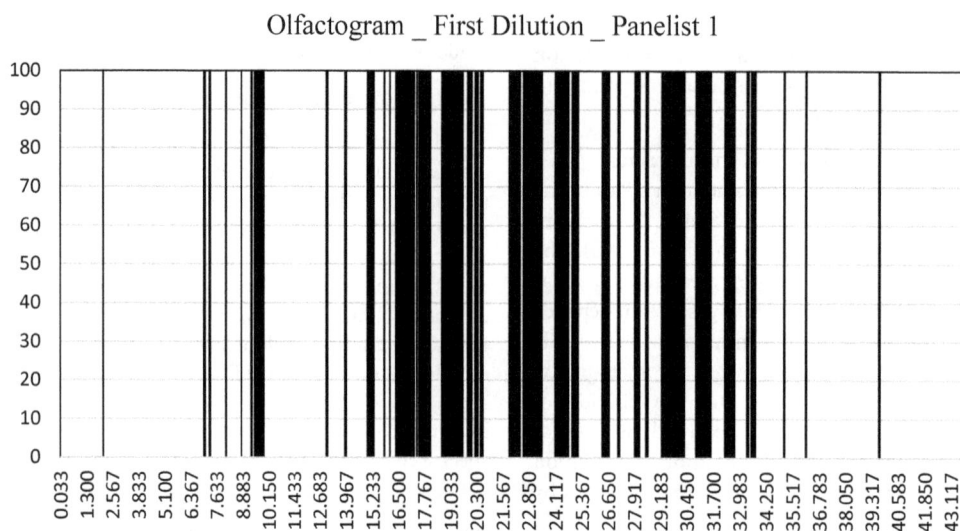

Figure 2: Olfactogram of first dilution of *Cinnamon cassia* obtained by Panelist1: retention time (X Axis) versus signal from GC-O joystick. When the panelist push the joystick, the instrument records a 100 signal that lasts until the panelist release the joystick.

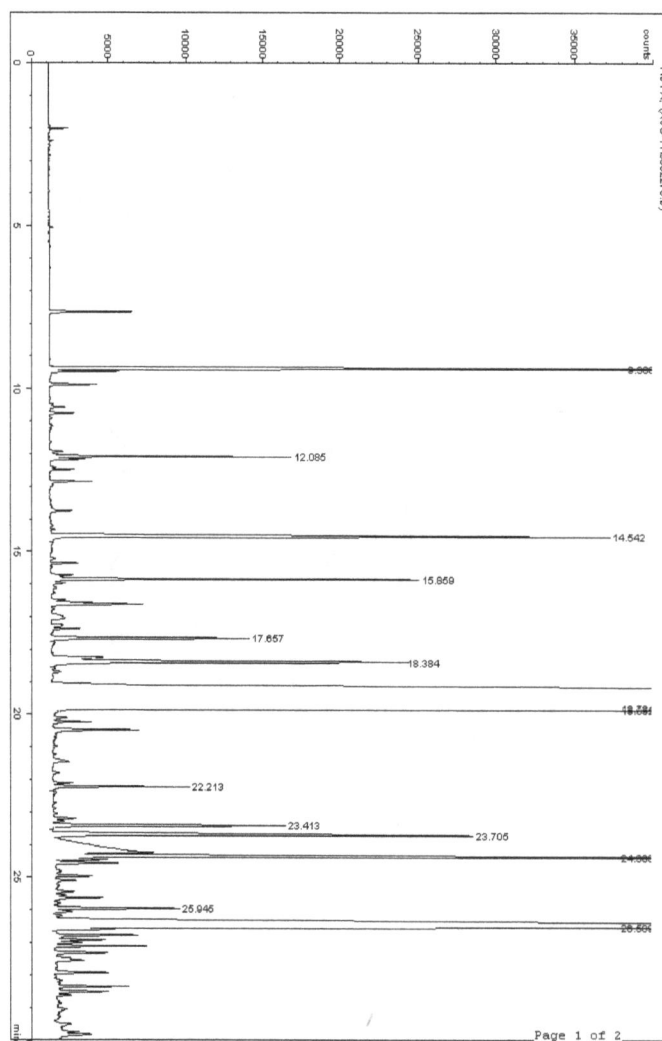

Figure 3: Chromatogram (FID) *Cinnamon cassia* essential oil (FID obtained by elution from column DB1).

determined and stored in an inland library for each molecule identified. The choice of benzyl benzoate as internal standard has been made after a preliminary evaluation confirmed negligible quantities of this component in our sample. The percentage of every component was then calculated using the response factors.

Response factor (K) is calculated (for each molecule, on both capillary column) using the following:

$$K=W(x) \times A(IS)/W(IS) \times A(x)$$

Where: W(x)=weight (g) of molecule "x"; W(IS)=weight (g) of internal standard (Benzyl benzoate); A (IS)=Peak area of internal standard; A (x)=peak area of molecule "x".

GC-O analysis

Cassia essential oil was analyzed with GC-O using an Agilent 5890 gas chromatograph equipped with DB-1 (30 m, 0.25 mm i.d., 0.25 μm film thickness) capillary column (Agilent J and W).

Chromatographic conditions were 50°C held for 3 min, 5°C/min to 220°C, 15°C/min to 280°C held for 10 min. The first temperature ramp was the same used for the determination of Kovats indices. Helium was used as carrier gas (flux 1.5 ml/min). The injector was in splitless mode, column flow was divided (70:30) by a splitter between the sniffing port and a flame ionization detector (F.I.D.).

A tailor-made software (Simplicius® from ABREG) was used to allow the recording of the odor intensity from the mechanical input of the panelists and to store the recorded signals that were further transformed into FD values or Log(FD).

GC-O analysis was performed using aroma extraction dilution analysis (AEDA) method. In order to find the most important odor-active compounds, which have the lowest odor thresholds, the amount of sample was reduced by ten dilution (w/v) of essential oil in ethyl alcohol (96°, food grade) factor dilution was 3. The sniffing started from the third dilution (1:27) due to peak overlapping (e.g., cinnamic aldehyde) and to overall too strong smell intensity of main eluting peak. According to AEDA method, samples were evaluated by the panelists in increasing dilution order and the impact of an odor-active component was measured as the last dilution value (FD).

The sniffing of cassia essential oil was carried-out by four trained assessors (1 male and 3 female, age 25-50 years), not smokers and without anosmia [15].

Results

In total 72 components were identified through GC-MS analysis and confirmed by Kovats indices. Out of them 41 were quantified using internal standard and response factor while 31 were found in traces (<0.01%).

The majority of volatiles identified belongs to oxygenated compounds (e.g., aldehydes) while non oxygenated terpenes represent about 18% of the oil. In the analyzed cassia oil sample, cinnamic aldehyde is confirmed as the major component (about 78.4%), followed by o-methoxy cinnamaldehyde (7.40%), cinnamyl acetate (1.70%) and benzaldehyde (1.13%) (Table 1).

Different publications on volatiles components of cassia oil have also been reviewed by VCF [16]; 69 components have been listed, with only 25% have been associated with a content range. In our analyses, we are able to quantify as much as 55% of the overall molecules including an high numbers of components usually found in traces such as eucalyptol (0.01%), 3- phenylpropanol (0.01%), 2,4-decadienal (0.01%), cinnamyl formate (0.01%).

Due to safety concerns related to coumarine presence in flavor and fragrance, it was important to detect its content in cassia oil. In our sample, coumarine content reached 0.88%, consistent with the previously reported 0-12.2% [5]. Phenylacetaldehyde was not found in quantitative data in literature but in our sample it was present in small quantity (0.01%); this compound has a relevant odor impression since it was been perceived for many dilution in GC-O analysis. The odor active components determined in our cassia oil sample were 26. The odorants were identified by comparing their mass spectra, retention index, Kovats index and odor description with those of authentic standards.

The data in Table 2 show that most intense odor active compounds were aldehydes: 3- phenylpropanal and cinnamic aldehyde were perceived until last dilution. Some compounds present in smaller quantity had stronger olfactory impression (e.g., guaiacol) than other compounds present in higher quantities (e.g., o-methoxy cinnamaldehyde, benzaldehyde). The odor quality was obtained by olfactory evaluation and description perceived during GC-O session. Among perceived odorants there were some compounds that have a typical spicy connotation: cinnamic aldehyde, methoxy cinnamic aldehyde, cinnamyl acetate, eugenol and cinnamyl alcohol.

Compound	FEMA	Kovats RI DB1a	Kovats RI DB1701[a]	Kovats RI DB1[b]	Kovats RI DB1701[b]	Quantitative data (%)
Ethanol	2419	500	554	513	550	0.02
Hexanal	2557	777	883	773	878	Traces
Styrene	3233	876	960	875	963	0.08
Benzaldehyde	2127	930	1085	927	1080	1.11
α-Pinene	2902	933	951	933	949	0.07
Camphene	2229	946	972	946	970	0.06c
Phenol	3223	964	1220	960	1218	Traces
6-Methyl-5-hepten-2-one	2707	966	1085	966	1084	0.03
β-Pinene	2903	972	1001	969	997	0.02
Benzofuran	-	974	-	-	-	Traces
(E,E)-2,4-Heptadienal	3164	977	1144	980	1144	Traces
2-Pentylfuran	3317	980	1034	980	1034	Traces
Octanal	2797	981	1090	981	1084	Traces
α-Phellandrene	2856	994	1033	994	1033	Traces

Benzyl alcohol	2137	1003	1207	1001	1202	0.02
2-Phenylacetaldehyde	2874	1004	1185	1004	1184	Traces
Salicylaldehyde	3004	1009	1171	1009	1176	0.41
p-Cymene	2356	1012	1075	1009	1072	0.03
Eucalyptol	2465	1019	1077	1016	1072	0.01
Limonene	2633	1022	1058	1020	1057	0.02
β-Ocimene	3539	1033	1085	1029	1085	Traces
Acetophenone	2009	1038	1200	1034	1203	0.05
m-Tolualdehyde	3068	1038	-	1038	-	Traces
γ-Terpinene	3559	1052	1085	1050	1083	Traces
Octanol	2800	1057	1183	1057	1184	Traces
Guaiacol	2532	1061	1236	1061	1231	0.04
Methyl benzoate	2683	1070	1204	1069	1204	Traces
α,p-Dimethylsty ene	3144	1074	1156	1073	1153	Traces
Terpinolene	3046	1078	1119	1079	1116	Traces
6-Methyl-3,5-heptadien-2-one	3363	1078	1238	1079	1238	Traces
2-Phenylethanol	2858	1084	1280	1084	1275	0.86
Fenchol	2480	1098	1227	1098	1227	Traces
Unknown	-	1104	-	-	-	0.01c
Unknown	-	1108	-	-	-	0.04c
Veratrole	3799	1110	1271	1111	1267	Traces
Camphor	2230	1115	1271	1115	1267	Traces
3-Phenylprop nal	2887	1125	1314	1125	1311	0.62
Pinocarvone	-	1138	-	1136	-	Traces
2-Phenylethyl formate	2864	1145	1302	1145	1302	Traces
Unknown	-	1148	-	-	-	0.06c
Borneol	2157	1149	1290	1147	1283	0.11
p-Methylacetophenone	2677	1151	1324	1155	1324	Traces
p-Mentha-1,8-dien-4-ol	-	1162	-	1155	-	Traces
Terpinen-4-ol	2248	1164	1271	1160	1273	0.02
Estragole	2411	1169	1302	1172	1291	0.03
Methyl salicylate	2745	1169	1302	1166	1302	0.06
α-Terpineol	3045	1173	1290	1170	1293	0.03
(Z)-Cinnamaldehyde	2286	1181	1377	1181	1369	0.31
Decanal	2362	1183	1291	1183	1291	Traces
3-Phenylpropanol	2885	1199	1396	1197	1395	0.05
Anisaldehyde	2670	1203	1415	1210	1427	0.86
Carvone	2249	1214	1373	1214	1377	0.02
(E)-Cinnamaldehyde	2286	1238	1462	1228	1463	78.20
o-Methoxyacetophenone	4163	1263	-	1263	-	Traces
Anethole	2086	1266	1396	1260	1393	0.05
Vitispirane	-	1264	-	1267	-	Traces
Cinnamic alcohol	2294	1275	1505	1270	1506	0.15
(E,E)-2,4-Decadienal	3135	1291	1428	1291	1430	0.01
Methyl o-methoxybenzoate	2717	1307	1502	1306	1502	Traces
3-Phenylpropionic acid	2889	1307	1581	1312	1581	0.11
Cinnamyl formate	2299	1319	1495	1319	1499	0.01
Eugenol	2467	1331	1512	1326	1506	0.11
Unknown	-	1335	-	-	-	0.22c
Methyl cinnamate	2698	1352	1514	1348	1519	Traces
α-Copaene	-	1377	1405	1377	1404	0.33c
Coumarin	2381	1388	1692	1388	1686	0.88
β-Elemene	-	1392	1435	1385	1436	Traces
Cinnamic acid	2288	1405	1702	1405	1697	0.74
Cinnamyl acetate	2293	1411	1593	1405	1586	1.70

α-Bergamotene		1434	1463	1434	1461	Traces
α-Curcumene		1465	1532	1464	1530	Traces
o-Methoxycinnamaldehyde	3181	1487	1756	1492	1752	7.40

[a]Observed Kovats RI; [b]Authentic chemicals analysis Kovats RI; [c]% of FID area calculated without using response factor

Table 1: Identifications and quantitative data of commercial sample of Cinnamon Cassia essential oil.

Molecule	Odor description	Mean dilution	Log FD
Styrene	Resinous, balsamic, pungent, plastic, ethereal	1.75	2.86
Benzaldehyde	Almonds, sweet cherry	5	9.07
6-Methyl-5-hepten-2-one	Oily, green, herbaceous	5	9.07
Octanal	Oily, fruity, sweet, citrus	3.5	6.20
2-Phenylacetaldehyde	Floral, hyacinth, rose, honey	5.75	10.50
Salicylaldehyde	Pungent, spicy, phenolic, almond, medicinal	3.75	6.68
Eucalyptol	Strong camphor, fresh	1.5	2.39
Limonene	Fresh, sweet, orange	1.5	2.39
Acetophenone	Sweet, pungent, cherry, sour	5.5	10.02
Guaiacol	Strong, sweet, smoky, vanilla, phenolic, medicinal	8.75	16.22
2-Phenylethanol	Flowers, rose	8	14.79
3-Phenylpropanal	Green, floral, hyacinth, balsamic, melon	9.75	18.13
Terpinen-4-ol	Fresh, green, lemon, fresh, spicy and musky with woody notes	2.75	4.77
Methyl salicylate	Warm, sweet, evergreen	2.5	4.29
Estragole	Sweet, herbaceous, anise-fennel, basil, licorice	3.5	6.20
(Z)-Cinnamaldehyde	Spicy, cinnamon, cassia, sweet, warm	4.25	7.63
Decanal	Strong, pungent, sweet, waxy, orange peel	2.5	4.29
(E)-Cinnamaldehyde	Spicy, cinnamon, cassia, sweet, warm	9.5	17.65
o-Methoxyacetophenone	Sweet, phenolic, evergreen, chemistry, medicine, musky, earthy	2.5	4.29
Cinnamyl alcohol	Balsamic, sweet with floral notes and spicy	3.75	6.68
Methyl-o-methoxybenzoate	Herbaceous, musk, anis, fruity, winey	4.25	7.63
Cinnamyl formate	Balsamic, herbaceous, fruity, apple, slightly bitter	3.75	6.68
Eugenol	Strong, spicy, dry, pungent, smoky, clove	1.25	1.91
Coumarin	Sweet, dried fruit, hay, tobacco	2.75	4.77
Cinnamyl acetate	Sweet, balsamic, floral, spicy	6	10.97
o-Methoxy cinnamaldehyde	Sweet, warm, spicy, cassia	4.5	8.11

Table 2: AEDA results. Mean dilution indicates the last dilution until the molecule had been identified. Odor description indicates the odor perception of panelist.

The identification of a large number of odorants, some of which found in trace and others such as cinnamaldehyde and methoxy cinnamaldehyde that made up to approximately 90% of the oil, confirm AEDA as a powerful technique to analyze the olfactory profile of complex mixtures [17-19].

Acknowledgements

The author would thank the entire panel group that took part to the sniffing sessions. The author would thank Kerry Ingredients and Flavors that supported the study with materials and experts.

References

1. Geng S, Cui Z, Huang X, Chen Y, Xu D, et al. (2011) Variations in essential oil yield and composition during Cinnamomum cassia bark growth. Industrial Crops and Products, p: 33.

2. Li YQ (2013) Analysis and evaluation of essential oil components of cinnamon barks using GC-MS and FTIR spectroscopy. Industrial Crops and Products, p: 41.

3. Wang R (2009) Extraction of essential oils from five cinnamon leaves and identification of their volatile compound compositions. Innovative Food Science and Emerging Technologies, p: 10.

4. Biniecka M, Caroli S (2011) Analytical methods for the quantification of volatile aromatic compounds. Trends in Analytical Chemistry 30: 11.

5. Lawrence L, Brian M (2012) Progress in essential oil. Perfumer & Flavorist, p: 37.

6. ESO 00 Database Report (1999).

7. Belitz HD, Grosch W, Schieberle P (2009) Food chemistry. Food science and nutrition.

8. Poole SK, Colin FP (1994) Thin layer chromatographic method for the determination of the principal polar aromatic flavor compounds of the cinnamons of commerce. Analyst 119: 113-120.

9. Miller KG, Colin FP, Tina MPC (1995) Solvent assisted supercritical fluid extraction for the isolation of semivolatile flavor compounds from the cinnamons of commerce and their separation by series-coupled column gas chromatographic. J High Resolut Chromatogr 18: 461-471.

10. D'Acampora Zellner B (2008) Gas chromatography-olfactometry in food flavor analysis. J Chromatogr A 1186: 123-143.

11. Van Ruth SM (2001) Methods for gas chromatography-olfactometry: a review. Biomolecular Engineering 17: 121-128.

12. Hempfling K, Fastowski O, Kopp M, Nikfardjan MP, Engel KH (2013) Analysis and sensory evaluation of Gooseberry (Ribes uva crispa L.) volatiles. Journal of Agricultural and Food Chemistry 61: 6240-6249.

13. Buttery G, Ronald R, Takeoka G (2013) Cooked carrot volatiles. AEDA and Odor activity comparisons. Identification of Linden Ether as an important aroma component. Journal of Agricultural and Food Chemistry 61: 9063-9066.

14. Kopka K (2006) Gas chromatography-mass spectrometry. Biotechnology in Agriculture and Forestry 57: 3-20.

15. Ferreira V, Petka J, Aznar M, Cacho J (2003) Quantitative gas chromatography-olfactometry. Analytical characteristics of a panel of judges using a simple quantitative scale as gas chromatography detector. Journal of Chromatography A 1002: 169-178.

16. VCF (Volatile Compounds in Food).

17. Benzo M, Gilardoni G, Gandini C, Caccialanza G, Finizi PV, et al. (2007) Determination of the threshold odor concentration of main odorants in essential oils using gas chromatography-olfactometry incremental dilution technique. J Chromatogr A 1150: 131-135.

18. Cho IH (2008) Volatiles and key odorants in the pileus and stipe of pine-mushroom (Tricholoma matsutake Sing.). Food Chemistry, p: 106.

19. Lan-Phi NT, Tomoko S, Hiroyuki U, Masayoshi S (2009) Chemical and aroma profiles of yuzu (Citrus junos) peel oils of different cultivars. Food Chemistry, p: 115.

3

Comparative Study on Separation of Proteins from Whey Waste by Solvent Sublation in Batch and Continuous Mode

Samadrita Saha[1], Nasim Sepay[2], Pranjal Das[3], Sarmisttha Kundu[3], Milan Kumar Maiti[3] and Goutam Mukhopadhyay[3]*

[1]Calcutta Institute of Pharmaceutical Technology and Allied Health Science, Howrah, West Bengal, India
[2]Department of Chemistry, Presidency University, West Bengal, India
[3]BCDA College of Pharmacy and Technology, Kolkata, West Bengal, India

Abstract

The work is based on investigation on the separation of proteins from whey waste collected from local confectionery. Separation phenomenon involves solvent sublation technique both in batch and continuous mode. The objective of this work was to make a comparative study of afore mentioned batch and continuous method. The effect of pH, initial feed, concentration of feed, nitrogen gas flow rate, surfactant concentration, ionic strength, concentration of organic solvent and temperature were investigated in details. The amount of protein recovered in single continuous mode after 5 hr was 3644 mg, %Rp (90%), total volume (8100 ml), effective concentration (49.5 mcg/ml). The VFR 14 cc/min, %Rp was nearly 1.3 times more in continuous solvent sublation technique in comparison to the batch mode at optimum pH and GFR.

Keywords: Volumetric flow rate; Solvent sublation; pH; Whey waste

Introduction

In cheese industry and sweet industry whey is produced as a by-product [1]. Whey contains number of essential proteins in about 5-10%, these proteins are either added to dairy products or animal fodder or are discharged, resulting in a high waste water burden. Whey becomes a strong pollutant when discharged into steams, its high organic matter enhances biochemical oxygen demand (BOD), ranging from 30 to 40 gm of oxygen per liter. There are various methods of utilizing or disposing of whey. Thus a costs effective method for the enrichment or isolation of such protein fractions is of high economic and ecological interest. Current applied techniques are ultrafiltration, gel filtration, ion exchange or precipitation and coagulation [2]. An alternative but still less investigative method for the enrichment of whey protein is called solvent sublation, belonging to Adsorptive bubble separation method [3]. The advantage of solvent sublation over adsorptive bubble separation method is that higher removal efficiencies are positive. The sublation process does mixer and phase separations, which is needed in the solvent extraction. Furthermore, the effluent water from a sublation column removes residual solvent. Handling is simple and expenses are cheap. So, the solvent sublation has the potential in the environmental pollution treatment. The whey proteins differs widely in their functional properties (solubility, gelation properties, dispersibility, water holding capacity, stability, adhesion, emulsification properties, film formation, foaming properties, organoleptic properties, viscosity, binding properties etc.) [4]. The functionality of protein is basically dependent on the molecular structure. There are various factors come variability in functional properties of whey proteins are the source of whey, season-dependent variation of its components, different conditions like pH, ionic strength, heat treatment, and the presence of minerals [5]. Among the three major proteins in whey like alpha-lactalbumin (α-la), beta-lactoglobulin (β-lg), and bovine serum albumin (BSA), α-la has good emulsifying properties but its gelation ability is poor. β-lg has excellent gelling and foaming properties. Whey waste in batch foaming process gives a small amount of whey waste recovery at a time [6,7]. Solvent sublation method is easier to operate than foam fractionation since latter involves the product as mixture of protein and surfactant. A further study is required to separate the protein from protein surfactant complex. In solvent sublation collector sodium dodecyl sulphate (SDS) was not used, so protein is in pure form had been accumulate at the interface and finally protein was settle in the solvent chamber as white soft material. It was easy to collet it. It did not require evaporation of solvent to collect product. This solvent acts as blanket so protein did not dry up at the top aqueous layer. Through solvent was used repeatedly but it is costly. Efficiency may be enhanced in solvent sublation method by dispersing coalescence intermittently throughout the liquid column, by using finer sparger, so that effluent concentration is sufficiently low.

A little investigation has been done by solvent sublation method. Few researchers investigated solvent sublation of pure proteins (BSA, β-lg, α-la, latoferrin, lectoperoxidase, lusozyme) in synthetic feed to see the effect of pH, and the volumetric flow rate with solvent concentration. Wu et al. [8] investigated on solvent sublation of L-lusine and found that it can be used to separate the low concentration L-lysine from the aqueous solution by using dodecyl benzene sulforc acid as surfactant and di-phosphoric acid as extractant, n-heptane as the extractant solvent. He found the separation efficiency of solvent sublation was higher than that of traditional foam fractionation and solvent extraction method. Separation of penicillin G from fermentation broth, was investigated by Dong et al. The flotation product was quantitatively analysed by HPLC and compared with traditional solvent extraction. The effect of pH of the solution, NaCl concentration, concentration of butyle acetate (BA) in organic phase were investigated in details. Solvent extraction, flotation complexation extraction was proposed first time by Dong et al. [9] This new technique was used to separate and purify L-phenylalanine from fermentations liquid with good results. The flotation product from the fermentation liquid under optimal conditions, after back-extraction and re-crystallization was characterized by FTIR and HPLC, and its

*Corresponding author: Goutam Mukhopadhyay, BCDA College of Pharmacy and Technology, Kolkata, West Bengal-700 127, India
E-mail: gmukhopadhyay8@gmail.com

purity was more than 98%. Martinez-Araganet et al. [10] thoroughly studied about the different solvents which can be used in extraction of proteins by solvent sublation method, they found out that such solvents which breaks the disulfide bond in potent that should not be used a stability this of proteins as may the bond cause helps to lower retain tertiary structure and thereby prevent unfolding of the proteins, so chloroform and alcohols are not advised to use. They also proposed that any organic solvents which is less toxic having higher interfacial tension value and with a logP value greater than 4 can be suitable used as solvent for carrier mediated extraction proteins.

The aim of the present work was to make a comparative study on separation of protein as a whole by solvent sublation batch and continuous method. Both the batch and continuous method of solvent sublation was performed under the effect of several parameters like; pH, volumetric flow rate (VFR) etc. [11,12]. But its application is still continuing. Its feasibility at various operating conditions had been elucidated. The role of pH and volumetric flow rate on the performance characteristics were the prime objective here on the comparative study of both batch and continuous mode of solvent sublation in separation of protein from whey waste.

Materials and Methods

Whey waste was supplied by a local confectionary. Octanol (i-octanol) was obtained from Merck Ltd. India. Double distilled water was prepared at the laboratory. The instruments used were: UV- spectrophotometer (UV-1700 Shimadzu), pH meter (Satorious), Peristaltic pump with volumetric flow rate controller (RIVOTEK, India). Ultrasonic cleaner (Takashi), Digital weighing balance (Sartorious). Solvent sublation glass column with fritte (Remco Ltd, local glass fabricator).

Preparation of whey filtrate

Commercial whey was obtained from local sweet shop, it was then filtrated using a cheese cloth and filtrate was then poured in centrifuged tube and was subjected to centrifugation at 500 g for 30 min. The filtrate was collect through whatman filter paper-4 and again centrifuged in the same manner for 15 min and then the process was repeated till it gives constant OD value at λ_{max} of 280 nm. The filtrate was kept in a glass container inside the refrigerator to be used for subsequent experiments. The filtrate whey sample (40 ml) was taken in a Petridis. The Petridis was kept in BOD at the temperature of 50°C over night for drying. After drying a solid layer was found on Petridis, which was taken by spatula and weight was taken. From the difference of weight the amount of powder whey was found. It was found that every 40 ml of whey gives 1.25 gm of dried powder.

Solvent sublation (Batch process)

The apparatus (Figure 1) consists of a long column of 1 meter height that is fitted by frit at the bottom and an enlarged solvent chamber at the top. The solvent chamber is fitted with a reflux condenser. The column shows inlet, outlet for feed and effluent. The solvent chamber shows also inlet and outlet for solvent. The main glass apparatus is assembled with a nitrogen cylinder, rotameter (gas flow rate controller) Feed of desired concentration was prepared then the pH of the feed was adjusted as per requirement. The column was then filled with feed solution. The level of aqueous feed reaches 1 cm below the top of this column. The enlarged part of the solvent chamber was then covered with the glass cover. Required volume of Octanol was poured inside the chamber and a clear feed/octanol interface was visible. Nitrogen gas was passed through the feed at desired gas flow rate (GFR). On the

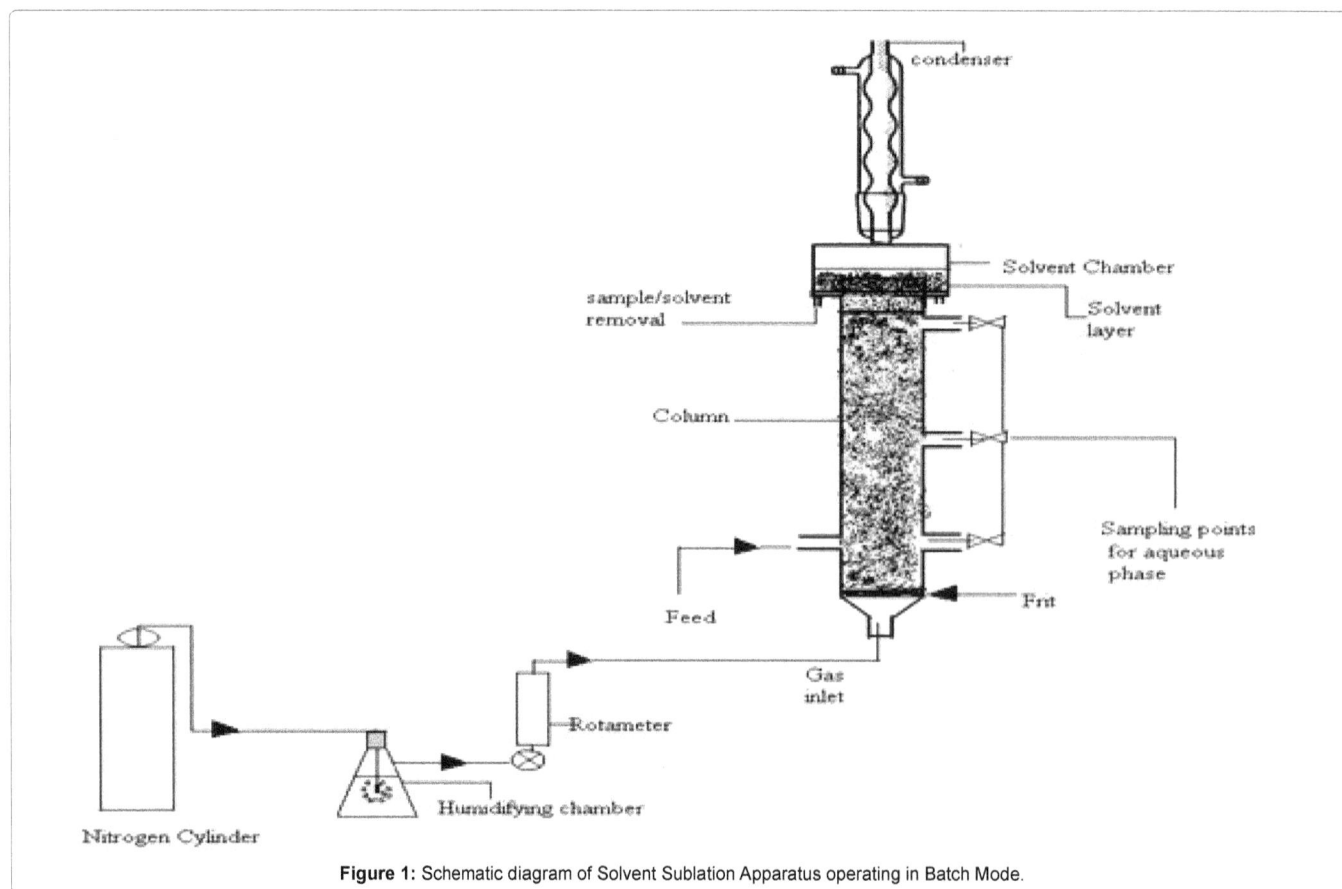

Figure 1: Schematic diagram of Solvent Sublation Apparatus operating in Batch Mode.

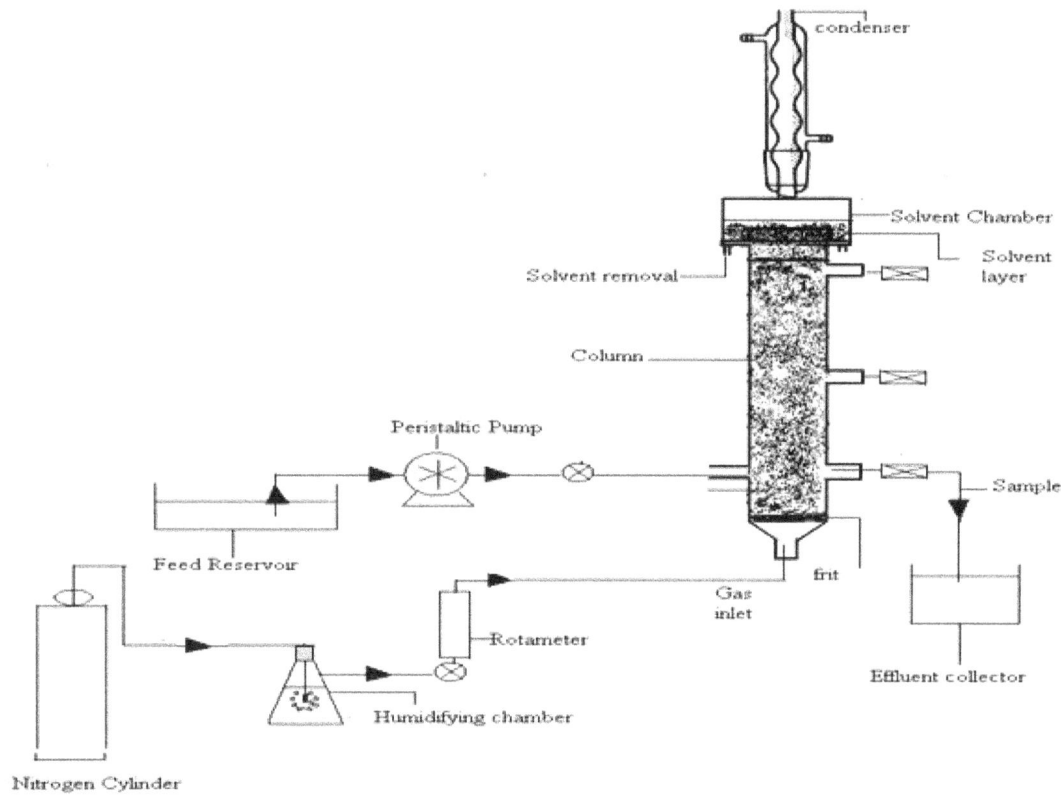

Figure 2: Schematic diagram of Solvent Sublation Apparatus operating in Continuous Mode.

Time (min)	Volume of Feed sample (ml)	Residual feed volume (ml)	Residual feed conc (mcg/ml)	Amount in residual feed (mg)	Amount separated (mg)	%RP	$t_{1/2}$ (min)
30	5	3895	449.5	1750.80	199.20	10.21	
60	5	3890	394.5	1534.60	415.40	21.30	
90	5	3885	342	1328.67	621.33	31.86	
120	5	3880	287	1113.56	836.44	42.89	146
150	5	3875	232	899	1051	53.89	
180	5	3870	177	684.99	1265.01	64.87	
210	5	3865	174.5	674.44	1275.56	65.41	

Initial volume of feed=3900 ml, Amount of protein in initial feed=1950 mg

Table 1: Performance criteria of whey in Batch mode under following condition (solvent sublation) Ci-500 mcg/ml, pH-2, GFR-175 cc/min.

top of the solvent chamber a condenser was connected to prevent the evaporation of solvent Residual liquid was collected from outlet at fixed intervals and was immediately analyzed. After sometime steady state concentration of effluent was observed. The experiment was continued upto 3 hours. At the end of run the total volume of residual liquid was collected and analyzed. The total input amount, output amount, loss amount, recovery %, were also calculated and tabulated [8].

Solvent sublation (Continuous process)

The glass column and its assembly description are same as Figure 2. a reservoir, peristaltic pump attached to supply feed to the column continuously at a desired volumetric flow rate (VFR). After loading the column with desired concentration of feed the top of the column was then covered with the receptacle (solvent chamber). Required volume of octanol was poured inside the chamber and a clear feed/octanol interface was visible. Feed was supplied from outside though a inlet in the column with the help of a peristaltic pump to maintain a constant volumetric flow rate, The flow rate of the outgoing effluent was same as the incoming feed. The effluent was continuously collected and

analyzed. Nitrogen gas was passed through the feed at desired gas flow rate (GFR). After sometime the effluent showed steady concentration. The experiment was continued till 5 hours. The total effluent volume. residual volume was measured at the end of the run and was analyzed. The total input amount, output amount, loss amount, recovery %, were also calculated and tabulated.

Results

The present work deals with the comparative study of the separation and removal of whey proteins from the commercial whey waste (which is more often discarded as a waste product in sweet industry) by both batch and continuous solvent sublation method. Both batch and continuous solvent sublation mode of operations were performed and effect of several parameters such as pH, volumetric flow rate (VFR), Gas flow rate (GFR) on the separation process were studied thoroughly.

Effect of pH

From the values obtained in Tables 1-3 in batch mode of solvent sublation and comparing them with Figure 3, in batch mode, it was

Time (min)	Volume of Feed sample (ml)	Residual feed volume (ml)	Residual feed conc (mcg/ml)	Amount in residual feed (mg)	Amount separated (mg)	%RP	$t_{1/2}$ (min)
30	5	3895	447	1741.06	208.94	10.71	
60	5	3890	392	1524.88	425.12	21.80	
90	5	3885	332	1289.82	660.18	33.85	
120	5	3880	272	1055.36	894.64	45.87	137
150	5	3875	214.5	831.19	1118.81	57.37	
180	5	3870	157	607.59	1342.41	68.84	
210	5	3865	154.5	597.14	1352.86	69.37	

Initial volume of feed=3900 ml, Amount of protein in Initial feed=1950 mg

Table 2: Performance criteria of whey in Batch mode under following condition (solvent sublation) Ci-500 mcg/ml, pH-5, GFR-175 cc/min.

Time (min)	Volume of Feed sample (ml)	Residual feed volume (ml)	Residual feed conc (mcg/ml)	Amount in residual feed (mg)	Amount separated (mg)	%RP	$t_{1/2}$ (min)
30	5	3895	449.5	1750.80	199.20	10.21	
60	5	3890	402	1563.78	386.22	19.80	
90	5	3885	352	1367.52	582.48	29.87	
120	5	3880	299.5	1162.06	787.94	40.40	157.7
150	5	3875	252	976.5	973.5	49.92	
180	5	3870	202	781.74	1168.26	59.91	
210	5	3865	199.5	771.07	1178.93	60.45	

Initial volume of feed=3900 ml, Amount of protein in initial feed=1950 mg

Table 3: Performance criteria of whey in Batch mode under following condition (solvent sublation) Ci-500 mcg/ml, pH-8, GFR-175 cc/min.

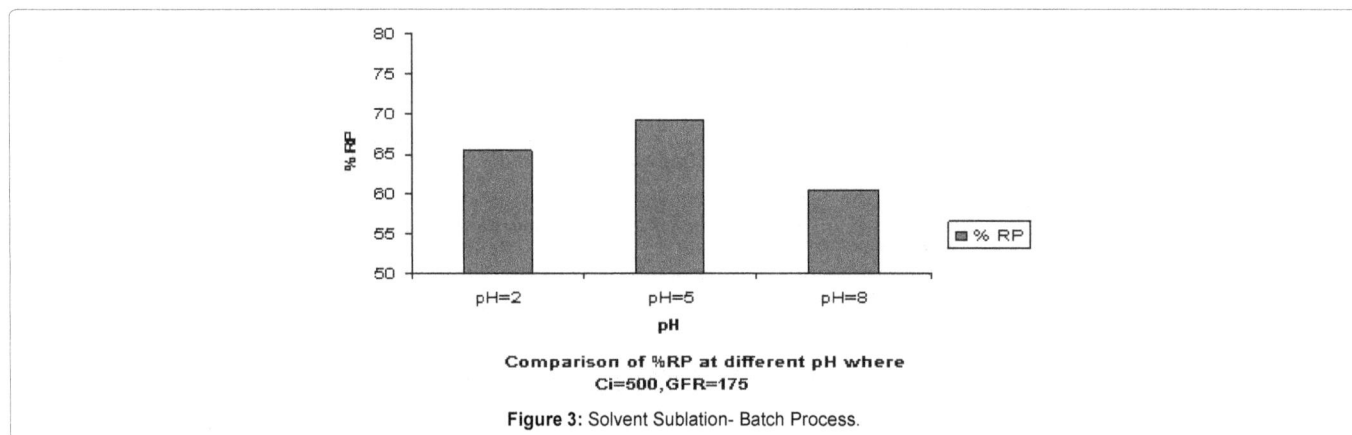

Comparison of %RP at different pH where Ci=500,GFR=175

Figure 3: Solvent Sublation- Batch Process.

Time (Min)	Effluent concentration(mcg/ml)	Eff+Res Vol (ml)	Input amount (mg)	Output amount (mg)	Loss amount (mg)	Separated amount of whey (mg)	%RP
30	104.5						79.10
60	99.5						80.10
90	92						81.60
120	94.5						81.10
150	89.5	7433	3750	3724.83	25.16	3074	82.10
180	89.5						82.10
210	89.5						82.10
240	87						82.60
270	89.5						82.10
300	87						82.60

Table 4: Performance criteria of whey in continuous mode under following condition (solvent sublation) Ci-500 mcg/ml, pH-5, GFR-175 cc/min, VFR=12 ml/min.

found that the recovery process was optimum at pH 5, at lower pH 2 at the performance criteria was lesser and still least at a higher pH 8 than the optimum. The enrichment ratio (E_R) (maximum at pH 5) increased with the increased in pH and then decreased above pH 5. This is due to the fact that the isoeletric pH of the whey protein is around 5, bellow which it becomes negative charge. Since it is no more ionic (net charge is zero), so it favours interface than the bulk liquid (aqueous). It seems that whey proteins may be separated from the bulk to adsorption of protein in the interface at IEP [13].

Effect of GFR

In continuous mode of solvent sublation range of GFR was 150-200

Time (Min)	Effluent concentration (mcg/ml)	Eff+Res Vol (ml)	Input amount (mg)	Output amount (mg)	Loss amount (mg)	Separated amount of whey (mg)	%RP
30	59.5						88.10
60	54.5						89.10
90	52						89.60
120	52						89.60
150	49.5	8030	4050	4044.0475	5.9525	3644	90.10
180	49.5						90.10
210	49.5						90.10
240	49.5						90.10
270	47						90.60
300	49.5						90.10

Table 5: Performance criteria of whey in continuous mode under following condition (solvent sublation) Ci-500 mcg/ml, pH-5, GFR-175 cc/min, VFR=14 ml/min.

Time (Min)	Effluent concentration (mcg/ml)	Eff +Res Vol (ml)	Input amount (mg)	Output amount (mg)	Loss amount (mg)	Separated amount of whey (mg)	%RP
30	149.5						70.10
60	139.5						72.10
90	137						72.60
120	139.5						72.10
150	134.5	8627	4350	4297.995	52.00	3174	73.10
180	134.5						73.10
210	134.5						73.10
240	129.5						74.10
270	132						73.60
300	129.5						74.10

Table 6: Performance criteria of whey in continuous mode under following condition (solvent sublation) Ci-500 mcg/ml, pH-5, GFR-175 cc/min, VFR=16 ml/min.

Time (Min)	Effluent concentration(mcg/ml)	Eff+Res Vol (ml)	Input amount (mg)	Output amount (mg)	Loss amount (mg)	Separated amount of whey (mg)	%RP
30	59.5						79.10
60	54.5						82.10
90	52						84.10
120	52						84.10
150	49.5	8026	4050	3758.8495	291.1505	3359	83.10
180	49.5						83.10
210	49.5						83.10
240	49.5						84.10
270	47						83.60
300	49.5						83.10

Table 7: Performance criteria of whey in continuous mode under following condition (solvent sublation) Ci-500 mcg/ml, pH-5, GFR-150 cc/min, VFR=14 ml/min.

Time (Min)	Effluent concentration(mcg/ml)	Eff +Res Vol (ml)	Input amount (mg)	Output amount (mg)	Loss amount (mg)	Separated amount of whey (mg)	%RP
30	74.5						85.10
60	62						87.60
90	64.5						87.10
120	64.5						87.10
150	67	8026	4050	4045.092	4.908	3504	86.60
180	67						86.60
210	67						86.60
240	62						87.60
270	67						86.60
300	67						86.60

Table 8: Performance criteria of whey in continuous mode under following condition (solvent sublation) Ci-500 mcg/ml, pH-5, GFR-200 cc/min, VFR=14 ml/min.

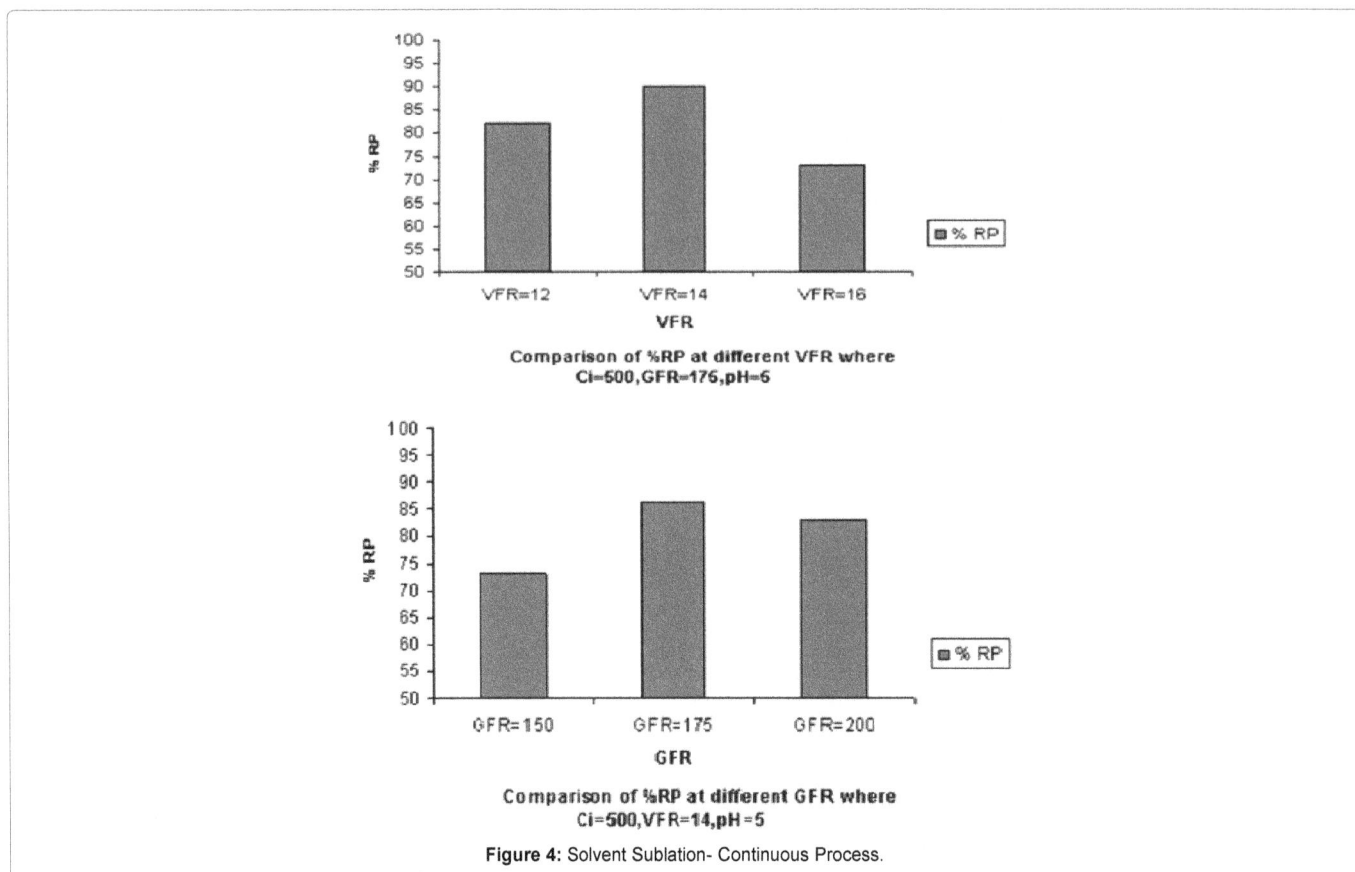

Figure 4: Solvent Sublation- Continuous Process.

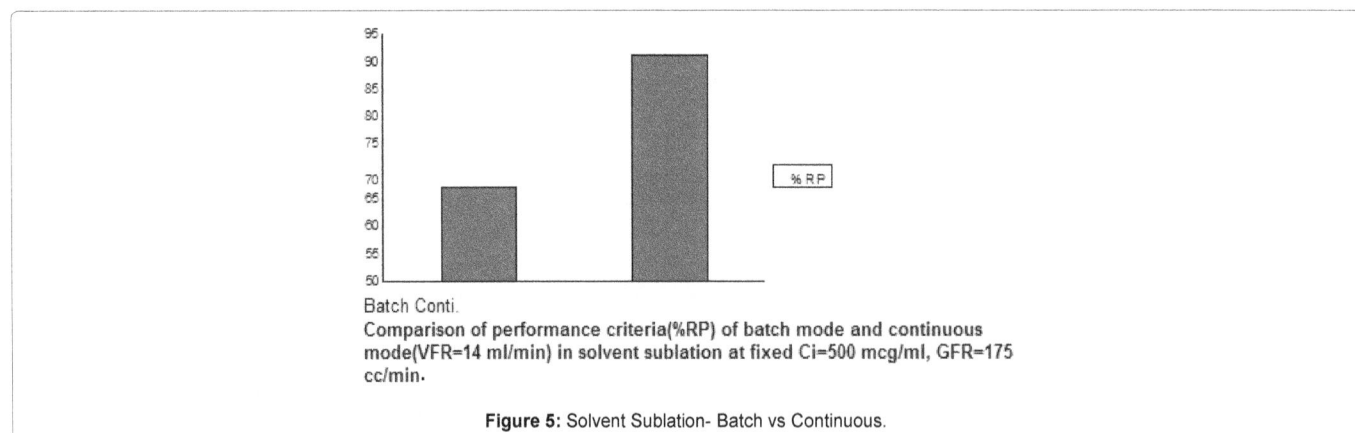

Figure 5: Solvent Sublation- Batch vs Continuous.

cc/min, %Rp increased with the increased of GFR up to 175 cc/min and then it decreased at 200 cc/min. At high GFR bubbles coalescence and decrease in interfacial area caused low %Rp. Therefore, GFR at 175 c/min was found suitable. From the values obtained in the Tables 4-8 in continuous mode of solvent sublation and compare them with Figure 4, its was found that the performance criteria was highest at a GFR 175 cc/min.

Effect of VFR

From the vales obtained in Tables 4-6 in continuous mode and comparing them with Figure 5, it was found that the optimum volumetric flow rate was 14 ml/min. It was observed that at a higher VFR the performance criteria showed lower value when compare to that of a lower VFR. %Rp increased with the increased of VFR up to 14

ml/min. With increased of input liquid %Rp increased up to a limit. At higher VFR (16 ml/min) residence time of input liquid is less and some of protein left the column without getting adsorbed. Therefore, effluent showed high concentration.

Effect of recovery process

From the Figure 4, it was found that in continuous mode at VFR 14 cc/min, %Rp was nearly 1.3 times more than that of batch mode at optimum pH and GFR. From the values obtained in the Tables 1-3 in batch process with various pH and Tables 4-6 at various VFR in continuous mode, it was observed that at any given condition the percentage recovery was greater in continuous operation mode at pH 5. The optimum operation conditions were found as GFR 175 cc/min, pH 5, VFR 14 ml/min. It is obvious that rate of separation of protein rom whey is found higher in continuous mode than the batch mode. In solvent sublation method, amount of protein recovered in single stage of continuous mode after 5 hours operation was 3644 mg, %Rp 90%, total volume treated was 8100 ml and effluent concentration was 49.5 mcg/ml, at the optimum condition (Table 5).

Conclusion

It was observed that solvent sublation is useful method either to concentrate protein from dilute solution of whey or to decrease the protein lable of waste solution buy both batch and continuous mode operation. The method was found more effective in continuous mode of VFR 14 ml/min, GFR 175 cc/min at pH 5. The enrichment ratio almost 1.5 times in continuous operation than that of batch mode operation. %Rp was also nearly 1.3 times more than that of batch mode at optimum pH and GFR. Optimisation of data adapting a suitable model may provide more precise operating conditions for the maximum recovery of proteins.

Acknowledgements

Authors are thankful to BCDA College of Pharmacy and Technology, Kolkata, India, for its technical support to perform this research work.

References

1. Dairy Council of California (2004) A monograph on whey.

2. Pomeranz Y (1985) Functional Properties of Food Components. Academia Press Inc., California, USA.

3. Lemlich R (1972) Adsorptive Bubble Separation Technique. Academic Press, New York, USA.

4. Casper JL, Wendorff WL, Thomas DL (1999) Functional Properties of Whey Protein from Caprine and Ovine Specialty Cheese Whey concentrates. Journal of Dairy Science 82: 265-271.

5. Krastanka G, Marinova M, Elka S, Basheva A, Boriana N, et al. (2009) Physico-chemical factors controlling the foamability and foam stability of milk proteins: Sodium caseinate and whey protein concentrates. Food Hydrocolloids 23: 1864-1876.

6. Geoffrey W, Krissansen K (2007) Emerging Health Properties of Whey Proteins and their Clinical implications. Journal of the American College of Nutrition 26: 713-723.

7. Daniel R, Rejean D, Yves P, Sylvie G, Patrice E, et al. (2009) Bovine Whey Protein Extract Can Enhance Innate Immunity by Priming Normal Human Blood Neutrophils. Journal of Nutrition 139: 386-393.

8. Zhaoliang W, Bo L, Bin H, Huijie Z (2009) Separation of l- lysine by solvent sublation. Separation and Purification Technology 66: 237-241.

9. Bi PY, Dong HR, Guo QZ (2009) Separation and Purification of Penicillin G from fermentation broth by solvent sublation. Separation and Purification Technology 65: 228-231.

10. Martinez AM, Burghoff S, Goetheer ELV, Haan DA (2009) (Extraction of biomolecules using) aqueous biphasic systems formed by ionic liquids and aminoacids. Separation and Purification Technology 65: 65-72.

11. Suzuki A, Yasuhara K (2002) Selective Foam Separation of Binary Protein Solution by SDS complexion Method. Journal of Colloid and Interface Science 253: 402-408.

12. Liping D, Prokop A, Robert DT (2003) Effect of pH on the startup of a continuous Foam Fractionation Process Containing Ovalumin. Separation Science and Technology 38: 1093-1109.

13. Ekici P, Backleh-sohrt M, Parlar H (2005) High efficiency enrichment of total and single whey proteins by pH controlled foam fractionation. International Journal of Food Science and Nutrition 56: 223-229.

4

Detection of Phytoconstituents of Medicinal Plant *Terminalia arjuna* Using Chromatographic Techniques

Chitte RR[1,2]*, Nagare SL[1,2], Date PK[2] and Shinde BP[2]

[1]Biotechnology, Vidya Pratishthan's School of Biotechnology, Vidyanagari, Baramati, Maharashtra, India
[2]Agriculture Biotechnology, Vidya Pratishthan's School of Biotechnology, Vidyanagari, Baramati, Maharashtra, India

Abstract

The plant extract revealed the presence of phytochemicals such as Phlobatannins, phenols, leucoanthocyanins, saponins, emodins, coumarins and quinones. Process of extraction of pure compound using column chromatography. The gradient of solvent eluted fraction has in pure form, tested and partial characterized. Thin layer chromatographic study was carried out by using various solvent system of varying polarity of which water, ethyl acetate and propanol system suited the best. *In vitro* anti-inflammatory activity was evaluated using albumin denaturation (50%) membrane stabilization assay (75%) and proteinase inhibitory activity (33.33%). For anti-inflammatory activity Aspirin (85.67%) used as standard drug. Using alfa amylase inhibition assay, *In vitro* antidiabetic activity was determined, fraction 5 (89.12%) and fraction 6 (80%) were showed at conc. 500 µg/mL. Antimicrobial efficiency of the plant extract fractions was determined using well diffusion method against *Pseudomonas sp.*, *Bacillus sp.*, *Protease* and *Staphylococcus aureus*, of which no activity was observed against *Pseudomonas sp.*

Keywords: Bioprospecting; *Terminalia arjuna*; Thin layer chromatography; Phytochemical analysis; Anti-inflammatory; Anti-diabetic; Antimicrobial

Introduction

Terminalia arjuna has been reported medicinal value plant for wide applications. The various parts have been reported for health benefit effects such as bark extract injection had increase the coronary flow in heart preparation of rabbit [1]. The bark powder also reported the anti hypocholesterolemic and anti-oxidant effects in human [2] while the methanol extract of *T. arjuna* leaves have moderate activity against *Aspergillus flavus*. There is antibiotically resistant strains of microbial pathogens, such as methicillin resistant *Staphylococcus aureus*, penicillin-resistant *Streptococcus pneumonia* is a problem and so search for better to develop new antimicrobial compounds are continued [3]. Rather than conventional antibiotics, the medicinal plants originated antimicrobial compounds may inhibit bacteria through different mechanisms and it has clinical values against resistant microbes had reported [4]. Modern research has discovered that *Terminalia arjuna* has antioxidant properties and may be clinically helpful in cardiovascular health. It has antibacterial [5] antimutagenic, hypolipidemic, antioxidant and hypocholesterolemic and anti-inflammatory effects. The aim of the present study was to deliver the literal studies of *T. arjuna* with its phytochemical and pharmacological characteristics. Arjuna regulates cholesterol by decreasing LDL levels in the liver and to be a natural liver tonic. Still today there is no single drug which showed definite and reliable protection or cure from atherosclerotic cardiovascular disorders.

The present paper aims to review Arjuna pharmacognostical, phytochemical, pharmacological, insecticidal, anthelmintic, immunomodulatory, antidiabetic and antioxidant Properties. There is ample evidence of its beneficial effect in coronary artery disease. The extracts of Arjuna used in strengthening the heart muscles, relieving stress, and hypertension. Arjuna is effective for a variety of heart related conditions like high blood pressure, heart palpitations, rapid heartbeat and high cholesterol. These reports are very encouraging and indicate that this should be studied more extensively for its therapeutic benefits.

Arjuna is a good hepatitis reliever and gives heart strength for human. Leaves of arjuna are best for recovery from hepatitis C and also

ideal tonic for heart disease. Glucoside reported in the bark of Arjuna is good for heart problem. Arjuna bark powder is also used on broken bone can joint of the fractured bone and also good remedy for swelling of gum and mouth. Bark is effectively used for controlling diarrhoea and dysentery. Arjuna bark is described as heart tonic and herbal preparations is used for treatment of cardiac disorders. World Health Organization, stated that the medicinal plants are best source of variety of drug [6].

Tannins, alkaloids, carbohydrates, terpenoids, steroids, flavonoids and phenols are present in medicinal plant. These bioactive substances of organic compounds play important role in body physiology of human. Secondary metabolites of plant are chemically and taxonomically extremely diverse compounds with unique function. They are widely used in the human therapy. We have extracted and purified these compounds using column chromatography and fractions are collected analyzed for purity and *in vitro* different assay condition of potency, isolated pure compounds have carried out.

Preparation of extract

1. Freshly collected Plant leaf sample were air dried at room temperature for 5-6 days.

2. Dried leaf sample was grinned to powder using mechanical grinder.

3. Dried leaf powder was homogenized in 10 mL (100%) methanol and was extracted on a rotatory shaker in a centrifuge tube at 350 rpm overnight [7].

4. Crude extract was filtered through Whatman No.1 filter paper and

*Corresponding author: Ratnakar Ravindra Chitte, Vidya Pratishthan's School of Biotechnology, Vidyanagari, Baramati, Maharashtra-413 133, India
E-mail: rrc10@rediffmail.com

evaporated on steam bath at 30 °C for 30 min (Figure 1).

5. Extract was stored at -20 °C for further study [8].

Column chromatography and TLC

1. Column chromatography was performed according to Patra et al. [9]. The solvent system that exhibited the best separation of compound was chosen for column chromatography. The methanol extract of *Terminalia arjuna* was absorbed onto silica gel.

2. The column 2 cm × 25 cm was packed with a solution of silica gel with water using the wet slurry method. This involve preparing a solution of silica gel with water in this case, in a beaker and subsequently adding this into the column till it is about three-fourth filled.

3. The solution was stirred for dispersal and quickly added to the column after the gel settles, this method was used to prevent the trapping of air bubble.

4. A boll of wool (glass wool) was pushed into the column to settle a top the packed silica gel.

The solvent system of water: ethyl acetate: propanol (2:5:3) was poured continuously into the column and allowed to drain and about 12 fractions of 5-6 mL was collected in sterile centrifuge tube. The fraction eluted on column was tested with same solvent system by TLC for the presence of active compounds. The column eluted fractions are lyophilized dried as power and stored at -20 °C (Figure 2).

Thin Layer Chromatography (TLC) [10] was performed

1. Fraction eluted on column was subjected to TLC as per conventional one-dimensional ascending method using silica gel (60F254 MERCK) pre-coated plate, plate was cut with ordinary household scissor.

2. Plate marking were made with soft pencil.

3. Glass capillary were used to spot the sample.

4. For TLC applied sample volume 1 μL by using capillary and solvent system was used is water: ethyl acetate: propanol.

5. After pre-saturation with mobile phase 20 min for development of band were used.

6. After running the plates, they are dried using dryer and plates were observed under various wavelength at 254 nm and 366 nm for band detection.

7. Colour of the spot and pattern were observed and RF value were calculated using formula:

$$RF \text{ (Retention factor)} = \frac{\text{Distance travelled by solute}}{\text{Distance travelled by solvent front}}$$

Phytochemical screening

Phytochemical studies were carried out for methanol extracts of *Terminalia arjuna* leaves to detect the presence of different phytochemicals (Table 1).

Estimation of protein content of eluted fractions of column chromatography using nano drop spectrophotometer

The protein content was measured using nanodrop, casein used as standard for estimation of protein Anti-inflammatory Activity was performed according to Leelaprakash and Dass [11].

Figure 1: Extraction and Filtration of Arjun leaves sample in Methanol as a solvent.

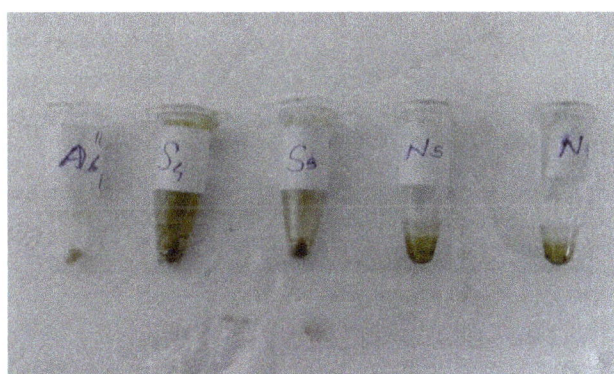

Figure 2: Lyophilisation of samples.

Albumin denaturation assay

1. Method of Mizushima and Kobayashi and Sakat et al. followed by minor modification [12,13].

2. The reaction mixture was consisting of test extract and 1% aqueous solution of bovine albumin fraction, pH of the reaction mixture was adjusted using small amount of 37 °C HCL

3. The sample extract was incubated at 37 °C for 20 min. and then heated to 51 °C for 20 min

4. After cooling the sample, the turbidity was measured spectrophotometrically at 660 nm

5. The experiment was performed in triplicate

6. Percent inhibition of protein denaturation was calculated as follow

$$\text{Percentage Inhibition} = \frac{Abs\ control - Abs\ sample}{Abs\ control} \times 100$$

Membrane stabilization test

Preparation of red blood cells (RBCs) suspension:

S. No.	Phytochemicals	Test	Inference	Observation
1	Alkaloid	Add 2 mL of extract to 2N HCL decand aqueous layer formed and few drops of mayers reagent	Cream precipitate observe indicating the presence of alkaloid	Cream precipitate was not observed
2	Phenolic compounds	Compounds-Add 3-5 drops of 5% FeCl₃ solution to 2 mL of extract	Formation of deep blue colour	Deep blue colour was observed
3	Flavonoids	In 2 mL of extract, add 2-5 drops of 1N NaOH	Formation of yellow orange colour	Yellow orange colour was not observed
4	Saponins	Add 2 mL of extract with 6 mL of water in a test tube	Observe for persistent foam	Observation of persistent foam
5	Tannins	Add 2 mL of aqueous extract with 2 mL of distilled water and few drops of FeCl₃	Formation of green precipitate	Green precipitate was observed
6	Leucoanthocyanin	Add 5 mL of aqueous extract to 5mL of isoamyl alcohol	Upper layer appears red in colour	Red colour was observed
7	Quinones	Add 2 mL of extract with concentrated HCl	Formation of yellow precipitate	Yellow precipitate was observed
8	Coumarin	Add 3 mL of 10% NaOH to 2 mL of aqueous extract	Formation of yellow colour	Yellow colour was observed
9	Steroid	Dissolve 1mL of extract in 10 mL of chloroform and add equal volume of concentrated H₂SO₄	The upper layer turns red and H₂SO₄ layer shows yellow green fluorescence	The upper layer was not red and H₂SO₄ layer was not yellow green fluorescence
10	Emodins	Add 2 mL of extract with concentrated HCl	Formation of yellow precipitate	Yellow precipitate was observed
11	Phlobatanin	Add 2 mL of aqueous extract to 2 mL of 1% HCl and boil the mixture	Deposition of red precipitate	Red precipitate was observed
12	Anthocyanin	Add 2 mL of aqueous extract to 2 mL of 2N HCl and Ammonia	Appearance of pink- red turns Blue-violet	Pink-red colour turns Blue-violet was not observed

Table 1: Phytochemical tests for secondary metabolites present in extract of Arjun.

1. Fresh whole human blood 10 mL was collected and transferred to the centrifuge tube.

2. The tubes were centrifuged at 3000 rpm for 10 min. and were washed three times with equal volume of normal saline.

3. The volume of blood was measured and constituted as 10% v/v suspension with normal saline [13,14].

Heat induced haemolytic:

1. The reaction mixture 2 mL consisted of 1 mL of test sample solution and 1 mL of 10% RBCs suspension

2. Instead of test sample only saline was added to the control test tube.

3. Aspirin was taken as a standard drug.

4. All the centrifuged tube containing reaction mixture were incubated in water bath at 56 °C for 30 min

5. At the end of the incubation the tube was cooled under running tap water

6. The reaction mixture was centrifuged at 2500 rpm for 5 min. and the absorbance of the supernatants was taken at 560 nm

7. The experiment was performed in triplicate for all the test sample, % membrane stabilization activity was calculated by formula [13,15].

$$\text{Percentage Inhibition} = \frac{Abs\ control - Abs\ sample}{Abs\ control} \times 100$$

Protein inhibitory action:

1. The test was performed according to the modified method of Oyedepo et al. and Sakat et al. [13,16].

2. The reaction mixture (2 mL) was containing 0.06 mg trypsin, 1 mL of 20 Mm Tris HCL buffer (pH 7.4) and 1 mL test sample of different concentration.

3. The reaction mixture was incubated at 37 °C for 5 min. and then 1 mL of 0.8% (W/V) casein was added

4. The mixture was inhibited for an additional 20 min., 2 mL of 70% perchloric acid was added to terminate the reaction

5. Cloudy suspension was read at 210 nm against buffer as blank

6. The experiment was performed in triplicate

7. Percentage protein inhibition activity was calculated by formula

$$\text{Percentage Inhibition} = \frac{Abs\ control - Abs\ sample}{Abs\ control} \times 100$$

Antidiabetic activity was performed according to Dhritiv et al. [17].

Inhibition of alpha amylase enzyme:

Standard maltose curve:

1. 0.2-1 mL of standard maltose (1 mg/mL) was taken into different tube.

2. Make the volume to 1 mL in each case with distilled water.

3. Added 1 mL of DNSA (Dinitro salicylic acid) reagent to each tube and then place all the tubes in boiling water bath for 15 mins.

4. Add 8 mL of distilled water in each tube and mix the content.

5. Then read the absorbance of the solution in Calorimeter at 570 nm against blank solution.

Alpha amylase inhibition assay:

1. 100-500 µL of extract was taken into different test tubes, make the volume 0.5 mL with phosphate buffer of pH 6.8

2. Blank was measured by taking 1 mL of phosphate buffer

3. Control was measured by taking 0.5 mL of phosphate buffer

4. The solution was taken treated with 0.5 mL of alpha amylase (0.5 mg/mL)

5. The solution was incubated at 25 °C for 10 mins

6. Added 0.5 mL of 1% starch solution in 0.02 M Sodium phosphate buffer of pH 6.9 to all tubes and then incubated at 25 °C for 10 mins

7. The reaction was stopped by DNSA and the reaction mixture was kept in boiling water bath for 5 mins cooled to RT

8. The solution was mixed with 8 mL distilled water

9. Read the absorbance of the solution in calorimeter at 570 nm against blank solution

10. Amount of maltose produced is calculated using standard maltose curve and enzyme activity is calculated by using formula

$$\text{Enzyme Activity} = \frac{\text{Amount of maltose formed}}{10 \times 342} \times 2$$

Antimicrobial activity was performed according to Narendra et al. [18].

Microorganism used

The bacterial strains were collected from microbial culture collection laboratory VSBT

1. The antimicrobial activity was performed by agar well diffusion method for solvent extract.

2. 20 ml of media Muller and Hinton (MH) agar was poured into the petri plates along with inoculum.

3. A well was prepared in the plate with the help of cork bores (6 mm).

4. 20 μL of the test sample was poured in each well using sterile micropipette.

5. For positive control standard, antibiotic tetracycline (30 mcg) was used.

6. The plates were incubated overnight at 37 °C in BOD incubator.

7. The microbial growth was determined by measuring the diameter of zone of inhibition.

8. The entire process was carried out aseptically in the laminar air flow.

Results and Discussion

Column chromatography and TLC studies

Thin layer chromatographic studies of partial purified methanol fraction of *Terminalia arjuna* was done by using silica gel 60 F254 (MERCK) aluminium plate. Solvent system water: ethyl acetate: propanol (2:5:3) was used for separation of compound. Partial purified fraction eluted on column chromatography showing different band pattern at 254 nm and 366 nm (Figure 3) Spot were characterized by Rf value under UV light (Figure 4).

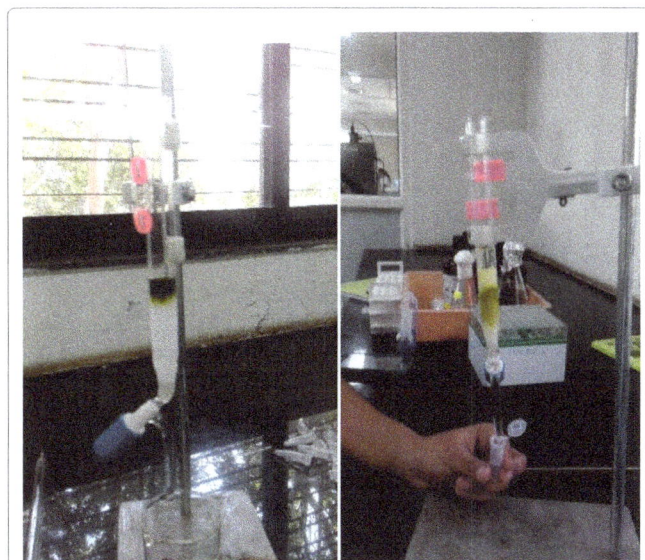

Figure 3: Column chromatography of Plant extract.

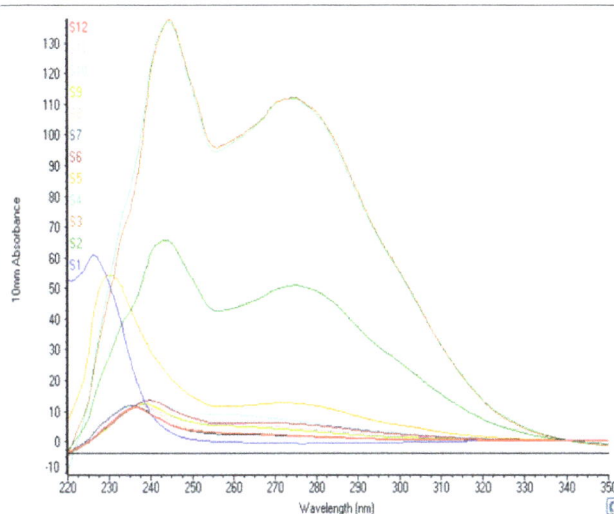

Figure 4: Determination of protein concentration of column eluted fractions by Nanodrop.

Figure 5: Quantitative analysis phytochemical present in the crude extract and partially purified fractions of Arjun.

Phytochemical analysis

Qualitative phytochemical investigation of extract of *Terminalia arjuna* leaves showed the presence of saponins, coumarins, emodins, phenols, quinones, leucoanthacyanins and phlobatannins. Methanolic extract of *Terminalia arjuna* leaves contain maximum number of phytoconstituents. The results of phytochemical screening and qualitative analysis were showed in the Tables 2-4 (Figures 5 and 6).

In vitro anti-inflammatory activity

Albumin denaturation assay: Denaturation of protein is a well-documented cause of inflammation. As a part investigation on the mechanism of the anti- inflammatory activity, ability of fraction inhibit denaturation was studied. It was effective in inhibiting head induced albumin denaturation. Maximum inhibition was observed in fraction No 4 and 8 as shown in Table 5. The percentage of albumin denaturation inhibition is given of respective column eluted fractions (Figure 7).

Membrane stabilization assay

The HRBC membrane stabilization has been used as a method to study the *in vitro* anti-inflammatory activity because the erythrocyte membrane is analogous to the lysosome membrane and its stabilization implies that the column eluted fraction may well stabilize lysosomal membranes. Stabilization of lysosomal is important in limiting the inflammatory response by preventing the release of lysosome constituents of activated neutrophil, such as bacterial enzymes and proteases, which causes further tissue inflammation and damage upon extra cellular release. The lysosomal enzymes released during inflammation produce a various disorder. purified fraction was effective in membrane stabilization at different concentration as shown in Table 6, maximum inhibition of fraction no.4 and 8 (75%) was observed at 500 μg/mL, followed by fraction 6 (50%) and fraction no.7 (50%) aspirin a standard drug shows maximum inhibition 81.32% at conc.500 μg/mL. Membrane stabilization assay of partially eluted sample was shown in the Figure 8.

Proteinase inhibitory activity

Proteinases have been implicated in arthritic reactions. Neutrophils are known to be a rich source of proteinase which carries in their lysosomal granules many serine proteinases. It was previously reported that leukocytes proteinase play an important role in the development of tissue damage during inflammatory reactions and significant level of protection was provided by proteinase inhibitors. *Terminalia arjuna* partial purified fraction exhibited significant anti proteinase activity at different concentrations as shown in Table 7. Maximum inhibition of fraction no.6 and 7 (33.33%) was observed at 500 μg/mL, followed by fraction 4 (25%) and fraction 8 (25%) aspirin a standard drug shows maximum inhibition 85.67% at conc.500 μg/mL. The protein denaturation of column eluted sample was shown in the Figure 9.

In vitro antidiabetic activity

Alpha amylase inhibition assay: The intestinal digestive enzyme alpha-amylase plays vital role in the carbohydrate digestion. Antidiabetic therapeutic approach reduces the post prandial glucose level in blood by the inhibition of alpha- amylase enzyme. The *in*

Fraction No.	Solvent system	No. of spot detected		Rf value	
		254 nm	366 nm	254 nm	366 nm
1	Water: ethyl acetate: propanol	-	-	-	-
2	Water: ethyl acetate: propanol	1	1	-	-
3	Water: ethyl acetate: propanol	1	1	-	-
4	Water: ethyl acetate: propanol	1	1	0.60	-
5	Water: ethyl acetate: propanol	1	1	0.58	-
6	Water: ethyl acetate: propanol	1	1	0.85	-
7	Water: ethyl acetate: propanol	1	1	0.67	-
8	Water: ethyl acetate: propanol	1	1	0.82	-
9	Water: ethyl acetate: propanol	1	1	0.59	-
10	Water: ethyl acetate: propanol	1	1	0.40	-
11	Water: ethyl acetate: propanol	1	1	-	-
12	Water: ethyl acetate: propanol	1	1	---	-

Note: (-) absence of zone.

Table 2: TLC investigation and banding pattern for column eluted fractions.

Fraction no	Protein concentration μg /mL
1 (solvent eluted)	-1.058
2 (solvent eluted)	48.788
3 (solvent eluted)	107.445
4 (solvent eluted)	106.418
5 (solvent eluted)	11.308
6 (solvent eluted)	4.89
7 (solvent eluted)	1.216
8 (solvent eluted)	1.418
9 (solvent eluted)	2.834
10 (solvent eluted)	5.535
11 (solvent eluted)	3.508
12 (solvent eluted)	1.232

Table 3: Determination of Protein concentration by Nano drop technique.

Name of the phytochemical	Result	Partially purified fraction No. 4, 6, 7 and 8
Phenolic compounds	+	+
Phlobatannins	+	+
Alkaloids	−	−
Saponins	+	−
Flavonoids	−	−
Coumarins	+	+
Anthocyanins	−	−
Terpenoids	+	+
Steroids	−	−
Fatty acid	−	−
Emodins	+	+
Quinones	+	+

Table 4: Phytochemical constituents present in methanol extracts of *Terminalia arjuna* leaves and partially purified fraction.

Test sample	Albumin Denaturation
Fraction 4	50 ± 0.003
Fraction 6	13.33 ± 0.003
Fraction 7	13.33 ± 0.003
Fraction 8	33.33 ± 0.003
Aspirin	76.54 ± 0.003

Table 5: Percentage inhibition of Albumin denaturation assay.

Test Sample	Membrane Stabilization assay
Fraction 4	75 ± 0.004
Fraction 6	50 ± 0.004
Fraction 7	50 ± 0.004
Fraction 8	75 ± 0.004
Aspirin	81.32

Table 6: Percentage inhibition of membrane stabilization assay.

Test sample	Proteinase inhibition
Fraction 4	25 ± 0.004
Fraction 6	33.33 ± 0.004
Fraction 7	33.33 ± 0.004
Fraction 8	25 ± 0.004
Aspirin (Control)	85.67

Table 7: Percentage inhibition of Proteinase denaturation of column eluted fractions of *Terminalia arjuna*.

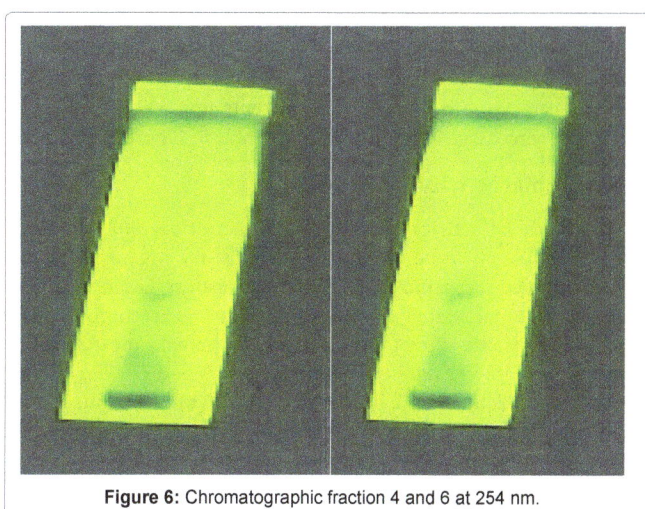

Figure 6: Chromatographic fraction 4 and 6 at 254 nm.

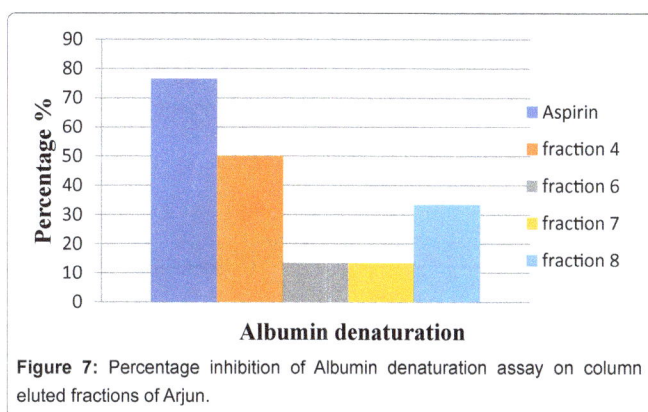

Figure 7: Percentage inhibition of Albumin denaturation assay on column eluted fractions of Arjun.

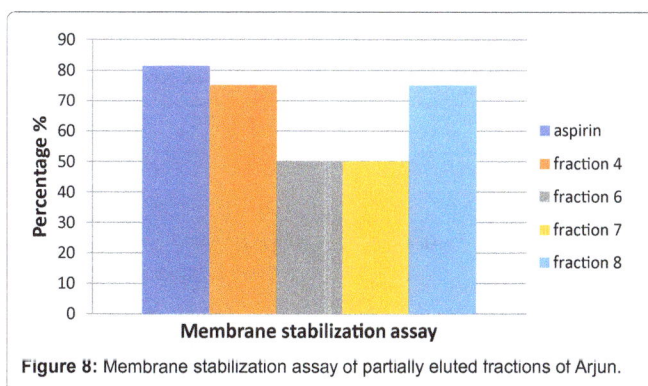

Figure 8: Membrane stabilization assay of partially eluted fractions of Arjun.

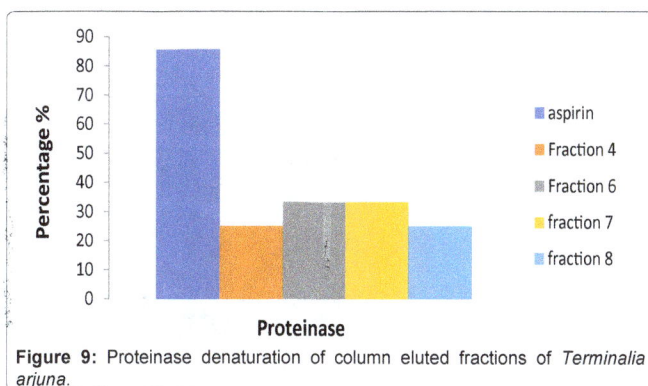

Figure 9: Proteinase denaturation of column eluted fractions of *Terminalia arjuna*.

vitro alpha amylase inhibitory studies demonstrated that *Terminalia arjuna* has well antidiabetic activity. Partially purified fraction showed maximum inhibition of fraction No.5 (84.98%) at concentration 500 µg/mL and fraction no.6 (70.71%) at concentration 500 µg/mL. conc. Dependent percentage inhibition listed in Table 8 (Figures 10 and 11).

Antimicrobial activity

Antimicrobial activity of methanol extract of *Terminalia arjuna* leaves gave different zone of inhibition against the organisms tested. The extract showed antimicrobial activity against Gram +ve and Gram -ve bacterial strains as shown in Figure 12. Fraction No. 3, 4 and 6 of methanolic extract of Terminalia arjuna was most sensitive against *Staphylococcus aureus*, *Protease mirabilis* and *Bacillus subtilis* with maximum zone of inhibition of diameter 15 mm, 12 mm, 15 mm respectively at conc. of 20 µg as shown in Table 9. Methanolic extract did not show resistance against *Pseudomonas aeruginosa*.

Results

Qualitative phytochemical investigation of extract of *T. arjuna* leaves showed the presence of alkaloids, aponins, coumaris, emodins, phenols, quinones, leucoanthacyonins and phlobatannins. A total of 12 different fractions were obtained from the extract of *T. arjuna* leaves through column chromatography. The most suitable solvent system out

of the three-different solvent system was Water: ethyl acetate: propanol (2:5:3) solvent system, which provided visible bands in fraction 4 and 5 only. The anti-inflammatory activity was most effectively depicted by fraction 4, 6, 7 and 8 in the respective tests conducted. The anti-diabetic activity was highest in fraction 5 and 6 at a concentration of 500 µg/mL. Methanolic extract of *T. arjuna* was most sensitive against *Staphylococcus aureus*, *Protease mirabilis* and *Bacillus subtilis* except *Pseudomonas aeruginosa*.

Discussion

In vitro assay of the extract of arjuna plant showed the various activities such as anti-inflammatory was valuated using albumin denaturation, membrane stabilization assay and proteinase inhibitory activity. *In vitro* antidiabetic activity was determined, fraction no 5 (89.12%) and fraction no 6 (80%) of the pure compound isolated using column chromatography techniques. These compounds showed pure band on TLC plate. Further identification of compounds and its structure need to be identified. These active Phyto constitutes would study further for targeted various spectrum of diseases and will determine its physiological and metabolic effects in animal's model, aspects of preclinical studies towards drug development process. Several medicinal values of the reported of genus Terminalia [19]. The preliminary phytochemical evaluation of flavonoids and alkaloids was carried out [20]. As the arjuna is reported historical medicinal value

Conc µg/mL	% Fraction 5		% Fraction 6		% Standard n	
	Abs.	% inhibition	Abs.	% inhibition	Abs.	% inhibition
100	0.186	69.12	0.044	55.8	0.027	67 ± 0.04
200	0.994	77.36	0.056	59.67	0.032	68 ± 0.04
300	1.026	85.08	0.658	34.24	0.053	81.42 ± 0.05
400	1.165	57.89	0.12	68.7	0.057	78.35 ± 0.05
500	0.426	89.12	0.128	75.08	0.075	80 ± 0.06

Table 8: *In vitro* alpha amylase inhibition method.

Fraction No. 20 µg/50 µL	Staphylococcus aureus	Protease mirabilis	Bacillus substilis	Pseudomonas aeruginosa
Fraction 4	15 mm	–	–	–
Fraction 9	–	12 mm	–	–
Fraction 6	–	–	15 mm	–
Fraction 7	–	–	14 mm	–

Microorganism zone of inhibition (mm), (-) absence of zone

Table 9: Antimicrobial activity *Terminalia arjuna* by well diffusion method.

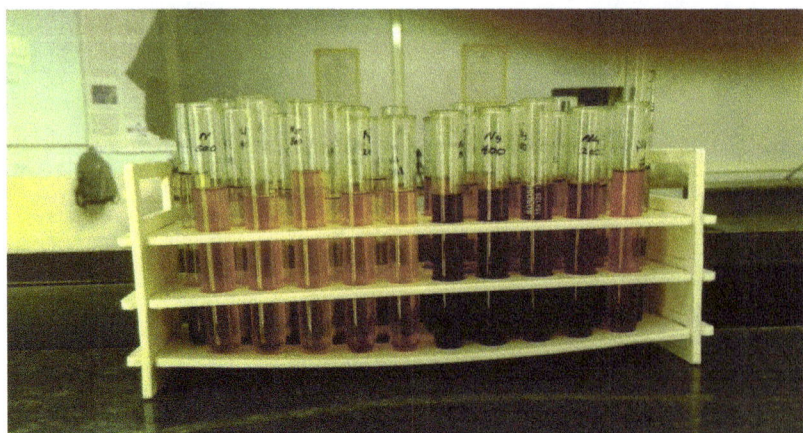

Figure 10: Proteinase denaturation of column eluted fractions of *Terminalia arjuna*.

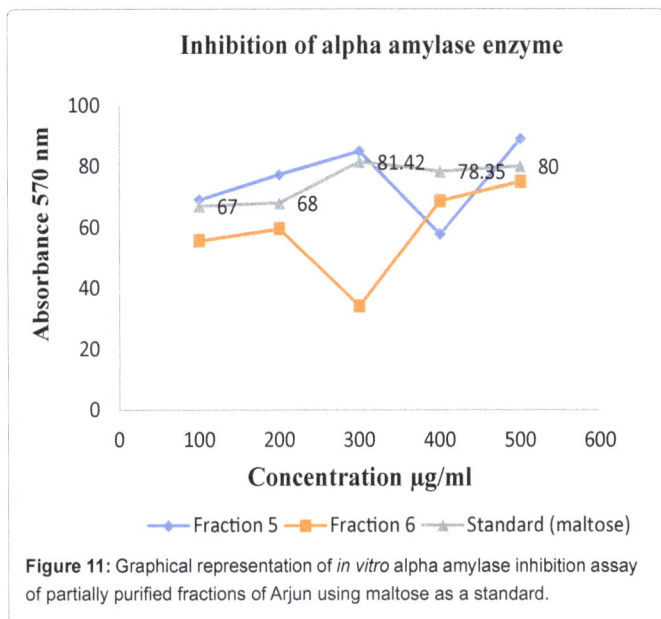

Figure 11: Graphical representation of *in vitro* alpha amylase inhibition assay of partially purified fractions of Arjun using maltose as a standard.

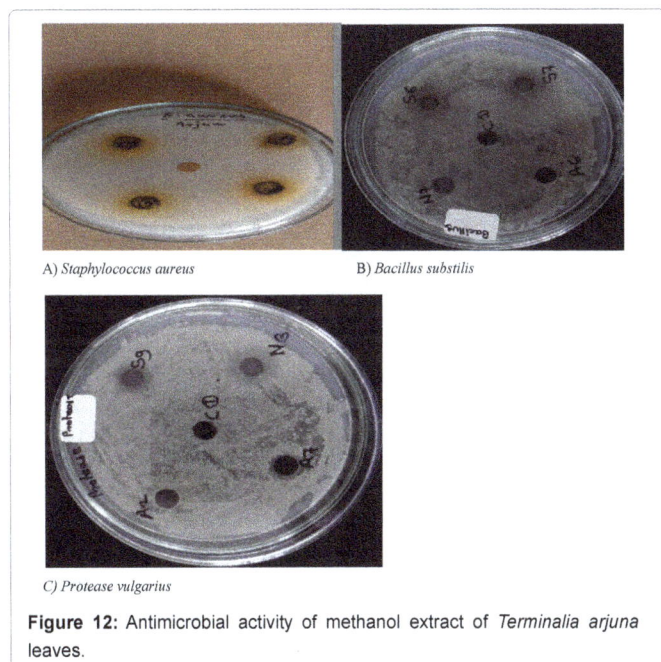

A) *Staphylococcus aureus*   B) *Bacillus substilis*

C) *Protease vulgarius*

Figure 12: Antimicrobial activity of methanol extract of *Terminalia arjuna* leaves.

and safe for consummation so trials will be low risk. The fraction no S6 and S7 showed good antimicrobial activities against *Staphylococcus aureus* and *Bacillus substilis* so the active component may be used as bactericidal agent.

Conclusion

On the basis of results, the following major conclusions were drawn. The presence of seven Phytochemicals in *T. arjuna* confirms its antioxidant activity. The results of column chromatography and thin layer chromatography enabled the purification of *T. arjuna* leaves extract. The anti-inflammatory effect of four fractions concludes the anti-inflammatory property of *Terminalia arjuna* and good anti-diabetic property which was confirmed by the anti-diabetic activity of two fractions obtained. *Terminalia arjuna* extracts has shown good activity against both the gram positive and gram-negative bacteria.

However, further research is required to determine the compound that has contributed to these activities of the extract and its exact mode of action.

References

1. Bhatia J (1998) Study of the possible cardioprotective role of *Terminalia arjuna* in experimental animals and its clinical usefulness in coronary artery disease. MD (Pharmacology) Thesis, India: University of Delhi.

2. Gupta R, Singhal S, Goyle A, Sharma V (2001) Antioxidant and hypocholesterolaemic effects of *Terminalia arjuna* tree-bark powder: a randomised placebo-controlled trial. The Journal of the Association of Physicians of India 49: 231-235.

3. Talley NJ, Stanghellini V, Heading RC, Koch KL, Malagelada JR, et al. (1999) Functional gastroduodenal disorders. Gut 45: II37-II42.

4. Eloff JN (1998) A Sensitive and Quick Microplate Method to Determine the Minimal Inhibitory Concentration of Plant Extracts for Bacteria. Planta Med 64: 711-713.

5. Perumalsamy P, Ignacimuthu S, Sen A (1998) Screening of 34 Indian medicinal plants for antibacterial properties. J Ethnopharmacol 62: 173-181.

6. Yadav RN, Agrawala M (2011) Phytochemical analysis of some medicinal plants. J Physiol 3: 10-14.

7. Daayf F, El Bellaj M, El Hassni M, J'aiti F, El Hadrami I (2003) Elicitation of soluble phenolics in date palm (Phoenix dactylifera L.) callus by Fusarium oxysporum f. sp. albedinis culture medium. Environmental and Experimental Botany 49: 41-47.

8. Vandana R, Upadhayaya K (2012) *In Vitro* antimicrobial activity and phytochemical screening of dry pets Roxburghii leaves extracts. IJRAP 3: 582-586.

9. Patra JK, Gouda S, Sahoo SK, Thatoi HN (2012) Chromatography separation, 1H NMR analysis and bioautography screening of methanol extract of Excoecaria agallocha L. from Bhitarkanika, Orissa, India. Asian Pacific Journal of Tropical Biomedicine, pp: S50-S56.

10. Rajendra Prasad G, Estari M (2013) Phytochemical screening and thin layer chromatographic studies of Aervallanata root extract. IJIRSET, p: 2.

11. Leelaprakash G, Dass SM (2011) *In vitro* anti-inflammatory activity of methanol extract of Enicostemma axillare. Int J Drug Dev Res 3: 189-196.

12. Mizushima Y, Kobayashi M (1968) Interaction of anti - inflammatory drugs with serum proteins, especially with some biologically active proteins. J Pharm Pharm 20: 169-173.

13. Sakat S, Juvekar A, Gambhire M (2010) *In vitro* antioxidant and antiinflammatory activity of methanol extract of Oxalis corniculata Linn. Int J Pharm Pharmaceut Sci 2: 146-155.

14. Sadique J, Al-Rqobah NA, Bughaith MF, El-Gindy AR (1989) The bioactivity of certain medicinal plants on the stabilization of RBC membrane system. Fitoterapia, pp: 525-532.

15. Shinde UA, Phadke AS, Nair AM, Mungantiwar AA, Dikshit VJ, et al. (1999) Membrane stabilizing activity - A possible mechanism of action for the antiinflammatory activity of Cedrus deodara wood oil. Fitoterapia 70: 251-257.

16. Oyedepo OO, Femurewa AJ (1995) Anti-protease and membrane stabilizing activities of extracts of Fagra zanthoxiloides, Olax subscorpioides and Tetrapleura tetraptera. Int J Pharmacogn 33: 65-69.

17. Dhriti V, Chowdary PVV, Rahul J, Vishank G, Bole SB (2014) Free radical scavenging and anti-diabetic activity of Kigelia pinnata. World J Pharm Pharmaceut Sci 3: 1249-1262.

18. Narender Prasad D, Ganga Rao B, Sambasiva Rao E, Mallikarjuna Rao T, Praneeth VSD (2012) Quantification of phytochemical constituents and in-vitro antioxidant activity of Mesua ferrea leaves. Asian Pacific Journal of Tropical Biomedicine, pp: S539-S542.

19. Kadam VB, Salve SB, Deore SV, Khandare KR, Kadam UB (2015) Estimation of lipid and alkaloid content in some medicinal plants of genus *Terminalia* (Combretaceae) of Marathwada Region in Maharashtra. International Journal of Medicine and Pharmaceutical Research 3: 1026-1029.

20. Nema R, Jail P, Khare S, Pradhan A (2012) Preliminary phytochemical evaluation and flavonoids quantification of Terminalia arjuna leaves extract. International Journal of Pharmaceutical and Phytopharmacological Research 1: 283-286.

5

Determination of Organochlorine Pesticides in Wildlife Liver and Serum Using Gas Chromatography Tandem Quadrupole Mass Spectrometry

Mark Bokhart[1,2,3], Andreas Lehner[2], Margaret Johnson[2] and John Buchweitz[2,3]*

[1]*College of Natural Science, Chemistry Department, Michigan State University, East Lansing, MI, USA*
[2]*Diagnostic Center for Population and Animal Health, Toxicology Section, Michigan State University, Lansing, MI, USA*
[3]*Department of Pathobiology and Diagnostic Investigation, College of Veterinary Medicine, Michigan State University, East Lansing, MI, USA*

Abstract

A single laboratory validation of a quantitative capillary gas chromatography tandem quadrupole mass spectrometer (GC-MS/MS) method utilizing a Quick, Easy, Cheap, Effective, Rugged, Safe (QuEChERS) approach for the extraction of 24 organochlorine analytes in liver and blood serum is presented here. The QuEChERS approach utilizes an acetonitrile extraction, partitioning facilitated by the addition of salts and a dispersive solid phase extraction cleanup. This method simultaneously monitored 24 organochlorine pesticide residues representing four different classes, including chlordecone, cyclodienes, dichlorodiphenylethanes, and hexachlorocyclohexanes. Calculated limits of detection (LOD) varied from 0.002 to 2.4 ppb and limits of quantification (LOQ) varied from 0.01 to 7.4 ppb. This multi-residue method proved to be a sensitive approach to the measurement of persistent organic pollutants in biological matrices.

Keywords: QuEChERS; Organochlorine pesticides; Veterinary toxicology; GC-MS/MS

Introduction

Organochlorine pesticides (OCPs) were widely used from the 1940's through 1970's and some are still used in developing countries today [1]. OCPs were highly effective in agricultural applications, improved crop production and enhanced public health through the reduction of diseases spread by insects [2]. However, OCPs are chemically stable, have high lipid solubility, low volatility and low rates of degradation leading to their classification as persistent organic pollutants (POPs). Organisms at higher trophic levels exhibit greater body burdens because OCPs bioconcentrate and biomagnify in the food chain [2]. When present in high concentrations, OCPs can produce a variety of toxic effects dependent on pesticide class [2]. Accordingly, nations observing the Stockholm Convention have eliminated or restricted the use and production of 12 POPs of high priority to protect the environment and human health [2]. Thus, it is imperative that veterinary diagnostic laboratories have the capability to identify and quantify low level traces of these compounds in animals that may be consumed by people, as well as in wildlife, particularly those animals with diets that lead to extensive accumulation such as certain avian species.

Traditional extraction methods for pesticide analysis in biological samples often use large quantities of solvent, are time consuming and may use expensive equipment to aid in the extraction process. Common pesticide extraction methods for a variety of matrices have been the subject of several reviews [3-7]. The QuEChERS extraction method was developed in 2003 by Anastassiades et al. in an effort to minimize time and analytical cost associated with multiresidue pesticide analysis in produce [8]. This approach to analyte extraction was a significant advancement over traditional multi-residue methods in that it utilized significantly less solvent and sample preparation time, making it a more economical approach for diagnostic labs to adopt. In addition to the economic advantages, the QuEChERS approach does not use harsh extraction conditions therefore making it is less likely to cause thermal breakdown of compounds in contrast to pressurized liquid extraction (PLE), supercritical fluid extraction (SFE) or microwave-assisted extraction (MAE) [7]. Additionally, the cleanup of samples prior to analysis by gas chromatography (GC) or liquid chromatography (LC) systems is performed by a dispersive solid phase extraction (dSPE) step.

The dSPE step uses sorbents such as PSA and C18 to remove interfering matrix components and this step is faster and has less chance of carryover compared to gel permeation chromatography (GPC), silica gel or solid phase extraction (SPE) cleanup methods [8].

Since the sentinel publication of this method in 2003, QuEChERS has been further developed, modified and adapted to many laboratory panels. The method was developed and utilized extensively by agricultural laboratories for pesticide residue testing of fruits and vegetables and environmental samples. The QuEChERS method has been the subject of recent review of pesticide analysis in food [9] and in environmental samples [10]. The extensive use of the method has led to commercial QuEChERS extraction kits from several vendors and an official AOAC method for analysis of pesticide residues in food [11]. Although the method has been widely used for pesticide analysis in a variety of food types, there have been few extensions to analysis of chemical residues in animal tissues [12,13]. The QuEChERS sample preparation method has been reported for the detection of pharmaceuticals in animal liver [14,15] and human whole blood [16]. To the best of the authors' knowledge, the method has not been applied to pesticide quantification in wildlife samples. With slight modifications to the extraction procedure, the method was readily adapted to pesticide analysis in biological samples of interest for veterinary diagnostic laboratories.

Analysis by gas chromatography tandem mass quadrupole spectrometry greatly enhances selectivity and sensitivity in comparison to single quadrupole techniques. Monitoring multiple precursor and

***Corresponding author:** John P. Buchweitz, Diagnostic Center for Population and Animal Health, Michigan State University, 4125 Beaumont Rd., Lansing, MI 48910-8104, USA, E-mail: buchweitzj@dcpah.msu.edu

product ions per analyte along with retention time provides greater selectivity nearly eliminating false positive results. With organic pesticides, the use of multiple reaction monitoring (MRM) with the tandem mass spectrometer provides sensitivity to at least the same order of magnitude as electron capture detection [17]. Previous reports have shown the validation of GC-MS/MS analysis of OCPs in biological samples [18,19].

The present work describes the development and single-laboratory validation of a multi-residue method for the detection of organochlorine pesticides in biological matrices using a QuEChERS based extraction approach and GC-MS/MS targeted analysis by MRM. This method allowed for the simultaneous monitoring of 24 organochlorine pesticide residues in liver and blood serum.

Materials and Methods

Pesticide standards

CLP organochlorine pesticide mix (alpha-chlordane, methoxychlor, gamma-chlordane, aldrin, alpha-BHC [benzene hexachloride or hexachlorocyclohexane], beta-BHC, gamma-BHC, delta-BHC, 4,4'-DDD [Dichlorodiphenyldichloroethane], 4,4'-DDE [Dichlorodiphenyldichloroethylene], 4,4'-DDT [dichlorodiphenyltrichloroethane], dieldrin, endosulfan I [alpha], endosulfan II [beta], endosulfan sulfate, endrin, endrin aldehyde, endrin ketone, heptachlor, heptachlor exo-epoxide isomer B) 2000 µg/mL in 50:50 toluene/hexane and trans-Nonachlor 100 µg/mL in hexane were obtained from Supelco Analytical (Sigma-Aldrich Corp. St. Louis, MO, USA); oxychlordane, chromatography purity, and decachlorobiphenyl (internal standard), 99.5% purity were purchased from Cerilliant Corporation (Round Rock, TX) and ChemService (West Chester, PA), respectively.

Solvents and reagents

Acetonitrile UV (chromatography grade) and Isooctane (chromatography grade) were purchased from Honeywell, Burdick and Jackson brand (Muskegon, MI). Glacial acetic acid was purchased from EMD Millipore USA (Billerica, MA). Anhydrous magnesium sulfate (Baker analyzed reagent) and sodium chloride (ACS reagent grade) were purchased from JT Baker (Phillipsburg, NJ). Bondesil-PSA, 40 µm was purchased from Varian (Division of Agilent, Palo Alto, CA). C18, octadecyl-functionalized silica gel, 200-400 mesh was purchased from Aldrich (Sigma-Aldrich Corp. St. Louis, MO, USA). Sodium acetate trihydrate (HPLC grade) was purchased from Fisher Scientific (Waltham, MA). 200-proof ethyl alcohol was purchased from Decon Labs (King of Prussia, PA). Solutions of 1% (v/v) acetic acid in acetonitrile UV and 80:20 isooctane:ethanol were prepared prior to extraction procedures.

Livers and serum

Control livers were obtained from bovine necropsy submissions to the veterinary diagnostic center and pooled control blood in ethylenediaminetetraacetic acid (EDTA) was obtained from canine clinical chemistry panel (CBC) submissions from the veterinary teaching hospital. Livers from bald eagles were obtained through submissions of birds found dead or moribund by the Michigan Department of Natural Resources (Lansing, MI).

Equipment

An Agilent Technologies 7890A GC system coupled with an Agilent 7000 GC/MS Triple Quad detector (Agilent, Santa Clara, CA) was used to perform the GC-MS/MS analysis. The GC separation was conducted with an Agilent 19091S-433UI fused silica capillary column (30 m × 250 µm × 0.25 µm) and was performed with a 4 µL injection into an inlet operated in splitless mode. The liner was an Agilent 5062-3587 split/splitless glass cylinder with single taper from which the glass wool was removed. The inlet was at 13.4 psi with an initial temperature of 210°C held 1 min, ramped 100°C/min to 280°C, held 10 min: then ramped 100°C/min to 210°C. Helium was used as the GC carrier gas at 1.5 mL/min. The GC oven initial temperature was 60°C, held 5 min, then ramped 8°C/min to 300°C, held 0.5 min for a total run time of 35.5 min. The MSD transfer line was held constant at 300°C.

The triple quadrupole mass analyzer was operated in negative EI mode at 70 eV with a source temperature of 230°C. Nitrogen gas was used as the collision gas at 1.5 mL/min and helium was used as the quench gas at 2.25 mL/min. MRM scanning was initiated after 3.75 min solvent delay. Acquisition and data processing were performed with MassHunter software (Agilent, Santa Clara, CA).

Backflushing

Installation of the T-union involved combination of two Agilent 19091S-431 fused silica capillary columns (each 15 m × 250 µm × 0.25 µm) with the first column in sequence operated at 1.2 mL/min constant He flow and the second at 1.4 mL/min constant flow. Backflush involved He flow introduced at the T-union at 17.3 mL/min with a backflush pressure of 100 psi for a total time of 0.52 min while the oven was held at 320°C.

GC/MS/MS Multi-mode Inlet

The Agilent Multi-mode Inlet (MMI) was temperature programmed as described above, but temperatures lower than the 210°C initial temperature were studied, specifically the range 50-75°C at 5-degree increments. Each temperature was held 0.3-min, and then ramped to 325°C at 600°C/min.

QuEChERS extraction procedure

Control bovine liver and serum was found to be uncontaminated with organochlorine pesticides. The liver and serum were stored at -17°C and were thawed before use. The following extraction was performed on all samples and in the preparation of matrix matched calibration standards. 2.0 g of liver was placed in a 50 mL glass test tube and spiked with 20 µL of 20 µg/mL decachlorobiphenyl (internal standard) and appropriate amount of stock or working solution of OC pesticide solution for calibration curve and QC samples. 5 mL of 1% acetic acid in acetonitrile was added to the test tube and the mixture was homogenized using a Brinkmann PTMR 3000 (Kinematica, Bohemia, NY) homogenizer. 0.5 g sodium chloride, 0.5 g sodium acetate, and 2.0 g of magnesium sulfate were added to the sample. The sample was vortexed for 30 seconds and centrifuged 5 min at 3000 rpm. 4 mL of the supernatant layer was transferred to a 20 mL glass test tube containing 0.60 g anhydrous magnesium sulfate, 0.10 g primary secondary amine, and 0.10 g C18. The extract was vortexed 30 seconds and centrifuged 5 min at 3000 rpm. A 2.5 mL aliquot of the supernatant was transferred to a clean test tube and dried on an N-EVAP (Organomation, Berlin, MA) just to dryness. The analytes were reconstituted in 0.5 mL 80:20 isooctane: ethanol and transferred to GC autosampler vials. The same procedure was followed for the serum samples with 1 mL being the initial sample volume.

Use of commercial QuEChERS tubes

The extraction method was simplified with the use of commercially available extraction and clean-up tubes available from Waters

Corporation (Milford, MA). The Waters extraction tube (PN 186004571) contained 1.5 g sodium acetate and 6.0 g magnesium sulfate. The Waters clean up tube (PN 186004834) contained 150 mg PSA, 150 mg C18 and 900 mg magnesium sulfate. The use of the commercially available tubes reduced possible contamination of glassware and streamlined the extraction process. No change in extraction efficiencies were observed in the transition to the commercial tubes.

Construction of calibration curve

Matrix matched standards were prepared from blank matrix samples that underwent the QuEChERS extraction and reconstituted with the addition of appropriate pesticide spikes. Neat standards in pure solvent were prepared from appropriate dilution of standard solution to make working solution of pesticide solution in same concentration as the matrix matched standards. Standard levels were 1, 10, 100, 200, 500 and 1000 ppb in 80:20 isooctane ethanol.

Results and Discussion

Analytical performance

The method's linearity, recovery, repeatability, selectivity, limit of detection and limit of quantitation were evaluated as measures of analytical performance [20]. Neat standards in pure solvent were prepared to examine matrix effects. Matrix may enhance the signal of a compound due to the binding of active sites in the inlet and chromatography column leading to higher amounts of analyte reaching the detector and producing significant increases in recovery [5,21,22]. The effects were apparent in serum samples with an average increase in response of 20.6%. Matrix effects were negligible in liver samples, but matrix matched standards were nevertheless utilized. For the purposes of this work, the general assumption was made that healthy organs such as liver or fluids such as serum provided roughly equivalent matrices on interspecific comparisons. By this reasoning, a commonly available or plentiful tissue such as bovine liver would be an appropriate matrix for determining avian liver pesticides. The authors are aware that there is a toxicology specialty dealing with extrapolation between matrices, usually between environmental categories such as air and water, and between organisms, such as from laboratory animals to humans [23]. Nevertheless, it seemed an appropriate assumption to make, particularly in light of the paucity of unadulterated eagle livers available for the study of subtle matrix effects.

Linearity

Matrix matched calibration curves were constructed in the range of 1 to 1,000 ppb. Calibration curves were constructed by a plot of peak area versus analyte concentration. Linearity was evaluated by the calculation of a six-point linear plot with three replicates, based on linear regression and coefficient of determination, r^2, which should be >0.990. 45 of the 48 calibration curves met this criterion. Experimental linear regression r^2 varied from 0.997 to 0.999 for liver and 0.985 to 0.996 for serum. These correlations improved slightly when calibration curves were fitted to quadratic regressions.

LOD and LOQ

The lower limits of detection (LOD) and of quantitation (LOQ) were calculated based on FDA guidance for bioanalytical method validation [20]. LOD was defined as 3.3 times the standard deviation of the blank divided by the slope of the calibration curve. The standard deviation was calculated from the responses of each analyte in 5 blank matrix samples. LOQ was defined as 10 times the standard deviation of the blank divided by the slope of calibration curve. The working upper

and lower LOQ were taken as the maximum and minimum of linear calibration curve, 1000 ppb and 10 ppb, respectively. The calculated LOD is likely to be far below that of the actual amount detectable. This is due to the way that the data analysis software calculates abundances for analytes in the blanks. A peak is identified in the RT window defined in the program for a specific analyte. In the blank sample, the algorithm for peak integration and start and end points are likely to be variable leading to very low peak areas on a very flat chromatogram. Because of this, a conservative LOD of the lowest consistently detected concentration in the calibration curve was used, 10 ppb (ng/mL). The r^2, LOD, and LOQ calculated values for the OCPs analyzed are summarized in Table 1; Figure 1 illustrates chromatographic separation with the mixed OCP standards.

Recovery, repeatability and stability

Recovery of analytes was assessed at 3 spike levels with 3 replicates: 10, 20 and 1000 ppb. Acceptable levels of recovery for method validation are 70 to 120% [20]. Experimental recovery averages ranged from 46.8 to 117%, with 94.4% of samples within the accepted range. Repeatability is expressed as the relative standard deviation (RSD) of the QC sample recoveries. The accepted RSD values are <15% or <20% at the LOQ [20]. Overall, 131 of 144 of the analyte determinations in the validation met this criterion. Hexachlorobenzene was the only analyte that had lower than acceptable recoveries in more than one spike concentration level. A summary of average recovery and RSD for each analyte is given in Table 2. Stability was measured for each analyte at spike concentrations of 10 and 100 ppb in serum and liver matrices. Extracts were taken from spiked serum and livers at 72 hours and 1 hour prior to analysis. The percent recovery was evaluated for each analyte tested (Table 3). The percent recovery ranged from 96.3% to 111.4% in the 10 ppb serum spike and 96.2% to 122% in the 100 ppb serum spike demonstrating good sample stability over 3 days in this matrix. Alternatively, the percent recovery ranged from 32.2% to 117% in the 10 ppb liver spike and 73.7% to 161.3% in the 100 ppb liver spike. The complexity of the liver matrix was most likely responsible for the increased variability observed between analytes.

Selectivity

The pesticides evaluated in this study were identified by comparing retention time (RT) and precursor ion to product ion transitions. Two additional qualifier ion transitions were chosen for each analyte. The qualifier response to quantifier response ratio should not deviate more than 20%. Ions used for pesticide detection in identification are presented in Table 4. Retention time stability was expressed as the RSD of the time. RT should be ± 1% or six seconds for GC, whichever is greater [20]. Experimental data shows excellent RT reproducibility. The RT was also compared to the average RT of several solvent standard runs to examine the matrix effects on RT. No significant change in RT was detected. Retention times for the 24 analytes are provided in Table 4.

Precision

Precision of the organochlorine pesticide assay was assessed by determination of average refit values of standards to their respective standard curves over the course of three separate determinations, and calculation of the relative standard deviation (RSD) as %RSD. Precision was reasonable, with the average %RSD <30% across all analytes for the 100, 200, 500 and 1000 ppb standards. Table 5 lists the %RSD for each compound and all six standards.

Application of the method to real-life samples

Bald eagles have become victims of the farming industry and its

Compound Name	Liver			Serum		
	r^2	LOD (ppb)	LOQ (ppb)	r^2	LOD (ppb)	LOQ (ppb)
Alpha- BHC	0.998	0.1	0.32	0.995	0.04	0.13
Hexachlorobenzene	0.998	0.09	0.27	0.996	0.02	0.05
Gamma- BHC	0.998	0.07	0.22	0.994	0.02	0.06
Beta-BHC	0.998	0.08	0.24	0.993	0.03	0.1
Delta- BHC	0.998	0.14	0.43	0.995	0.06	0.17
Heptachlor	0.999	0.04	0.13	0.993	0.01	0.04
Aldrin	0.998	1.16	3.53	0.995	0.14	0.42
Oxychlordane	0.999	0.7	2.13	0.995	0.01	0.02
Heptachlor epoxide	0.999	0.1	0.32	0.993	0.01	0.03
Endosulfan I	0.999	1.83	5.56	0.993	0.01	0.03
Gamma-chlordane	0.999	0.07	0.21	0.993	0.01	0.03
Alpha-chlordane	0.999	0.03	0.09	0.993	0.01	0.03
Trans-nonachlor	0.999	0.08	0.25	0.993	0.01	0.02
4,4'-DDE	0.999	0.32	0.95	0.994	0.03	0.09
Dieldrin	0.999	2.44	7.4	0.991	0.14	0.42
Endrin	0.998	0.87	2.63	0.996	0.22	0.65
Endosulfane II	0.998	0.22	0.66	0.994	0.03	0.09
4,4'-DDD	0.998	0.14	0.43	0.993	0.02	0.04
Endosulfane Sulfate	0.998	0.05	0.15	0.989	0.02	0.07
4,4'-DDT	0.998	0.36	1.08	0.985	0.06	0.19
Endrin ketone	0.997	0.3	0.91	0.994	0.04	0.13
Endrin Aldehyde	0.999	0.32	0.97	0.994	0.1	0.29
Methoxychlor	0.998	0.31	0.93	0.985	0.05	0.16
Mirex	0.998	0.04	0.11	0.993	0.01	0.02

Table 1: Determined limits of detection (LOD) and quantitation (LOQ) in liver and serum matrices, including coefficient of determination (r^2) of corresponding standard curves.

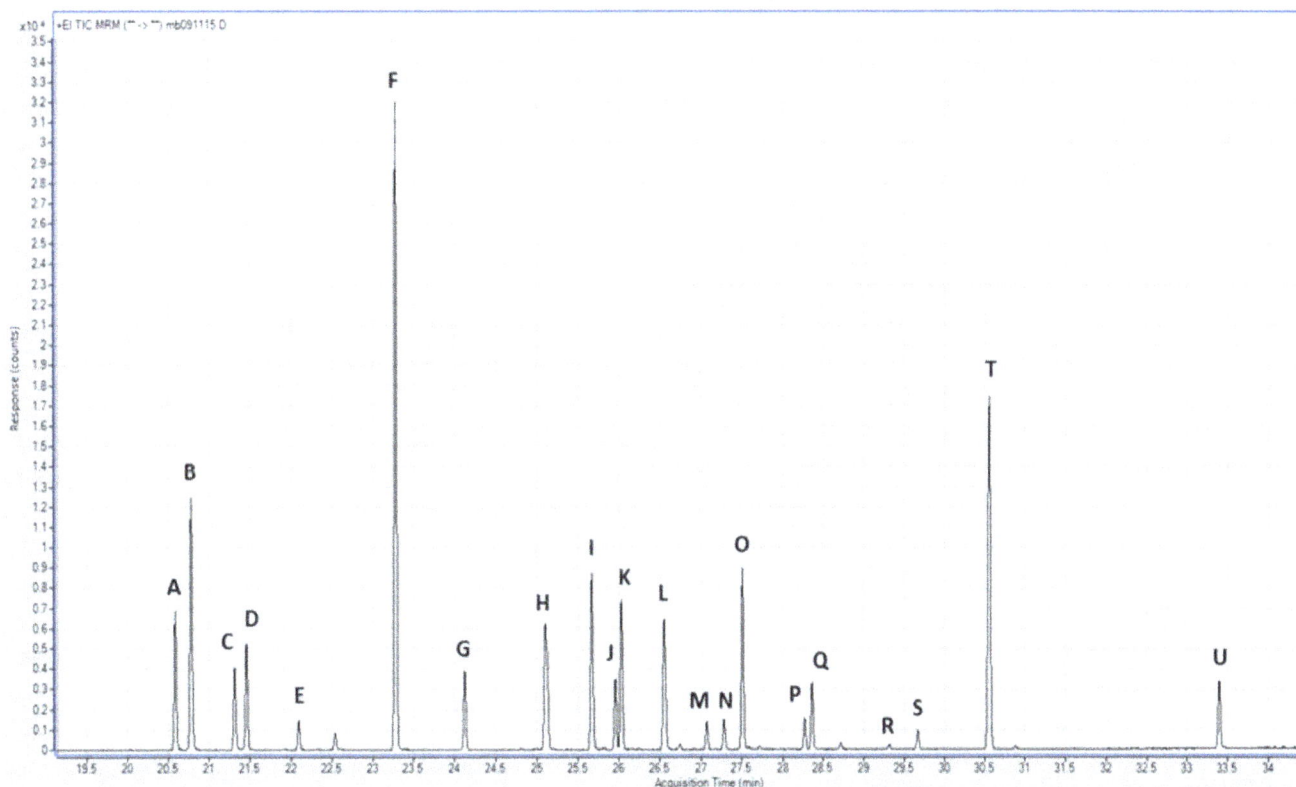

Figure 1: TIC MRM of 24 organochlorine analytes of study. Chromatograph peaks were assigned as follows: A) Alpha-BHC; B) Hexachlorobenzene; C) Gamma-BHC; D) Beta-BHC; E) Delta-BHC; F) Heptachlor; G) Aldrin; H) Oxychlordane, Heptachlor Epoxide; I) Endosulfan I; J) Alpha-Chlordane, Gamma-Chlordane; K) Trans-Nonachlor; L) 4,4'-DDE, Dieldrin; M) Endrin; N) Endosulfan II; O) 4,4'-DDD; P) Endosulfan Sulfate; Q) 4,4'-DDT; R) Endrin Ketone, Endrin Aldehyde; S) Methoxychlor; T) Mirex; U) Decachlorobiphenyl (internal standard).

Compound Name	Liver			Serum		
	10 ppb	200 ppb	1000 ppb	10 ppb	200 ppb	1000 ppb
Alpha-BHC	93.5 (2.4)	87.4 (9.0)	92.4 (7.4)	73.5 (9.9)	73.0 (15.9)	62.4 (20.5)
Hexachlorobenzene	69.6 (2.6)	75.2 (5.7)	77.4 (5.7)	56.9 (2.1)	60.8 (14.2)	46.8 (9.4)
Gamma-BHC	88.2 (1.9)	92.2 (6.4)	96.3(8.9)	88.3 (10.8)	87.7 (12.1)	83.2 (11.8)
Beta-BHC	94.8 (0.9)	88.0 (0.9)	94.2 (7.7)	83.2 (7.4)	80.4 (14.6)	70.7 (18.4)
Delta-BHC	87.8 (1.9)	94.7 (7.0)	94.6 (9.0)	91.3 (12.8)	91.7 (16.3)	77.3 (10.5)
Heptachlor	96.9 (3.6)	87.9 (9.3)	91.2 (9.6)	78.8 (8.8)	73.9 (12.7)	67.0 (15.9)
Aldrin	84.3 (3.9)	85.2 (7.6)	89.1 (8.0)	87.2 (5.6)	77.9 (10.1)	68.7 (12.9)
Oxychlordane	84.1 (0.8)	91.8 (9.0)	93.5 (9.9)	85.1 (6.9)	82.4 (9.7)	80.4 (11.5)
Heptachlor epoxide	89.3 (6.2)	92.7 (9.2)	95.0 (9.4)	87.9 (14.0)	83.5 (9.3)	81.4 (11.5)
Endosulfan I	80.1 (16.7)	89.1 (5.3)	94.4 (9.0)	73.7 (11.5)	81.0 (12.9)	81.8 (12.2)
Gamma-Chlordane	86.5 (2.0)	90.3 (9.7)	94.2 (9.1)	79.6 (3.7)	83.2 (10.0)	85.4 (9.1)
Alpha-Chlordane	80.6 (5.0)	92.1 (8.6)	94.5 (10.0)	88.8 (6.9)	85.3 (8.0)	85.9 (8.1)
Trans-Nonachlor	86.3 (2.3)	92.5 (8.3)	91.9 (11.3)	89.6 (11.4)	85.0 (8.3)	85.6 (8.9)
4,4'-DDE	81.0 (2.0)	90.0 (10.0)	92.7 (9.9)	83.9 (12.5)	82.6 (6.7)	86.3 (6.1)
Dieldrin	83.8 (3.2)	93.6 (8.0)	94.6 (9.5)	77.9 (4.1)	85.3 (8.0)	92.7 (1.7)
Endrin	94.7 (9.9)	93.0 (10.6)	95.8 (10.1)	83.6 (3.1)	84.6 (10.3)	78.3 (4.6)
Endosulfan II	98.0 (8.7)	93.6 (8.3)	97.5 (8.7)	91.9 (13.4)	85.9 (9.3)	89.8 (8.0)
4,4'-DDD	117.3 (1.8)	93.0 (6.2)	96.7 (8.1)	86.0 (4.4)	94.1 (8.0)	83.0 (5.6)
Endosulfan Sulfate	86.2 (0.6)	92.5 (6.2)	90.4 (6.4)	92.3 (18.6)	92.2 (23.8)	91.4 (10.0)
4,4'-DDT	85.2 (3.7)	90.1 (12.1)	93.6 (10.8)	82.9 (16.0)	79.3 (10.6)	87.5 (10.7)
Endrin Ketone	109.4 (15.4)	94.5 (8.1)	96.7 (9.6)	80.2 (15.1)	87.8 (16.7)	82.6 (6.8)
Endrin Aldehyde	56.7 (35.4)	94.6 (2.2)	96.3 (8.0)	84.3 (8.8)	86.3 (14.3)	82.3 (6.7)
Methoxychlor	84.6 (3.9)	92.5 (5.7)	96.4 (8.7)	76.9 (10.2)	88.5 (11.7)	81.5 (5.0)
Mirex	69.3 (3.0)	78.3 (7.2)	80.8 (6.5)	75.4 (8.0)	80.5 (4.5)	75.1 (4.0)

Table 2: Percent recovery and percent RSD (in parentheses) summary of spiked samples in two matrices at three spiked levels.

heavy reliance on chlorinated pesticides. For example, organochlorine chemicals appear as significant stressors on Great Lakes bald eagle populations when compared with stresses on successful populations of bald eagles continent-wide [24], despite the compounds' discontinuance in the US since the 1970-80s. Livers from three bald eagles submitted to the Diagnostic Center for Population and Animal Health at Michigan State University (DCPAH) were randomly chosen for assessment of organochlorine burden by the described GC-MS/MS method, and quantitative results are shown in Table 6. In addition, serum samples from three bald eagles, a common raven and three turkey vultures were also assessed (Table 7). Figure 2 illustrates the quantitative standard curves and confirmatory quantifier and qualifier ion chromatograms for two representative compounds in one of the eagles. There were significant amounts of the DDT metabolites 4,4'-DDE and 4,4'-DDD, as well as hexachlorobenzene, trans-nonachlor and mirex. DDT was banned by the EPA in 1976; hexachlorobenzene was used as a pesticide until 1965 and has been banned globally under the Stockholm Convention on persistent organic pollutants, 2001; trans-nonachlor is a bioaccumulating component of the insecticide chlordane, banned in 1983; and mirex has been banned since 1976 [25]. This finding is an unfortunate testament to the classification of the chlorinated pesticides as persistent organic pollutants.

Improvements to the method

The method as described has been validated; however, improvements were sought to provide increased robustness for daily operation and the best sensitivity possible. The approach took the tact of 1) introducing a T-union into the chromatographic column in order to reduce inlet contaminant buildup by enabling backflushing, 2) making use of an installed multi-mode inlet for temperature programming and large volume injection and 3) use of commercial extraction tubes. Chromatography with backflushing post-run had relatively minor

effects, in general with some increase in retention; for example, alpha-BHC increased from 20.5 to 22.8 min RT, and methoxychlor increased from 29.6 to 31.8 min RT (not shown). This was principally an effect of the lower flow rate. The principal advantages of introducing backflush were improvements in chromatographic column and inlet cleanliness, as judged by decreased need for inlet cleaning or column maintenance. Its mention here was crucial in affirming that no deleterious effects were seen on the organochlorine pesticide chromatography or sensitivity.

The Agilent multi-mode inlet was studied with the same GC oven conditions but with decreased initial inlet temperatures substantially lower than 210°C. This offered advantages in being able to invoke larger injection volumes if desired as well as some chromatographic advantages from solvent focusing in cases where the initial oven temperature could also be lowered [26]. Initial inlet temperatures of 50, 55, 60, 65, 70 and 75°C were studied, and the lower temperatures did offer improvement on overall TIC area relative to the original starting conditions of 210°C. As shown in Table 8 for 55°C initial temperature, there was substantial increase in areas for compounds G-S, with little change in relative retention time, but some relatively minor decreases for the initial group of compounds, particularly the BHC family. Compound chromatograms are compared in Figure 3. Although it was difficult to determine differences within the relatively narrow range of inlet temperatures from 50 to 75°C, it was found that matching the inlet temperature to a reduced oven starting temperature was more important in improving sensitivity. Figure 4 compares the MMI 55°C inlet matched to a 45 to 300°C oven program in comparison to the original 210°C splitless inlet matched to the 60 to 300°C program. Note the shift in RTs accompanied by substantial increases in TIC peak areas.

The Agilent vapor volume software program [27] enables the chromatographer to judge whether a given injection volume will overload the injection liner volume and thereby shunt injected

Compound Name	Liver		Serum	
	10 ppb	100 ppb	10 ppb	100 ppb
Alpha- BHC	94.1%	75.6%	100.4%	117.5%
Hexachlorobenzene	ND	ND	ND	ND
Gamma- BHC	96.9%	91.3%	100.7%	122.7%
Beta-BHC	102.3%	89.9%	99.5%	108.4%
Delta- BHC	112.3%	83.8%	98.8%	111.5%
Heptachlor	78.8%	108.1%	101.4%	113.4%
Aldrin	94.1%	96.6%	105.2%	108.0%
Oxychlordane	74.7%	112.5%	97.9%	107.5%
Heptachlor epoxide	37.5%	118.7%	102.6%	111.2%
Endosulfan I	90.1%	106.2%	100.9%	104.1%
Gamma-chlordane	32.2%	101.6%	99.5%	106.8%
Alpha-chlordane	74.4%	97.0%	101.4%	104.4%
Trans-nonachlor	74.4%	95.2%	100.5%	107.6%
4,4'-DDE	104.2%	104.4%	99.9%	99.4%
Dieldrin	72.3%	161.3%	101.3%	102.9%
Endrin	78.2%	80.0%	101.9%	102.8%
Endosulfane II	99.3%	73.7%	100.9%	112.8%
4,4'-DDD	110.7%	86.2%	100.0%	101.0%
Endosulfane Sulfate	48.7%	115.8%	97.1%	98.4%
4,4'-DDT	117.0%	99.6%	96.3%	97.0%
Endrin ketone	78.3%	99.9%	101.0%	109.7%
Endrin Aldehyde	ND	ND	ND	ND
Methoxychlor	85.0%	81.4%	99.2%	96.2%
Mirex	ND	ND	ND	ND

Table 3: Stability of OCP analytes in liver and serum matrices over three days represented by percent recovery. ND: Not determined.

Compound Name	RT (min)	Quantitation MRM (CE)	Qualifier MRM 1 (CE)	Qualifier MRM 2 (CE)
Alpha- BHC	20.55	180.9→145 (15)	217 -> 181 (5)	219→183 (5)
Hexachlorobenzene	20.74	283.8→213.9 (30)	283.8→248.8 (15)	281.8→211.9 (30)
Gamma- BHC	21.28	180.9→145 (15)	217 -> 181 (5)	219→183 (5)
Beta-BHC	21.42	180.9→145 (15)	217 -> 181 (5)	219→183 (5)
Delta- BHC	22.06	180.9→145 (15)	217 -> 181 (5)	219→183 (5)
Heptachlor	23.33	272→236.9 (15)	236.8→118.9 (25)	273.9→238.9 (15)
Aldrin	24.09	263→193 (35)	263→191 (35)	263→263 (0)
Heptachlor epoxide	25.07	272→236.9 (15)	262.9→192.9 (35)	273.9→238.9 (15)
Oxychlordane	25.09	114.9→51.1 (25)	184.9→121 (15)	236.9→142.9 (25)
Gamma-chlordane	25.63	272→236.9 (15)	273.9→238.9 (15)	277→241 (5)
Endosulfan I	25.91	195.1→159 (5)	272→236.9 (15)	273.9→238.9 (15)
Alpha-chlordane	26.00	272→236.9 (15)	273.9→238.9 (15)	277→241 (5)
Trans-nonachlor	26.12	409→409 (0)	407→407 (0)	411→411 (0)
4,4'-DDE	26.53	246.1→176.2 (30)	315.8→246 (15)	317.8→317.8 (0)
Dieldrin	26.54	263→191 (35)	277→241 (5)	263→193 (35)
Endrin	27.03	263→193 (35)	263→191 (35)	263→263 (0)
Endosulfan II	27.25	195.1→159 (5)	206.9→172 (15)	180.9→145 (15)
4,4'-DDD	27.48	235.1→165.2 (20)	237→165.1 (27)	237→237 (0)
Endrin Aldehyde	27.70	184.9→121 (15)	206.9→172 (15)	195.1→159 (5)
Endosulfan Sulfate	28.23	272→236.9 (15)	273.9→238.9 (15)	277→241 (5)
4,4'-DDT	28.33	235.1→165.2 (20)	237→237 (0)	237→165.1 (27)
Endrin ketone	29.27	180.9→145 (15)	114.9→51.1 (25)	195.1→159 (5)
Methoxychlor	29.64	227→141.1 (40)	227→169.1 (30)	227→212 (17)
Mirex	30.51	272→236.9 (15)	273.9→236.9 (15)	273.9→238.9 (15)

Table 4: Mass spectrometer settings for the OCP GC-MS/MS method, including precursor ion→product ion m/z values and collision energies, kV, (CE) in parentheses, with compounds arranged in retention time (RT) order.

Compound Name	Calibrator levels					
	1000 ppb	500 ppb	200 ppb	100 ppb	10 ppb	1 ppb
Alpha-BHC	1%	4%	13%	21%	11%	104%
Hexachlorobenzene	2%	4%	32%	23%	42%	124%
Gamma-BHC	1%	4%	10%	27%	33%	116%
Beta-BHC	7%	10%	15%	11%	64%	97%
Delta-BHC	8%	16%	10%	25%	3%	65%
Heptachlor	0%	2%	10%	19%	18%	87%
Aldrin	1%	0%	7%	13%	20%	106%
Heptachlor epoxide	1%	8%	5%	10%	10%	19%
Oxychlordane	1%	6%	6%	10%	10%	56%
Trans-Chlordane	2%	9%	5%	9%	11%	12%
Endosulfan I	1%	7%	3%	10%	61%	49%
Cis-Chlordane	2%	9%	4%	5%	12%	26%
4,4'-DDE	1%	6%	5%	8%	6%	23%
Dieldrin	2%	7%	6%	6%	6%	19%
Trans-Nonachlor	1%	5%	5%	13%	16%	46%
Endrin	1%	6%	7%	9%	14%	40%
Endosulfan II	2%	9%	7%	14%	54%	88%
4,4'-DDD	3%	6%	10%	13%	25%	64%
Endrin aldehyde	14%	22%	28%	18%	65%	142%
Endosulfan sulfate	5%	10%	9%	10%	38%	88%
4,4'-DDT	5%	7%	14%	22%	33%	23%
Endrin ketone	2%	6%	7%	13%	39%	141%
Methoxychlor	5%	5%	15%	18%	29%	47%
Mirex	2%	6%	7%	8%	10%	21%
Average across all compounds	3%	7%	10%	14%	26%	67%

Table 5: Average percent RSD for all OCPs measured as a refit of calibrators to their respective standard curves across three separate calibrations performed on different days.

Compound (RT order)	Bald Eagle 1	Bald Eagle 2	Bald Eagle 3
Alpha-BHC	<10	<10	<10
Hexachlorobenzene	311	355	407
Gamma-BHC	<10	<10	<10
Beta-BHC	<10	<10	<10
Delta-BHC	<10	<10	<10
Heptachlor	<10	<10	<10
Aldrin	<10	<10	<10
Heptachlor epoxide	<10	<10	<10
Oxychlordane	<10	<10	<10
Trans-Chlordane	<10	<10	<10
Endosulfan I	<10	<10	<10
Cis-Chlordane	<10	<10	<10
4,4'-DDE	1950	977	2600
Dieldrin	<10	<10	<10
Trans-Nonachlor	812	132	803
Endrin	<10	<10	<10
Endosulfan II	<10	<10	<10
4,4'-DDD	<10	<10	62
Endrin aldehyde	<10	<10	<10
Endosulfan sulfate	<10	<10	<10
4,4'-DDT	<10	<10	<10
Endrin ketone	<10	<10	<10
Methoxychlor	<10	<10	<10
Mirex	2111	570	464

Table 6: OCP contaminants found in three separate bald eagle liver samples. All values expressed in ppb.

Compound (RT order)	Bald Eagle 1	Bald Eagle 2	Bald Eagle 3	Common Raven	Turkey Vulture 1	Turkey Vulture 2	Turkey Vulture 3
Alpha-BHC	<10	<10	<10	<10	<10	<10	<10
Hexachlorobenzene	ND	ND	ND	ND	ND	ND	ND
Gamma-BHC	<10	<10	<10	<10	<10	<10	<10
Beta-BHC	<10	<10	<10	<10	<10	<10	<10
Delta-BHC	<10	<10	<10	<10	<10	<10	<10
Heptachlor	<10	<10	<10	<10	<10	<10	<10
Aldrin	<10	<10	<10	<10	<10	<10	<10
Heptachlor epoxide	<10	<10	<10	<10	<10	<10	<10
Oxychlordane	<10	<10	<10	<10	<10	<10	<10
Trans-Chlordane	<10	<10	<10	<10	<10	<10	<10
Endosulfan I	<10	<10	<10	<10	<10	<10	<10
Cis-Chlordane	<10	<10	<10	<10	<10	<10	<10
4,4'-DDE	62	288	133	121	15	10	22
Dieldrin	11	10	<10	<10	<10	<10	<10
Trans-Nonachlor	<10	41	12	<10	<10	<10	<10
Endrin	<10	<10	<10	<10	<10	<10	<10
Endosulfan II	<10	<10	<10	<10	<10	<10	<10
4,4'-DDD	<10	<10	<10	<10	<10	<10	<10
Endrin aldehyde	<10	<10	<10	<10	<10	<10	<10
Endosulfan sulfate	<10	<10	<10	<10	<10	<10	<10
4,4'-DDT	<10	<10	<10	<10	<10	<10	<10
Endrin ketone	<10	<10	<10	<10	<10	<10	<10
Methoxychlor	<10	<10	<10	<10	<10	<10	<10
Mirex	ND	ND	ND	ND	ND	ND	ND

Table 7: OCP contaminants identified in the serum of three avian species: bald eagle, common raven, and turkey vulture. All values expressed in ppb. ND = not determined

Compound	Inlet 55°C RT, min	Inlet 210°C RT, min	TIC area ratio, t55/t210	RT ratio, t55/t210
A	22.84	22.87	0.1986	0.9985
C	23.52	23.62	1.5822	0.9958
D	23.69	23.73	0.4915	0.9982
E	24.28	24.38	0.4663	0.9962
F	25.53	25.56	0.4214	0.9990
G	26.39	26.42	4.2794	0.9991
H	27.34	27.39	2.1271	0.9982
J	27.89	27.95	10.2979	0.9979
I	28.20	28.25	3.2446	0.9982
K	28.26	28.30	7.1228	0.9986
L	28.70	28.76	8.8798	0.9980
O	29.65	29.73	5.9753	0.9974
Q	30.50	30.57	3.4425	0.9978
S	31.76	31.82	3.2488	0.9982

Table 8: Study of effects of inlet temperature in the MMI. Shown are retention times (RT) at 55 and 210°C for 14 of the compounds under study, and the relative changes in total ion chromatogram (TIC) area and RT on comparing temperature 55 divided by temperature 210 (t55/t210).

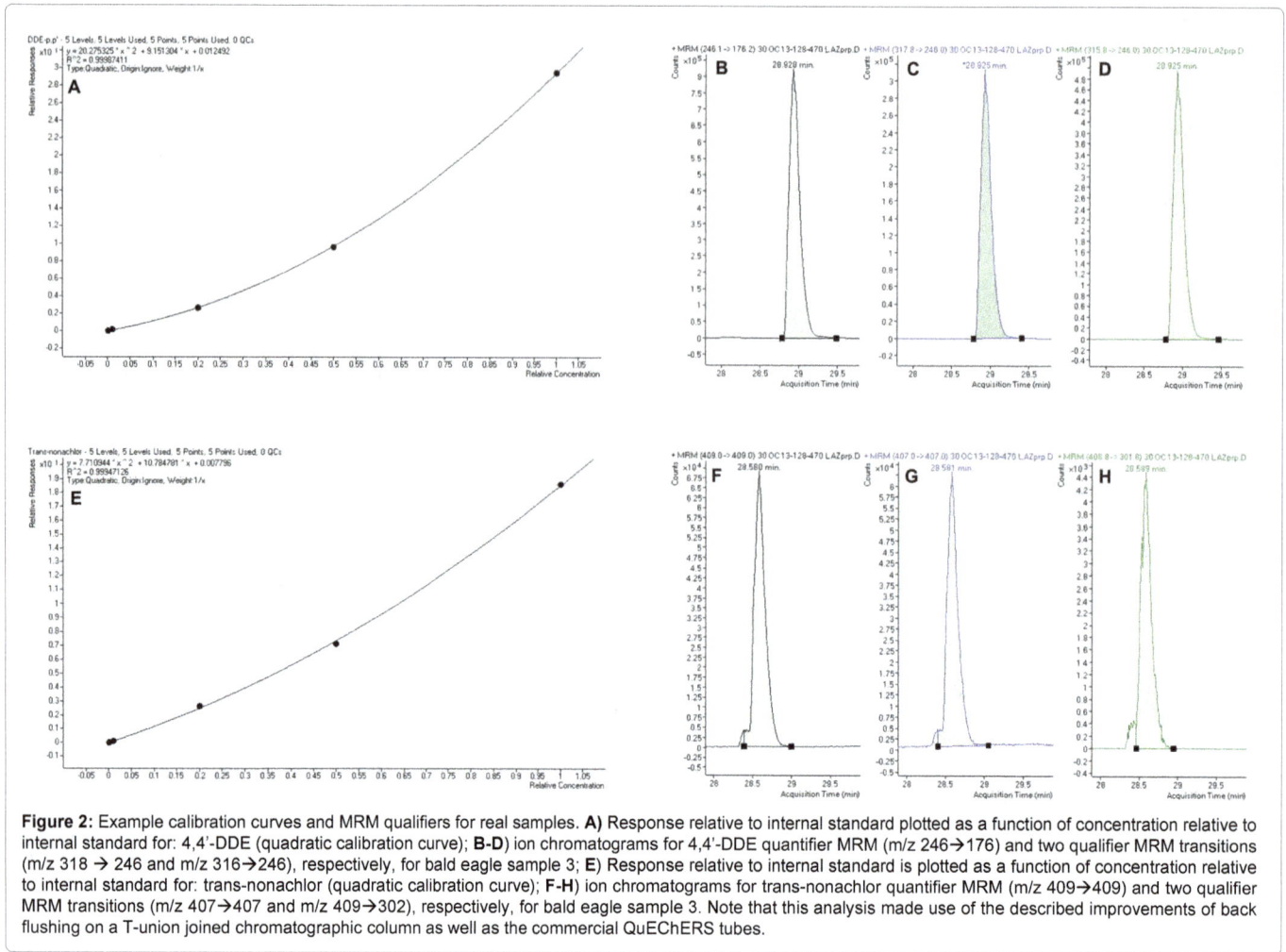

Figure 2: Example calibration curves and MRM qualifiers for real samples. **A)** Response relative to internal standard plotted as a function of concentration relative to internal standard for: 4,4'-DDE (quadratic calibration curve); **B-D)** ion chromatograms for 4,4'-DDE quantifier MRM (m/z 246→176) and two qualifier MRM transitions (m/z 318 → 246 and m/z 316→246), respectively, for bald eagle sample 3; **E)** Response relative to internal standard is plotted as a function of concentration relative to internal standard for: trans-nonachlor (quadratic calibration curve); **F-H)** ion chromatograms for trans-nonachlor quantifier MRM (m/z 409→409) and two qualifier MRM transitions (m/z 407→407 and m/z 409→302), respectively, for bald eagle sample 3. Note that this analysis made use of the described improvements of back flushing on a T-union joined chromatographic column as well as the commercial QuEChERS tubes.

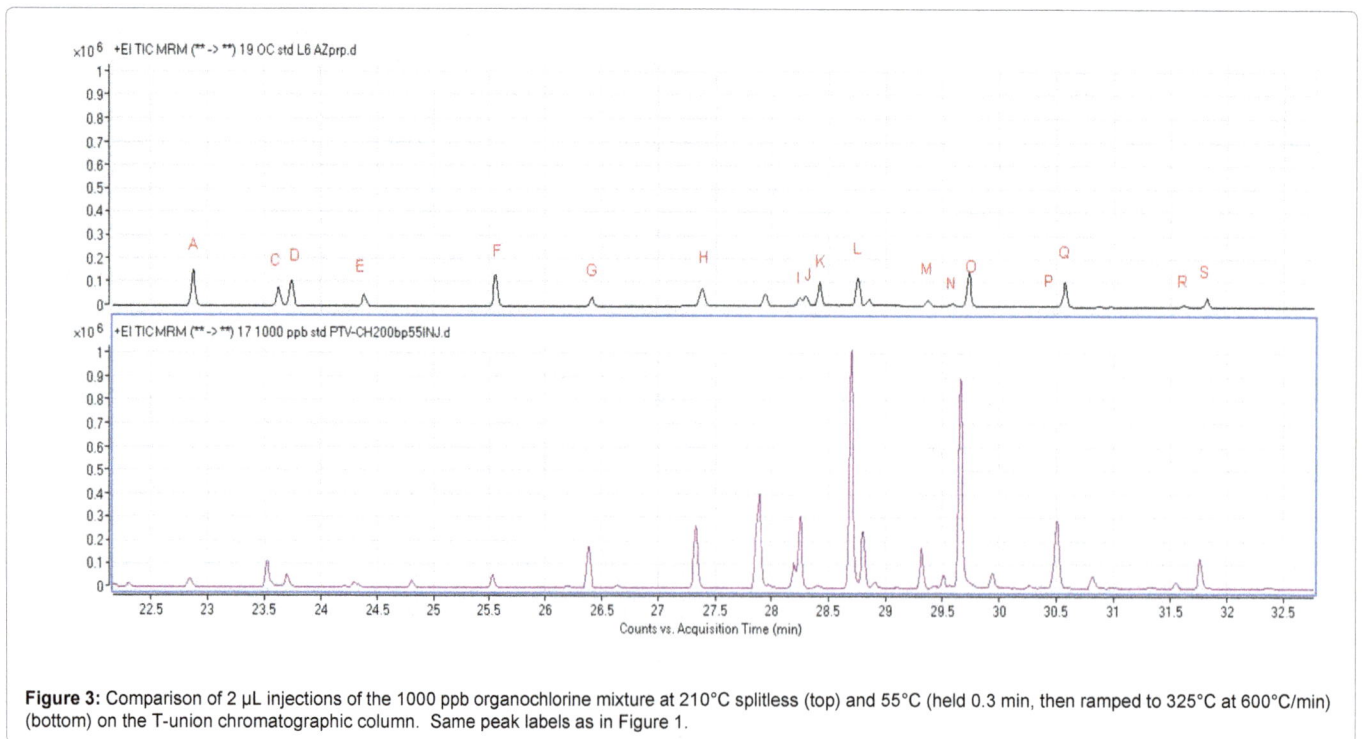

Figure 3: Comparison of 2 µL injections of the 1000 ppb organochlorine mixture at 210°C splitless (top) and 55°C (held 0.3 min, then ramped to 325°C at 600°C/min) (bottom) on the T-union chromatographic column. Same peak labels as in Figure 1.

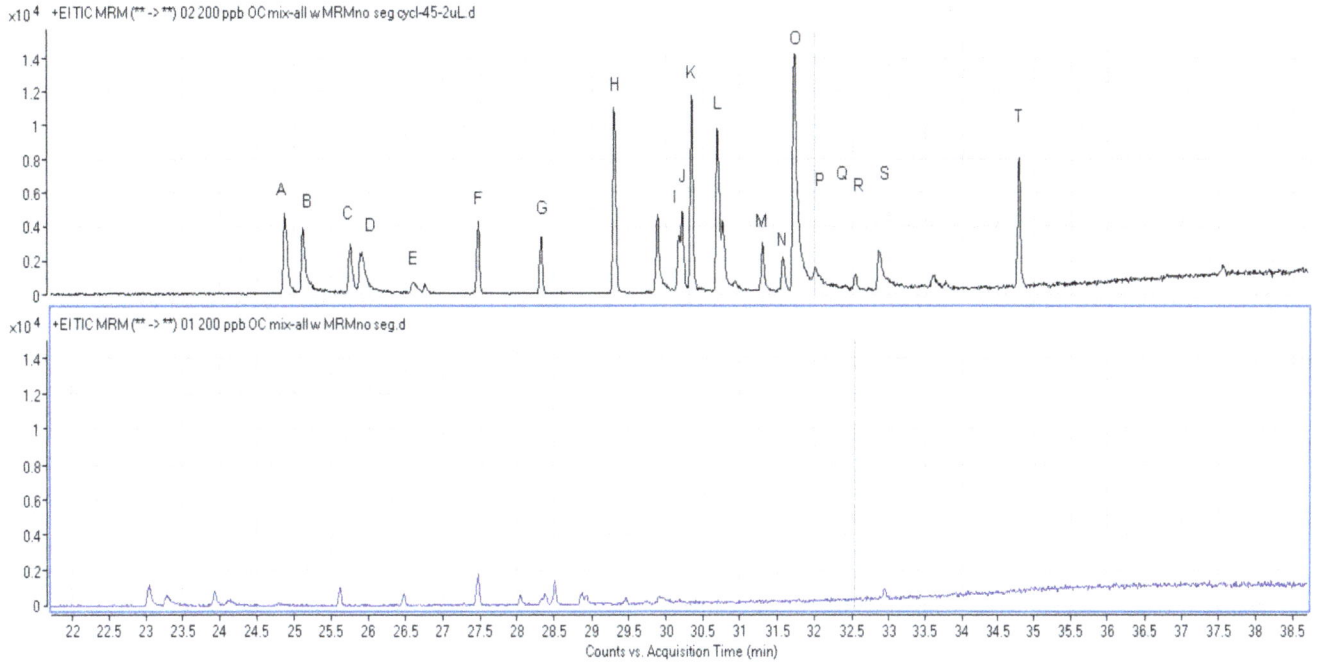

Figure 4: Comparison of 2 µL injections of the 200 ppb organochlorine mixture at 55°C (held 0.3 min, then ramped to 325°C at 600°C/min) with the oven temperature program starting at 45°C [top] with the original oven temperature program and splitless 210°C injector (bottom), both on the T-union chromatographic column. Same peak labels as in Figure 1. Note that peak retention times in the top TIC trace have increased by 2 min uniformly.

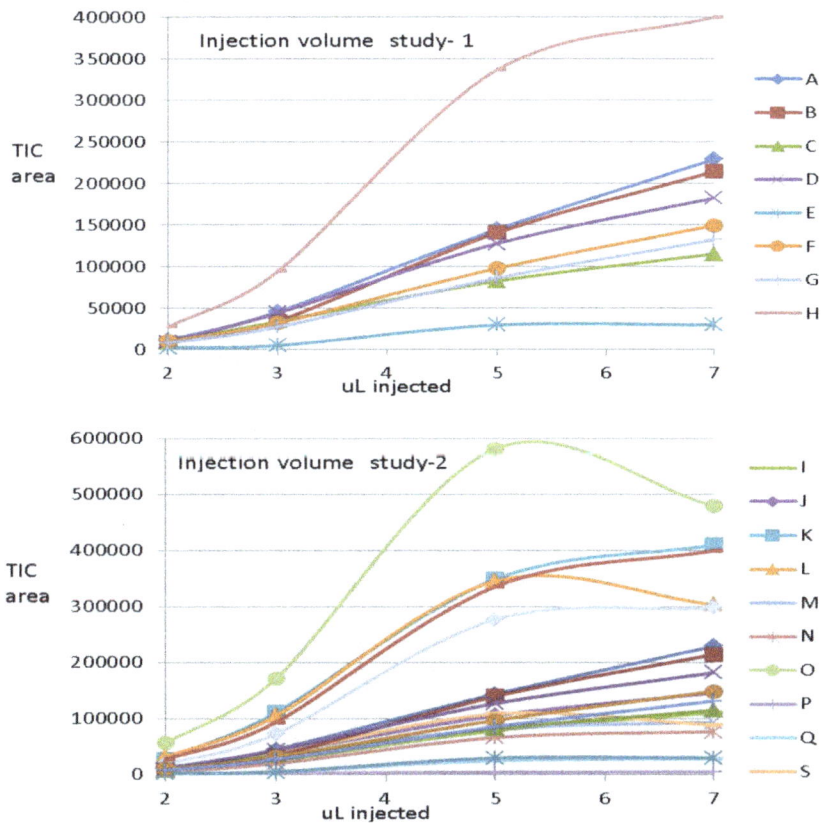

Figure 5: TIC area as a function of number µL injected for compounds A-H [top] and I-S [bottom]. The oven was temperature programmed from 45 (held 5 min) to 300°C at 8°C/min, and the MMI inlet was programmed from 55 to 320°C at 600°C/min.

material to the exhaust valve rather than onto the chromatographic column. By this criterion, 100% iso-octane theoretically appears suitable for a 4 µL injection at an estimated 94% of the 900 µL total liner volume at 210°C inlet temperature and 12 psi inlet pressure, whereas 100% ethanol would be unsuitable at 267% the inlet capacity. Reducing the inlet temperature to 55°C reduces these high percentages to 64% and 181% respectively, making it likely that large volume injection should not be a problem with the iso-octane:ethanol 80:20 solvent mixture. Injection volume capacity on the instrument was assessed with the optimized combination of 55°C programmed inlet with the 45 to 300°C oven temperature program, and the results are displayed in Figure 5. Most of the compounds demonstrated increase in response across the entire 2 to 7 µL range, whereas a few (E, O, L and S) were at maximum response around 5 µL.

Conclusion

The modified QuEChERS extraction approach coupled with GC-MS/MS analyte detection improved analytical turn-around time and increased sensitivity for analyte detection of organochlorine pesticides having a limit of detection of 10 ppb. The implementation of matrix matched calibration curves allowed for enhanced sensitivity demonstrated by the increased instrument response to the same concentration of analyte in matrix and solvent. This approach can clearly be extended to a variety of veterinary diagnostic panels that similarly use less efficient or lower specificity methodologies for assessing animal matrices for toxicant exposure. Finally, application of the optional temperature-programmable multi-mode inlet revealed techniques for improving sensitivity of the method as applied to analytes extracted from matrix, should greater sensitivity be required due to small sample size and/or a requirement for maximum detectability.

Acknowledgements

The authors would like to thank Dr. Tom Cooley from the Michigan Department of Natural Resources for extending the opportunity to evaluate liver samples from bald eagles submitted to the veterinary diagnostic center for routine post-mortem analysis. In addition, we would like to thank Dr. Dan Varland with Coastal Raptors for his contribution of avian serum for organochlorine analysis.

References

1. Centers for Disease Control and Prevention (CDC) (2005) Fourth National Report on Human Exposure to Environmental Chemicals. Atlanta (GA).

2. Resource Futures International for the World Bank and CIDA (2001). Persistent Organic Pollutants and the Stockholm Convention: A Resource Guide.

3. Margariti MG, Tsakalof AK, Tsatsakis AM (2007) Analytical methods of biological monitoring for exposure to pesticides: recent update. Ther Drug Monit 29: 150-163.

4. Barr DB, Needham LL (2002) Analytical methods for biological monitoring of exposure to pesticides: a review. J Chromatogr B Analyt Technol Biomed Life Sci 778: 5-29.

5. LeDoux M (2011) Analytical methods applied to the determination of pesticide residues in foods of animal origin. A review of the past two decades. J Chromatogr A 1218: 1021-1036.

6. Brown P, Charlton A, Cuthbert M, Barnett L, Ross L, et al. (1996) Identification of pesticide poisoning in wildlife. J Chromatogr A 754: 463-478.

7. Lambropoulou DA, Albanis TA (2007) Methods of sample preparation for determination of pesticide residues in food matrices by chromatography-mass spectrometry-based techniques: a review. Anal Bioanal Chem 389: 1663-1683.

8. Anastassiades M, Lehotay SJ, Stajnbaher D, Schenck FJ (2003) Fast and easy multiresidue method employing acetonitrile extraction/partitioning and "dispersive solid-phase extraction" for the determination of pesticide residues in produce. J AOAC Int 86: 412-431.

9. Wilkowska A and Biziuk M (2011) Determination of pesticide residues in food matrices using the QuEChERS methodology. Food Chemistry 125: 803-812.

10. Bruzzoniti MC, Checchini L, De Carlo RM, Orlandini S, Rivoira L, et al. (2014) QuEChERS sample preparation for the determination of pesticides and other organic residues in environmental matrices: a critical review. Anal Bioanal Chem 406: 4089-4116.

11. Lehotay SJ (2011) QuEChERS sample preparation approach for mass spectrometric analysis of pesticide residues in foods. Methods Mol Biol 747: 65-91.

12. Jia F, Wang W, Wang J, Yin J, Liu Y, et al. (2012) New strategy to enhance the extraction efficiency of pyrethroid pesticides in fish samples using a modified QuEChERS (Quick, Easy, Cheap, Effective, Rugged and Safe) method. Analytical Methods 4: 449-453.

13. Stubbings G, Bigwood T (2009) The development and validation of a multiclass liquid chromatography tandem mass spectrometry (LC-MS/MS) procedure for the determination of veterinary drug residues in animal tissue using a QuEChERS (QUick, Easy, CHeap, Effective, Rugged and Safe) approach. Anal Chim Acta 637: 68-78.

14. Whelan M, O'Mahony J, Moloney M, Cooper KM, Furey A, et al. (2013) Maximum residue level validation of triclabendazole marker residues in bovine liver, muscle and milk matrices by ultra high pressure liquid chromatography tandem mass spectrometry. J Chromatogr A 1275: 41-47.

15. Zhang MX, Li C, Wu YL (2012) Determination of phenylethanolamine A in animal hair, tissues and feeds by reversed phase liquid chromatography tandem mass spectrometry with QuEChERS. J Chromatogr B Analyt Technol Biomed Life Sci 900: 94-99.

16. Usui K, Hayashizaki Y, Hashiyada M, Funayama M (2012) Rapid drug extraction from human whole blood using a modified QuEChERS extraction method. Leg Med (Tokyo) 14: 286-296.

17. Fernandes VC, Domingues VF, Mateus N, Delerue-Matos C (2012) Analysing organochlorine pesticides in strawberry jams using GC-ECD, GC-MS/MS and QuEChERS sample preparation. Food additives & contaminants Part A 29: 1074-1084.

18. Moreno Frías M, Jiménez Torres M, Garrido Frenich A, Martínez Vidal JL, Olea-Serrano F, et al. (2004) Determination of organochlorine compounds in human biological samples by GC-MS/MS. Biomed Chromatogr 18: 102-111.

19. Medina CM, Pitarch E, Portolés T, López FJ, Hernández F (2009) GC-MS/MS multi-residue method for the determination of organochlorine pesticides, polychlorinated biphenyls and polybrominated diphenyl ethers in human breast tissues. J Sep Sci 32: 2090-2102.

20. U.S. Department of Health and Human Services, Food and Drug Administration, (2001) Center for Drug Evaluation and Research (CDER), Center for Veterinary Medicine (CVM) Guidance for Industry, Bioanalytical Method Validation.

21. Erney DR, Gillespie AM, and Gilvydis DM (1993) Explanation of the matrix-induced chromatographic response enhancement of organophosphorus pesticides during open tubular column gas chromatography with splitless or hot on-column injection and flame photometric detection. Journal of Chromatography A 638: 57-63.

22. Lehotay SJ, Son KA, Kwon H, Koesukwiwat U, Fu W, et al. (2010) Comparison of QuEChERS sample preparation methods for the analysis of pesticide residues in fruits and vegetables. J Chromatogr A 1217: 2548-2560.

23. Extrapolation Practice for Ecotoxicological Effect Characterization of Chemicals (2008) edited by Keith RS, Theo CM Brock, Dick De Zwart, Scott D Dyer, Leo Posthuma, Sean Richards, Hans Sanderson, Paul Sibley, Paul J. van den Brink. CRC Press.

24. Colborn T (1991) Epidemiology of Great Lakes bald eagles. J Toxicol Environ Health 33: 395-453.

25. United States Environmental Protection Agency, accessed December 2013.

26. Klee (1990) M.S.GC Inlets-An Introduction. Hewlett-Packard Co.

27. http://www.agilent.com/en-us/support/gas-chromatography/gccalculators

6

A Simple and Fast Solid-Phase Extraction GC-ECD Method for the Routine Assessment of Atrazine Residues in Agricultural Produces

Sharma DK*, Anil Kumar and Mahender

Department of Chemistry, Himachal Pradesh University, Shimla, Himachal Pradesh, India

Abstract

A simple and rapid solid-phase extraction (SPE) supported Gas Chromatographic-Electron Capture Detection (GC-ECD) method has been presented for the routine determination of atrazine in some common agricultural produces. Equivalent to USP G-42 GC stationary phase, DB-35 ms capillary column and ECD detector was selected on the basis of physical and structural characteristics of atrazine. Temperature programming of 20 minutes with 0.8 mL/min optimized gas flow gave maximum number of the theoretical plates ($N=10^4$) and high resolution (R_s~2.1 min) with LOD and LOQ values of 0.001 µg/L and 0.1 µg/L respectively. LOQ achieved by the present method is almost below the maximum residue levels recommended by the European legislation. Satisfactory spiked recoveries of the atrazine were obtained from agricultural produces ranging from 82.2 to 97.8% with relative standard deviations below 3.8%. The minimum consumption of solvent (5 mL ethyl acetate) in SPE for residual workup in 5 minutes, with subsequent 20 minutes GC analysis makes the proposed method simple and fast having promise as an excellent economical alternative for the routine analysis of atrazine in environmental samples.

Keywords: Atrazine; Solid-phase extraction; GC-ECD; Residue analysis

Introduction

Atrazine (2-chloro-4-ethylamino-6-isopropylamino-1,3,5-triazine), structure shown in Figure 1, is one of the most heavily used and less expensive pre and post emergence herbicide and used against a broad spectrum of weeds on many crops including corn, potato, maize, wheat, apple, rice and forestry [1-4]. The residue due to spraying of high dosage of atrazine is one of the threats to human health [5]. Atrazine exhibits endocrine disruption both in human and animals and is responsible for adverse effects on reproductive system, hypothalamus, pituitary and thyroid glands [6,7]. Considerable concentration of this herbicide in agricultural produces is increasing and frequently exceeds the maximum residual limit (MRL) of 0.1 µg/L. To prevent adverse effects on health, assessment of atrazine residues in agricultural produces is required which necessitates the development of convenient, reliable and economic analytical method for screening its quantitative presence in agricultural produces viz. cereals, fruits, vegetables etc. Of the various methods viz. UV-Vis spectrophotometry [8,9], electrochemical method [10,11], thin layer chromatography (TLC) [12,13], and immunoassay [14], the methods based on Gas chromatographic (GC) [15-19] and high performance liquid chromatographic (HPLC) [20-22] techniques are generally used for the determination of atrazine. GC methods however, are largely used for quantification studies due to thermal stability of atrazine [23]. Solid phase micro extraction (SPME) and gas chromatography with mass spectrometry (GC-MS) methods increase the sensitivity in the determination of atrazine but are relatively costlier

and time consuming in comparison to the methods employing other detectors like flame ionization detector (FID), nitrogen phosphorus detector (NPD) and thermal conductivity detector (TCD). The greater selectivity and sensitivity of electron capture detector (ECD) detector for halogenated compounds including atrazine compared with FID and TCD has not largely been reported in the determination of atrazine and thus prompted us to use it in the present work. The determination also involves extraction and clean up for matrix interference removal. Liquid-Liquid extraction has a long history for the purpose and is still useful. Solid phase extraction (SPE) involves minimum consumption of solvent and represents a convenient alternative to conventional extraction reported. A cost effective SPE (Silica based, 60 mesh~250 micron, 1.2 cm i.d. homogeneous glass column) with ethyl acetate elution at room temperature has been fabricated for residual workup [24,25] which completes the extraction process in 5 minutes. The greater selectivity of ECD detector compared with FID and TCD, with optimized temperature programming of 20 minutes makes the method rapid and sensitive.

Experimental

Reagents and materials

Standard Atrazine (Product No-34053, GC purity, 99.0%) was purchased from Sigma-Aldrich, Bangalore, India. Ethyl acetate and Methanol (HPLC Grade; Merck, Germany) were used. AR grade chemicals were used in the analysis.

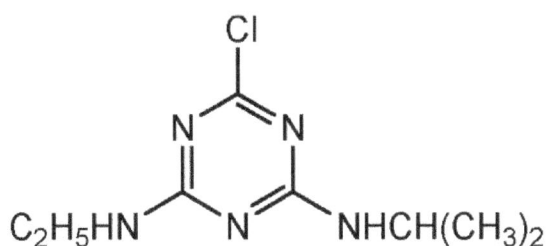

Figure 1: Structure of atrazine.

*Corresponding author: Sharma DK, Department of Pharmacognosy, Department of Chemistry, Himachal Pradesh University, Shimla, Himachal Pradesh, India
E-mail: dksharma_dk@rediffmail.com

Dilution Number	Concentration (µg/L)	Retention Time (Min)	GC Peak Area	Limits	S/N Ratio
1	10^3	14.8	19705		26.2
2	10^2	14.8	16837		23.8
3	10^1	14.9	14156		21.1
4	10^0	14.9	11340		17.2
5	10^{-1}	14.9	8569	LOQ	12.7
6	10^{-2}	14.9	5888		8.2
7	10^{-3}	14.9	3364	LOD	4.9
8	10^{-4}	14.9	608		2.7

Table 1: GC Quantization Limits for Standard Atrazine Solutions.

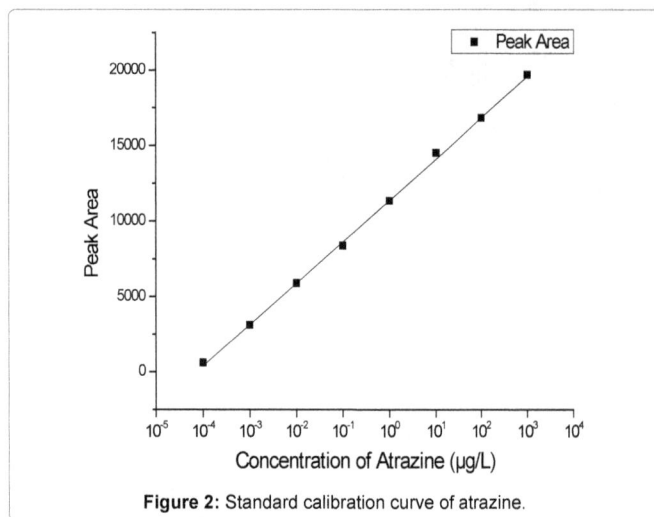

Figure 2: Standard calibration curve of atrazine.

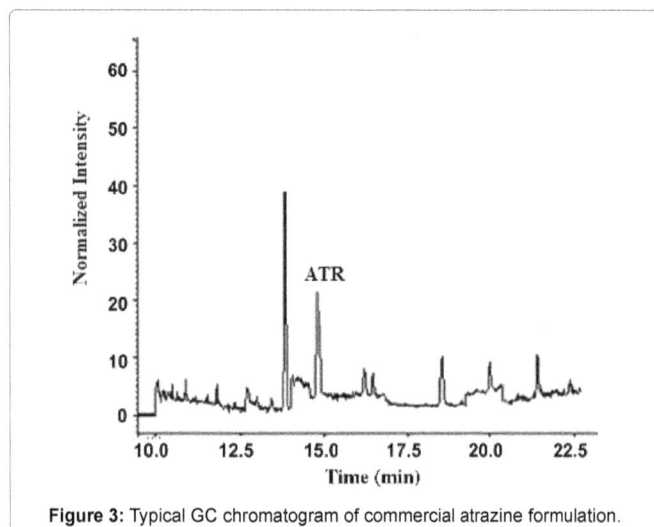

Figure 3: Typical GC chromatogram of commercial atrazine formulation.

Instrumentation and method

An Agilent 7890 A GC system (Germany) with Chemstation B.02.01 version software using ECD detector was used to perform GC analysis. The GC chromatographic Capillary Column DB-35 ms, consisted of 35%-phenyl-methyl polysiloxane, (length 25 m, i.d. 0.20 mm with 0.33 µm film thickness) and is equivalent to USP G-42, connected with the splitless injector with extended temperature 50-375°C with the injector liner of i.d.0.75 mm.

Career gas

Helium (99.9%) with Nitrogen as make up gas with constant flow of 0.8 mL/min was optimized and set.

Temperature gradient

The initial Temperature was 50°C (hold for 2 min) and then was raised to 230°C at a speed of 10°C/min; finally it was programmed to 370°C (hold for 2 min) at a speed of 10°C/min with initial injector injector temperature at 225°C and 1 µL (splitless) atrazine volume injected for each sample.

Column preparation for SPE

Glass column of 15 cm length and 1.2 cm in diameter was used, silica 8 g (Silica gel 60-120 mesh) for column chromatography (Sisco Research Laboratory Pvt. Ltd., Mumbai) was taken in beaker added 15 mL of ethyl acetate and column was filled with this slurry slowly so that silica spread homogeneously in the column [25].

Preparation of standard calibration curve

For the preparation of GC calibration curve standard 10^4 µg/L solution of atrazine was prepared in methanol and then diluted to 10^{-5} µg/L with the same solvent and calibration curve was built for quantitative measurements using the solutions prepared according to Table 1. GC analysis was performed in a 20 min run time method following above temperature gradient and 1 µL of atrazine sample was injected in each run and peak area was measured from ECD response peaks. Calibration curve was constructed by plotting peak area versus concentration of standard atrazine solutions as presented in Figure 2.

Formulation analysis

A wettable powder (WP) formulation of Atrazine purchased from Dhanuka Agritech Ltd, Gurgaon (An authorized dealer with State Govt. Haryana, India), containing 50% active atrazine ingredient was used. Solution of higher concentration was prepared in methanol and then solution of 100 µg/L was prepared by dilution with same solvent. Aliquots of 100 µg/L solution were taken and diluted to 5 ml with methanol and processed for GC analysis in the same manner as described above. The makers specification has also been established by a reference method [26].

Recovery experiment

Standard solution of 200000 µg/L of atrazine was prepared in methanol. This solution was diluted to get the final concentration of 100 µg/L with the same solvent and aliquots of this solution were added to 5 g portion of each grain material. The samples were mixed thoroughly and extracted with 10 mL of ethyl acetate and subsequent extracts was purified by passing through the silica column extractor at a flow rate of 0.7 mL/min. The eluates collected were dried with nitrogen gas drier and the remainder was dissolved in10 ml methanol and analyzed with GC-ECD method developed above.

Results

Linearity of the method was investigated with correlation of

Atrazine Concentration (µg)	Recovery (% age)[a]				
	Maize	Rice	Wheat	Apple	Potato
0.10	95.49 ± 2.39	86.79 ± 2.90	91.67 ± 2.61	82.27 ± 2.25	89.04 ± 2.49
0.15	96.43 ± 2.80	88.95 ± 3.43	92.69 ± 2.36	83.15 ± 2.76	89.26 ± 1.26
0.20	96.61 ± 3.09	90.00 ± 3.04	94.43 ± 2.88	83.64 ± 2.63	90.22 ± 2.92
0.25	97.85 ± 2.33	91.54 ± 3.15	95.06 ± 2.79	84.34 ± 2.80	91.42 ± 2.43

[a]Values are mean ± standard deviation for five determination.

Table 2: Recovery data of Atrazine from spiked grains, fruit, and vegetable matrices.

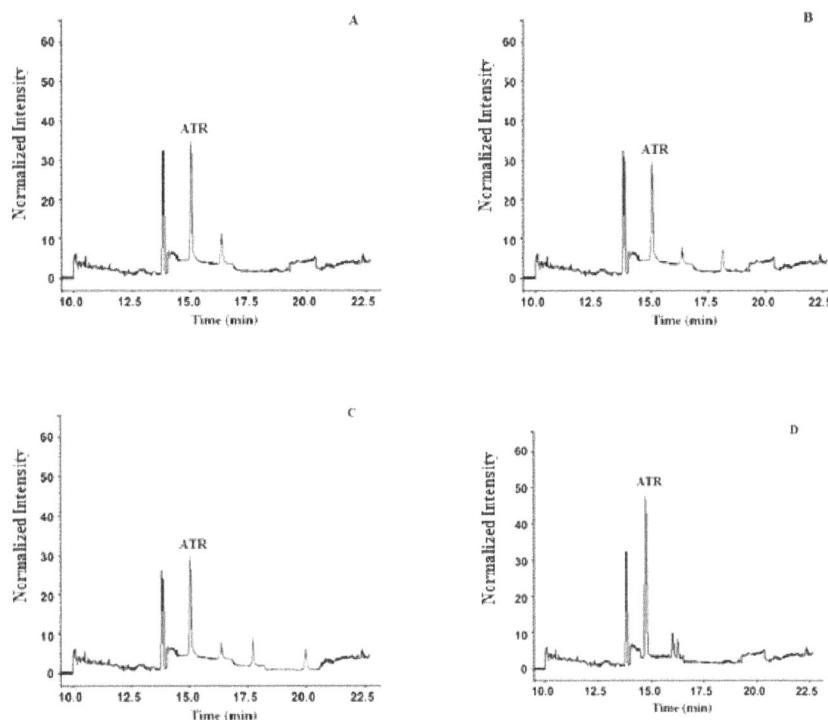

Figure 4: GC chromatograms for atrazine extracted from spiked Maize (a), Rice(b), Wheat(c), Potato(d) and Apple(e) respectively; one of five recovery results.

determination (r^2) values of 0.99 calculated from Regression equation y=2721.65 × +19584.16. Temperature programming of 20 minutes with 0.8 mL/min optimized gas flow gave maximum number of the theoretical plates (N=10^4) and high resolution (R_s~2.1 min) thus permitting the determination of atrazine below its maximum residual limit fixed by the European legislation (Council Directives 90/642/EEC, 1990) as 0.1 µg/L with the estimated LOD and LOQ ranged from 0.001 µg/L and 0.1 µg/L respectively. The method was applied to the analysis of a formulation of atrazine for its active ingredient content. The recovery of the atrazine (active ingredient) from the wettable powder (WP) formulation comprising 50% of standard active ingredient, has been found in the range 49.7-50.3% with relative standard deviation in the range 3.2% and its GC chromatogram is presented in Figure 3. A cost effective SPE column (Silica based, 60 mesh, 1.2 cm i.d homogeneous glass column) with ethyl acetate elution at room temperature has been fabricated for pre-concentration workup. The samples were extracted with ethyl acetate then purified with a silica SPE column, and finally, detected by GC–ECD. Recovery experiments of four fortified concentrations were carried out on maize, rice, wheat, potato and apple samples which did not contain the target compound. The recovery and precision data is shown in Table 2 and chromatograms are presented in Figures 4A-4D. Average recovery and maximum standard deviation of the analytical method applied were 82.27 to 97.85% of spiked amount and 3.43% respectively.

Discussion

Selection of Stationary Phase and ECD system was based on the high boiling point, structure with long hydrocarbon chain with chloride substituent and maximum residue levels (MRL) 0.1 µg/L values of atrazine. All data were subject to strict quality control procedures, including the analysis of procedural blanks and spiked samples with each set of samples analyzed. Though MS/MS[n] detectors are more quantitative and sensitive over ECD, higher selectivity for halogenated analyte like atrazine and comparative low cost of ECD over Mass detectors expertise their preferred use for excellent Limit of Detection (LOD), Limit of Quantification (LOQ) and Limit of Linearity (LOL). Silica gel adsorbents are widely accepted as one of the best and cheaper adsorbents used in chromatography and the important criteria for opting for silica gel in the present study is that it remains neutral and does not interact with atrazine residues that are passed through it and also maintains its own stable structure throughout a process. Importantly, it can be regenerated or reused many times, cutting the costs of purification considerably. All it needs is to be heated to a specific temperature (about 150°C), when it releases all the substances absorbed by it. The method has successfully applied to the analysis of atrazine in its commercial formulation and residues in agricultural produces. Atrazine was not detected in the procedural blanks and method performance was assessed using spiked food samples and the method was shown to have good precision and high recoveries. The former is essential not only to ensure the quality of marketed products but also to get reliable residue data.

Conclusion

A new simple and rapid SPE-GC-ECD method for the determination of atrazine in some agricultural samples has been described. The samples were extracted with ethyl acetate and purified with a cost effective SPE Silica based glass column with ethyl acetate elution at room temperature in five minutes, with subsequent 20 minutes GC-ECD detection makes the proposed method fast and simple and could be potentially extended to other matrices. The selectivity, linear range, recovery, precision, and limit of quantification were all evaluated and verified. The method is accurate, sensitive, and convenient. The limit of detection is 0.001 µg/L and keeps pace with the advances of international technology. LODs and LOQs obtained by this research can meet the European Union standards for MRLs of atrazine in agricultural stuffs and have a promising application ahead.

Acknowledgment

Authors are thankful to Council of Scientific and Industrial Research (CSIR), India for Junior Research Fellowship (CSIR-JRF-09/237-0154/2015-EMR-1) and necessary research grants to Mr. Mahender.

References

1. Agrawal H, Shrivastav AM, Gupta BD (2016) Surface plasmon resonance based optical fiber sensor for atrazine detection using molecular imprinting technique. Sensors and Actuators B: Chemical 227: 204-211.

2. Mansour SA, Mohamed DA, Sutra JF (2014) Which exposure stage (gestation or lactation) is more vulnerable to atrazine toxicity? Studies on mouse dams and their pups. Toxicology Reports 1: 53-68.

3. Brender JD, Weyer PJ (2016) Agricultural compounds in water and birth defects. Current Environmental Health Reports 3: 144-152.

4. Liu T, Zhang Z, Chen D, Wang L, Yao H, et al. (2013) Effect of atrazine and chlorpyrifos exposure on heat shock protein response in the brain of common carp (Cyprinus carpio L.). Pesticide Biochemistry and Physiology 107: 277-283.

5. Syed JH, Alamdar A, Mohammad A, Ahad K, Shabir Z, et al. (2014) Pesticide residues in fruits and vegetables from Pakistan: a review of the occurrence and associated human health risks. Environmental Science and Pollution Research 21: 13367-13393.

6. Stoker TE, Guidici DL, Laws SC, Cooper RL (2002) The effect of atrazine metabolites on puberty and thyroid in the male wistar rat. Toxicological Sciences 67: 198-206.

7. Cooper RL, Soker TE, Tyrey L, Goldman JM, McElroy WK (2000) Atrazine disrupts the hypothalamic control of pituitary-ovarian function. Toxicological Sciences 53: 297-307.

8. Liu G, Yang X, Li T, Yu H, Du X, et al. (2015) Spectrophotometric and visual detection of the herbicide atrazine by exploiting hydrogen bond-induced aggregation of melamine- modified gold nanoparticles. Microchim Acta 182: 1983-1989.

9. Assaker K, Rima J (2012) Improvement of spectrophotometric method for the determination of atrazine in contaminated water by inducing of mannich reaction. Journal of Food Research 1: 17-26.

10. Borras N, Oliver R, Arias C, Brillas E (2010) Degradation of atrazine by electrochemical advanced oxidation processes using a boron-doped diamond anode. Journal of Physical Chemistry A114: 6613-6621.

11. Norouzi P, Larijani B, Ganjali MR, Faridbod F (2012) Admittometric electrochemical determination of atrazine by nano-composite immune-biosensor using FFT- square wave voltammetry. International Journal of Electrochemical Science 7: 10414-10426.

12. Kumar V, Upadhyay N, Singh S, Singh J, Kaur P (2013) Thin layer chromatography: Comparative estimation of soil's atrazine. Current World Environment 8: 469-472.

13. Broszat M, Ernst H, Spangenberg B (2010) A simple method for quantifying triazine herbicides using thin-layer chromatography and a CCD camera. Journal of Liquid Chromatography & Related Technologies 33: 948-956.

14. Qie Z, Bai J, Xie B, Yuan L, Song N, et al. (2015) Sensitive detection of atrazine in tap water using TELISA. Analyst 140: 5220-5226.

15. Pecek G, Pavlovic DM, Babic S (2013) Development and validation of a SPE-GC-MS method for the determination of pesticides in surface water. International Journal of Environmental Analytical Chemistry 93: 1311-1328.

16. Sanagl MM, Muhammad SS, Hussain I, Ibrahim WAW, Ali I (2015) Novel solid-phase membrane tip extraction and gas chromatography with mass spectrometry methods for the rapid analysis of triazine herbicides in real waters. Journal of Separation Science 38: 433-438.

17. Williams DBG, George MJ, Marjanovic L (2014) Rapid detection of atrazine and metachlor in farm soils: Gas chromatography-mass spectrometry-based analysis using the bubble-in-drop single drop microextraction enrichment method. Journal of Agricultural and Food Chemistry 62: 7676-7681.

18. Leyva-Morales JB, Voldez-Torres JB, Bastidas-Bastidas PJ, Betancourt-Lozano M (2015) Validation and application of a multi-residue method, using accelerated solvent extraction followed by gas chromatography, for pesticides quantification in soil. Journal of Chromatographic Science 53: 1623-1630.

19. Pereira A, Silva E, Cerejeira MJ (2014) Applicability of the new 60 µm polyethylene glycol solid-phase microextraction fiber assembly for the simultaneous analysis of six pesticides in water. Journal of Chromatographic Science 52: 423-428.

20. Akdogan A, Divrikl U, Elci L (2013) Determination of triazine herbicides and metabolites by solid phase extraction with HPLC analysis. Analytical Letters 46: 2464-2477.

21. Li X, Wang Y, Sun Q, Xu B, Yang Z, et al. (2016) Molecularly imprinted dispersive solid-phase extraction for the determination of triazine herbicides in grape seeds by high-performance liquid chromatography. Journal of Chromatographic Science 54: 871-877.

22. Wang C, Ji S, Wu Q, Wu C, Wang Z (2011) Determination of triazine herbicides in environmental samples by dispersive liquid-liquid microextraction coupled with high performance liquid chromatography. J Chromatographic Sci 49: 689-694.

23. Watson DH (2004) Pesticide, veterinary and other residues in food. woodhead publishing limited, Cambridge, England, pp: 400.

24. Tadeo JL, Sanchez-Brunete C, Perez RA (2000) Fernandez, M.D.; Analysis of herbicide residues in cereals, fruits and vegetable. Journal of Chromatography A882: 175-191.

25. Supelco (1998) Guide to Solid Phase Extraction, Bulletin, 910.

26. Wang L, Li C, Peng C, Li X, Xu C (2008) A rapid multi residue determination method of herbicides in grain by GC-MS-SIM. Journal of Chromatographic Science 46: 424-429.

7

Development of Solid-Phase Extraction Using Molecularly Imprinted Polymer for the Analysis of Organophosphorus Pesticides-(Chlorpyrifos) in Aqueous Solution

Binsalom A[1], Chianella I[2], Campbell K[1] and Zourob M[3]

[1]Institute for Global Food Security, School of Biological Sciences, Queen's University, UK
[2]Cranfield Health, Cranfield University, Cranfield, Bedfordshire, MK43 0AL, UK
[3]Department of Chemistry, Alfaisal University, Riyadh, KSA

Abstract

A new and selective sorbent for molecularly imprinted solid-phase extraction (MISPE) was prepared to extract chlorpyrifos (CPF) residue from solutions. The extracted analyte was analyzed by high performance liquid chromatography (HPLC) coupled with photodiode array detection. To synthesize the molecularly imprinted polymers, four different pyrogens (acetonitrile, toluene, dichloromethane and chloroform) were initially studied. CPF was used as the template molecule, methacrylic acid as the functional monomer, ethylene glycol dimethacrylate as the cross-linker. Thermo-polymerization method was used to produce bulk polymers. In order to determine the medium that enhances the best molecular recognition, the adsorption study of CPF to the MIPs was investigated. Both organic solvents and water were utilized as media. The acetonitrile solvent was finally selected as pyrogen for the synthesis of the polymers and water was chosen as the medium for loading the analytes into the polymers. The selectivity of the MISPE method for CPF and other pesticides in aqueous solution was also assessed.

Keywords: Molecularly imprinted polymer; Solid phase extraction; Chlorpyrifos pesticide

Introduction

Sample preparation plays a key role for a fruitful and accurate analysis of pesticide residues in food [1]. The ultimate objective of the sample preparation is to pre-concentrate and purify the desired compounds of a complex matrix. Solid Phase Extraction (SPE) is amongst the most common techniques used to extract, purify and concentrate analytes. This technique relies on the repartition of the analyte between a solid phase (usually a polymeric sorbent such as C18 packed in a tube) and a mobile phase namely the solvent used for loading, washing and finally recovering the analyte. Standard sorbents like C18 do not have the require selectivity and binding capacity. Molecularly imprinted polymers (MIPs), which are synthetic polymeric materials that contain receptor sites able to recognize a target compound or similar structures [2], have shown to be an excellent sorbent for molecularly imprinted solid-phase extraction (MISPE). MISPE usually is used to purify and pre-concentrate a target analyte [3-5]. MIPs are produced by a polymerization process following a self-assembly step between the target analyte and functional monomers. Figure 1 shows the synthetic process; which involves a functional monomer, a cross-linker, an imprinted molecule (template), an organic solvent (or pyrogen) and an initiator.

MIPs are highly cross-linked polymers, with binding sites specific and selective for the target molecule template), which was used during the polymerization. MIPs have several advantages over natural bio-molecules, with stability being the most important. In fact they can be used in harsh conditions, such as high temperature, pressure, extreme pH, and organic solvents [6].

Organophosphate (OP) is a major class of pesticides and has a number of applications in agriculture, public health and home use. Chlorpyrifos (CPF) is used as an insecticide to control insects in the field, indoor and outdoor. MIPs have been already successfully developed for organophosphorus pesticides. For example, MIPs were synthesized by bulk polymerization for the chromatographic determination of OP pesticides, using disulfoton as a template [7].

Liu and his colleagues produced MIP microspheres (MIPMs) via emulsifier-free polymerization method for CPF [8]. Also, MIP utilizing quinalphos as a template was developed and applied as SPE sorbent for sample enrichment of organophosphorus pesticides including diazinon and CPF [9]. Besides, CPF was used to synthesize magnetic MIP (MMIPs) by surface imprinted polymerization [10].

This work reports the synthesis of different MIPs synthesized using various pyrogens: acetonitrile (ACN), dichloromethane (DCM), chloroform (CHCl$_3$) and toluene (TOL). The four different MIPs were all prepared in bulk format by thermal polymerization using CPF as template, methacrylic acid (MAA) as monomer and ethylene glycol dimethacrylate (EGDMA) as cross linker. Among all the pyrogens tested for preparation of MIPs, ACN has shown to produce the best MIP for extraction of CPF. The resulting polymer has the potential to be applied as MISPE for selective extraction for CPF pesticide from aqueous solutions.

Materials and Methods

Materials

Chlorpyrifos (CPF), chlorpropham (CIPC), propham (IPC), 3-chloroanaline (3-CA), 3,5,6-trichloropyridinol (TCP), methacrylic acid (MAA), 2,2'-azobisisobutyronitrile (AIBN), ethylene glycol dimethacrylate (EGDMA), and silicon oil were purchased from

---

*Corresponding author: Binsalom A, Institute for Global Food Security, School of Biological Sciences, Queen's University, UK
E-mail: abinsalom01@qub.ac.uk

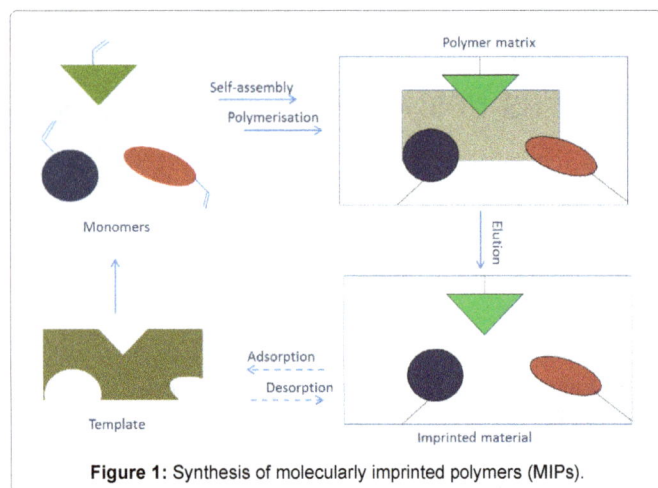

Figure 1: Synthesis of molecularly imprinted polymers (MIPs).

MIPs	T	FM	XL	Molar ratio*	Solvent	Polymerisation condition
MIP1	CPF	MAA	EDGMA	1:4:20	ACN	Thermal 65°C
MIP2	CPF	MAA	EDGMA	1:4:20	CHCL₃	Thermal 65°C
MIP3	CPF	MAA	EDGMA	1:4:20	DCM	Thermal 65°C
MIP4	CPF	MAA	EDGMA	1:4:20	TOL	Thermal 65°C

*Molar ratio: Template (T); Functional monomer (FM); Crosslinking monomer (XL).

Table 1: Different pyrogens were used to develop the MIPs and NIPs.

Sigma Aldrich (UK). Acetonitrile (ACN), methanol (MeOH), toluene, dichloromethane (DCM), acetic acid, mortar and pestle with 199 mL capacity were purchased from Fisher Scientific (Dorset, UK). The 75 and 125 μm mesh size sieves used were from Retsch. All chemicals and solvents were HPLC-analytical grade and used as received.

HPLC chromatography method

The quantification of the target analyte was conducted in a isocratic mode using an HPLC (Agilent Technology, 1200 series) equipped with UV/Vis detector and coupled with a Gemini C18 (150 × 4.6 mm, 5 μm) column (Phenomenex, UK). A sample volume of 20 μl was injected using a mobile phase consisting of 85% MeOH and 15% H₂O, adapted from [11] the mobile phase flow-rate was set at 1 ml/ min. Chromatograms were recorded for 10 minutes at the wavelength of 240 nm. A calibration curve for the target analyte was performed for every quantification experiment using standards over a concentration range of 10 to 500 ng/ml.

Polymer preparation

All CPF-MIPs were prepared under the same conditions except for the pyrogen. The synthetic compositions are represented in Table 1. The polymerization was achieved with the ratio (1:4:20) for template, MAA and EGDMA respectively as follow: CPF (0.5 mmol), MAA (2 mmol) as functional monomer, EGDMA (10 mmol) as a cross-linker. These were dissolved in 3.017 mL of different pyrogens namely ACN, DCM, CHCl₃ and TOL in a 15 mL glass vial. After mixing, the initiator AIBN (30 mg) was added into the solution, which was purged with nitrogen for 5-7 min, and then polymerization was performed at 65°C for 24 h in a thermostat-controlled oil bath. Then the resulting monoliths were ground and sieved using 75-125 μm sieves. After that, MIPs were washed with a mixture of MeOH-acetic acid (9:1, v/v) in a Soxhlet extractor for 24 h to remove any trapped template molecule and were dried at 55°C in an HPLC oven (Shimadzu CTO-10AC). For each MIP, a corresponding non-molecularly imprinted polymer (NIP) was prepared and handled in the same way but without the presence of the template at the polymerization step.

After the synthesis, portions of the polymers (30 mg of both MIPs and NIPs) were packed into empty 1 mL SPE columns (polypropylene, from Sigma) and these were fitted into a solid phase extraction manifold with 12 positions (Supelco) connected to a pump (KNF Laboport® mini-pump) for the vacuum.

Affinity assessment of MIP by SPE

First, an attempt was made to assess whether the same organic solvents used as pyrogens could be utilized as media for loading the target analyte into the MIPs. For these experiments, solutions of CPF (1 mL of 100 ng/mL), prepared in the four different organic solvents, were loaded into the MISPE and NISPE tubes. The filtrates were collected and injected into the HPLC to quantify the residual amount of CPF.

To assess the affinity of the polymers in aqueous solutions, after conditioning MISPE tubes with MeOH (3 mL) and water (3 mL), CPF (30 mL of 500 ng/mL prepared in water) was loaded into all MISPE tubes. These were then washed with several different amounts and mixtures of ACN-water and MeOH-water followed by elution with MeOH-acetic acid (90:10, v/v). The filtrates of all the steps were collected and injected into the HPLC CPF quantification. Based On the results of the washing step optimization, the final SPE protocol was as follow: loading 30 mL of CPF sample (500 ng/mL prepared in water) into MISPE and NISPEs, followed by washing with 5 mL of methanol-water (50:50 v:v) and elution with 5 mLs MeOH-acetic acid (90:10; v/v). All filtrates were collected for HPLC analysis.

Polymer binding capacity

After packing SPE tubes with the CPF-MIP (30 mg), the binding capacity was determined by continuously loading a CPF standard solution prepared in water into the MISPE tubes. After conditioning the tube with MeOH (3 mL) and water (3 mL), 1 L of 500 ng/mL of CPF solution prepared in water was loaded on the tubes, a total amount of CPF (500 μg). MeOH-water (50:50, v/v) (5 mL) and MeOH-acetic acid (90:10, v/v) (5 mL) were used for washing and elution steps. All filtrates in the loading, washing and elution steps were collected for HPLC analysis. The binding capacity of the MISPE reported as μg of CPF bound / g polymer was calculated with the following formula: binding capacity=[(500-X)/0.03 g], where 500 (μg) is the total loaded amount of CPF, X (μg) is the amount of CPF-residue-found in the filtrates, and 0.03 (g) is the polymer mass of a MISPE. The value was then extrapolated for 1 g of MIP.

Selectivity test

To investigate the MIP selectivity in water, analytical standards of analogues of CPF and other pesticides and herbicides (CIPC, IPC, 3-CA and TCP), whose chemical structures are shown in Figure 2, were tested. For the experiments, after conditioning the MISPE and NISPE tubes with MeOH (3 mL) and water (3 mL), aqueous solutions of CPF, CIPC, IPC, 3-CA and TCP (30 mL of 500 ng/mL) were loaded into MISPEs. This was followed by washing with 5 mL MeOH-water (50:50) and recovery with 5 mL of MeOH and acetic acid (9:1; v/v). The filtrates of all the steps were collected and injected into the HPLC for quantification of CPF and the other tested compounds. The experiments were performed in triplicate.

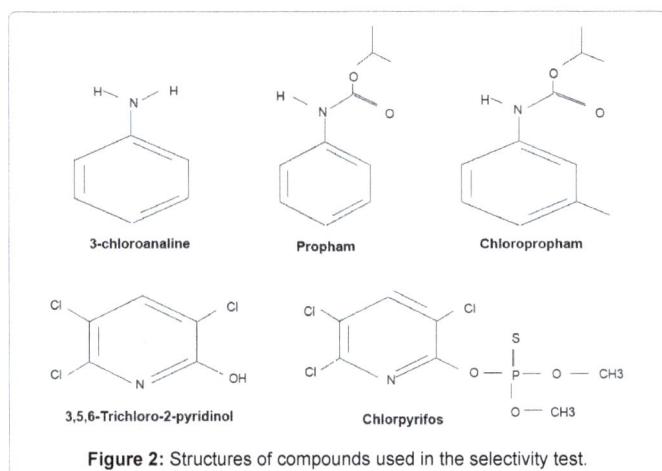

Figure 2: Structures of compounds used in the selectivity test.

Results and Discussion

Polymer preparation

The creation of polymeric binding cavities that capable to recognize a target analyte can be achieved via non-covalent interactions (e.g., hydrogen bonds, electrostatic and ionic interactions) or by covalent and semi-covalent interactions [12]. Regardless of the method, the configuration and alignment of functional groups in the polymeric network are crucial for a specific and selective recognition. In this work, the non-covalent method was preferred over the others, because of its wide applicability and flexibility. Molecularly imprinted polymers can be prepared in many formats (e.g., bulk, films, microparticles and nanoparticles), with the bulk format being the most straightforward, simple and suitable to produce an SPE sorbent. Therefore, bulk polymerization with the optimum ratio of 1:4:20 of template, monomer and cross-linker respectively was selected for this work. The molar ratio for reagents has been proven to produce good monomer-template interactions in the polymeric network of bulk polymers [13]. MAA was chosen here as it is a widely used functional monomer which can behave both as a hydrogen bond donor and acceptor if it is used with a suitable pyrogen [14]. The proposed interaction between MAA and CPF takes place via the oxygen atoms presents in the structure of both compounds [15,16]. EGDMA was used as the cross-linking agent to strengthen the imprinted binding sites and provide mechanical stability to the polymer matrix. In this method, a high proportion of cross-linking agent was used to give the polymeric network specificity and rigidity as well as strengthen the binding site [17].

In this work polymerization was performed by a thermal method to produce polymers with a microporous structure, which maximize surface area and the number of binding sites and consequently to achieve materials with high binding capacity [18,19].

The strength of the assembly between a template and a functional monomer is governed by the physical and chemical characteristics of the solvent. The solvent can impart an effective imprinting particularly in a non-covalent method. In addition, it can also change the polymer morphology and porosity, which subsequently affects the interactions during the rebinding experiments. Therefore, initially, four different pyrogenic solvents were used to synthesize the polymers and then to identify an appropriate pyrogen which could generate a good imprinting for CPF. As a result, aprotic and non-polar pyrogens solvents such as ACN, DCM, CHCl$_3$, and toluene were selected for this study [20]. Such solvents were preferred due to their capability to promote non-covalent interactions without interfering with the radical polymerization.

MIPs and NIPs were therefore successfully synthesized using the four pyrogens.

MIP affinty assessment

The basis of molecular recognition is to preserve the interactions established between a template and MAA in the pre-polymerization mixture. It has been indicated that CPF binds MAA via hydrogen bonds. Thus, the four different pyrogenic solvents used in this work were selected to maximize such interactions and impart affinity and selectivity to the resulting MIPs. Therefore, MIPs and NIPs prepared with the four pyrogens were packed in empty SPE tubes and screened for their affinity towards CPF. Initially, the organic solvents; ACN, DCM, CHCl$_3$ and TOL, used for the synthesis, were investigated as solvent for loading CPF (100 ng/ml) into the tubes. In fact, it has been stated that employing the same solvent as a pyrogen for the polymerization and as media for analyte rebinding is beneficial as there can be a memory effect during rebinding [21]. The results indicated none of the MIPs and NIPs showed sufficient affinity and binding capacity for the target analyte in any of the organic solvents tested (data not shown). One reason for these results might be that the highly hydrophobic CPF preferred to establish interactions with the aprotic and low polar organic solvents rather than with the more hydrophilic polymer matrix containing MAA groups [22]. Therefore, the MIP affinity study was repeated using water as a loading solvent.

Changing the loading solvent to water alters the chemistry of interactions between CPF and the MIPs, which in such environment is mainly due to nonspecific hydrophobic interactions and in part to electrostatics interactions [23]. Thus a high absorption (specific and nonspecific) of CPF to the polymers was expected. In this conditions, the washing step following the analyte loading becomes the crucial step capable to evidence the selectivity of the MIPs by disrupting the weakest nonspecific interactions [24]. Even though none of the MIPs synthesized here demonstrated sufficient affinity for CPF in organic solvents, the MIP prepared in ACN was the only one consistently showing better binding than the corresponding NIP in all the testing conditions. Therefore, the washing step was firstly optimised using ACN-MIP and ACN-NIP, polymers that prepared in ACN, a standard solution of CPF (30 mL of 500 ng/ml in water) was loaded and totally bound to the MIPs and N, then different mixtures of methanol or acetonitrile and water (5 mL) were tested as washing. The amount of CPF in the filtrates was quantified by HPLC and the results are reported in Figure 3. Clearly, all mixtures containing ACN at 10, 20 and 50% in

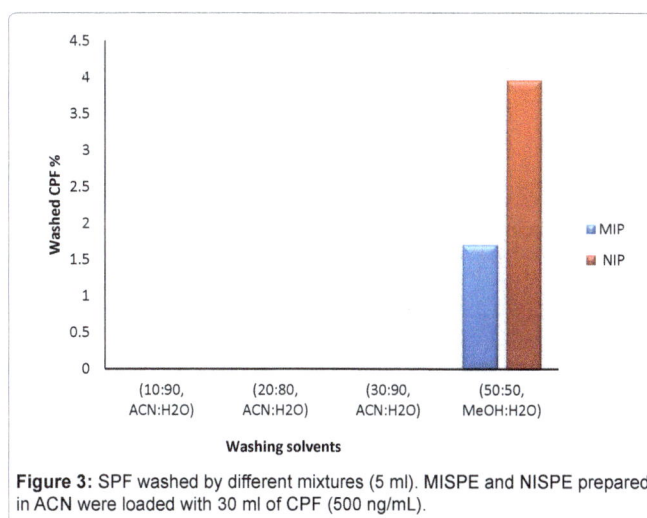

Figure 3: SPF washed by different mixtures (5 ml). MISPE and NISPE prepared in ACN were loaded with 30 ml of CPF (500 ng/mL).

water were not capable to disrupt the nonspecific binding in either the MIP or NIP. This could be explained by the fact that ACN was the solvent used to synthesise the polymers and therefore it promotes interactions between the polymers and template. In addition, ACN is slightly more polar than MeOH (polarity index for 5.8 for ACN and 5.1 for MeOH) [25] and therefore less capable to wash out a non-polar compound such as CPF. Conversely, MeOH seemed to be capable to disrupt some of the weak interactions between polymers and CPF to achieve an acceptable washing. Therefore, the mixture MeOH: water (50%: 50%), was then able to disrupt in part the non-specific interactions and remove a greater amount of CPF from the NIP than the MIP. The mixture used for the subsequent step, the elution consisting in 5 mL MeOH containing 10% acetic acid, was adapted from a comparable study [27] and was capable of eluting CPF from both polymers [data not shown].

After this optimisation, the washing experiments were repeated using all the MIPs and NIPs synthesized with the different four pyrogens. Once again there was no detection of CPF in the filtrates at the loading steps, demonstrating that all polymers (MIPs and NIPs) were capable of retaining the analyte completely. This could be attributed to the nature of aqueous solutions, which promote binding between analyte and polymer through hydrophobic and electrostatic interactions.

The optimized washing solution (5 mL of MeOH/water) (50:50%) was then used to wash the cartridges. The percentage of CPF found in the filtrates at the washing step for all the MIPs and NIPs is shown in Figure 4. It can be seen that with the exception of DCM-MIP, a higher amount of CPF was removed from the NIPs than the MIPs. This indicates that the MIPs have a higher affinity for CPF when compared to the corresponding NIPs [26]. Among all the MIPs, ACN-MIP showed the greatest difference with the smallest standard deviations, which makes it the most promising material for an effective MISPE. The large standard deviations shown in Figure 4 were due to the fact that CPF concentrations were close to the LOD of the HPLC measurements.

Following the washing step, all the MISPE cartridges were eluted with 5 mL of the elution mixture, methanol-acetic acid (9:1 v:v). The results, summarized in Figure 5, illustrate that CPF recovered from ACN-MIP, CL-MIP and DCM-MIP was higher than that eluted from their corresponding NIPs. However only the difference between ACN-MIP and ACN-NIP is high enough to be considered significant. In fact, for the other polymers the differences fall inside the standard deviations of the measurements, thus they are not significant. This includes TOL-

MIP and NIP, where TOL-NIP seemed to release a higher (but not significant) amount of CPF. One common result for all the polymers is that CPF was not fully recovered with the 5 mL of elution mixture, even considering the amount lost at the washing step. A subsequent 5 mL of elution mixture was passed through the cartridges, but the amount of CPF released was below the LOD of the HPLC measurement and therefore it was not quantifiable. Nevertheless, the MIP prepared with ACN, in addition to possessing the best affinity (greatest differences between MIP and NIP) also showed the highest amount of CPF recovered at the elution step (around 75%), demonstrating to be the best performing MIP. The CPF recovered was also pre-concentrated 6 times, as 30 mL were used to load CPF into the cartridges and 5 mL of elution mixture was used to recover it.

Although other studies in literature have reported the use of DCM, toluene and chloroform as pyrogens for producing MIPs for other pesticides [27,28], this finding agrees with several studies, where MIPs were produced for organophosphate pesticide similar to CPF using ACN as a pyrogen [29].

Breakthrough volume and mass capacity

The binding capacity of the best performing MIP, ACN-MIP, was assessed by a breakthrough experiment where the breakthrough volume, which is the volume of the loading solution at which total analytes adsorption can no longer be attained due to the saturation of polymer binding sites, was determined [30]. In this study, for the experiment, even though a large volume of CPF solution (1 L of 500 ng/mL of CPF prepared in water) was loaded into the MIP cartridges (30 mg of MIP), traces of CPF in the filtrate were still not observed. Therefore, the binding capacity for the MIP was calculated to be higher than 16.7 mg/g polymer.

The polymer performance cannot be evaluated only by investigating the affinity and capacity; the selectivity to rebind and distinguish the target template from other related compounds is also essential [31]. The MIP binding selectivity towards CPF was evaluated here by testing other pesticide analogues of CPF such as CIPC, IPC, 3-CA and TCP. The same protocol explained above was applied again by loading CPF and the pesticides at a concentration of 500 ng/mL in water (30 mL), followed by the established washing and elution mixtures. The results of the experiments showed that, not surprisingly, all compounds were retained completely at the loading step by the MIP. Conversely the

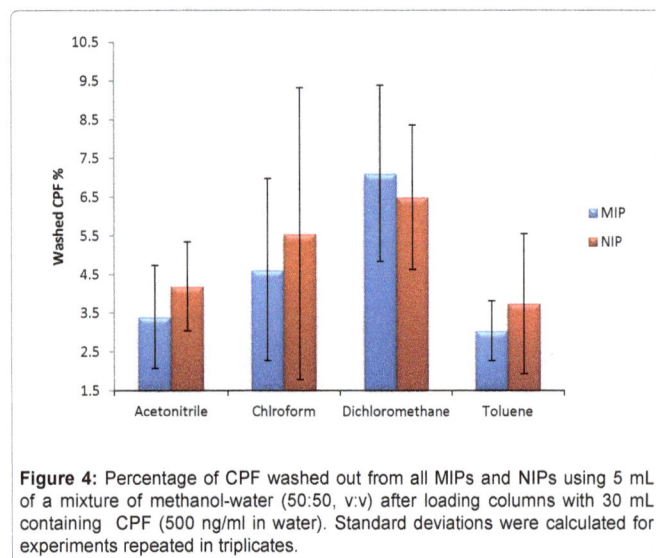

Figure 4: Percentage of CPF washed out from all MIPs and NIPs using 5 mL of a mixture of methanol-water (50:50, v:v) after loading columns with 30 mL containing CPF (500 ng/ml in water). Standard deviations were calculated for experiments repeated in triplicates.

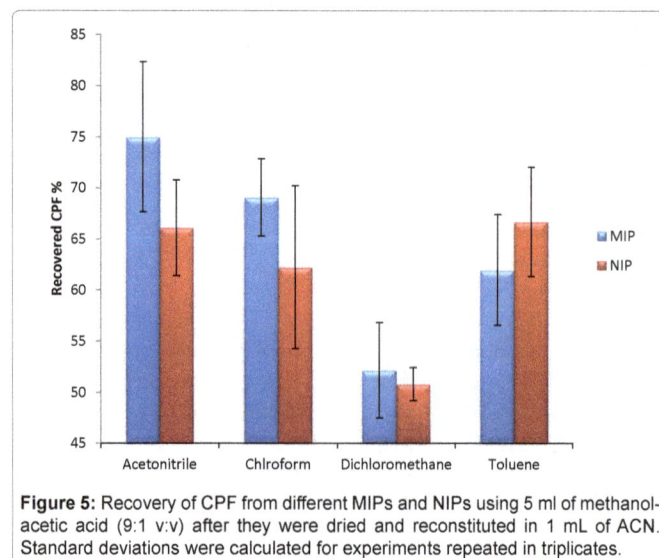

Figure 5: Recovery of CPF from different MIPs and NIPs using 5 ml of methanol-acetic acid (9:1 v:v) after they were dried and reconstituted in 1 mL of ACN. Standard deviations were calculated for experiments repeated in triplicates.

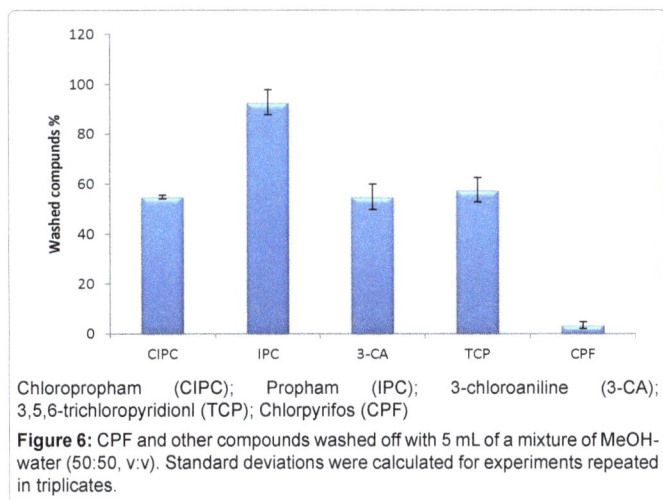

Chloropropham (CIPC); Propham (IPC); 3-chloroaniline (3-CA); 3,5,6-trichloropyridionl (TCP); Chlorpyrifos (CPF)

Figure 6: CPF and other compounds washed off with 5 mL of a mixture of MeOH-water (50:50, v:v). Standard deviations were calculated for experiments repeated in triplicates.

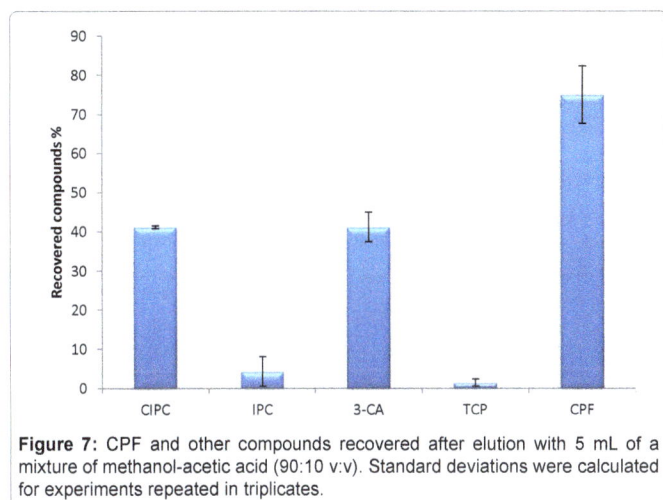

Figure 7: CPF and other compounds recovered after elution with 5 mL of a mixture of methanol-acetic acid (90:10 v:v). Standard deviations were calculated for experiments repeated in triplicates.

amount of CPF and other compounds found in the filtrates after the washing and the elution steps are shown in Figures 6 and 7 respectively. Even though all these compounds were retained by the polymers, about 55-85% of the amount were removed by a 5 mL mixture at the washing step (Figure 6). However, the MIP still selectively retained CPF. The result indicated that the MIP has low affinity for these tested compounds compared to CPF, which possibly contained higher molecular recognition towards its template. In comparison with CPF, the other compounds showed lower affinities for the MIP as larger amounts were removed at the washing step. Significant lower recovery were obtained at the final elution step (Figure 7), the possible explanation for this is that the interaction with these compounds took place mainly via non-specific binding sites available in the polymer. Indeed the MIP can distinguish its target molecule due to the presence of cavities with the right shape, size and specific binding groups suitable to interact with the target molecule [29]. Another explanation for this phenomena could be attributed to the different pKa values for these analytes varying (from 13.3 to - 4.1) as 13.3, 4.55, 3.521, -0.94 and -4.1 [30] for CIPC, TCP, 3-CA, IPC and CPF respectively, which indicate that CPF is the least polar. Nevertheless, the behavior observed does not strictly follow the values of the pKa. In fact, IPC, which is the least polar after CPF, is nearly all lost at the washing step and very little is recovered at the end. Therefore, selectivity could be in part influenced under the applied conditions by the polarity of these analytes [31], but there is definite indication of an imprinting effect.

This result demonstrates that in the tested conditions, ACN-MIP is sufficiently selective for analysis of real samples, which might contain the tested interfering compounds. The main implicit mechanism for the template selectivity of our MIP could be inferred for a shape recognition of the binding sites contained inside the polymeric structure [32].

Conclusion

A MISPE sorbent for the selective binding of CPF was synthesized here by thermo-polymerization. An investigation on how the type of solvent used during polymerization influences the affinity of the polymers was carried out. The MIP prepared with ACN as a pyrogen showed the highest affinity to CPF when tested in water in comparison with MIPs made with other organic solvents. Successful extraction and pre-concentration (6 times) of CPF from water was achieved using ACN MIP. When the MIP selectivity towards CPF was assessed by testing other pesticides, it was found that with the optimized SPE protocol the MIP was capable to selectively recognize CPF. The sorbent developed here is therefore could be a good candidate for extraction CPF from environmental samples.

References

1. Ferrer C, Gomez MJ, Garcia Reyes JF, Ferrer I, Thurman EM, et al. (2005) Determination of pesticide residues in olives and olive oil by matrix solid-phase dispersion followed by gas chromatography/mass spectrometry and liquid chromatography/tandem mass spectrometry. Journal of Chromatography A 1069: 183-194.

2. Xu Z, Fang G, Wang S (2010) Molecularly imprinted solid phase extraction coupled to high-performance liquid chromatography for determination of trace dichlorvos residues in vegetables. Food Chemistry 119: 845-850.

3. Alexander C, Andersson HS, Andersson LI, Ansell RJ, Kirsch N, et al. (2006) Molecular imprinting science and technology: a survey of the literature for the years up to and including 2003. Journal of molecular recognition 19: 106-180.

4. Piletska EV, Navarro Villoslada F, Chianella I, Bossi A, Karim K, et al. (2008) Extraction of domoic acid from seawater and urine using a resin based on 2-(trifluoromethyl) acrylic acid. Analytica Chimica Acta 610: 35-43.

5. Chianella I, Karim K, Piletska EV, Preston C, Piletsky SA (2006) Computational design and synthesis of molecularly imprinted polymers with high binding capacity for pharmaceutical applications-model case: Adsorbent for abacavir. Analytica Chimica Acta 550: 73-78.

6. Mirsky VM, Yatsimirsky A (2010) Artificial receptors for chemical sensors. John Wiley & Sons.

7. Baldim IM, de Oliveira Souza MC, Da Cunha Souza JCJ, Figueiredo EC, Martins I (2012) Application of the molecularly imprinted solid-phase extraction to the organophosphate residues determination in strawberries. Analytical and bioanalytical chemistry 404: 1959-1966.

8. Liu J, Yang M, Huai LF, Chen YT (2010) Removal of Chlorpyrifos from contaminated water using molecularly imprinted polymeric microspheres. In Bioinformatics and Biomedical Engineering (iCBBE). 4th International Conference.

9. Sanagi MM, Salleh S, Ibrahim WAW, Naim AA, Hermawan D, et al. (2013) Molecularly imprinted polymer solid-phase extraction for the analysis of organophosphorus pesticides in fruit samples. Journal of food composition and analysis 32: 155-161.

10. Ma G, Chen L (2014) Determination of chlorpyrifos in rice based on magnetic molecularly imprinted polymers coupled with high-performance liquid chromatography. Food Analytical Methods 7: 377-388.

11. Zhang C, Wu YJ, Jin SF, Yang H (2013) Analysis of chlorpyrifos and chlorpyrifos-methyl residues in multi-environmental media by cloud-point extraction and HPLC. Analytical Methods 5: 3089-3095.

12. Li S, Ge Y, Piletsky SA, Lunec J (2012) Molecularly imprinted sensors: overview and applications. Elsevier.

13. Andersson HS, Karlsson JG, Piletsky SA, Koch Schmidt AC, Mosbach K, et al. (1999) Study of the nature of recognition in molecularly imprinted polymers, II: influence of monomer–template ratio and sample load on retention and selectivity. Journal of Chromatography A 848: 39-49.

14. Zhu X, Su Q, Cai J, Yang J, Gao Y (2006) Molecularly imprinted polymer membranes for substance-selective solid-phase extraction from aqueous solutions. Journal of Applied Polymer Science 101: 4468-4473.

15. Yao W, Fang Y, Gao GL, Cheng Y (2008) Adsorption of carbaryl using molecularly imprinted microspheres prepared by precipitation polymerization. Polymers for Advanced Technologies 19: 812-816.

16. Sanagi MM, Salleh S, Ibrahim WAW, Naim AA (2011) Determination of organophosphorus pesticides using molecularly imprinted polymer solid phase extraction. Malaysian Journal of Analytical Sciences 15: 175-183.

17. Mayes AG, Whitcombe MJ (2005) Synthetic strategies for the generation of molecularly imprinted organic polymers. Advanced drug delivery reviews 57: 1742-1778.

18. Sellergren B (2000) Molecularly imprinted polymers: man-made mimics of antibodies and their application in analytical chemistry. Elsevier.

19. Turner NW, Holdsworth CI, Donne SW, McCluskey A, Bowyer MC (2010) Microwave induced MIP synthesis: comparative analysis of thermal and microwave induced polymerisation of caffeine imprinted polymers. New Journal of Chemistry 34: 686-692.

20. Vasapollo G, Sole RD, Mergola L, Lazzoi MR, Scardino A, et al. (2011) Molecularly imprinted polymers: present and future prospective. International journal of molecular sciences 12: 5908-5945.

21. Yan M (2004) Molecularly imprinted materials: Science and technology. CRC press.

22. Beltran A, Borrull F, Marce RM, Cormack PAG (2010) Molecularly-imprinted polymers: useful sorbents for selective extractions. Trends in Analytical Chemistry 29: 1363-1375.

23. He C, Long Y, Pan J, Li K, Liu F (2007) Application of molecularly imprinted polymers to solid-phase extraction of analytes from real samples. Journal of biochemical and biophysical methods 70: 133-150.

24. Siemann M, Andersson LI, Mosbach K (1996) Selective recognition of the herbicide atrazine by noncovalent molecularly imprinted polymers. Journal of Agricultural and Food Chemistry 44: 141-145.

25. Zhu QH, He JF, Feng JY (2007) Optimization of the process parameters of synthesis of vinblastine imprinted polymer. European Polymer Journal 43: 4043-4051.

26. Schirmer C, Meisel H (2006) Synthesis of a molecularly imprinted polymer for the selective solid-phase extraction of chloramphenicol from honey. Journal of Chromatography A 1132: 325-328.

27. Zhu X, Yang J, Su Q, Cai J, Gao Y (2005) Selective solid-phase extraction using molecularly imprinted polymer for the analysis of polar organophosphorus pesticides in water and soil samples. Journal of Chromatography A 1092: 161-169.

28. Xin J, Qiao X, Xu Z, Zhou J (2013) Molecularly imprinted polymer as sorbent for solid-phase extraction coupling to gas chromatography for the simultaneous determination of trichlorfon and monocrotophos residues in vegetables. Food Analytical Methods 6: 274-281.

29. Wang X, Qiao X, Ma Y, Zhao T, Xu Z (2013) Simultaneous determination of nine trace organophosphorous pesticide residues in fruit samples using molecularly imprinted matrix solid-phase dispersion followed by gas chromatography. Journal of agricultural and food chemistry 61: 3821-3827.

30. Chapuis F, Pichon V, Lanza F, Sellergren B, Hennion MC (2004) Retention mechanism of analytes in the solid-phase extraction process using molecularly imprinted polymers: Application to the extraction of triazines from complex matrices. Journal of Chromatography B 804: 93-101.

31. Bitar M, Cayot P, Bou Maroun E (2014) Molecularly imprinted polymer solid phase extraction of fungicides from wine samples. Analytical Methods 6: 6467-6472.

32. De Barros LA, Martins I, Rath S (2010) A selective molecularly imprinted polymer-solid phase extraction for the determination of fenitrothion in tomatoes. Analytical and bioanalytical chemistry 397: 1355-1361.

8

Determination of Major Polyphenolic Components in *Euphoria longana* Lam. by Validated High Performance Thin-Layer Chromatography Method and Direct Analysis in Real Time Mass Spectrometry

Atul S Rathore, Sathiyanarayanan L and Kakasaheb R Mahadik*

Centre for Advanced Research in Pharmaceutical Sciences, Poona College of Pharmacy, Bharati Vidyapeeth Deemed University, Pune, Maharashtra, India

Abstract

In the current study, a simple, sensitive, precise, and specific method has been developed and validated for the determination of four major polyphenolic compounds in *Euphoria longana* Lam. seeds based on high performance thin-layer chromatography (HPTLC) and confirmed by direct analysis in real time mass spectrometry (DART-MS). The chromatographic separation was accomplished on Merck HPTLC aluminum plates precoated with silica gel G60 F_{254} (20 cm × 10 cm) with 250 µm thickness as stationary phase using a mobile phase composed of n-butanol: water: methanol: formic acid (7.59: 1.27: 0.13: 1.01, v/v/v/v). Densitometric measurement was performed in the absorbance mode at 280 nm. The compounds were resolved satisfactorily with R_f values of 0.40 ± 0.01, 0.57 ± 0.02, 0.69 ± 0.01, and 0.79 ± 0.01 for corilagin, ellagic acid, epicatechin, and gallic acid, respectively. The method developed was validated with acceptable linearity (r^2>0.995), sensitivity, precision (RSD ≤ 1.60%), robustness, and recovery (RSD ≤ 1.77%). The proposed method was successfully applied for the determination of active components for comprehensive quality control of *E. longana* products. Furthermore, peak identities of compounds from each band were confirmed by direct analysis in real time mass spectrometry.

Keywords: *Euphoria longana* Lam; Polyphenols; HPTLC; Validation; DART-MS

Introduction

Euphoria longana Lam. (*E. longana*) belongs to Sapindaceae family, is extensively grown in South of China, India and Southeastern Asia in which Thailand is a major exporter of *E. longana* fruit [1,2]. Dried seed extract of *E. longana* is a source of high levels of polyphenolic compounds such as corilagin, gallic acid, and ellagic acid [3]. Also, previous reports have confirmed the presence of epicatechin (Figure 1) in *E. longana* seeds as other major phenolic compound [3,4]. As *E. longana* seeds contain significant amounts of polyphenolic compounds, seed extract exhibits excellent antioxidant [1,4-7], anti-tyrosinase [1], antifungal [8], anti-inflammatory [9], antigelatinase, anti-angiogenesis, and anticancer [2,10,11] activities. Also, *E. longana* seed extract has been investigated concerning its beneficial effects on learning and memory impairment in mice [12]. As a matter of fact, these polyphenolic compounds are usually considered as the markers for quality control of *E. longana* [3,4,13]. Therefore, for the purpose, a suitable and preferred method for analysis of these polyphenolic compounds is necessary.

So far, a few methods such as high performance liquid chromatography (HPLC) [1,3,4,6] have been developed for the determination of polyphenolic compounds in *E. longana*. These liquid chromatography methods reported for quantification of only either two or three major polyphenolic compounds. Also, few other liquid chromatography tandem mass spectrometry (LC-MS) methods [4,13,14] primarily focus on isolation and characterization of polyphenolic compounds in *E. longana* seed/pericarp extracts. However, these methods suffer from longer analysis time, time-consuming sample preparation and a large amount of organic solvent consumption.

In particular, HPTLC serves as the preferred technique nowadays for the detection of analyte at a very low operating cost with advantages of the need of minimum sample cleanup, high sample throughput, and lower solvent consumption as the matter of fact several samples can be run simultaneously unlike HPLC, thus lowering analysis time and cost per analysis [15,16].

Further, a DART-MS technique was established to confirm peak identities of compounds, is a fast, reliable and non-invasive high throughput technique that involves the ionization of organic molecules in diverse samples directly from the surface. Electronically excited species such as metastable helium and nitrogen atoms ionize atmospheric water molecules that transfer protons to the sample molecules resulted in soft ionization. As the DART ion source can ionize molecules directly from the surface, compounds can be analyzed as such without sample preparation [17]. DART-MS is, therefore, an appropriate technique for the rapid confirmation of analytes from each band on HPTLC plate based on exact mass spectra acquired in negative ion mode being directly analyzed from sample bands using glass capillary without sample preparation.

Nevertheless, to the best of our knowledge, no single method has been reported for the simultaneous determination of four major polyphenols namely, gallic acid, corilagin, epicatechin, and ellagic acid in plant matrices of *E. longana*. Moreover, no method has been reported till date for quantification of these compounds using HPTLC technique. Also, this is the first ever report of HPTLC determination of corilagin in any plant matrices till date.

In this study, the authors establish cost-effective, rapid, and less time consuming analytical strategy that combines simple, precise

***Corresponding author:** Kakasaheb R Mahadik, Centre for Advanced Research in Pharmaceutical Sciences, Poona College of Pharmacy, Bharati Vidyapeeth Deemed University, Pune-411 038, Maharashtra, India
E-mail: krmahadik@rediffmail.com

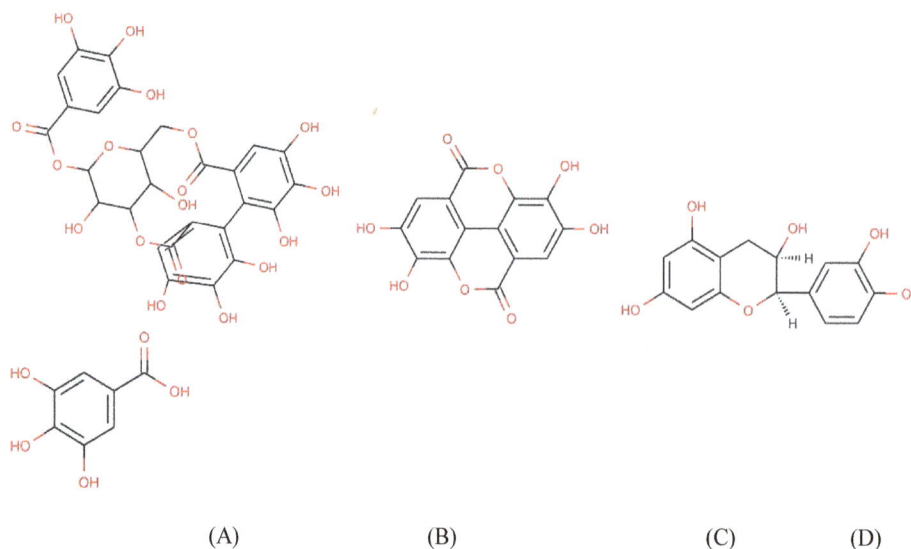

Figure 1: Chemical structures of corilagin (A), ellagic acid (B), epicatechin (C), and gallic acid (D).

and accurate HPTLC method and DART-MS for simultaneous determination and confirmation of four major polyphenolic components in *E. longana* seeds.

Materials and Methods

Reagents and chemicals

Standard substances of gallic acid, ellagic acid and epicatechin (purity ≥ 95.5%) were purchased from Sigma-Aldrich Co. (St. Louis, MO, USA). The analytical standard of corilagin (purity ≥ 98.3%) was purchased from Chengdu Biopurify Phytochemicals Co., Ltd. (Chengdu, China). Methanol (HPLC grade), n-butanol and formic acid (analytical grade) were of from Merck (Darmstad, Germany). Ultra high purity water was prepared using a Milli-Q water purification system (Millipore Corporation, Bedford, MA, USA).

E. longana fruits (Sample 1-5) were collected from the plants grown in Muzaffarpur district, Bihar, India between the months of June to August. A voucher specimen of *E. longana* (ASR-1) was deposited in the Botanical Survey of India, Western Region Centre, Pune, India. The seeds were sundried and ground to powder using a stainless-steel grinder.

Instruments and conditions

Chromatographic analysis was performed on HPTLC plates prewashed with methanol and activated at 110°C for 5 min prior to chromatography. The samples were spotted in the form of bands 6 mm width with a Camag 100 microlitre sample syringe (Hamilton, Bonaduz, Switzerland) on silica gel precoated HPTLC aluminum plate 60 F$_{254}$, (20 cm × 10 cm) with 250 μm thickness; E. Merck, Darmstadt, Germany, supplied by Anchrom Technologies, Mumbai] using a Camag Linomat V applicator (Switzerland). A constant application rate of 0.1 μL/s was used and the space between two bands was 6 mm. Linear ascending development was carried out in 20 cm × 10 cm twin trough glass chamber (Camag, Muttenz, Switzerland) saturated with the mobile phase. The mobile phase was consisted of n-butanol: water: methanol: formic acid (7.59: 1.27: 0.13: 1.01, v/v/v/v) and 20 mL was used per chromatography run. The optimized chamber saturation time with mobile phase was 30 min using saturation pads at room temperature (25°C ± 2). The length of chromatogram run was 80 mm and run time

was 40 min. Densitometric scanning was performed using a Camag TLC scanner III in the reflectance-absorbance mode and operated by winCATS software (V1.1.4, Camag). The slit dimension was kept at 5 mm × 0.45 mm and the scanning speed was 10 mm/s. The source of radiation used was a deuterium lamp emitting a continuous UV spectrum between 200 and 400 nm. All determinations were performed at ambient temperature with a detection wavelength of 280 nm. Concentrations of the compound chromatographed were determined from the intensity of the diffused light. Evaluation was by peak areas with linear regression.

Mass spectrometry confirmation

For analytes confirmation, all samples were analyzed using the same HPTLC conditions mentioned above and mass spectrometry detection was conducted on a JMS T100 LC (Accutof) atmospheric pressure ionization time of flight mass spectrometer (Jeol, Tokyo, Japan) fitted with a DART ion source. The mass spectrometer was operated in negative ion mode with resolving power of 6000 (full width at half maximum). The orifice 1 potential was set to 15 V, resulting in minimal fragmentation. The ring lens and orifice 2 potential were set to 13 and 5 V, respectively. Orifice 1 was set to a temperature of 100°C. The RF ion guide potential was 300 V. The DART ion source was operated with helium gas flowing at approximately 4.0 L/min. The gas heater was set to 300°C. The potential on the discharge needle electrode of the DART source was set to 3000 V, electrode 1 was 100 V and the grid was at 250 V. The samples were collected from each band of developed plate using glass capillary and positioned in the gap between the DART source and mass spectrometer for measurements. Data acquisition was from m/z 50.0 to 1000.0. Exact mass calibration was accomplished by including a mass spectrum of neat polyethylene (PEG) glycol (1:1 mixture PEG 200 and PEG 600) in the data file. m-Nitrobenzyl alcohol was also used for calibration. The mass calibration was accurate to within 0.002 u. It was found that the mass error is a linear function of the logarithm of the signal intensity adjusted to the associated lock-mass intensity. When applied to all mass data points, the correction function reduced the mass error for the majority of the tested compounds to ≤ 1 ppm over a wide range of signal intensities. The systematic error in mass measurements using TDC-based TOF-MS can only be avoided by careful and time-consuming manual analysis, limiting the exact mass calculations to those chromatographic scans that display an analyte mass intensity similar to the lock-mass intensity. The elemental composition could be determined on selected peaks using the Mass Center software. Molecular ions of polyphenolic compounds present in the samples were found to be same with that of standard analytes at same R_f value.

Preparation of standard solutions

The appropriate amount of each standard was weighed and dissolved in methanol to make individual stock solutions. A mixed standard stock solution containing corilagin, ellagic acid, epicatechin, and gallic acid was prepared in methanol. The working standard solutions were prepared by diluting the mixed standard solution with methanol to a series of proper concentrations. Each concentration was applied six times on the HPTLC plate. The plate was then developed using the previously described mobile phase.

Preparation of sample solutions

E. longana seed powder (1 g) was weighed accurately, and the extraction process was performed as described previously [3,7] with some modification. Ultrasound-assisted extraction was implemented by an ultrasonic cleaner (Equitron, Medica Instrument Mfg. Co., Mumbai, India) using 25 mL 70% methanol for 30 min at a frequency of 53 KHz and 52°C temperature. The extracted solution was centrifuged at 12000 rpm for 10 min at 25°C, and the supernatant was transferred to the volumetric flask. The residues were then re-extracted with 25 mL 70% methanol using the same conditions. The extraction solution was collected, pooled and filtered through a 0.45 μm PTFE syringe filter. The filtrate was diluted with methanol to final working solutions and analyzed by HPTLC. All samples were stored in a refrigerator at 4°C until analysis.

Method validation

The HPTLC method was validated in terms of linearity, sensitivity, precision (intra- and inter-day variability), robustness, specificity, and accuracy in accordance with International Conference on Harmonization (ICH) guidelines on analytical method validation [18].

A series of mixed standard solutions containing appropriate concentrations of the four standards were analyzed to construct the calibration curves by plotting the peak area (y) of individual standard versus the concentration (x) of each analyte with least square linear regression of slope (m) and intercept (c) (y=mx+c). Then, the correlation coefficient (r^2) and linear range of each analyte were calculated. Limit of detection (LOD) and limit of quantification (LOQ) was determined on the basis of response and slope of each regression equation at signal-to-noise ratio (S/N) of 3:1 and 10:1, respectively for each analyte to determine the sensitivity of the method. The precision of the developed HPTLC method was analyzed by intra-day and inter-day variation. For intra-day variability test, the mixed standard solutions were analyzed for six replicates within one day, while for inter-day variability test, the solutions were examined in duplicates on three consecutive days. Variations in the precision were expressed by relative standard deviation (RSD) of the peak area of the standard. To assess the stability of the target analytes in the final extraction solution, the randomly selected E. longana extracted sample (S2) solution was stored at room temperature and analyzed at 0, 2, 4, 6, 8, 12 and 24 h. Robustness of the method was checked by making intentional changes in the parameters. Small change in the mobile phase composition was tried (formic acid ± 0.01 ml). The amount of mobile phase was varied in the range of ± 5%. The plates were prewashed with methanol and activated at 110°C ± 5 for 5, 10, 15 min respectively prior to chromatography. Time from spotting

to chromatography and from chromatography to scanning was varied from 0, 30, 60 and 90 min. Robustness was done at three different concentration levels of each analytes. The specificity of the method was determined by analyzing standard and extract samples. The peak for polyphenols in E. longana extract samples was confirmed by comparing the R_f value and the spectrum of the peak with that of the standard. The peak purity of compounds was determined by comparing the spectrum at three different regions of the spot, i.e., peak start (S), peak apex (M) and the peak end (E). Moreover, peak identity and purity was confirmed by mass spectrometry. The accuracy of the method was evaluated by the standard addition method. In the accuracy experiment, known amounts of the standards at low (80% of the known amounts), medium (100% of the known amounts), and high (120% of the known amounts) levels was spiked into E. longana sample (S2). Three replicates were performed and the extraction recovery of each analyte was calculated using the following equation:

Recovery (%)=100 × (found amount-original)/Spiked amount

Results and Discussion

Selection of analytical wavelength

HPTLC spectrum of corilagin, ellagic acid, epicatechin, and gallic acid showed maximum absorbance at 293 nm, 278 nm, 279 nm, and 274 nm, respectively. Further, in situ HPTLC spectral overlain of corilagin, ellagic acid, epicatechin, and gallic acid was taken. Isoabsorptive point was found at 280 nm and was selected as scanning wavelength (Figure 2).

Optimization of chromatographic conditions

Several solvent systems were assessed to attain the optimum resolution of four polyphenolic compounds. Initially, the mobile phase was selected on the basis of previous reports of gallic acid and ellagic acid. The first combination of mobile phase was composed of toluene, ethyl acetate, methanol, and formic acid (3: 3: 0.2: 0.8, v/v/v/v) [19]. The mobile phase resulted in good resolution of four peaks with same R_f value of gallic acid and ellagic acid as described in reported method however corilagin peak did not migrate considerably from the spotting zone and recorded at an R_f value of 0.03. In an attempt to achieve the desired R_f value in the range (0.2-0.8), methanol in the same mobile phase was increased to 0.5, 1, and 1.5 while other components were kept constant. Despite increasing methanol, the peak of corilagin was remain constant at the starting point that indicates increasing polarity using methanol has no effect on the elution of corilagin and on the contrary this resulted in complete elution of gallic acid at the solvent front more than the desired R_f range due to high polarity difference among analytes. Any change in formic acid yielded broad peak shape of corilagin. So, it was kept constant in later trials. Further, methanol was replaced by chloroform and then acetone in different ratio, resulted in similar results with no movement of corilagin. Afterwards, trials were directed towards elution of corilagin with n-butanol-water system that had been tried in different ratio along with methanol and formic acid, which finally resulted in the movement of corilagin. Later on, fine improvements in the ratio of solvents were made to obtain optimum mobile phase composed of n-butanol: water: methanol: formic acid (7.59: 1.27: 0.13: 1.01, v/v/v/v) for required elution of corilagin and simultaneously maintaining the adequate retention of gallic acid. The optimum mobile phase resulted in well resolved peaks with good shape with R_f values of 0.40 ± 0.01, 0.57 ± 0.02, 0.69 ± 0.01, and 0.79 ± 0.01 for corilagin, ellagic acid, epicatechin, and gallic acid, respectively (Figure 3).

Method validation

The proposed HPTLC method for quantitative analysis was validated by determining the linearity, LOD, LOQ, intra-day and

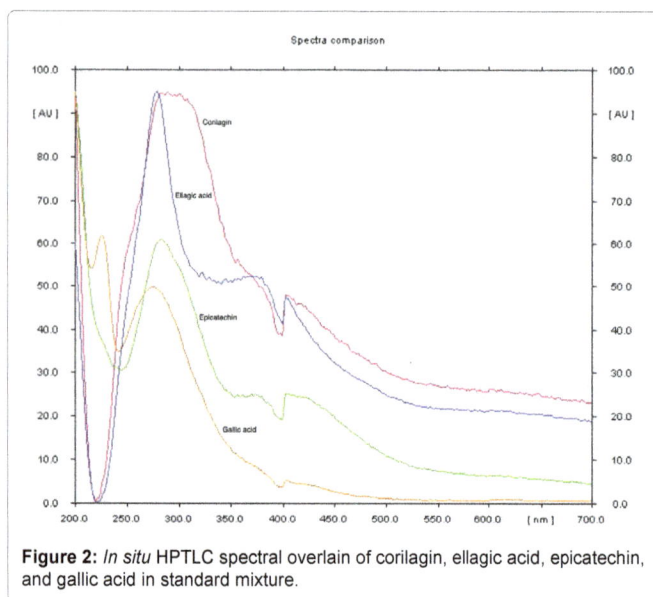

Figure 2: *In situ* HPTLC spectral overlain of corilagin, ellagic acid, epicatechin, and gallic acid in standard mixture.

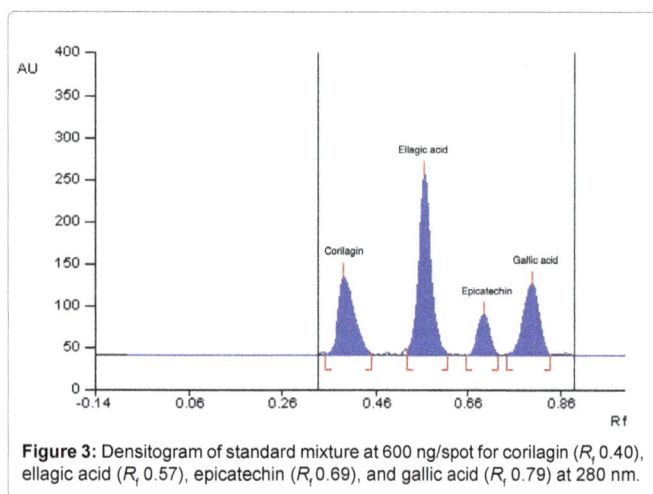

Figure 3: Densitogram of standard mixture at 600 ng/spot for corilagin (R_f 0.40), ellagic acid (R_f 0.57), epicatechin (R_f 0.69), and gallic acid (R_f 0.79) at 280 nm.

inter-day precisions, robustness, specificity, and accuracy. The linear regression equation, linearity ranges, together with the corresponding correlation coefficient (r^2) for the four analytes are listed in Table 1. All of the analytes expressed satisfactory linearity ($r^2>0.995$) over a relatively wide concentration range. The LOD and LOQ for the four compounds were <189.5 and 574.3 ng/spot, respectively, indicating a high sensitivity of the method with these chromatographic conditions. As shown in Table 1, the RSD values were calculated to express the intra- and inter-day precisions of the investigated compounds, ranged from 0.12 to 1.17% for the intra-day precision, and from 0.16 to 1.60% for the inter-day precision, illustrating good precision of the established method. Through analyzing the *E. longana* sample solution within 24 h, it was found to be stable as the RSD values of the peak areas were lower than 1.29%. Robustness of the proposed method was evaluated through investigating its ability to remain unaffected by small but deliberate variations. The standard deviation of the peak areas were calculated for each parameter and the RSD was found to be less than 2%. The low values of the RSD, indicated the robustness of the method. The peak purity of corilagin, ellagic acid, epicatechin, and gallic acid was assessed by comparing their respective spectra at the peak start, apex and peak end positions of the spot i.e., r (S, M)=0.999180, r (M, E)=0.998236, r (S, M)=0.999876, r (M, E)=0.999790, r (S, M)=0.999484, r (M,

E)=0.998909, and r (S, M)=0.999666, r (M, E)=0.999995, respectively. A good correlation (r^2=0.998, r^2=0.999, r^2=0.998 and r^2=0.999) was obtained between the standard and *E. longana* sample spectra of corilagin, ellagic acid, epicatechin, and gallic acid, respectively (Figure 4). A further confirmation of adequate specificity and resolution was established by DART-MS. The average recoveries of the four polyphenolic compounds were in the range of 98.42-101.73% with RSD values <1.77%, indicating that the proposed method is accurate and reproducible.

Quantitative analysis of samples

The developed HPTLC method was subsequently applied for the quantitative analysis of four major polyphenolic compounds (corilagin, ellagic acid, epicatechin, and gallic acid) in *E. longana* seeds. The contents of four major components were calculated with an external standard method based on the respective calibration curves. HPTLC densitogram of *E. longana* extract sample S2 shown in Figure 5.

As per results are shown in Table 2, corilagin content ranged from 3.94 mg/g to 4.98 mg/g (sample, S5 and S1) was markedly highest among all other polyphenolic components make it considered as the most prevalent compound in *E. longana* seeds. Ellagic acid content ranged from 1.08 mg/g to 1.76 mg/g (sample, S5 and S1) was found as the other dominant polyphenolic component. Similarly, gallic acid content ranged from 1.06 mg/g to 1.61 mg/g (sample, S5 and S1). The lowest content was of epicatechin correspond to others, which ranged from 0.48 mg/g to 1.22 mg/g (sample, S5 and S3). The highest amount of total polyphenolic content was found in sample S1. The various alterations of the quality of *E. longana* samples might be due to intrinsic factors such as genetic variation, plant origin, and extrinsic factors, such as geographic location, environmental conditions, cultivation techniques, etc. [20-23]. The quantitative analysis indicated that this method has significant importance in the comprehensive evaluation of selected polyphenolic compounds, which could be used in the quality control of *E. longana* and its related preparations.

DART-MS confirmation

Representative DART-MS spectra of HPTLC bands of *E. longana* extract (60000 ng/spot) correspond to corilagin (A), ellagic acid (B), epicatechin (C), and gallic acid (D) at their respective R_f are shown in Figure 6. Elemental compositions of major peaks were calculated by a built-in software based on the exact mass numbers of the elements and deprotonated molecules [M-H]⁻ were recorded under conditions of negative DART ionization. Based on the published literature and comparison with standards, four major polyphenolic compounds were confirmed in HPTLC bands of *E. longana* sample are shown in Table 3. According to the previously reported phytochemical studies, the mass to charge ratio of m/z 633.1488 [M-H]⁻ was identified as deprotonated corilagin [13]. Similarly, ion peak of m/z 300.9771 [M-H]⁻, 289.0802 [M-H]⁻ and 169.0337 [M-H]⁻ attributed to ellagic acid, epicatechin, and gallic acid, respectively [24-26].

To summarize, in the present study, the authors have established a suitable HPTLC method for the simultaneous determination of four major polyphenolic compounds in *E. longana*. The validated results demonstrated that the linearity, sensitivity, precision, robustness, specificity and accuracy of the proposed HPTLC method were satisfactory for the determination of polyphenolic compounds in *E. longana*. The authenticity and identification of the compounds were confirmed by DART-MS. The proposed method based on quantification and identification confirmation of bioactive components using a combination of two rapid analytical techniques was successfully applied

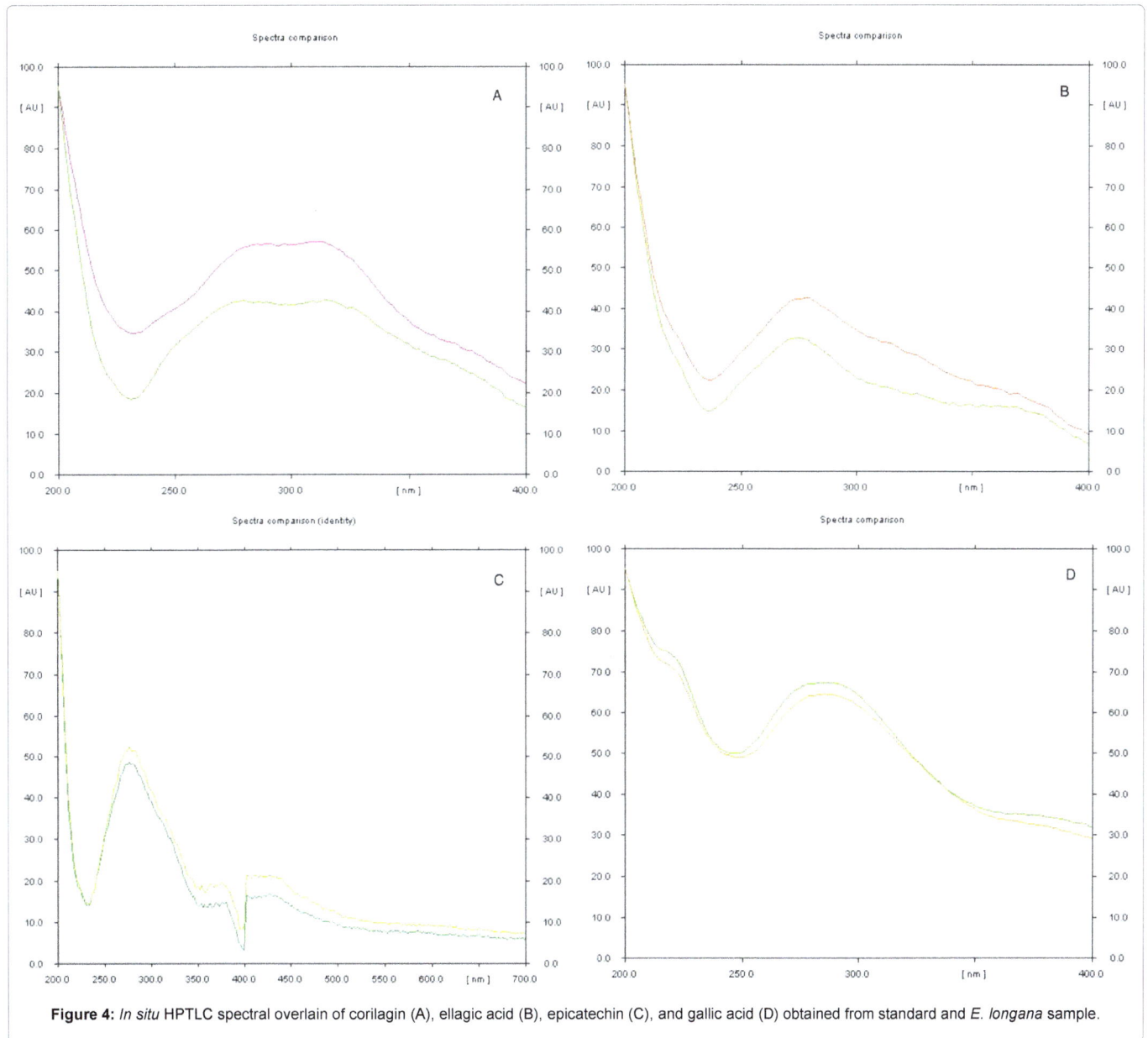

Figure 4: *In situ* HPTLC spectral overlain of corilagin (A), ellagic acid (B), epicatechin (C), and gallic acid (D) obtained from standard and *E. longana* sample.

	Corilagin	Ellagic acid	Epicatechin	Gallic acid
Regression equation	y=15.465x+187.2	y=15.987x-296.6	y=2.6583x+163.93	y=6.888x+2651.4
r^2	0.99813	0.99770	0.99941	0.99836
Linear range (ng/spot)	200-1200 ng/spot	200-1200 ng/spot	600-3600 ng/spot	200-1200 ng/spot
Sy.x*	279.6	321.1	80.9	116.8
LOD (ng/spot)	63.6	59.2	189.5	50.3
LOQ (ng/spot)	192.9	179.6	574.3	151.6
Precision RSD % (Intra-day, n=6)	0.81	0.77	1.17	0.12
Precision RSD % (Inter-day, n=6)	0.62	0.16	0.79	1.6
Stability RSD %, n=5	0.34	1.29	0.36	0.59
Recovery RSD %	1.77	0.61	1.21	0.48

*Standard deviation of residuals from line

Table 1: Method validation parameters: regression equation, correlation coefficients (r^2), limits of detection (LOD) and quantification (LOQ), precisions intra and inter-day, stability, and recovery.

for comprehensive quality control of *E. longana* products. This method could be an appropriate alternative for determination of bioactive components in plant matrices/ phytopharmaceuticals.

Acknowledgements

The authors thank Council of Scientific and Industrial Research (CSIR), New Delhi for research fellowship [08/281(0022)/2011-EMR-I] to Atul Singh Rathore.

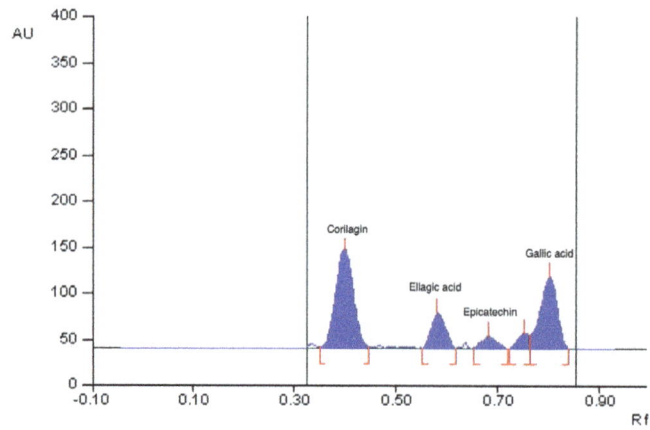

Figure 5: Densitogram of *E. longana* extract (60000 ng/spot) for corilagin (R_f 0.40), ellagic acid (R_f 0.58), epicatechin (R_f 0.68), and gallic acid (R_f 0.80) at 280 nm.

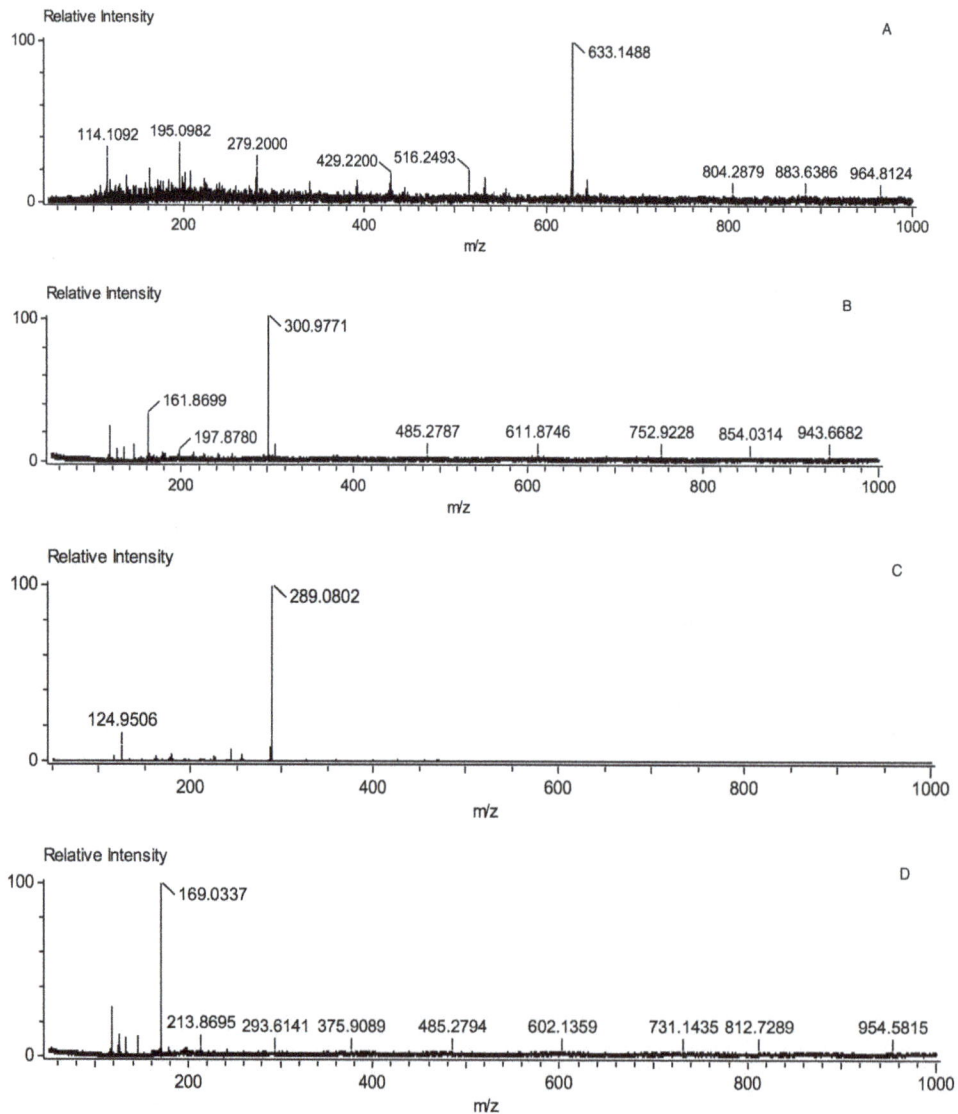

Figure 6: DART-MS spectra of HPTLC bands of *E. longana* extract (60000 ng/spot) correspond to corilagin (A), ellagic acid (B), epicatechin (C), and gallic acid (D) in negative ion mode.

Sample	Corilagin	Ellagic acid	Epicatechin	Gallic acid	Total
S1*	4.98 ± 0.02*	1.76 ± 0.05	0.92 ± 0.01	1.61 ± 0.06	9.27 ± 0.14*
S2	4.44 ± 0.15	1.61 ± 0.07	1.16 ± 0.03	1.28 ± 0.12	8.49 ± 0.37
S3	4.34 ± 0.04	1.52 ± 0.02	1.22 ± 0.16	1.57 ± 0.17	8.65 ± 0.39
S4	4.07 ± 0.10	1.13 ± 0.08	0.73 ± 0.04	1.20 ± 0.05	7.12 ± 0.27
S5	3.94 ± 0.21	1.08 ± 0.09	0.48 ± 0.10	1.06 ± 0.06	6.56 ± 0.46

*Highest content

Table 2: Content (mg/g) ± SD of polyphenolic compounds in *E. longana* seed (n=3).

Identification	R_f	Molecular formula	Measured mass	Theoretical mass	Error (mmu)
Corilagin	0.40	$C_{27}H_{22}O_{18}$	633.1488 [M-H]⁻	633.1506 [M-H]⁻	-0.18
Ellagic acid	0.58	$C_{14}H_6O_8$	300.9771 [M-H]⁻	300.9732 [M-H]⁻	0.39
Epicatechin	0.68	$C_{15}H_{14}O_6$	289.0802 [M-H]⁻	289.0790 [M-H]⁻	0.12
Gallic acid	0.80	$C_7H_6O_5$	169.0337 [M-H]⁻	169.0375 [M-H]⁻	-0.38

*a*Retardation factor

Table 3: DART-MS exact mass measurement of polyphenolic compounds in HPTLC bands of *E. longana* extract.

The authors are thankful to Principal, Poona College of Pharmacy, Bharati Vidyapeeth Deemed University, Pune and Dr. Brijesh Kumar, Sophisticated Analytical Instrument Facility (SAIF), CSIR-CDRI, Lucknow for providing facilities for carrying out this research work.

References

1. Rangkadilok N, Sitthimonchai S, Worasuttayangkurn L, Mahidol C, Ruchirawat M, et al. (2007) Evaluation of free radical scavenging and antityrosinase activities of standardized longan fruit extract. Food Chem Toxicol 45: 328-336.

2. Panyathep A, Chewonarin T, Taneyhill K, Vinitketkumnuen U, Surh YJ (2013) Inhibitory effects of dried Longan (Euphoria longana Lam.) seed extract on invasion and matrix metalloproteinases of colon cancer cells. J Agric Food Chem 61: 3631-3641.

3. Rangkadilok N, Worasuttayangkurn L, Bennett RN, Satayavivad J (2005) Identification and quantification of polyphenolic compounds in longan (Euphoria longana Lam.) fruit. J Agric Food Chem 53: 1387-1392.

4. Sudjaroen Y, Hull WE, Erben G, Würtele G, Changbumrung S, et al. (2012) Isolation and characterization of ellagitannins as the major polyphenolic components of Longan (Dimocarpus longan Lour) seeds. Phytochemistry 77: 226-237.

5. Zheng G, Xu L, Wu P, Xie H, Jiang Y, et al. (2009) Polyphenols from longan seeds and their radical-scavenging activity. Food Chem 116: 433-436.

6. Soong YY, Barlow PJ (2006) Quantification of gallic acid and ellagic acid from longan (Dimocarpus longan Lour.) seed and mango (Mangifera indica L.) kernel and their effects on antioxidant activity. Food Chem 97: 524-530.

7. Wen L, Yang B, Cui C, You L, Zhao M (2012) Ultrasound-assisted extraction of phenolics from longan (Dimocarpus longan Lour.) fruit seed with artificial neural network and their antioxidant activity. Food Anal Method 5: 1244-1251.

8. Rangkadilok N, Tongchusak S, Boonhok R, Chaiyaroj SC, Junyaprasert VB, et al. (2012) In vitro antifungal activities of longan (Dimocarpus longan Lour.) seed extract. Fitoterapia 83: 545-553.

9. Kunworarath N, Rangkadilok N, Suriyo T, Thiantanawat A, Satayavivad J (2016) Longan (Dimocarpus longan Lour.) inhibits lipopolysaccharide-stimulated nitric oxide production in macrophages by suppressing NF-κB and AP-1 signaling pathways. J Ethnopharmacol 179: 156-161.

10. Panyathep A, Chewonarin T, Taneyhill K, Surh YJ, Vinitketkumnuen U (2013) Effects of dried longan seed (Euphoria longana Lam.) extract on VEGF secretion and expression in colon cancer cells and angiogenesis in human umbilical vein endothelial cells. J Funct Foods 5: 1088-1096.

11. Chung YC, Lin CC, Chou CC, Hsu CP (2010) The effect of Longan seed polyphenols on colorectal carcinoma cells. Eur J Clin Invest 40: 713-721.

12. Losuwannarak N, Pasadhika C, Tantisira MH, Tantisira B (2009) Effects of longan seed extract on scopolamine-induced learning and memory deficit in mice. Thai J Pharm Sci 33: 31-38.

13. Soong YY, Barlow PJ (2005) Isolation and structure elucidation of phenolic compounds from longan (Dimocarpus longan Lour.) seed by high-performance liquid chromatography–electrospray ionization mass spectrometry. J Chromatogr A 1085: 270-277.

14. Li L, Xu J, Mu Y, Han L, Liu R, et al. (2015) Chemical characterization and anti-hyperglycaemic effects of polyphenol enriched longan (Dimocarpus longan Lour.) pericarp extracts. J Funct Foods 13: 314-322.

15. Rathore AS, Sathiyanarayanan L, Mahadik KR (2012) Stability-indicating high-performance thin-layer chromatographic method for quantitative estimation of emtricitabine in bulk drug and pharmaceutical dosage form. ISRN Chromatog 275883: 7.

16. More S, Tandulwadkar S, Nikam A, Rathore A, Sathiyanarayanan L, et al. (2013) Separation and determination of lamivudine, tenofovir disoproxil fumarate and efavirenz in tablet dosage form by thin-layer chromatographic-densitometric method. J Planar Chromatogr-Mod TLC 26: 78-85.

17. Chandra P, Bajpai V, Srivastva M, Kumar KR, Kumar B (2014) Metabolic profiling of Piper species by direct analysis using real time mass spectrometry combined with principal component analysis. Anal Method 6: 4234-4239.

18. Guideline IHT: Validation of analytical procedures: text and methodology (2005) Q2 (R1).

19. Bagul M, Srinivasa H, Padh H, Rajani M (2005) A rapid densitometric method for simultaneous quantification of gallic acid and ellagic acid in herbal raw materials using HPTLC. J Sep Sci 28: 581-584.

20. Valipour M (2015) Handbook of Environmental Engineering Problems. OMICS Press. USA.

21. Valipour M (2014) Handbook of drainage engineering problems. OMICS Group International.

22. Yannopoulos SI, Lyberatos G, Theodossiou N, Li W, Valipour M, et al. (2015) Evolution of water lifting devices (pumps) over the centuries worldwide. Water 7: 5031-5060.

23. Mahdizadeh KM, Gholami SMA, Valipour M (2015) Simulation of open-and closed-end border irrigation systems using SIRMOD. Arch Agron Soil Sci 61: 929-941.

24. Fang N, Yu S, Prior RL (2002) LC/MS/MS characterization of phenolic constituents in dried plums. J Agric Food Chem 50: 3579-3585.

25. Chandra P, Pandey R, Srivastva M, Kumar B (2015) Quality control assessment of polyherbal formulation based on a quantitative determination multimarker approach by ultra high performance liquid chromatography with tandem mass spectrometry using polarity switching combined with multivariate analysis. J Sep Sci 38: 3183-3191.

26. Gu D, Yang Y, Bakri M, Chen Q, Xin X, et al. (2013) A LC/QTOF-MS/MS Application to Investigate Chemical Compositions in a Fraction with Protein Tyrosine Phosphatase 1B Inhibitory Activity from Rosa Rugosa Flowers. Phytochem Anal 24: 661-670.

9

Electrospray Tandem Mass Spectrometric Study of a Furo-Furan Lactone in *Heliotropium eichwaldi*

Ali A[1], Uddin J[1], Ansari HN[2], Firdous S[2] and Musharraf SG[1]*

[1]*HEJ Research Institute of Chemistry, International Center for Chemical and Biological Sciences, University of Karachi, Karachi, Pakistan*
[2]*Department of Chemistry, University of Karachi, Karachi, Pakistan*

Abstract

Detailed gas phase fragmentation behavior of a furo-furan lactone **1** with a novel skeleton of three fused furan rings isolated from medicinally important *Heliotropium eichwaldi* plant was studied. Elucidation of the major fragmentation routes was performed using CID-MS/MS analysis, insource-MS^n fragmentation, parent ion scan analysis and exact mass measurements using electrospray ionization quadropole time-of-flight mass spectrometry (ESI-QqTOF-MS/MS) hybrid instrument in negative and positive ion modes. Analysis showed the presence of [M+H]$^+$ and [M-H]$^-$ at *m/z* 275 and *m/z* 273, respectively. Low-energy collision induced dissociation tandem mass spectrometric analysis (CID-MS/MS) in both ion modes showed several characteristics losses of carbon monoxide and water molecules along with characteristic losses of acetaldehyde, formaldehyde, formic acid and carbondioxide. Moreover, compound **1** showed significant insource-MS^n behavior during in-source fragmentation studies. LC-MS identification of **1** was performed in crude extract of *H. eichwaldi*. The knowledge of the fragmentation pattern and key fragment ions of furo-furan lactone, will be useful for further exploration of new members of this class of compounds in the *H. eichwaldi* and related species of family Boraginaceae utilizing very limited plant material.

Keywords: Furo-furan lactone; LC-MS/MS; Mass spectrometry; *Heliotropium eichwaldi*

Introduction

Plants have been used for the treatment of various diseases and are the source of unprecedented lead molecules since several decades. In the search for novel natural products in plants, particularly those possessing potential bioactivity, it is vitally important to efficiently distinguish between previously reported compounds and the novel natural products, a process termed as dereplication. Techniques capable of quick and accurate identification of previously reported compounds are therefore of great value [1] thus saving time and money spent on the re-isolation, purification and spectrometric identification of previously known natural products.

Mass spectrometry (MS) has emerged as a complementary method with high sensitivity and selectivity for the high-throughput analysis of natural products coupled with various chromatographic techniques [2,3]. Electrospray ionization tandem mass spectrometry (ESI-MS/MS) using low-energy collision-induced dissociation (CID) has been the technique of choice for such studies [4] owing to its ability to analyze compounds of medium to high polarities [5]. Several natural products have been investigated through ESI-MS/MS and this approach in combination with liquid chromatography (LC) offers a rapid dereplication of plant extract. In order to characterize a compound from its MS/MS data, previous knowledge of the fragmentation pathway of homologous compounds exhibiting a conserved structural core is required for confident identification [6] Such information is not only helpful in identifying characteristic fragments which provide useful information about the structures of compounds but also can be projected to infer the new and similar chemical structures [4]. Moreover, mass spectral database search has now turn out to be the well-accepted approach for the identification of unknown chemicals thus minimizing the need for authentic reference material [7].

The genus *Helitropium* belongs to the family Boraginaceae, is known to possess a number of medicinal properties which are predominantly attributed to pyrrolizidine alkaloids [8]. *Helitropium eichwadi* is herbaceous weed, widely distributed in the state of Punjab,

Haryana and Rajasthan of India [9] In Pakistan this annual herb grows in cultivated fields, gardens and shallow lands of the province of Sind. The plant is emetic, antidote, antiseptic and is used for the treatment of scorpion stings, bee stings, mad dog bite, snake bite, and ear ache. It is also known to be used for cleaning and healing of wounds, warts and ulcers [10,11]. The plant has demonstrated hypotensive effect, [12] antimicrobial activity and nephrotoxicity [13].

In continuation of our studies on the gas phase fragmentation behavior and the identification of diagnostic fragments ions produced during CID-MS/MS analysis of various unique naturally occurring structures and elucidation of their fragmentation pathways, [14-17] this paper describe the ESI-QqTOF-MS (+ve ion and -ve ion mode) and the CID-MS/MS analysis of the novel furo-furan lactone (**1**) isolated from *Helitropium eichwadi* [18] The elucidation of the gas-phase fragmentation pathways and the rapid characterization of this complex structure will be helpful for the identification of its congeners in other plant species. Present study is the first ESI-MS/MS report of this novel class of compound.

Experimental

Standard and reagents

Chemicals and solvents were of analytical and HPLC grades, respectively, and were purchased from Aldrich-Sigma (USA).

*Corresponding author: Syed Ghulam Musharraf, HEJ Research Institute of Chemistry, International Center for Chemical and Biological Sciences, University of Karachi, Karachi-75270, Pakistan, E-mail: musharraf1977@yahoo.com

Deionized water (Milli-Q) was used in the study. Isolation and characterization details of standard furo-furan lactone (4a,6b-Dihydroxy-5-(hydroxymethyl)-2a-isopropyl-2-methylhexahydro-4H-1,3,6-trioxa-cyclopenta[cd] pentalen-4-one) **1** has already been reported [18].

ESI-QqTOF-MS analysis

Compound (**1**) was dissolved in methanol (0.5 mg/mL) and analyzed by electrospray ionization (ESI) and collision-induced dissociation (CID) on QqTOF-MS/MS instrument (QSTAR XL mass spectrometer Applied Biosystems/ MDS Sciex, Darmstadt, Germany) at laboratory conditions. Working dilutions (10 μg/mL) were prepared in 1:1 acetonitrile-water containing 0.1% formic acid and 1:1 acetonitrile-water containing 4 mM ammonium acetate for the analysis in positive and negative ion modes, respectively. High-purity nitrogen gas was used as the curtain gas and collision gas delivered from Peak Scientific nitrogen generator. The ESI interface conditions for positive ion modes were as follows: Ion spray capillary voltage of 5500 V, curtain gas flow rate 20 L min^{-1}, nebulizer gas flow rate 30 L min^{-1}, DP1 60 V, DP2 15 V, and FP 265 V. Collision energy was swept from 05 to 45 eV for MS/MS analysis in positive modes. Calibration was performed by using 10 pM reserpine solution as internal calibrant. For the generation of product ions in-source, DP1 was optimized at 90 while curtain gas flow was set at 35 L min^{-1}. Sample was introduced into the mass spectrometer using a Harvard syringe pump (Holliston, MA) at a flow rate of 5 μL/ min during all analysis.

ESI interface conditions for the ESI-MS and ESI-MS/MS analysis in negative ion mode were optimized as follows: Ion spray voltage -4200 V, curtain gas flow rate 20 L, nebulizer gas flow rate 20 L min^{-1} DP1 -55, DP2 -15 V, and FP 25V. Collisional energy, during CID-MS/MS analysis, was swept from 5 to 40 eV and instrument was calibrated by using 0.2 ng/uL taurocholic acid solution as internal calibrant.

Plant sample preparation and LC-MS analysis

The plant material of *H. eichwaldi* was collected and identified according to already published report. The shade-dried and chopped whole plant of *H. eichwaldi* (5 g) was extracted by sonication with MeOH (15 mL) for 30 min at laboratory temperature and then centrifuged. The supernatant was then partitioned with EtOAc (5 mL × 3). EtOAc extract was evaporated to gummy mass (0.6 g). MeOH was then used to reconstitute the gummy mass to 1 mg/mL solution. After necessary dilution resulting solution was then directly injected to mass spectrometer for the ESI-MS/MS analysis of *H. eichwaldi* plant extract.

Chromatographic separation and mass spectrometric detection was achieved using UPLC QqQ-MS system (Series 1260, Agilent Technologies, Wilmington, DE, USA), and on Agilent ZORBAX SB-C18 column (2.1 × 50 mm, 1.8 μm). Injection volumes were 2 μL and the mobile phases were as follows: eluent A, H_2O (0.1% formic acid) and eluent B, MeOH (0.1% formic acid). The flow rate was 0.6 mL min^{-1}, and a gradient elution program was used. The chromatographic procedure was initialized and maintained at 10% B till 2 min, followed by linear increase to 90% B in 8 min. This composition was hold for 1 min and at 9.1 min system returns to initial conditions.

Results and Discussion

Compound **1** possesses three fused furan rings and has highly functionalized molecular structure. The structure comprises of seven oxygen functionalities present within the frame of twelve carbon atoms which include primary and tertiary hydroxyls, lactone and ether functions. Characteristic fragments in this novel furo furan lactone

(**1**) are formed due to the subsequent losses of water and carbon monoxide molecules. Other characteristic fragments include losses of acetaldehyde, formaldehyde, and formic acid.

Optimization of collision energy and in-source parameters

ESI-QqTOF-MS analysis of **1** showed [M+H]$^+$ at *m/z* 275.1307 corresponding to the protonated molecular formula $C_{12}H_{19}O_7$ (calcd 275.1125). MS/MS analysis of [M+H]$^+$ ion exhibiting interesting fragmentation pattern and the product ion abundance were found to be significantly influenced by the variation of collision energy after the optimization of all major parameters. MS/MS spectra of furo-furan lactone were screened against laboratory collision energies ranging from 5 to 40 eV in both negative and positive ion modes due to the significant changes in fragmentation pattern (Figure 1). It was observed that the fragment ions resulted due to the sequential losses were best appeared at collision energy 20 eV in positive ion mode (Figure 2A) whereas the collision energy of 25 eV was found to be the optimum energy for the fragment ions formed in the negative ion mode in order to observe intensity of all major fragment ions (Figure 2B). By optimizing declustering potential (DP) at 90 and curtain gas flow (CUR) at 35 Lmin^{-1} we were able to yield all major product ions via in-source with considerable intensity (Figure 3). These product ions were then successfully fragmented further using CID-MS/MS option and were utilized in proposing fragmentation routes (Supplementary Figures 1-6).

Characteristics fragmentation in the ESI-MS/MS

Major characteristic peaks were observed by the sequential removals of CO and water molecules from the [M+H]$^+$ in positive ion mode. A characteristic simultaneous removal of methanol and carbon monoxide followed by the loss of two molecules of water i.e., [M+H-MeOH-2H$_2$O -CO]$^+$ appeared as a base peak at *m/z* 179 with collision energy of 25 eV. Other characteristic peaks were observed due to the loss of multiple hydroxyls as water and carboxylic group as carbon monoxide molecules at *m/z* 257 [M+H-H$_2$O]$^+$, 239 [M+H-2H$_2$O]$^+$, 197 [M+H-H$_2$O-MeOH-CO]$^+$, 179 [M+H-2H$_2$O-MeOH-CO]$^+$, 133 [M+H-MeOH-3H$_2$O-2CO]$^+$, and 105 [M+H-MeOH-3H$_2$O-3CO]$^+$.

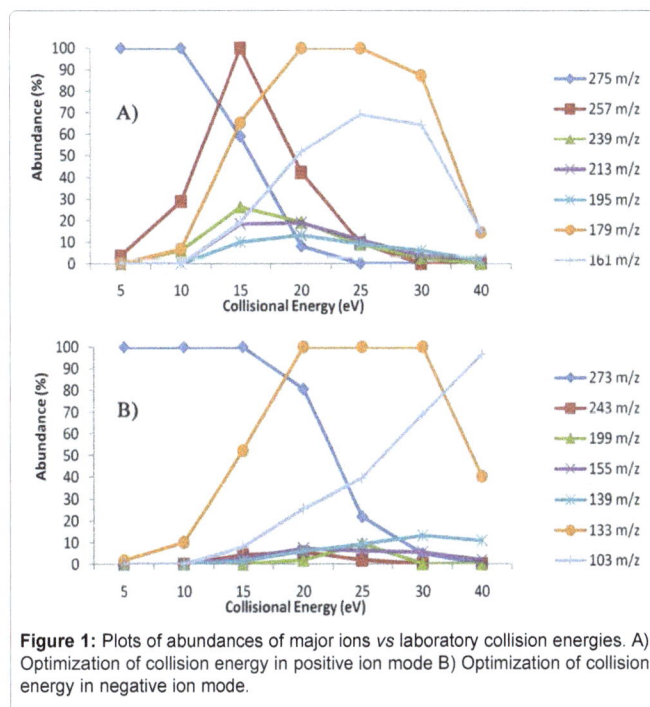

Figure 1: Plots of abundances of major ions *vs* laboratory collision energies. A) Optimization of collision energy in positive ion mode B) Optimization of collision energy in negative ion mode.

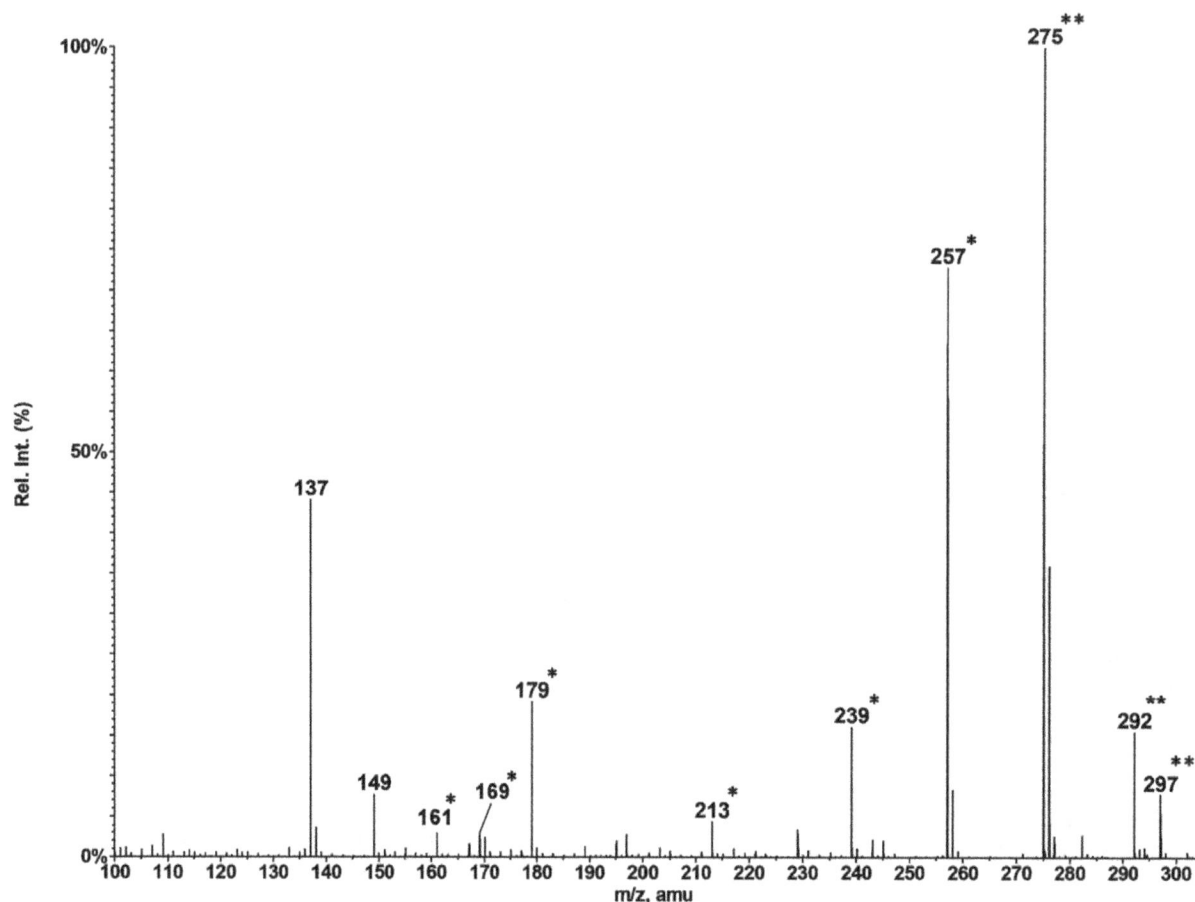

Figure 2: Insource MSMS of *m/z* 275 observed during TOF-MS analysis of Furo-Furan lactone standard. [*Fragment ions formed during insource-MSMS analysis of *m/z* 275, **Molecular ions of furo-furan lactone appearing at *m/z* 275 [M+H]⁺, *m/z* 292 [M+NH₄]⁺ and *m/z* 297 [M+Na]⁺ respectively].

The lactone and hydroxyl moieties have a significant influence on the fragmentation pattern of this new class of compound. In the second pathway, fragments were observed at m/z 229 [M+H-H_2O-CO]⁺, 169 [M+H-2H_2O-CO-CH_2CO]⁺, 151 [M+H-3H_2O-CO-CH_2CO]⁺, 141 [M+H-2H_2O-2CO-CH_2CO]⁺, 123 [M+H-2H_2O-2CO-CH_2CO]⁺, 95 [M+H-3H_2O-3CO-CH_2CO]⁺. In another pathway, loss of acetaldehyde was observed preceding removal of water molecule at m/z 213 [M+H-H_2O-CH_3CHO]⁺ and 195 [M+H-2H_2O-CH_3CHO]⁺ (Figure 3A).

In negative ion mode, parent ion was observed at m/z 273 [M-H]⁻. Few characteristic fragments were observed including m/z 243 [M-H-CH_2O]⁻, 133 [M-H-CH_2O-C_7H_7O]⁻, 103 [M-H-2CH_2O-C_7H_7O]⁻ and 59 [M-H-2CH_2O-C_7H_7O-CO_2]⁻ (Figure 2B). A second series of product ions were observed at m/z 199 [M-H-CH_2O-CO_2]⁻, 155 [M-H-CH_2O-CO_2-CH_3CHO]⁻, 127[M-H-CH_2O-CO_2-CH_3CHO-CO]⁻ and 99 [M-H-CH_2O-CO_2-CH_3CHO-2CO]⁻ (Figure 3B).

Evaluation of the mechanistic pathways in positive ion mode

Mechanistic pathways for the furo-furan lactone (**1**) fragmentation was proposed using different techniques including exact mass measurement for predicting elemental composition, *in silico* studies to predict possible protonation sites from where fragmentation possibly have initiated, MS/MS, in-source fragmentation and parent ion scan to elucidate fragmentation pathway. The HR-ESI-MS data of characteristic fragments in both positive and negative modes is summarized in Supplementary Tables 1 and 2, respectively.

In silico calculation of HOMO by DFT at 6-31G* basis set using Spartan 08 (Wavefunction, CA, USA) suggested carbonyl oxygen to be the most favorable protonation site (Figure 4) that may tautomarize through a proton transfer to lactone oxygen forming specie **A**. Loss of water molecule was observed forming product ion **B** at m/z 257. Parent ion scan of product ion m/z 179 showed its formation from two ions i.e., m/z 239 and m/z 197 (Supplementary Figure 7). Therefore, simultaneous loss of carbon monoxide and methanol was suggested to produce fragment **C** (m/z 197) a highly stabilize aromatic ion. Alternatively, loss of water molecule is also preceded by simultaneous loss of methanol and carbon monoxide forming fragment ion **E** (m/z 179) via fragment **D** at m/z 239. Rearrangement of fragment **E** to **F** is proposed through ring expansion that may tautomarize to phenol-like form. Ion form **F** on releasing water molecule may forms fragment **G** (m/z 161), structure that readily undergo removal of carbon monoxide to produce fragment **H** at m/z 133. Loss of another carbon monoxide from fragment **I** resulted in the formation of fragment ion I with m/z 105 (Scheme 1).

Parent ion scan of product ion at m/z 229 suggested that it was formed directly from parent ion at m/z 275 (Supplementary Figure 8). It revealed that another pathway is initiated with simultaneous loss of water and carbon monoxide molecule from parent ion providing fragment ion **J** at m/z 229 that may undergo rearrangement via six member transition state to stable vinyl cation **K** with m/z 169 by simultaneously losing ketene and water molecules, such reactions are

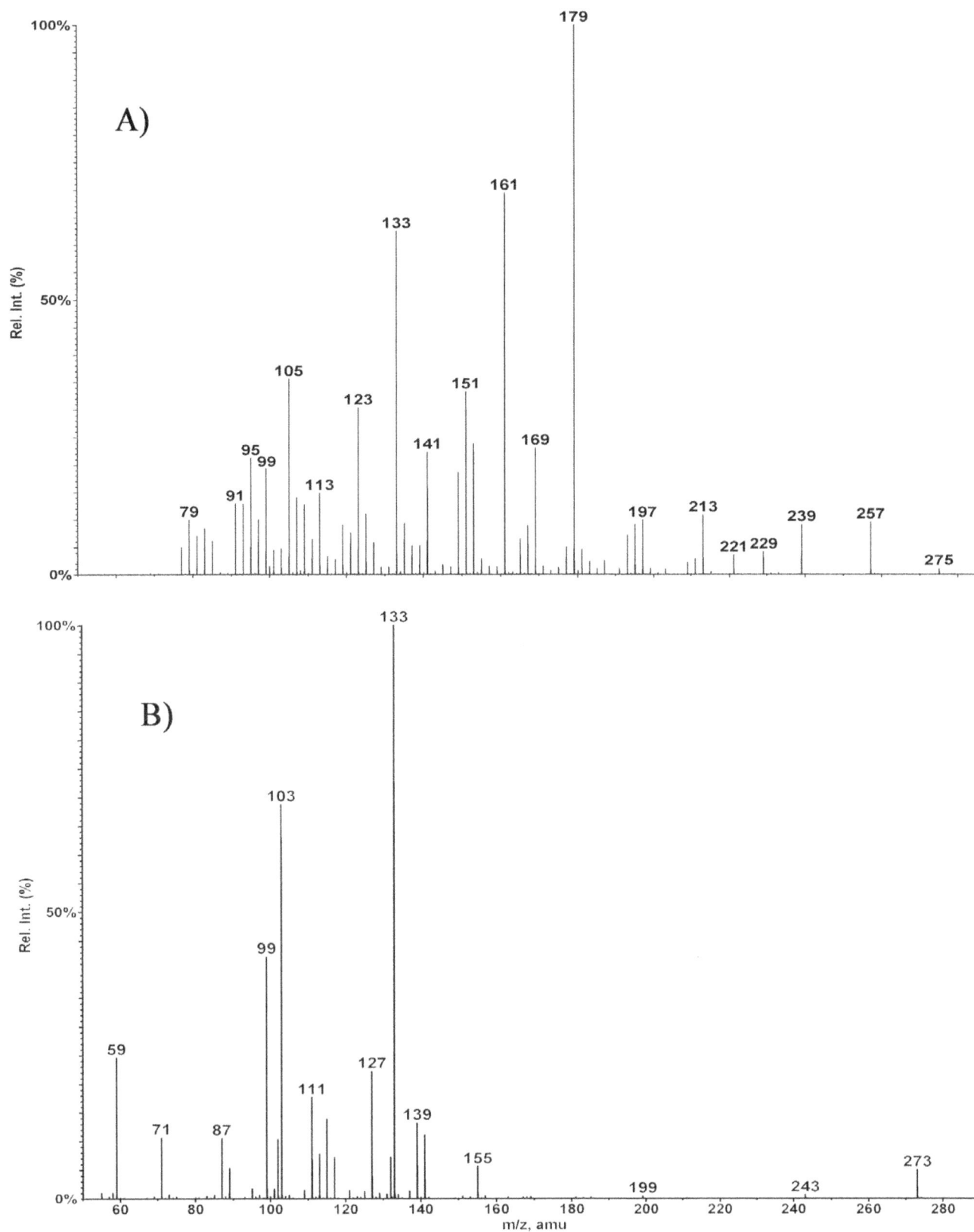

Figure 3: Product ion spectra of compound **1**. A) MS/MS spectrum of [M+H]+ (*m/z* 275) in positive ion mode. B) MS/MS spectrum of [M-H]- *m/z* 273 in negative ion mode.

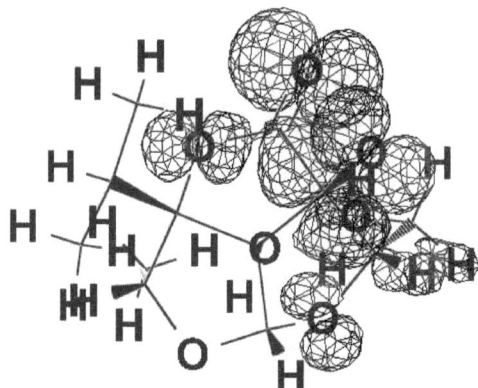

Figure 4: HOMO of furo-furan lactone, calculated by DFT at basis set 6-31G, showing largest lobes located at carbonyl oxygen.

entropically favored. Fragment ion K on further loss carbon monoxide molecule form fragment **L** with m/z 141. Loss of a water molecule from product ion at m/z 169 produces fragment ion **M**, a conjugated ion with m/z 151. Product ion at m/z 151 sequentially losses two carbon monoxide molecules to give rise to product ions N and O at m/z 123 and 95, respectively. Loss of ethene molecule is observed from product ion at m/z 95 to form product ion at m/z 67 (Scheme 2).

Parent ion scan of product ion at m/z 213 confirms its formation from product ion at m/z 257 (Supplementary Figure 9). Therefore, a different pathway is suggested, instead of multiple carbon monoxide losses, loss of an acetaldehyde molecule from fragment ion Q (m/z 257) was observed followed by loss of water molecule forming fragment ions R and S with m/z 213 and 195, respectively (Scheme 3).

Evaluation of mechanistic pathway in negative ion mode

It was suggested that fragmentation in the negative ion mode initiated from deprotonation of primary alcohol forming [M-H]⁻ ion

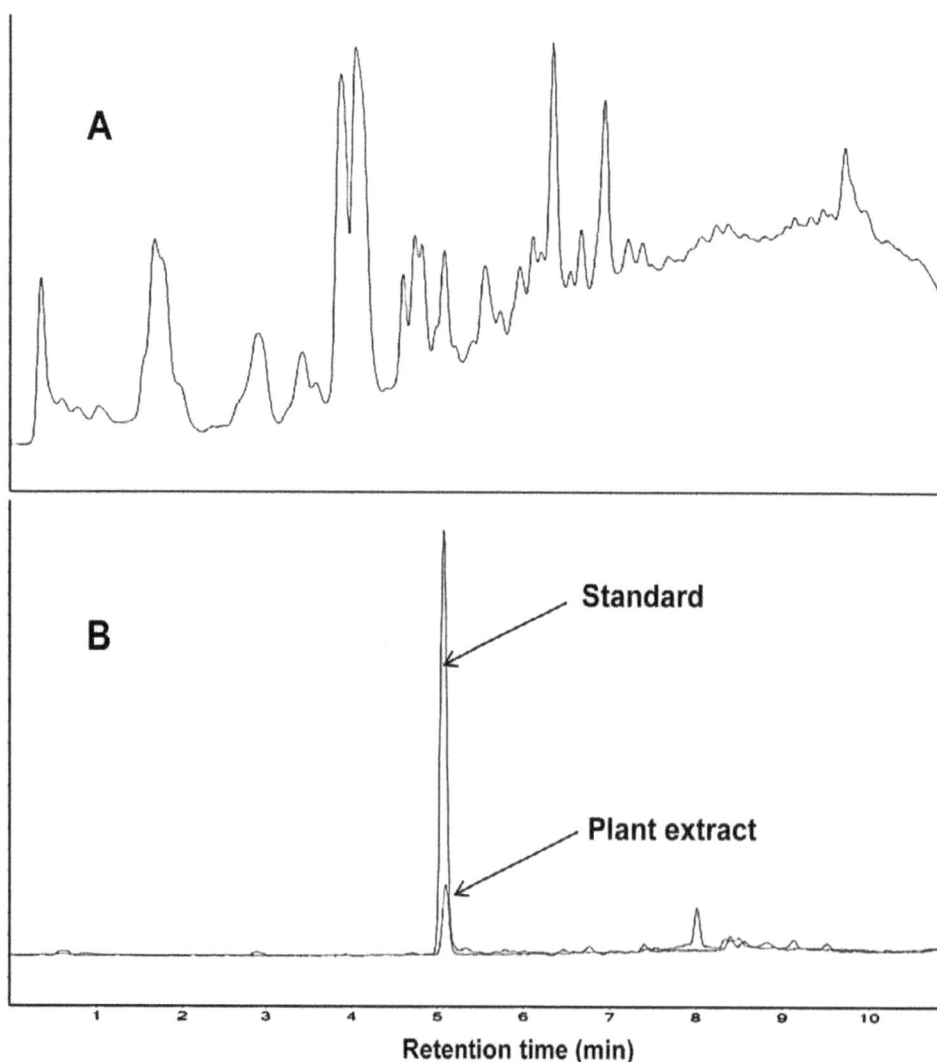

Figure 5: LC-MS analysis of *Heliotropium eichwaldi* plant extract. A) Total ion chromatogram (TIC). B) Extracted ion chromatogram of m/z 275 in standard and plant extract *H. eichwaldi* plant extract.

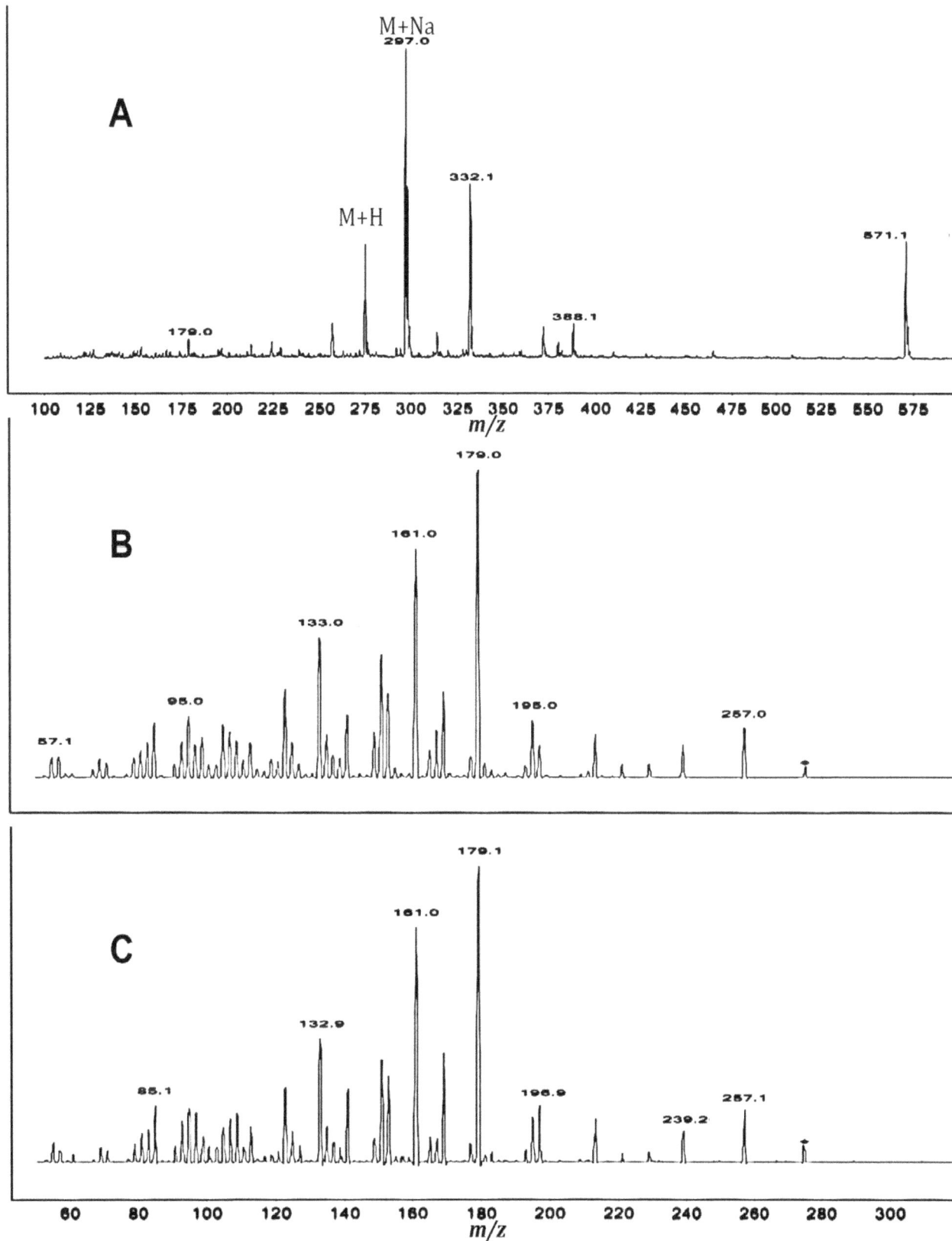

Figure 6: MS and MS/MS scans at Rt 5.1 min in plant extract *vs.* standard solution of **1**. A) MS scan of *H. eichwaldi* plant extract. B) MS/MS scan of *m/z* 275 appearing in plant extract. C) MS/MS scan of *m/z* 275 appearing in standard solution.

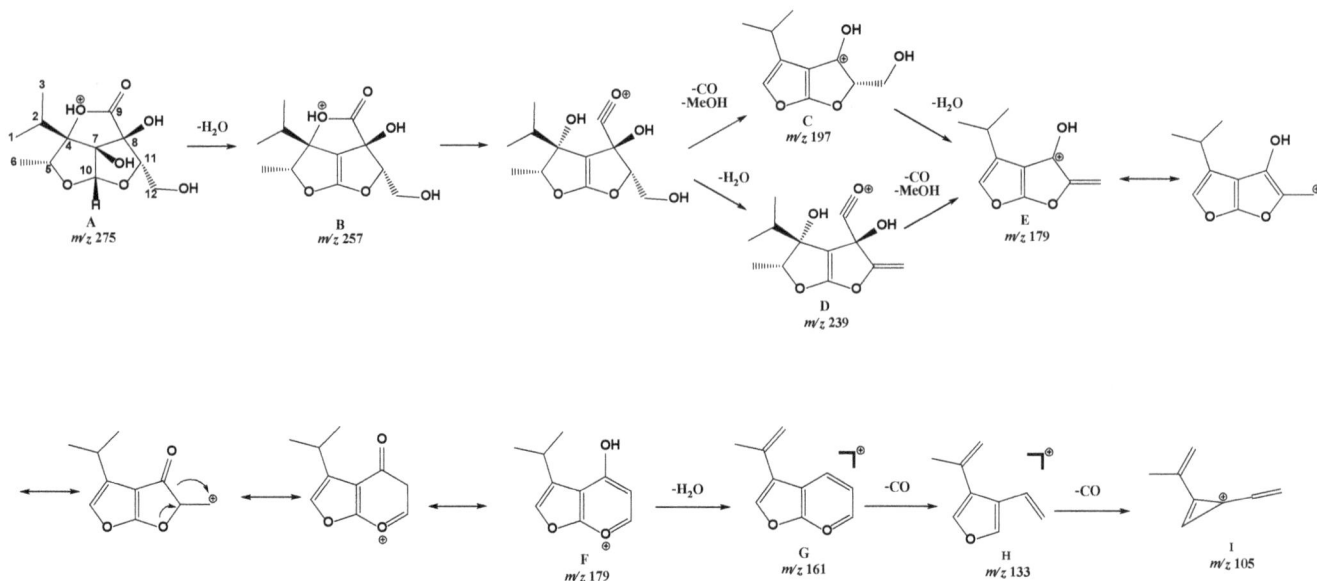

Scheme: 1

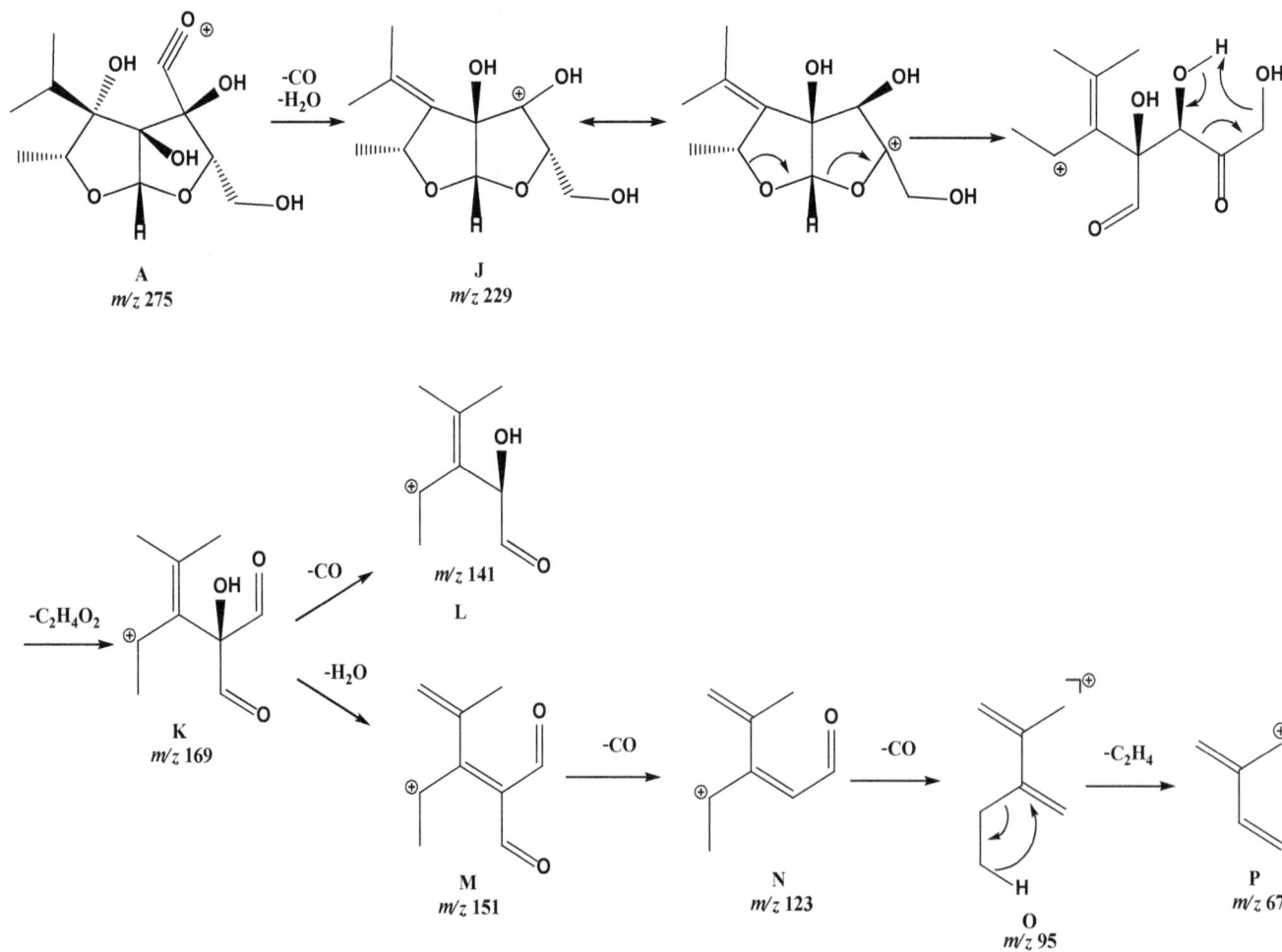

Scheme 2

a at *m/z* 273. The other two possible sites of deprotonation are the two relatively more sterically hindered hydroxyls at C-7 and C-8. Parent ion scans of product ions at *m/z* 103 and 133 shows that these ions are formed from parent [M-H]⁻ ion (*m/z* 273) via fragment ion *m/z* 243 (Supplementary Figure 10). On this basis it is suggested that ion **a** readily losses formaldehyde molecule to give rise to a weak signal of fragment ion **b** at *m/z* 243. Fragment ion **b** simultaneously loses alkenyl part and carbon monoxide molecule to form base peak at *m/z*

133 of fragment ion **c** (for CE 25-35 eV). Fragment **c** may dissociate into fragment **f** at *m/z* 59 via two possible routes, either via product ion **d** (*m/z* 103) by sequentially losing formaldehyde and carbon dioxide molecules, respectively or via product ion **e** (*m/z* 87) through sequentially giving out formic acid and carbon monoxide molecules, respectively (Scheme 4).

Following alternative dissociation pathway, fragment ion **b** may rapidly release carbon dioxide molecule to form fragment ion **g** at *m/z*

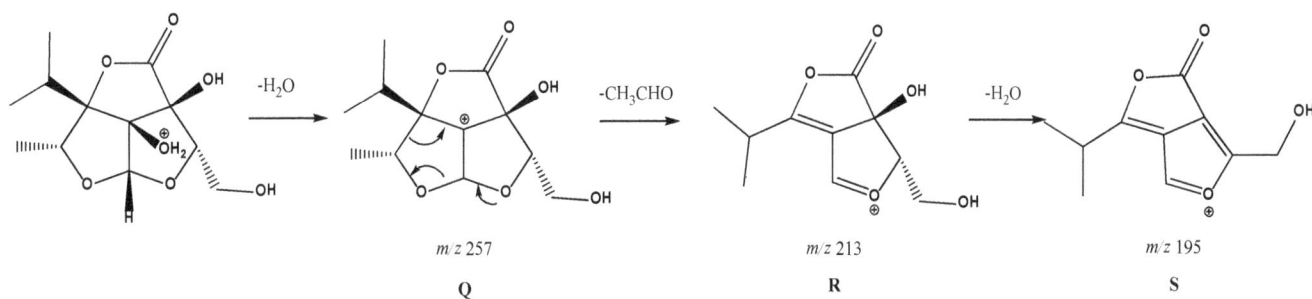

Scheme: 3

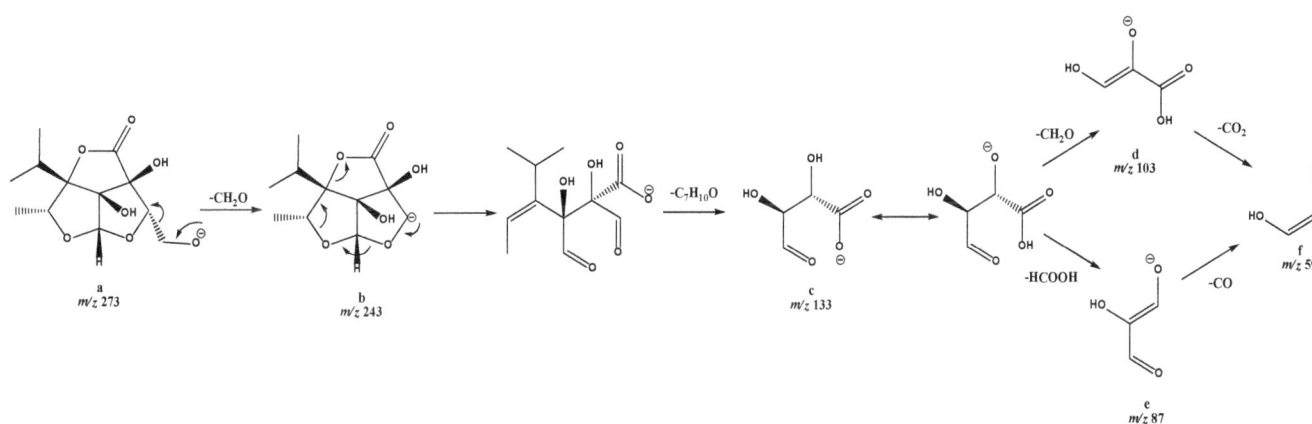

Scheme: 4

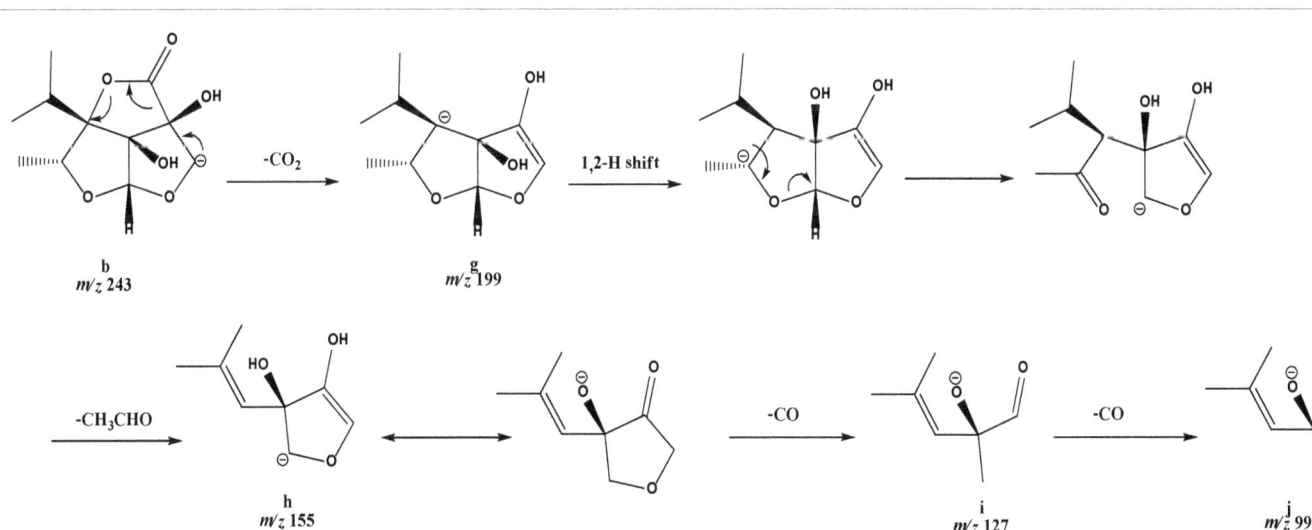

Scheme: 5

199. This product ion **g** may rearrange into an open structure that loses acetaldehyde molecule to give rise peak at m/z 155 of the product ion **h**. Product ion **h** upon sequential removal of two carbon monoxide molecules form product ions **i** and **j** at m/z 127 and 99, respectively (Scheme 5).

ESI-MS/MS analysis of *Heliotropium eichwaldi* extract

The furo-furan lactone in the extract of *Heliotropium eichwaldi* was investigated by LC-ESI-MS/MS. The full LC-MS scan in positive mode showed the presence of $[M+H]^+$ and $[M+Na]^+$ ions of novel furo furan lactone (m/z 275) at Rt 5.1 min (Figure 5). The Rt and MS/MS data of this protonated ion matches the standard furo furan data, which showed several neutral losses of water and carbon monoxide molecules giving rise to characteristic peaks such as at m/z 257, 195, 179, 161, 133, and 95 etc. (Figure 6). Therefore, the compound was identified as novel furo furan lactone (**1**).

Conclusion

This study showed that CID-MS/MS fragmentation pattern of furo-furan lactone (**1**) is highly dominated by neutral losses of several carbon monoxide molecules in positive ion mode and losses of formaldehyde and carbon monoxide is predominant in negative ion spectra. It was found that due to highly functionalized nature of compound (**1**) fragmentation pattern is very complicated and several parallel pathways guide the dissociation of parent ion. Fragmentation mechanism has been proposed using several tools including exact masses and linked scanning data such as MS/MS and parent ion scanning data etc. The information acquired from MS/MS data and parent ion scanning of compound **1** standard was successfully applied to identify it in *Heliotropium eichwaldi* crude extract without rigorous sample preparation method. Such studies will be potentially useful to rapidly identify this class of compounds in complex mixtures such as plant extracts of other species and herbal product based on them.

Acknowledgments

The authors acknowledge the OPCW (Project No. L/ICA/ICB/173673/12) for providing the partial funding for this study. The authors are also thankful to Mr. Junaid ul Haq for kind assistance in LC-MS analysis.

References

1. McGarvey BD, Liao H, Ding K, Wang X (2012) Dereplication of known pregnane glycosides and structural characterization of novel pregnanes in Marsdenia tenacissima by high-performance liquid chromatography and electrospray ionization-tandem mass spectrometry. J Mass Spectrom 47: 687-693.

2. Crotti AE, Lopes JL, Lopes NP (2005) Triple quadrupole tandem mass spectrometry of sesquiterpene lactones: a study of goyazensolide and its congeners. J Mass Spectrom 40: 1030-1034.

3. Motilva MJ, Serra A, Macià A (2013) Analysis of food polyphenols by ultra-high-performance liquid chromatography coupled to mass spectrometry: an overview. J Chromatogr A 1292: 66-82.

4. Fredenhagen A, Derrien C, Gassmann E (2005) An MS/MS library on an ion-trap instrument for efficient dereplication of natural products. Different fragmentation patterns for $[M+H]^+$ and $[M+Na]^+$ ions. J Nat Prod 68: 385-391.

5. Crotti AEM, Vessecchi R, Lopes JLC, Lopes NP (2006) Electrospray ionization mass spectrometry: chemical processes involved in the ion formation from low molecular weight organic compounds. Quim Nova 29: 287-292.

6. Furtado NA, Vessecchi R, Tomaz JC, Galembeck SE, Bastos JK, et al. (2007) Fragmentation of diketopiperazines from Aspergillus fumigatus by electrospray ionization tandem mass spectrometry (ESI-MS/MS). J Mass Spectrom 42: 1279-1286.

7. Subramaniam R, Östin A, Nygren Y, Juhlin L, Nilsson C, et al. (2011) An isomer-specific high-energy collision-induced dissociation MS/MS database for forensic applications: a proof-of-concept on chemical warfare agent markers. J Mass Spectrom 46: 917-924.

8. Reina M, GonzalezColoma A, Gutierrez C, Cabrera R, Henriquez J, et al. (1997) Bioactive saturated pyrrolizidine alkaloids from Heliotropium floridum. Phytochem 46: 845-853.

9. Jain SC, Sharma R (1987) Antimicrobial activity of pyrrolizidine alkaloids from Heliotropium ellipticum. Chem Pharm Bull (Tokyo) 35: 3487-3489.

10. Bhakuni DS, Dhar ML, Dhar MM, Dhawan BN, Mehrotra BN (1969) Screening of Indian plants for biological activity. II. Indian J Exp Biol 7: 250-262.

11. Jain R, Arora R (1997) Phytochemical and Pharmacological Investigation of Heliotropium ellipticum Ledeb. J Indian Chem Soc 74: 430-431.

12. Gupta SK, Mathur IS (1972) The effect of Arnebia nobilis and its naphthaquinones in rat Walker carcinosarcoma 256. Indian J Cancer 9: 50-55.

13. Sharma SK, Goyal N (2012) Protective effect of Heliotropium eichwaldi against cisplatin-induced nephrotoxicity in mice. Zhong Xi Yi Jie He Xue Bao 10: 555-560.

14. Musharraf SG, Goher M, Ali A, Adhikari A, Choudhary MI, et al. (2012) Rapid characterization and identification of steroidal alkaloids in Sarcococca coriacea using liquid chromatography coupled with electrospray ionization quadrupole time-of-flight mass spectrometry. Steroids 77: 138-148.

15. Ghulam MS, Goher M, Hussain A, Choudhary MI (2012) Electrospray tandem mass spectrometric analysis of a dimeric conjugate, salvialeriafone and related compounds. Chem Cent J 6: 120.

16. Musharraf SG, Goher M, Wafo P, Kamdem RST (2012) Electrospray tandem mass spectrometric analysis of duboscic acid, exploring the structural features of a new class of triterpenoids, dubosane. Int J Mass Spectrom 310: 77-80.

17. Musharraf SG, Ali A, Ali RA, Yousuf S, Rahman AU, et al. (2011) Analysis and development of structure-fragmentation relationships in withanolides using an electrospray ionization quadrupole time-of-flight tandem mass spectrometry hybrid instrument. Rapid Commun Mass Spectrom 25: 104-114.

18. Firdous S, Ansari NH, SWNg, Yousu S, Malik A, et al. (2012) Crystal Structure of a Novel Furo-Furan Lactone from Heliotropium eichwaldi. Zeitschrift für Naturforschung B 67: 269-271.

10

Comparison of Purification Strategies of Three Horseradish Peroxidase Isoenzymes Recombinantly Produced in *Pichia pastoris*

Rajamanickam V[1,2], Winkler M[2], Flotz P[2], Meyer L[2], Herwig C[1,2] and Spadiut O[1,2]*

[1]*Research Division Biochemical Engineering, Institute of Chemical Engineering, Vienna University of Technology, Vienna, Austria*
[2]*Christian Doppler Laboratory for Mechanistic and Physiological Methods for Improved Bioprocesses, Institute of Chemical Engineering, Vienna University of Technology, Vienna, Austria*

Abstract

Horseradish peroxidase (HRP), a versatile heme-containing glycoprotein frequently used in medical diagnostics, is primarily isolated from plant. This process involves filtration, salt precipitation and several chromatography steps, which is expensive, renders low yields and gives isoenzyme mixtures. Although single isoenzymes can be recombinantly produced in *Pichia pastoris*, they get hyper-glycosylated in the yeast rendering the downstream process cumbersome and thus not competitive.

In this study, we analyzed the purification of three HRP isoenzymes differing in the number of N-glycosylation sites recombinantly produced in *P. pastoris*. We wanted to 1) determine potential correlations between the mode of protein production and the resulting product quality and downstream process, 2) investigate correlations between the number of N-glycosylation sites of HRP and the mode of purification, and 3) find the optimal purification strategy for the recombinantly produced HRP isoenzymes.

Thus, we applied different cultivation strategies and tested downstream processes employing both particle-based resins as well as monolithic supports. We showed that the mode of cultivation affected product purity but not the subsequent downstream process. Purification of each of the three HRP isoenzymes using monolithic supports was successful and independent of the number of N-glycosylation sites, whereas purification by particle-based resins was highly affected by glycosylation.

Summarizing, we demonstrated a novel feasibility of using monoliths operated in flow through mode for the purification of comparatively small biomolecules.

Keywords: Horseradish peroxidase; *Pichia pastoris*; Downstream process; Monolith; Particle-based resin

Introduction

Due to advances in recombinant DNA technology the market for recombinantly produced biopharmaceuticals is booming [1-3]. Amongst the different recombinant host organisms [4-6], the methylotrophic yeast *Pichia pastoris* can be used for the recombinant production of complex proteins due to its ability of performing post-translational modifications [7-11]. However, it is well known that glycoproteins get hyperglycosylated in this yeast [12], which renders the subsequent downstream process difficult and limits the use of the respective product. For some applications, like medical diagnostics, the degree of glycosylation is not as crucial, leaving the problem of purifying hyperglycosylated product from *P. pastoris* as major bottleneck for the routine use of this organism as production platform.

Horseradish peroxidase (HRP; E.C. 1.11.1.7) is a heme-containing glycoprotein frequently used in medical diagnostics and coupled enzyme assays [13-20]. HRP originates from the horseradish root, where 28 different HRP isoenzymes were identified [21]. Although recombinant production of single isoenzymes in various host organisms has been investigated [22], commercially available HRP is still isolated from its native source. However, extraction from plant is cumbersome, only gives low yields and the final enzyme preparation describes a mixture of isoenzymes rather than single isoenzyme species [23-25] which disagrees with Quality by Design guidelines [26]. Unfortunately, studies on the recombinant production and subsequent purification of this versatile enzyme are scarce. In a recent study, we recombinantly produced 19 individual HRP isoenzymes in *Pichia pastoris* in shake flasks and developed a 2-step purification comprising a conventional particle-based hydrophobic charge induction chromatography (HCIC)

followed by anion exchange chromatography (AEX) using a monolith [27]. We obtained satisfactory results for several recombinantly produced HRP isoenzymes (rHRP) when both purification steps were operated in flowthrough mode. We speculated that hyperglycosylation caused a masking effect of the physico-chemical properties of rHRP and thus prevented interactions with the stationary phases. Less glycosylated contaminant host cell proteins (HCPs) on the other hand interacted with the resins and were retained. Interestingly, we observed a correlation between the number of N-glycosylation sites and the success of flowthrough purification [27]. In fact, these observations motivated us to analyze the purification of recombinant glycoproteins from yeast in greater detail.

Here, we chose three HRP isoenzymes differing in the number of N-glycosylation sites. We recombinantly produced these HRP isoenzymes extracellularly in *P. pastoris* either in shake flasks using a pulse-wise feeding or in a bioreactor applying a constant feed. Subsequently we compared different protein purification strategies

*****Corresponding author:** Oliver Spadiut, Institute of Chemical Engineering, Research Area Biochemical Engineering, Vienna University of Technology, Gumpendorfer Strasse 1a, 1060 Vienna, Austria
E-mail: oliver.spadiut@tuwien.ac.at

based on either particle-based resins or monolithic supports for purification of the produced isoenzymes. The goals of this study were to 1) determine potential correlations between the mode of protein production and the resulting product quality and downstream process, 2) investigate correlations between the number of N-glycosylation sites and the mode of purification, and 3) find the optimal purification strategy for the rHRP isoenzymes.

Materials and Methods

HRP isoenzymes and *P. pastoris* strains

The three HRP isoenzymes were selected based on their differences in the number of N-glycosylation sites (Table 1; [27]).

All three HRP isoenzymes were recombinantly expressed extracellularly in *P. pastoris* CBS7435 Mut^S. The recombinant strains were provided by Prof. Anton Glieder (TU Graz, Austria). A comprehensive description of the identification of the HRP isoenzymes and subsequent generation of the *P. pastoris* strains has been published before in Ref. [28]. In short, the codon-optimized genes were N-terminally fused to the pre-pro peptide of the alpha-factor from *Saccharomyces cerevisiae* enabling product secretion and the expression of rHRP isoenzymes was regulated by the methanol inducible AOX1 promoter.

Recombinant production of HRP isoenzymes

Recombinant production was done both in shake flasks and in a bioreactor. Three strains, namely rHRP_A2A, rHRP_1805 and rHRP_5508, were cultivated using two different feeding strategies: a dynamic pulse-wise feeding was done during shake flask cultivations, whereas a constant feeding based on the specific methanol uptake rate ($q_{s\,MeOH}$; [29]) was applied during bioreactor cultivations (Figure 1).

Shake flask cultivations: The three rHRP isoenzymes were expressed in 6 × 2.5 L Ultra Yield flasks (BioSilta; Finland). All media were prepared according to the Pichia protocols (Invitrogen; [30]). An overnight culture (ONC) of 50 mL YPD+zeocin in 250 mL Erlenmeyer shake flasks was cultivated at 25°C and 230 rpm for 24 hr. The ONC

HRP isoenzyme	pI	MW [kDa]	N-glycosylation sites	GenBank	UniProt
A2A	4.84	32.09	9	HE963806.1	K7ZW28
1805	5.75	35.96	5	HE963809.1	K7ZW05
5508	8.22	31.35	3	HE963815.1	K7ZWW9

Table 1: Physico-chemical properties of the three HRP isoenzymes used in this study [27].

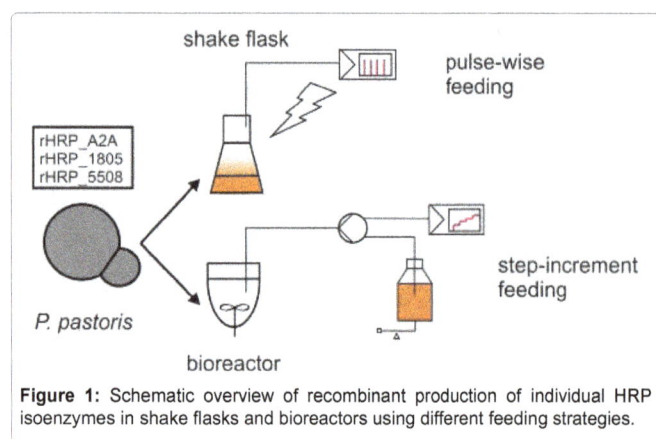

Figure 1: Schematic overview of recombinant production of individual HRP isoenzymes in shake flasks and bioreactors using different feeding strategies.

was transferred to Ultra Yield flasks and a batch phase in 470 mL BMGY+zeocin was done at 30°C and 230 rpm for 24 hr. After that the cofactor hemin was added to a final concentration of 1 mM [31] and an adaptation pulse of 50 mL BMMY was added to each flask. The temperature was reduced to 20°C and 1% (v/v) pure methanol was pulsed every day. Samples were taken every 24 hours to analyze volumetric activity, total protein and RNA concentration in the cell-free cultivation broth. After 120 hours of induction, the cultivation was stopped and the fermentation broth was harvested.

Bioreactor cultivations: Cultivations were done in a 5 L lab scale bioreactor (Infors; Switzerland) and comprised of a batch on glycerol, a fed-batch on glycerol to generate biomass and an induction phase on methanol where the feed was controlled based on $q_{s\,MeOH}$ [29,32]. To identify $q_{s\,max\,MeOH}$ and thus the upper limit of the feeding strategy, dynamic batch cultivations with repeated methanol pulses were performed for each recombinant *P. pastoris* strain. We have repeatedly described this dynamic strategy to evaluate strain specific parameters before (e.g., [29,33,34]).

Bioreactor cultivations were performed as following: a preculture was grown in YNB+zeocin at 230 rpm and 30°C for 24 hr. Then the preculture was aseptically transferred to 1.5 L sterile 2-fold BSM medium in the bioreactor [30]. The inoculation volume was 10% (v/v). The following process parameters were controlled throughout cultivation: pO$_2$ above 30%, pH at 5.0, temperature at 30°C for batch, fed batch and adaptation phase and at 20°C during induction. Complete consumption of glycerol in the batch phase was indicated by a sharp increase in pO$_2$ accompanied by a drop in CO$_2$ in the offgas signal. A glycerol fed batch was initiated upon batch end with an exponential feeding profile controlled at μ=0.1 h^{-1}. The fed batch phase on glycerol was terminated when the cell density reached 80-90 g·L^{-1} dry cell weight (DCW). Before adaptation the precursor hemin was added to a final concentration of 1mM [31]. An adaptation pulse of 0.5% (v/v) pure methanol was added to the bioreactor. Upon complete consumption of the pulsed methanol, again monitored in the offgas, the methanol fed batch was started. In analogy to our previous studies [29,35], we performed a fed batch where we stepwise increased the feeding rate corresponding to increasing $q_{s\,MeOH}$. The first step was at 25%, the second at 50%, the third at 75% and the last step at 90% of $q_{s\,max\,MeOH}$. Each step was held for 24 hr. This dynamic feeding strategy was shown to result in high productivity before [29].

Processing of cultivation broth

Tthe cultivation broth was centrifuged at 3,500 rpm and 4°C for 30 min. The supernatant containing the rHRP was collected, concentrated 10 to 15-fold and buffer was exchanged for subsequent purification by diafiltration using a 10 kDa cut-off membrane (Omega T-Series; PALL; Austria). Prior to purification, all samples were filtered through a 0.2 μm filter (GE Healthcare; Sweden).

Purification of rHRP isoenzymes

All purifications were conducted on an Äkta Pure25 system (GE Healthcare) at room temperature. We compared two strategies based on either particle-based resins or monolithic supports. For the particle-based strategy we applied MEP HyperCel™ (PALL, Austria) for HCIC followed by a HiLoad™16/600 Superdex™ 75 pg column (GE Healthcare) for size exclusion chromatography (SEC). For the monolith-based strategy we used CIMmultus™ C4-HLD for hydrophobic interaction chromatography (HIC) and CIMmultus™ DEAE for AEX, both purchased from BIA separations (Slovenia). For the latter strategy we also investigated the order of the steps. In total, eighteen two-step purification runs were carried out. An overview of the respective experiments is given in Figure 2. For all purification steps the amount of total protein loaded onto the column was kept 20% below the maximum binding capacity of the respective column.

Particle-based purification strategy: The HCIC column was equilibrated with 20 column volumes (CV) HCIC-A (20 mM sodium acetate, 1 M sodium chloride, pH 8). After 5 CV post-load wash, bound proteins were eluted with HCIC-B (20 mM sodium acetate, pH 8) with a 100% step gradient. The flow velocity during all steps was 100 cm·h^{-1}. The HCIC flowthrough was then concentrated to a volume of 1 mL with centrifugal filters (Ultracel-30 K; Millipore, Ireland) and diafiltrated in SEC buffer (50 mM Bis-Tris, 150 mM NaCl, pH 6.5). The SEC column was equilibrated with 5 CV SEC buffer. The flowrate was 15 cm·h^{-1}. Fractions were collected and pooled by following the absorbance signal at 404 nm indicating the presence of rHRP.

Monolith-based purification strategy: A full factorial multivariate screening with the three factors "type of salt" (NaCl and $(NH_4)_2SO_4$), "molar strength" (between 1 M and 3 M) and "temperature" (room temperature and 4°C) using the programme MODDE (Umetrics; Sweden) revealed the most suitable buffer compositions. For HIC, the sample was loaded with HIC-A (50 mM Tris-HCl, 2 M NaCl, pH 8) and bound impurities were eluted with HIC-B (50 mM Tris-HCl, pH 8). For AEX, the samples were loaded in AEX-A (50 mM Tris-HCl, pH 8) and bound contaminants were eluted with AEX - B (50 mM Tris-HCl, 2 M NaCl, pH 8). We tested two different ways of using the monolithic supports. The first approach, hereafter called monolith approach A, comprised of AEX as primary purification followed by HIC for polishing. After AEX we simply added salt to the flowthrough and loaded it onto the HIC monolith. In monolith approach B we switched the order of the steps. We used HIC as the first purification step, removed the salt from the flowthrough by diafiltration and loaded the salt-free sample onto the AEX monolith. The flow velocity was 66 cm·h^{-1} in all these experiments.

Data analysis

Volumetric enzyme activity of rHRP isoenzymes A2A and 5508 was automatically measured by an ABTS assay in a Cubian XC photometric robot (OptoCell, Germany) as described before [36]. Since the maximum reaction rate for HRP isoenzyme 1805 was very low [27], enzyme activity was measured manually using a spectrophotometer (UV-1601; Shimadzu, Long Beach, CA, USA). The manual measurement was carried out at 37°C. 100 μL of 20 mM hydrogen peroxide solution were mixed with 600 μL of 75 mM potassium phosphate buffer, pH 6.5 and 100 μL rHRP. The reaction was started by adding 200 μL of 50 mM ABTS, prepared in 50 mM potassium phosphate buffer (pH 6.5). The change in absorbance was recorded at 420 nm for 600 s. The volumetric activity, a_{HRP} [U·mL^{-1}], was calculated according to equation 1.

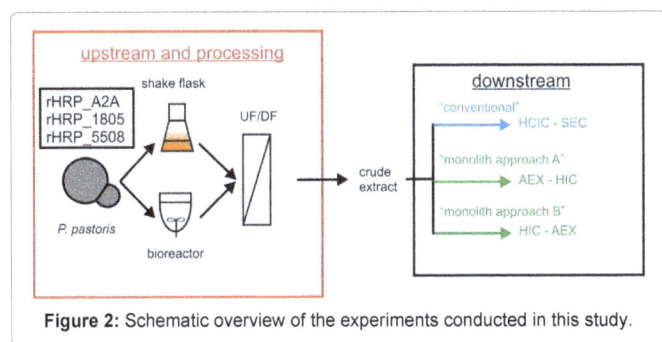

Figure 2: Schematic overview of the experiments conducted in this study.

$$a_{HRP} = \frac{V_{total} \cdot \frac{\Delta A}{min}}{V_{sample} \cdot \varepsilon \cdot l} \tag{1}$$

Protein concentration was determined by the Bradford assay. The purification factor in the flowthrough (PF) and the recovery of rHRP in percentage (R_a %) were calculated according to equations 2 and 3.

$$PF = \left(\frac{\text{specific activity}_{post}}{\text{specific activity}_{pre}} \right) \tag{2}$$

$$R_a\% = 100 \cdot \left(\frac{\text{volumetric activity}_{post}}{\text{volumetric activity}_{pre}} \right) \cdot \left(\frac{\text{volume}_{post}}{\text{volume}_{pre}} \right) \tag{3}$$

RNA concentration was measured using NanoDrop 1000 (Thermo Scientific; USA). Two μL of sample were loaded on the sample pedestal and RNA concentration was measured at 260 nm. Removal of RNA in percentage (RNA_{rem} %) was calculated following equation 4.

$$RNA_{rem}\% = 100 \cdot \left(100 - \left(\frac{\text{RNA concentration}_{post}}{\text{RNA concentration}_{pre}} \right) \right) \tag{4}$$

All measurements were at least done in duplicates.

Results and Discussion

Monolithic supports are usually used for the purification of large biomolecules, like viruses, virus like particles [37-39] and monoclonal antibodies [40-43]. Owing to their large channels, convective flow allowing high mass transfer is possible (e.g., [44]). In a previous study we have shown that monoliths operated in flowthrough mode can also be used for polishing of HRP isoenzymes recombinantly produced in the yeast *P. pastoris*. We found a linear correlation between the number of N-glycosylation sites, which get hyperglycosylated in the yeast, and the success of the flowthrough purification strategy. In the present study we chose three HRP isoenzymes differing in the number of N-glycosylation sites and recombinantly produced them in *P. pastoris* either in shake flasks or in the bioreactor. We wanted to determine potential correlations between the mode of protein production and the resulting product quality and downstream process, investigate correlations between the number of N-glycosylation sites and the mode of purification, and finally find the optimal purification strategy for the recombinant rHRP isoenzymes.

Recombinant production of HRP isoenzymes

As shown in Figure 1 we cultivated the recombinant *P. pastoris* strains either in shake flasks or in a bioreactor. During shake flask cultivations we daily pulsed methanol, whereas in the bioreactor we constantly fed methanol following a dynamic feeding strategy [29,34]. For the latter we had to know the maximum specific uptake rate of methanol ($q_{s\,max\,MeOH}$) to avoid overfeeding the cells. Thus, we performed batch cultivations with dynamic methanol pulses [45] to determine strain specific parameters on methanol (Table 2).

As shown in Table 2, the three recombinant *P. pastoris* strains had a comparable methanol metabolism, which is why it was possible to apply the same dynamic feeding strategy. In the subsequent fed-batch cultivations we fed corresponding to $q_{s\,MeOH}$=0.01, 0.02, 0.03 and 0.036 g·g^{-1}·h^{-1} for 24 hours each. In Table 3 we compared the volumetric activity of HRP, the total protein concentration and the total amount of RNA in the cell free cultivation broths at the time of harvest.

As shown in Table 3 the total protein concentrations in cell-free cultivation broths from shake flasks and bioreactor were similar. However, the amount of active rHRP was 2- to 3-fold higher in the bioreactor and the amount of extracellular RNA was around 6-fold lower. Apparently, production in uncontrolled shake flasks applying a pulse-wise feeding is more stressful for the cells resulting in a significant higher amount of contaminants. This is also shown in the specific activity, which can be regarded as a measure of enzyme purity (Table 3).

Purification of rHRP isoenzymes

We tested three different purification strategies using particle-based resins and monoliths and discussed the results in an orthogonal manner, as we compared 1) how the mode of cultivation influenced the DSP, 2) how the number of N-glycosylation sites affected the success of purification, and finally 3) how successful the two different purification strategies were.

Particle-based purification strategy: In a previous study, we developed a simple two-step flowthrough purification strategy for the well-studied isoenzyme rHRP_C1A providing nine N-glycosylation sites. HCIC in flowthrough mode was used for primary purification followed by SEC for polishing [36]. Here, we applied the same strategy for the three rHRP isoenzymes differing in the number of N-glycosylation (N-gly) sites produced either in shake flasks or in the bioreactor (Table 4).

• Effect of mode of cultivation on DSP

We expected a higher PF for shake flask cultures than for bioreactor broths, owing to uncontrolled conditions, pulse-wise feeding and complex medium and the resulting higher amount of contaminating proteins. However, we did not observe such a trend. In fact, the mode of cultivation did not affect the success of purification, since the amount of rHRP recovered in the flowthrough as well as the amount of removed RNA were comparable independent of the cultivation strategy (Table 4).

• Effect of N-glycosylation sites

As shown in Table 4, the flowthrough purification strategy with particle-based resins worked best for rHRP_A2A providing 9 N-glycosylation sites. There was a clear correlation between the number of N-gylcosylation sites and the amount of product recovered in the flowthrough regardless of the cultivation strategy (Figure 3). The less N-glycosylation sites, the more was the respective rHRP retained

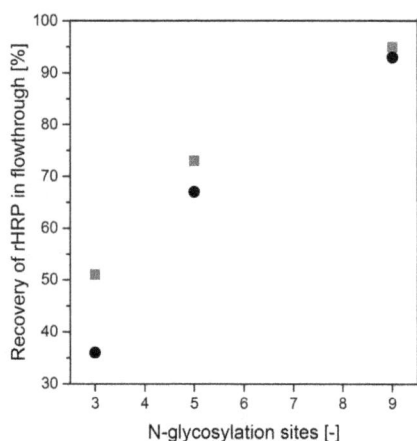

Figure 3: Correlation between the number of N-glycosylation sites and the amount of rHRP recovered in the flowthrough using particle-based resins. Gray squares, rHRP produced in shake flasks; Black circles, rHRP produced in bioreactor.

Strain	$q_{s\ adapt}$ [g·g⁻¹·h⁻¹]	$q_{s\ max\ MeOH}$ [g·g⁻¹·h⁻¹]
rHRP_A2A	0.006	0.042
rHRP_1805	0.008	0.044
rHRP_5508	0.009	0.038

Table 2: Physiological strain-specific parameters. $q_{s\ adapt}$, methanol uptake rate during adaptation, $q_{s\ max\ MeOH}$, maximum methanol uptake rate.

rHRP	Vol. activity [U·mL⁻¹]	Protein conc. [mg·mL⁻¹]	RNA conc. [mg·mL⁻¹]	Specific activity [U·mg⁻¹]
		Shake flask		
A2A	3.92	0.12	2.38	32.7
1805	0.45	0.10	2.28	4.50
5508	11.5	0.10	2.29	115
		Bioreactor		
A2A	11.5	0.16	0.39	71.9
1805	0.62	0.12	0.38	5.17
5508	21.5	0.12	0.33	180

Table 3: Production of three different rHRP isoenzymes either in shake flasks or in the bioreactor.

rHRP	N-gly sites	Shake flask PF	R_a [%]	RNA$_{rem}$[%]	Bioreactor PF	R_a [%]	RNA$_{rem}$[%]
A2A	9	1.7	95	91	3.0	93	96
1805	5	1.5	73	98	1.0	67	94
5508	3	n.a.	51	91	n.a.	36	86

Table 4: Purification of three rHRP isoenzymes produced in either shake flask or bioreactor using particle-based resins.

on the column, a phenomenon which we had observed before [27]. For rHRP_5508 providing 3 N-glycosylation sites, more than 50% of the product was recovered in the eluate. Therefore, the purification factor in the flowthrough could not be calculated. We concluded that purification of HRP isoenzymes recombinantly produced in *P. pastoris* using particle based resins in flowthrough mode is only efficient and successful, if the product provides a high number of N-glycosylation sites.

Monolith-based purification strategy: We used two monoliths, a HIC and an AEX column, for the purification of the three rHRP isoenzymes in this study. We tested two different ways of using the monolithic supports. In monolith approach A we used AEX for primary purification followed by HIC for polishing. After AEX we simply added salt to the flowthrough and loaded it onto the HIC monolith. However, results from initial screening experiments indicated addition of salt to play a dominant role in removal of contaminant proteins. Therefore, in monolith approach B we switched the order of the steps. We used HIC as the first purification step, removed the salt from the flowthrough by diafiltration and loaded the salt-free sample onto the AEX monolith.

Monolith approach A: In Table 5 the purification of the three rHRP isoenzymes using AEX followed by HIC is summarized.

• Effect of mode of cultivation on DSP

Also when using monoliths the mode of cultivation did not seem to affect the success of purification as the amount of rHRP recovered in the flowthrough as well as the amount of removed RNA was comparable independent of the cultivation strategy.

• Effect of N-glycosylation sites

In contrast to particle-based resins (Table 4), all three rHRP isoenzymes could be purified by monoliths independent of the number of their N-glycosylation sites (Table 5). At least 85% of the total amount of rHRP was found in the flowthrough for each of the rHRP isoenzymes. We speculate that the accelerated convective flow through the monoliths [44] impeded product-resin interactions for the less glycosylated rHRP isoenzymes, whereas the slower diffusion in particle-based resins allowed retention.

Monolith approach B: In monolith approach B we changed the order of the monolithic steps and performed HIC followed by AEX (Table 6). In fact, the results were similar to monolith approach A. However, for ease of operation we recommend performing monolith approach A in flowthrough mode to purify rHRP isoenzymes from *P. pastoris.*

Particle-based vs. Monolith-based purification strategy

The overall PFs for each of the three rHRP isoenzymes were similar when using particle-based resins or monoliths operated in flowthrough mode. However, when particle-based resins were used, rHRP was retained on the column in dependence of the number of N-glycosylation sites (Figure 4).

Summarizing using the monolith-based purification strategy HRP isoenzymes recombinantly produced in the yeast *P. pastoris* could be purified independently of the number of N-glycosylation sites. We speculate that the accelerated convective flow through the monoliths

[44] impeded product-resin interactions for the less glycosylated rHRP isoenzymes. Another huge advantage is the high flow and thus the short process times using monoliths [44]. Using monolith approach A allows an integrated, continuous DSP since the two monoliths can be connected and run in series without a holding step in between, as demonstrated for another product before.

Conclusions

Motivated by previous observations we investigated the use of both particle-based resins and monoliths for the purification of recombinant glycoproteins derived from the yeast *P. pastoris*. The results of this study can be summarized as:

1)   recombinant production in the controlled environment of a bioreactor yielded higher specific activity and lower extracellular RNA concentrations in comparison to shake flasks.

2)   the number of N-glycosylation sites played an important role for purification with particle-based resins, whereas it was irrelevant for purification with monolithic supports.

3) with respect to process time, ease of operation and success of purification monoliths outperformed particle-based columns.

In this study we showed that monoliths cannot only be used for the purification of large molecules but also for comparatively small glycoproteins. This purification strategy might describe a platform tool for the purification of recombinant glycoproteins from yeast.

rHRP	N-glysites	Shake flask			Bioreactor		
		PF	R_a [%]	RNA_{rem} [%]	PF	R_a [%]	RNA_{rem} [%]
A2A	9	1.8	97	84	2.7	87	82
1805	5	1.4	85	99	1.7	91	84
5508	3	1.9	93	76	3.4	92	74

Table 5: Purification of three rHRP isoenzymes produced in either shake flask or bioreactor using monolith approach A.

rHRP	N-glysites	Shake flask			Bioreactor		
		PF	R_a [%]	RNA_{rem} [%]	PF	R_a [%]	RNA_{rem} [%]
A2A	9	3.2	90	94	2.8	99	81
1805	5	1.1	89	92	1.5	92	87
5508	3	1.6	100	84	3.2	96	89

Table 6: Purification of three rHRP isoenzymes produced in either shake flask or bioreactor using monolith approach B.

Acknowledgements

The authors would like to thank BIA separations (Slovenia) for providing the columns and their technical support.

Author Contributions

Vignesh Rajamanickam, Maximillian Winkler, Peter Flotz and Lorena Meyer performed experiments. Christoph Herwig and Oliver Spadiut supervised the work. Vignesh Rajamanickam and Oliver Spadiut did the literature search and wrote the article.

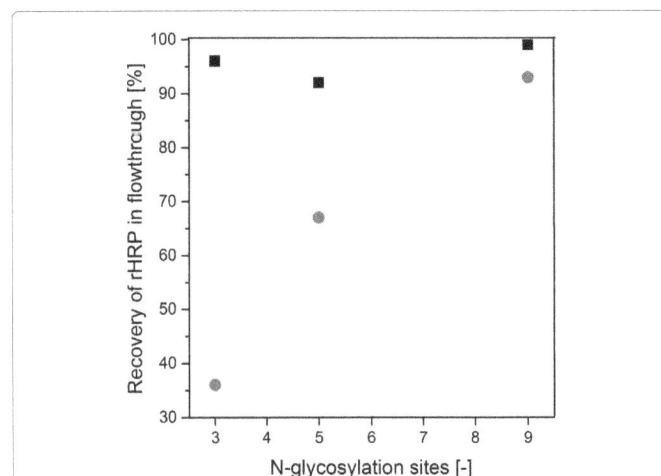

Figure 4: Correlation between the number of N-glycosylation sites and the amount of rHRP (produced in the bioreactor) recovered in the flowthrough. Gray circles, particle-based purification strategy; Black squares, monolith-based purification strategy.

References

1. Bhopale GM, Nanda RK (2005) Recombinant DNA expression products for human therapeutic use. Curr Sci 89: 614-622.

2. Reichert JM, Paquette C (2003) Therapeutic recombinant proteins: trends in US approvals 1982 to 2002. Curr Opin Mol Ther 5: 139-147.

3. Kamionka M (2011) Engineering of therapeutic proteins production in Escherichia coli. Curr Pharm Biotechnol 12: 268-274.

4. Fernández FJ, Vega MC (2013) Technologies to keep an eye on: alternative hosts for protein production in structural biology. Curr Opin Struct Biol 23: 365-373.

5. Overton TW (2014) Recombinant protein production in bacterial hosts. Drug Discov Today 19: 590-601.

6. Bonander N, Bill RM (2012) Optimising yeast as a host for recombinant protein production (review). Methods Mol Biol 866: 1-9.

7. Sun W, Lai Y, Li H, Nie T, Kuang Y, et al. (2016) High level expression and purification of active recombinant human interleukin-15 in Pichia pastoris. J Immunol Methods 428: 50-57.

8. Dai M, Yu C, Fang T, Fu L, et al. (2015) Identification and Functional Characterization of Glycosylation of Recombinant Human Platelet-Derived Growth Factor-BB in Pichia pastoris. PLoS One 10: e0145419.

9. Wang TY, Tsai YH, Yu IZ, Chang TS (2015) Improving 3'-hydroxygenistein production in recombinant Pichia pastoris by a periodic hydrogen peroxide-shocking strategy. J Microbiol Biotechnol.

10. Facchinetti de Castro Girao L, Goncalves da Rocha SL, Sobral RS (2015) Saccharomyces cerevisiae asparaginase II, a potential antileukaemic drug: purification and characterization of the enzyme expressed in Pichia pastoris. Protein Expr Purif 120: 118-125.

11. Krainer FW, Darnhofer B, Birner-Gruenberger R, Glieder A (2016) Recombinant production of a peroxidase-protein G fusion protein in Pichia pastoris. J Biotechnol 219: 24-27.

12. Romanos MA, Scorer CA, Clare JJ (1992) Foreign gene expression in yeast: a review. Yeast 8: 423-488.

13. Azevedo AM, Martins VC, Prazeres DM, Vojinovi V, Cabral JM, et al. (2003) Horseradish peroxidase: a valuable tool in biotechnology. Biotechnol Annu Rev 9: 199-247.

14. Krainer FW, Glieder A (2015) An updated view on horseradish peroxidases: recombinant production and biotechnological applications. Appl Microbiol Biotechnol 99: 1611-1625.

15. Veillette M, Coutu M, Richard J, Batraville LA, Désormeaux A, et al. (2014) Conformational evaluation of HIV-1 trimeric envelope glycoproteins using a cell-based ELISA assay. J Vis Exp: 51995.

16. Chen H, Jiang C, Yu C, Zhang S, Liu B, et al. (2009) Protein chips and nanomaterials for application in tumor marker immunoassays. Biosens Bioelectron 24: 3399-3411.

17. Kim J, Rho THD, Lee JH (2014) Rapid chemiluminescent sandwich enzyme immunoassay capable of consecutively quantifying multiple tumor markers in a sample. Talanta 129: 106-112.

18. Patris S, De Pauw P, Vandeput M, Huet J, Van Antwerpen P, et al. (2014) Nanoimmunoassay onto a screen printed electrode for HER2 breast cancer biomarker determination. Talanta 130: 164-170.

19. Barber JEB, Damry AM, Calderini GF, et al. (2014) Continuous colorimetric screening assay for detection of d-amino acid aminotransferase mutants displaying altered substrate specificity. Anal Biochem 463: 23-30.

20. Xu Y, Hou J, Liu Z, Yu H, Sun W, et al. (2011) Gene therapy with tumor-specific promoter mediated suicide gene plus IL-12 gene enhanced tumor inhibition and prolonged host survival in a murine model of Lewis lung carcinoma. J Transl Med 9: 39.

21. Näätsaari L, Krainer FW, Schubert M, Glieder A, Thallinger GG (2014) Peroxidase gene discovery from the horseradish transcriptome. BMC Genomics 15: 227.

22. Spadiut O, Herwig C (2013) Production and purification of the multifunctional enzyme horseradish peroxidase. Pharm Bioprocess 1: 283-295.

23. Hoyle MC (1977) High resolution of peroxidase-indoleacetic Acid oxidase isoenzymes from horseradish by isoelectric focusing. Plant Physiol 60: 787-793.

24. Lavery CB, Macinnis MC, Macdonald MJ, Williams JB, Spencer CA, et al. (2010) Purification of peroxidase from Horseradish (Armoracia rusticana) roots. J Agric Food Chem 58: 8471-8476.

25. Aibara S, Kobayashi T, Morita Y (1981) Isolation and properties of basic isoenzymes of horseradish peroxidase. J Biochem 90: 489-496.

26. International Conference on Harmonisation Draft Guidance: Q8(R2) Pharmaceutical Development Revision.

27. Krainer FW, Pletzenauer R, Rossetti L, Herwig C, Glieder A, et al. (2014) Purification and basic biochemical characterization of 19 recombinant plant peroxidase isoenzymes produced in Pichia pastoris. Protein Expr Purif 95: 104-112.

28. Krainer F, Naeaetsaari L, Glieder A (2013) Horseradish peroxidase isoenzymes.

29. Spadiut O, Zalai D, Dietzsch C, Herwig C (2014) Quantitative comparison of dynamic physiological feeding profiles for recombinant protein production with Pichia pastoris. Bioprocess Biosyst Eng 37: 1163-1172.

30. Invitrogen (2010) Pichia Expression Kit. Integr Vlsi J.

31. Krainer FW, Capone S, Jäger M, Vogl T, Gerstmann M, et al. (2015) Optimizing cofactor availability for the production of recombinant heme peroxidase in Pichia pastoris. Microb Cell Fact 14: 4.

32. Dietzsch C, Spadiut O, Herwig C (2011) A fast approach to determine a fed batch feeding profile for recombinant Pichia pastoris strains. Microb Cell Fact 10: 85.

33. Capone S, Ćorajević L, Bonifert G, Murth P, Maresch D, et al. (2015) Combining Protein and Strain Engineering for the Production of Glyco-Engineered Horseradish Peroxidase C1A in Pichia pastoris. Int J Mol Sci 16: 23127-23142.

34. Dietzsch C, Spadiut O, Herwig C (2011) A dynamic method based on the specific substrate uptake rate to set up a feeding strategy for Pichia pastoris. Microb Cell Fact 10: 14.

35. Spadiut O, Rossetti L, Dietzsch C, Herwig C (2012) Purification of a recombinant plant peroxidase produced in Pichia pastoris by a simple 2-step strategy. Protein Expr Purif 86: 89-97.

36. Burden CS, Jin J, Podgornik A, Bracewell DG (2012) A monolith purification process for virus-like particles from yeast homogenate. J Chromatogr B Analyt Technol Biomed Life Sci 880: 82-89.

37. Rupar M, Ravnikar M, Tušek-Žnidarič M, Kramberger P, Glais L, et al. (2013) Fast purification of the filamentous Potato virus Y using monolithic chromatographic supports. J Chromatogr A 1272: 33-40.

38. Urbas L, Jarc BL, Barut M, Zochowska M, Chroboczek J, et al. (2011) Purification of recombinant adenovirus type 3 dodecahedric virus-like particles for biomedical applications using short monolithic columns. J Chromatogr A 1218: 2451-2459.

39. Tscheliessnig A, Jungbauer A (2009) High-performance monolith affinity chromatography for fast quantitation of immunoglobulin G. J Chromatogr A 1216: 2676-2682.

40. Leblebici P, Leblebici ME, Ferreira-da-Silva F, Rodrigues AE, Pais LS (2014) Separation of human immunoglobulin G subclasses on a protein A monolith column. J Chromatogr B Analyt Technol Biomed Life Sci 962: 89-93.

41. Mönster A, Hiller O, Grüger D, Blasczyk R, Kasper C (2011) Isolation and purification of blood group antigens using immuno-affinity chromatography on short monolithic columns. J Chromatogr A 1218: 706-710.

42. Rajak P, Vijayalakshmi MA, Jayaprakash NS (2012) Purification of monoclonal antibodies, IgG1, from cell culture supernatant by use of metal chelate convective interaction media monolithic columns. Biomed Chromatogr 26: 1488-1493.

43. Rajamanickam V, Herwig C, Spadiut O (2014) Monoliths in Bioprocess Technology. Chromatography 2: 195-212.

44. Spadiut O, Herwig C (2014) Dynamics in bioprocess development for Pichia pastoris. Bioengineered 5: 401-404.

45. Gerster P, Kopecky EM, Hammerschmidt N, Klausberger M, Krammer F, et al. (2013) Purification of infective baculoviruses by monoliths. J Chromatogr A 1290: 36-45.

11

Anti-Human IgG-Horseradish Peroxidase Conjugate Preparation and its Use in ELISA and Western Blotting Experiments

Ramesh Kumar K*, Sneha Xiavour, Swarna Latha, Vijay Kumar and Sukumaran

Bhavans Vivekananda College, Department of Biochemistry, Osmania University, Hyderabad, India

Abstract

Studies of our experiment are related to the binding of enzymes to antibodies. This involves the formation of a stable, covalent linkage between an enzyme & antibody. We used different Enzymes for this studies which are altered in binding action. Enzymes we used-e.g. Horseradish peroxidase (HRPO), urease, or alkaline phosphatase. Studies of Enzyme and an antigen-specific monoclonal or polyclonal antibody in which neither the antigen-combining site nor the active site of the enzyme is functionally altered. The enzyme most commonly used in the immunoreagent (the antibody enzyme conjugate) preparation is horseradish peroxides. This enzyme is cheap and can be attached to the immunoreagent by a variety of methods. Moreover many chromogenic substrates for it are also available. Our Experiment studies on conjugation of Horseradish Peroxidase (HRPO) to antibody (anti human IgG) will be carried out by Periodate oxidation method. We have prepared Anti human IgG-HRP conjugate in our laboratory. This will be used for ELISA and western blot experiments and IgG purified from different human serum samples would be used as the antigen in the above experiments. This experiment is carried out in two different steps mentioned as: Preparations of anti-human IgG-Horseradish peroxidase conjugate, Characterization of isolated human IgG.

Keywords: HRPO; Immunoglobulin's; Enzyme conjugation; Blotting techniques; Periodate oxidation method; Binding of enzymes to antibodies

Introduction

Conjugation of enzymes to antibodies

Conjugation of enzymes to antibodies involves the formation of a stable, covalent linkage between an enzyme [e.g. Horseradish Peroxidase (HRPO), urease, or alkaline phosphatase] and an antigen-specific monoclonal or polyclonal antibody in which neither the antigen-combining site nor the active site of the enzyme is functionally altered. The chemistry of cross-linking HRPO or urease to immunoaffinity purified monoclonal or polyclonal antibodies (IgG) is presented in the chemistry of cross-linking alkaline phosphatase to antibodies is presented in Figure 1.

The enzyme most commonly used in the immunoreagent (the antibody enzyme conjugate) preparation is horseradish peroxides. This enzyme is cheap and can be attached to the immunoreagent by a variety of methods. Moreover many chromogenic substrates for it are also available [1,2].

Background information

The conjugation of HRPO to antibody is dependent on the generation of aldehyde groups by periodate oxidation of the carbohydrate moieties on HRPO. Combination of these active aldehydes with amino groups on the antibody forms Schiff base upon reduction by sodium borohydride, become stable. For urease conjugation, cross-linking the enzyme and antibody with MBS (m-maleimidobenzoyl N-hydroxysuccinimide ester) is achieved through benzoylation of free amino groups on antibody. This is followed by thiolation of the maleimide moiety of MBS by the cysteine sulfhydryl groups of urease. The advantages of urease conjugates are their stability in solution at normal working dilutions, the rapid turnover rate of the enzyme, the easily discernible color change when substrate is added, and the fact that urease is not found in most mammalian or bacterial systems. The disadvantage is that since no precipitable substrate is available, urease conjugates cannot be used for immunohistology or western blotting. Alkaline phosphatase conjugates are useful for all types of

immunological assays depending on the alkaline phosphatase substrate used (i.e., p-nitrophenyl phosphate in diethanolamine is the preferred substrate for ELISA with colorimetric detection, 4-methylumbelliferyl phosphate is useful for ELISA with fluorimetric detection, and nitrobluetetrazolium/5-bromo-4-chloro-3-indolyl phosphate is the preferred substrate for western blotting). Alkaline phosphatase conjugates are as stable as urease conjugates and more stable than HRPO conjugates. Endogenous phosphatases can cause false positive reactions. However, levamisole will inhibit alkaline phosphatase in many mammalian tissues but not the alkaline phosphatase (i.e., bovine intestinal) used in the conjugates and for this reason levamisole may be added to the substrate solution. The one-step glutaraldehyde method is the simplest available procedure for preparing alkaline phosphatase–antibody conjugates. Various alternative procedures for preparing alkaline phosphatase conjugates have been compared. The sensitivity that can be achieved with HRPO, urease, or alkaline phosphatase conjugates is comparable and between 1 ng/ml and 10 ng/ml of antigen can be detected [3,4].

Critical parameters

The most critical parameters of both conjugation methods are the quality of enzyme and the cross-linking reagents. These reagents should be tested as described in the protocol before conjugating to larger quantities of antibodies. It is imperative that the m-maleimidobenzoyl N-hydroxysuccinimide ester (MBS), sodium periodate ($NaIO_4$) and

*Corresponding author: Ramesh Kumar K, Department of Biochemistry, Bhavans Vivekananda College, Osmania University, Hyderabad, India
E-mail: ramesh_4k@rediffmail.com

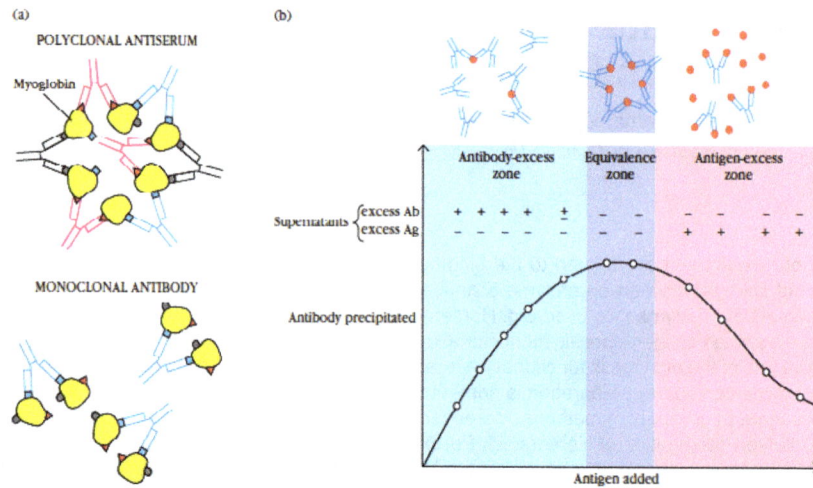

Figure 1a: (Precipitation reactions description: (a) Polyclonal antibodies can form lattices, or large aggregates, that precipitate out of solution. However, if each antigen molecule contains only a single epitope recognized by a given monoclonal antibody, the antibody can link only two molecules of antigen and no precipitate is formed. (b) A precipitation curve for a system of one antigen and its antibodies. This plot of the amount of antibody precipitated versus increasing antigen concentrations (at constant total antibody) reveals three zones: a zone of antibody excess, in which precipitation is inhibited and antibody not bound to antigen can be detected in the supernatant; an equivalence zone of maximal precipitation in which antibody and antigen form large insoluble complexes and neither antibody nor antigen can be detected in the supernatant; and a zone of antigen excess in which precipitation is inhibited and antigen not bound to antibody can be detected in the supernatant).

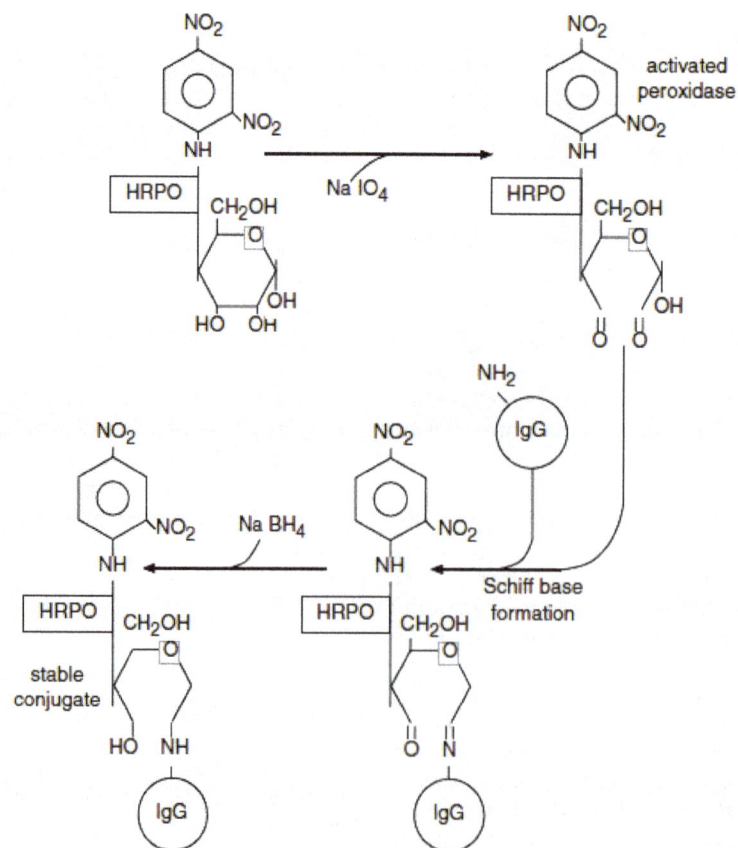

Figure 1b: (Reaction Description: Conjugation of Horseradish Peroxidase (HRPO) to antibody (IgG) using the periodate oxidation method. The method involves three chemical steps: (1) sodium periodate (NaIO$_4$) oxidation of the carbohydrate side chains of HRPO, (2) Schiff base formation between activated peroxidase and amino groups of the antibody, and (3) sodium borohydride (NaBH$_4$) reduction of the Schiff base to form a stable conjugate. Contributed by Scott E. Winston, Steven A. Fuller, Michael J. Evelegh, and John G.R. Hurrell Current Protocols in Molecular Biology (2000) 11.1.1-11.1.).

sodium borohydride ($NaBH_4$) be stored in a desiccator and that solutions containing these chemicals be prepared immediately prior to use. The method described is applicable to most antibodies and should produce conjugates that are useful for developing an ELISA for detecting sensitively and specifically for a given antigen. However, not all antibodies conjugate in an identical manner. It may be necessary to vary the ratio of MBS/antibody or urease/antibody for the urease conjugation and the $NaIO_4$/HRPO and HRPO/antibody ratios for a given HRPO conjugation. The quality and grade of alkaline phosphatase is crucial to the generation of effective conjugates. Immunoassay grade material is recommended over lower grades, and the enzyme should not be conjugated beyond its expiration date. In the case of polyclonal antisera, the specificity and titer of the antiserum will be reflected in the conjugate and any purification procedures that increase these values, such as immunoaffinity chromatography will enhance conjugate performance. The selection of an optimal conjugation time for preparing alkaline phosphatase–antibody conjugates varies for different antibodies, in particular when monoclonal antibodies are used. In contrast, polyclonal antibodies may be reliably conjugated in 120 min.

Trouble shooting

There are several factors that may contribute to the production of poor enzyme-antibody conjugates. It is important to determine first whether a poor conjugate is the result of inactivation of either the antibody or the enzyme (or both) or the result of insufficient or excessive cross-linking. The affinity of the antibody for substrate can be measured by determining the presence of bound antibody with another immunoassay employing anti-antibody conjugated to a different enzyme. Enzyme activity can be measured by cleavage of substrate at different enzyme concentrations. Precipitation of material in the conjugate solution or opaque solutions, are indicative of excessive cross-linking. Sodium dodecyl sulfate–polyacrylamide gel electrophoresis is useful for monitoring the extent of cross-linking by determining the Mr of the cross-linked species. Insufficient cross-linking usually results from the use of inactive or poor-quality crosslinking agents. Excessive cross-linking and inactivation of antibody or enzyme can be eliminated by either reducing the concentration of antibody and enzyme or by reducing the time of reaction. It may not be possible to generate effective alkaline phosphatase conjugates with all antibodies using the one-step glutaraldehyde method. An alternative is to try a different conjugation technique. Another alternative is to use an anti-species antibody–alkaline phosphatase conjugate to detect the antibody in question. These reagents may be purchased or prepared using the above technique.

Anticipated results

The yield and titer of the resultant conjugate will depend on the original antibody's properties and specific application. It is difficult to estimate the yield or working dilution of the conjugates, as it is dependent on numerous factors such as antibody affinity, type of ELISA, and quality of antigen. In general, the working dilutions range from 1:100 to 1:10,000.

Aim

In our studies conjugation of horseradish peroxidase (HRPO) to antibody (anti human IgG) will be carried out by periodate oxidation method. Anti-human IgG-HRP conjugate prepared in our laboratory. This will be used for ELISA and western blot experiments (IgG purified from different human serum samples would be used as the antigen in the above experiments). Thesis work will be carried out in two parts for

the sake of clarity and convenience.

Preparations of anti-human IgG-HRP conjugate

Prepared anti human IgG HRP conjugate will be used for the following experiments

Qualitative ELISA for detection of human IgG

Identification of IgG by western blotting

Toward this objective, IgG from different human serum samples will be isolated and it will be used for the above experiments.

Characterization of isolated human IgG

1. Estimation of protein content

2. SDS-PAGE analysis: Determination of purity, Determination of molecular weight

Preparations of anti-human IgG-HRP conjugate

Chemicals and reagents: Antibody labeling teaching kit was purchased from Bangalore Genie Pvt Ltd, Bangalore India.

Materials provided in the kit are given below:

1. Antigen spotted strip (Human IgG spotted)

2. Antibody for coupling (Anti human IgG)

3. Carbonate buffer

4. Desalting column G 25 (2 ml)

5. Oxidation tubes

6. Reductant solution

7. Stabilizer

8. 10 X ELISA buffer

9. 10 X Phosphate buffer saline (PBS)

10. 10 X TMB/H_2O_2

Other requirements

1. Micropipettes and tips

2. Measuring cylinder, test tubes and beakers

3. 2,5 and 10 ml pipettes

4. Burettestand

5. Distilled water.

Method

Oxidation of HRP

0.2 ml of distilled water was added to the oxidation tube containing HRP and sodium Meta-periodate. The solution was stirred for 20 min at room temperature. (Note: colour changes from orange to green).

Coupling

0.2 ml of carbonate buffer was added to antibody coupling tube and the contents were mixed thoroughly. Immediately after mixing, oxidized HRP was added to the antibody solution and the contents were stirred for 2 h at room temperature. After 2 h, 10 μl of reductant solution (borohydride) was added and the solution was mixed for 10 min at room temperature (Figure 2).

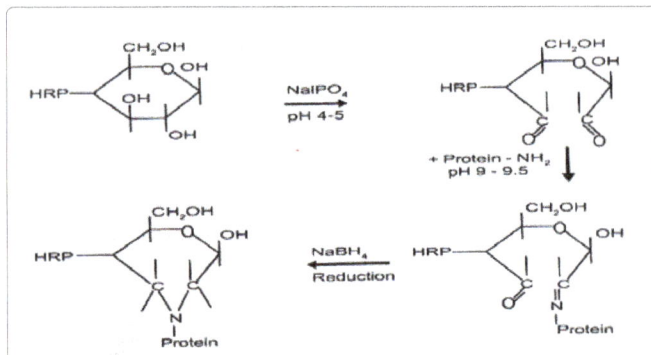

Figure 2: The most commonly used method for labeling IgG molecules with HRP exploits the glycoprotein nature of the enzyme. The saccharide residues of HRP are oxidized with sodium meta-periodate to produce aldehyde groups that can react with the amino groups of the IgG molecule and the Schiff's bases formed are then reduced by borohydride to give a stable conjugate.

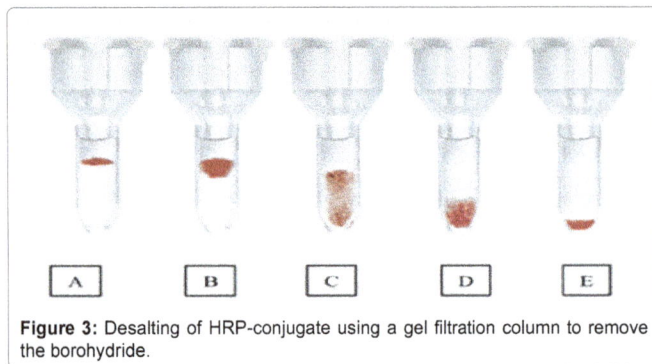

Figure 3: Desalting of HRP-conjugate using a gel filtration column to remove the borohydride.

Desalting

The desalting column was first equilibrated with 20 ml of 1X PBS. After allowing the reaction mix to settle down the supernatant was loaded on to the desalting column that was previously equilibrated with PBS. After running down the reaction mixture into the column 0.1 ml of 1X PBS was added along the walls of the column to wash down the reaction mixture sticking onto the column. Once the PBS entered the bed of the column, 1 ml of 1X PBS, was added at a time till the entire colored fraction (HRP conjugate) eluted out of the column completely (Figure 3) Immediately after collecting the HRP conjugate 0.5 ml of stabilizer was added to the HRP conjugate and the conjugate was stored at 4°C until use. After eluting the HRP conjugate, the column was washed with 5 ml of 1X PBS and stored at 4°C.

Titration

The HRP conjugate was diluted with 1X assay buffer. Four antigen (Fraction no. 2 from serum sample 1, 2 and 3) spotted nitrocellulose strips were labeled 1:1000, 1:2000, 1:4000 and 1:8000 and then the strips were agitated in appropriately diluted conjugate (i.e. 1:1000, 1:2000, 1:4000 and 1:8000 diluted anti human IgG HRP conjugate (Figure 4).

The strips were then incubated at room temperature for 30 min with constant shaking. After incubation, the strips were washed 5 times with 2-3 ml of rinse buffer (1 min each wash). After each wash the old buffer was replaced with fresh buffer. After the fifth wash the strips were washed with distilled water once. After washing, the strips were agitated in 2 ml of substrate solution (TMB/H2O2) till blue/ grey colored spots were observed. Soon after the spots developed the strips were removed before the background turned dark. The highest dilution at which the spot are seen is the titer value of the conjugate [6-8].

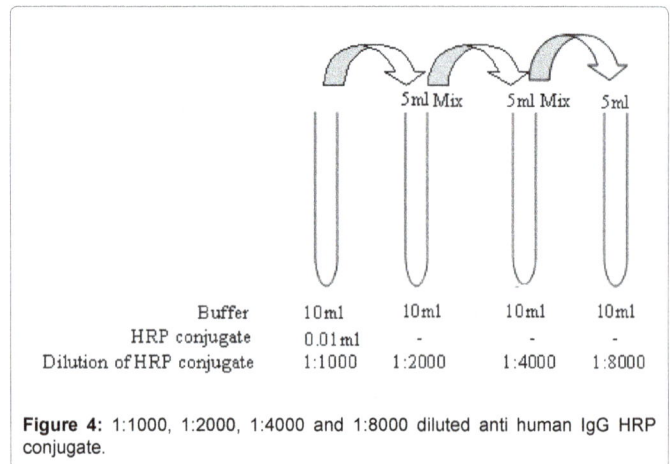

Figure 4: 1:1000, 1:2000, 1:4000 and 1:8000 diluted anti human IgG HRP conjugate.

Purification of Human IgG Employing protein-A Affinity Chromatography

Chemicals and reagents

10N sodium hydroxide

a. 4 g sodium hydroxide

b. 10 ml of distilled water

4 g sodium hydroxide was dissolved in about 8 ml of distilled water and then the volume was made up to 10 ml with distilled water.

Phosphate buffer saline (PBS) pH 7.4 (equilibration buffer):(20x/1000 ml)

a. Sodium chloride=180 g

b. Potassium dihydrogen orthophosphate=2.88 g

c. Disodium hydrogen ortho phosphate $2H_2O$=17.78 g

d. 1000 ml of distilled water

180 g sodium chloride, 2.88 g of dipotassium hydrogen orthophosphate and 17.78 g of disodium hydrogen orthophosphate was dissolved in about 800 ml of distilled water. After adjusting the pH to 7.4 with 10 N sodium hydroxide, the volume was made up to 1000 ml with distilled water (2.1 ml 0f 10N NaOH was required to adjust the pH to 7.4). To prepare 1X buffer, the 20X buffer was diluted 20 times (e.g. 50 ml of 20 X PBS was diluted to 1000 ml with distilled water (+950 ml of distilled water).

Elution buffer: (0.1M citrate buffer pH=4.6)

0.1 M citric acid=2.1 g

2.1 g citric acid was dissolved in about 80 ml of distilled water and then the volume was made up to 100 ml with distilled water.

0.1 M sodium citrate=2.94 g

2.94 g sodium citrate was dissolved in about 80 ml of distilled and then the volume was made up to 100 ml with distilled water.

To prepare 0.1 M citrate buffer pH 4.6, 25.5 ml of solution (a) was mixed with 24.5 ml of solution (b) and then the volume was made up to 100 ml with distilled water.

Neutralization buffer: (2 M Tris (hydroxy methyl) amino methane)

Tris (hydroxy methyl) amino methane=2.42 g, 10 ml of distilled water

2.42 g of Tris (hydroxy methyl) amino methane was dissolved in about 8 ml of distilled water and then the volume was made up to 10 ml with distilled water.

Storage buffer: (equilibration buffer containing 0.05% (w/v) sodium azide)

25 mg sodium azide, 50 ml of equilibration buffer, 25 mg of sodium azide was dissolved in 50 ml of equilibration buffer.

Protein A affinity column: (column size-2 ml): Protein-A affinity column was purchased from Bangalore Genie Pvt Ltd Bangalore India.

Serum samples: Human serum samples 1, 2 and 3.

Other requirements

1. Test tubes and test tube stand

2. 2, 5 and 10 ml glass pipettes

3. Glass beakers and conical flasks

4. 10, 50 and 100 ml measuring cylinder

5. Micro pipettes and tips; 2-20, 20-200 and 100-1000 μl

6. Eppendorf tubes (1.5 ml)

7. Refrigerated centrifuge

8. Burettestand

9. Tissue paper

10. Distilled water

11. 1ml quartz cuvettes

12. UV-VIS spectrophotometer

13. Graph sheet, pencil and eraser

Method

Sample preparation

Serum sample was mixed with equilibration buffer in 1:1 ratio in an eppendorftube. After mixing, the sample was subjected to centrifugation at 6000 rpm at 4°C for 10 min to remove particulate matter.

Protein A Column Chromatography (Bed volume 2 ml)

After the column was brought to ambient temperature, the column was equilibrated with 10 bed volumes of equilibration buffer. After equilibration, 0.9 ml of the 1:1 diluted clear serum sample was loaded on to the column. While loading, the flow rate of the column was adjusted to 1 ml/5 min. Once the serum sample passed through the column, the column was washed with equilibration buffer till the A_{280} of the equilibration buffer was less than 0.1 O.D. (About 10 bed volumes of equilibration buffer was required for washing). After the unbound materials were removed by washing, the bound proteins were eluted with 15 bed volumes of elution buffer. 1ml fractions were collected in tubes containing neutralizing buffer (25 μl/ml) (Figure 5).

After collecting the fractions, A_{280} for each fraction was measured at 280 nm in a UV-VIS spectrophotometer. Serum samples 2 and 3 were treated similarly. After taking the absorbance value a graph was plotted with fraction number on x-axis vs. absorbance values on y-axis.

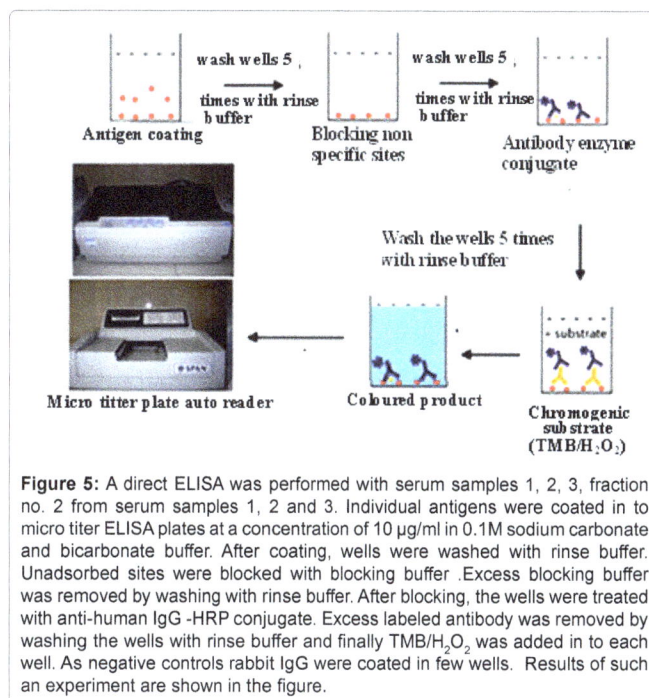

Figure 5: A direct ELISA was performed with serum samples 1, 2, 3, fraction no. 2 from serum samples 1, 2 and 3. Individual antigens were coated in to micro titer ELISA plates at a concentration of 10 μg/ml in 0.1M sodium carbonate and bicarbonate buffer. After coating, wells were washed with rinse buffer. Unadsorbed sites were blocked with blocking buffer .Excess blocking buffer was removed by washing with rinse buffer. After blocking, the wells were treated with anti-human IgG -HRP conjugate. Excess labeled antibody was removed by washing the wells with rinse buffer and finally TMB/H_2O_2 was added in to each well. As negative controls rabbit IgG were coated in few wells. Results of such an experiment are shown in the figure.

Enzyme Linked Immunosorbent Assay (ELISA)

Chemicals and reagents

Antigen:

a. Pure human IgG was purchased from Bangalore Genie Pvt Ltd, Bangalore India.

b. Pure Rabbit IgG was purchased from Bangalore Genie Pvt Ltd, Bangalore India.

c. Serum samples 1, 2, 3

d. Fraction no 2 (serum samples 1, 2 and 3)

Samples were prepared at a concentration of 5μg/ml individually in coating buffer (0.1M sodium carbonate and bicarbonate buffer pH 9.6).

Antigen coating buffer: (0.1 M sodium carbonate bicarbonate buffer pH 9.6)

0.1 M sodium carbonate=1.06 g/100 ml: 1.06 g of sodium carbonate was dissolved in about 80 ml of distilled water and then the volume was made up to 100 ml with distilled water.

0.1M sodium bicarbonate=1.26 g/150 ml:

1.26 g of sodium bicarbonate was dissolved in about 80 ml of distilled water and then the volume was made up to 150 ml with distilled water.

To prepare 0.1 M sodium carbonate bicarbonate buffer pH 9.6, 84 ml of (a) was added to 150 ml of (b) and mixed.

Phosphate buffer saline (PBS) pH 7.4 (equilibration buffer): (20x/1000 ml)

a. Sodium chloride=180 g

b. Potassium dihydrogen orthophosphate=2.88 g

c. Disodium hydrogen orthophosphate 2H₂O=17.78 g

d. 1000 ml of distilled water

180 g sodium chloride, 2.88 g of dipotassium hydrogen orthophosphate and 17.78 g of disodium hydrogen ortho phosphate was dissolved in about 800 ml of distilled water. After adjusting the pH to 7.4 with 10N sodium hydroxide, the volume was made up to 1000 ml with distilled water. (2.1 ml of 10 N NaOH was required to adjust the pH to 7.4). To prepare 1x buffer, the 20X buffer was diluted 20 times (e.g. 50 ml of 20X PBS was diluted to 1000 ml with distilled water (+950 ml of distilled water).

Rinse buffer: 50 ml of 20X PBS was diluted to 1000 ml with distilled water. After mixing, 1 ml of tween 20 was added to it.

Serum diluent buffer: 1 g of egg albumin was dissolved in 100 ml of rinse buffer.

Labeled antibody: Anti-human IgG HRP conjugate prepared in our laboratory was used.

Substrate solution: Tetramethylbenzidine/hydrogen peroxide (TMB/H$_2$O$_2$):

TMB/H$_2$O$_2$ was purchased from Bangalore Genie Pvt Ltd, Bangalore India. TMB is a noncarcinogenic chromogen used in assays involving Horseradish Peroxidase (HRP) enzyme. TMB/H$_2$O$_2$ for ELISA produces a soluble blue colored product with HRP. The reagent is supplied as 20X concentration. Just before use 1 volume of the substrate solution was mixed with 19 volumes of distilled water.

Other requirements

1. 96 well micro titer plates.
2. Micro pipettes 2-20 and 200-1000 µl and tips.
3. Glass pipettes 5 ml and 10 ml.
4. Glass beakers and conical flasks
5. Measuring cylinders
6. Wash bottle
7. Incubator set at 37°C.
8. Moist chamber (plastic box with a layer of wet cotton)
9. ELISA Reader.

Method

All assays were carried out in 96 well flat bottomed polystyrene plates.

Antigen coating: 100 µl of antigen at a concentration of 5 µg/ml in 0.1 M sodium carbonate–bicarbonate buffer, pH 9.6 was coated onto each well and the plate was incubated for 30 min at 37°C (Few wells coated with rabbit IgG was used as negative control). After incubation, excess antigen was removed from the well by washing 5 times with rinse buffer (PBS with 0.1% tween 20).

Blocking non specific sites: After washing, non-specific sites were blocked by addition of 300 µl serum diluent buffer (SDB, containing PBS with 0.1% tween-20 and 1% egg albumin) into each well and plate was incubated for 30 min at 37°C. After incubation, plate was washed 5 times with rinse buffer.

Labeled antibody (Anti-human IgG HRP-conjugate and goat anti-human IgG HRP conjugate): After washing, 100 µl of 1:2000 diluted anti-human IgG-HRP conjugate was added to each well and the plate was incubated for 30 min at 37°C. Excess antibody enzyme conjugate was removed by washing the well 5 times with rinse buffer.

Substrate solution: After washing, 100 µl of the chromogenic substrate (TMB/H$_2$O$_2$) was added to each well and the plate was incubated for 10 min at room temperature till colour developed. Colour developed was read at 450 nm in a micro titter plate auto reader ELISA reader (Figure 6).

Western Blotting

Chemicals and reagents

Blotting buffer:

a. 2.42 g of tris base
b. 10.25 g of glycine
c. 200 ml of 20% methanol
d. 800 ml distilled water

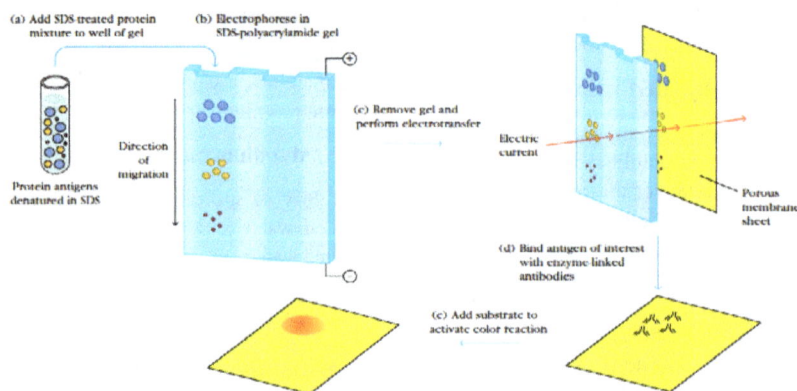

Figure 6: In Western blotting, a protein mixture is (a) treated with SDS, a strong denaturing detergent, (b) then separated by electrophoresis in an SDS polyacrylamide gel (SDS-PAGE) which separates the components according to their molecular weight; lower molecular weight components migrate farther than higher molecular weight ones. (c) The gel is removed from the apparatus and applied to a protein-binding sheet of nitrocellulose or nylon and the proteins in the gel are transferred to the sheet by the passage of an electric current. (d) Addition of enzyme-linked antibodies detects the antigen of interest, and (e) the position of the antibodies is visualized by means of a reaction that generates a highly colored insoluble product that is deposited at the site of the reaction.

2.42 g of tris base, 10.25 g of glycine was dissolved in about 700 ml of distilled water and then 200 ml of 20% methanol was added to it and then the volume was made up to 1000 ml with distilled water.

Phosphate buffer saline (PBS) pH 7.4 (equilibration buffer): (20x/1000 ml)

a. Sodium chloride=180 g

b. Potassium dihydrogen orthophosphate=2.88 g

c. Disodium hydrogen orthophosphate $2H_2O$=17.78 g

d. 1000 ml of distilled water

180 g sodium chloride, 2.88 g of dipotassium hydrogen ortho phosphate and 17.78 g of disodium hydrogen ortho phosphate was dissolved in about 800 ml of distilled water. After adjusting the pH to 7.4 with 10N sodium hydroxide, the volume was made up to 1000 ml with distilled water (2.1 ml of 10N NaOH was required to adjust the pH to 7.4). To prepare 1x buffer, the 20x buffer was diluted 20 times (e.g. 50 ml of 20 X PBS was diluted to 1000 ml with distilled water (+950 ml of distilled water).

Rinse buffer: 50 ml of 20X PBS was diluted to 1000 ml with distilled water. After mixing;

1 ml of tween 20 was added to it.

Serum diluent buffer: 1 g of egg albumin was dissolved in 100 ml of rinse buffer.

Labeled antibodies: Anti-human IgG HRP conjugate was prepared in our laboratory.

Substrate solution: Tetramethylbenzidine/hydrogen peroxide (TMB/H_2O_2) was purchased from Bangalore Genie India. TMB is a noncarcinogenic chromogen used in assays involving Horseradish Peroxidase (HRP) enzyme. The reagent is supplied as 20 X concentration. Just before use 1 volume of the substrate solution was mixed with 19 volumes of distilled water.

Fraction no 2 from serum samples 1, 2 and 3: Fraction no 2 from serum samples 1, 2 and 3 were spotted on nitrocellulose membrane strip. Unadsorbed (non-specific) sites on the membrane were blocked by agitating the membrane in 10 ml blocking buffer. After washing, the membrane was agitated in 1:200 diluted goat anti human IgG-FITC. Excess goat anti human IgG-FITC conjugate was removed by washing. After air drying, the membrane was placed in a UV cabinet to observe the fluorescence spot developed.

Other requirements

1. Nitro cellulose membrane (pore size 0.22-0.45 micro meter)

2. Glass pipettes

3. Beakers

4. Conical flask

5. Petri dish

6. Scalpel blade

7. Glass tray or Plastic tray

8. Glass plate platform 25x20 cm

9. Tissue paper rolls

10. Whatman filter paper

11. cissors

12. Forceps

13. Gloves

14. Small electrophoresis unit

15. Electrotransfer unit or small electro blotting system

16. Power supply unit

17. Connecting cords

18. Shaker.

Method

Western blotting is essentially a combination of three techniques electrophoresis (PAGE), blotting (protein blotting) and immunochemical detection (blot development).

Stage I: Separation of proteins on SDS-PAGE

The proteins to be analyzed (fraction no 2 from serum samples 1, 2 and 3) by western blotting was first subjected to separation on a 10-12% separating gel and 6% stacking gel. Samples were electrophoresed at 80-100 volts.

Stage II: Electro transfer of the separated protein onto the nitro cellulose membrane

Blotting is the transfer of resolved proteins from the gel to the surface of a suitable membrane. The separated proteins are transferred out of the gel either by the capillary action of the buffer or in an electric field (known as electro blotting).

After separation of proteins by SDS-PAGE, the stacking gel was discarded and the required portion of the gel was cut with a scalpel blade, if the whole gel was not to be blotted. Similarly, the nitro cellulose membrane (one layer) and the filter paper (six layers) were cut to the exact size of the gel. The gel, the NC membrane and the filter papers were then soaked in a tray containing the blotting buffer for 10 min. After 10 min, the blotting sandwich was assembled. During sandwich assembly preparation care was taken to avoid air bubbles between the gel and NC membrane. After assembling the blotting sandwich, the cassette was inserted into the apparatus filled with the blotting buffer. The gel was placed towards the cathodic end; since proteins are negatively charged (due to presence of SDS) their migration will be towards anode. After inserting the cassette into the apparatus lid was placed over the buffer tank and with the help of connecting cords the apparatus was connected to power supply and electro transfer was carried out overnight at 50 volts. The presence of SDS facilitates the migration of proteins. Once out of the gel, the proteins come in contact with the nitrocellulose membrane, which binds the protein very strongly on to the surfaces as a band, thus producing a replica of original gels. However, the protein location and detection can only be assessed after immune detection (Figures 6 and 7).

Stage III: Immunodetection

The transferred proteins are bound to the surface of the nitrocellulose membrane and are accessible for reaction with immunochemical reagents.

1. Blocking nonspecific sites: After electrotransfer, the NC membrane was placed in a petridish and the non-specific sites i.e. the unabsorbed sites on NC membranes were blocked by agitating the membrane for 1 h in serum diluent buffer. After incubation, the

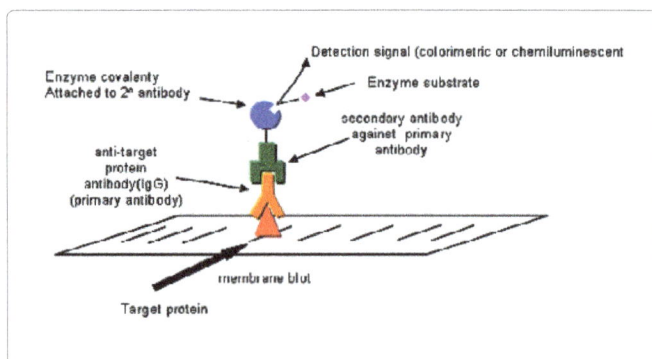

Figure 7: Identification of proteins separated by gel electrophoresis is limited by the small pore size of the gel, as macromolecular probes for protein analysis cannot permeate the gel. The separated proteins are buried in the polyacrylamide gels and therefore further analysis of proteins or their recovery is cumbersome. However, the proteins can be effectively transferred from the gel to the supporting medium by blotting. The presence of SDS facilitates the migration of proteins. Once out of the gel, the proteins come in contact with the nitrocellulose membrane, which binds the protein very strongly on to the surfaces as a band, thus producing a replica of original gels. A variety of analysis involving immuno-chemical analysis can be carried out for the detection of protein bands.

membrane was washed thrice with rinse buffer [8,9].

2. Labeled antibody (anti-human IgG HRP-conjugate): After washing, the membrane was agitated for 1 h in 10 ml of 1:1000 diluted anti-human IgG HRP conjugate Excess antibody enzyme conjugate was removed by washing the membranes thrice with rinse buffer.

3. Substrate solution: After washing, the membrane was agitated in substrate solution (TMB/H_2O_2) till blue/grey colored bands were observed. The blot was photographed for permanent record.

Immunofluorescence (dot blot method)

Chemicals and reagents

Antigen: Fraction no 2 (serum samples1, 2 and 3)

Phosphate buffer saline (PBS) pH 7.4 (equilibration buffer):(20X/1000 ml):

a. Sodium chloride=180 g

b. Potassium dihydrogen orthophosphate=2.88 g

c. Disodium hydrogen orthophosphate 2H_2O=17.78 g

d. 1000 ml of distilled water

180 g sodium chloride, 2.88 g of dipotassium hydrogen orthophosphate and 17.78 g of disodium hydrogen ortho phosphate was dissolved in about 800 ml of distilled water. After adjusting the pH to 7.4 with 10 N sodium hydroxide, the volume was made up to 1000 ml with distilled water. At the time of use the buffer was diluted 20 times (e.g. 50 ml of 20 X PBS+950 ml of distilled water).

Rinse buffer: 50 ml of 20X PBS was diluted to 1000 ml with distilled water. After mixing, 1 ml of tween 20 was added to it.

Serum diluent buffer: 1 g of egg albumin was dissolved in 100 ml of rinse buffer.

Labeled antibody: Goat anti-human fluorescein isothiocyanate (FITC) was purchased from Bangalore Genie Pvt Ltd Bangalore India.

Other requirements:

1. Nitrocellulose membrane strips

2. Micro pipettes 2-20 and 200-1000 µl and tips.

3. Glass pipettes 5 ml and 10 ml.

4. Glass beakers and conical flasks

5. Measuring cylinders

6. Wash bottle

7. Petridish

8. Shaker

9. UV cabinet

Method

2-5 µl of the antigen (Fraction no.2, serum samples 1, 2 and 3) was spotted on nitrocellulose membrane strips. After air drying, the unadsorbed (non-specific) sites on the membrane was blocked by agitating the membrane in 10ml blocking buffer (taken in a petridish) for 1 h on a shaker. After incubation, excess blocking buffer was removed by washing the membrane 3 times with rinse buffer (3×5 min each). After washing with rinse buffer, the membrane was agitated in 1:200 diluted goat anti-human IgG-FITC for 30 min on a shaker. Excess goat anti-human IgG-FITC conjugate was removed by washing the membranes thrice with rinse buffer (3×5 min each). After air drying, the membrane was placed in a UV cabinet to observe the fluorescence spot developed [10-12].

Determination of Purity and Relative Molecular Weight of Purified IgG

Sodium dodecyl sulphate-polyacrylamide gel electrophoresis (SDS-PAGE)

Chemicals and Reagents:

2% Agarose in 0.9% Saline:

a. 2 g of agarose

b. 0.9 g of sodium chloride

c. 100 ml distilled water

2 g of agarose and 0.9 g of sodium chloride were added to 100 ml of distilled water and the contents were boiled till agarose dissolved completely and the solution was clear.

6N HCl:

a. 48.7 ml distilled water

b. 51.3 ml of conc. HCl

To 48.7 ml distilled water, 51.3 ml of conc. HCl was added.

1.5M Tris-HCl pH 8.8:

a. 18.171 g of Tris base

b. 100 ml distilled water

18.171 g of Tris base was dissolved in about 70 ml of distilled water. After adjusting the pH to

8.8 with 6N HCl, the volume was made up to 100 ml with distilled water(*Initial pH of Tris-base was 10.68; a total of 4.6 ml of 6N HCl was required to get down the pH from 10.68 to 8.8).

0.5M Tris-HCl pH 6.8:

a. 6.057 g of Tris base

b. 100 ml distilled water

6.057 g of Tris base was dissolved in about 70 ml of distilled water. After adjusting the pH to 6.8 with 6N HCl, the volume was made up to 100 ml with distilled water (*Initial pH of Tris-base was 10.47; a total of 9.1 ml of 6N HCl was required to get down the pH from 10.47 to 6.8).

Acrylamide: N, N'-methylene bis acrylamide (normally referred to as bisacrylamide)

a. 30 g of acrylamide

b. 0.8 g of bis- acrylamide

c. 100 ml distilled water

30 g of acrylamide and 0.8 g of bis-acrylamide was dissolved in about 60 ml of distilled water and then the volume was made up to 100 ml with distilled water. The solution was filtered through a Whatman filter paper and then stored in brown bottle.

Note: Acrylamide and Bis-acrylamide are slowly converted during storage to acrylic acid, Bis -acrylic acid. This deamination reaction is catalyzed by light and alkali. Check the pH of the solution, it should be 7.0 or less, and store the solution in dark bottles. Fresh solution should be prepared for every few months. Pre packed, pre mixed stock solutions are commercially available.

10% Sodium dodecyl sulfate (SDS):

a. 10 g of SDS

b. 100 ml distilled water

10 g of SDS was dissolved in about 80 ml of distilled water and then the volume was made up to 100 ml with distilled water. Note: SDS is an irritant, while weighing avoid contact with skin or inhalation.

10% ammonium per sulfate (APS):

a. 1 g of Ammonium per sulfate

b. 10 ml distilled water

1 g of Ammonium per sulfate was dissolved in about 8 ml of distilled water and the volume was made up to 10 ml with distilled water. Aliquots of 0.5 ml were pipetted out in eppendorf tube. The vials were frozen at -20°C. Just before use, one vial was thawed and used for experiment.

Note: APS has to be prepared freshly. An alternative method is to prepare and store frozen.

N, N, N', N'-tetramethylethylene di amine (TEMED): Commercially available.

0.05% Bromophenol Blue

a. 50 mg of bromophenol Blue

b. 100 ml distilled water

50 mg of bromophenol Blue was dissolved in 100 ml of distilled water.

β-Mercaptoethanol: Commercially available.

Glycerol: Available commercially. Note: As an alternative to glycerol 40% sucrose can be used.

Sample buffer:

a. 0.5M Tris -HCl pH 6.8=1 ml

b. Glycerol=1 ml

c. 10% SDS=1 ml

d. β-Mercaptoethanol=0.2 ml

e. 0.05% Bromo phenol Blue=0.2 ml

Ingredients a-e were mixed together and then stored in a screw cap vial.

Leveler:

To 1 ml of distilled water 10 μl of 10% SDS was added.

Electrode buffer (reservoir buffer): Tris-Glycine pH 8.3

a. 6 g of Tris-base

b. 28.4 g glycine

c. 10 ml 10% SDS

d. 1000 ml distilled water

6 g of Tris-base was dissolved in about 700 ml of distilled water. The pH of the solution was adjusted to 8.3 with gradual addition of solid glycine. After adjusting the pH, 10 ml of 10% SDS was added and the volume made up to 1000 ml with distilled water. (*28.4 g of glycine was required to adjust the pH to 8.3).

Note: Buffer used in the apparatus with small volume buffer components should be discarded after each run because pH changes occur as a result of the electrolysis of water during electrophoresis. When compartment volumes are >700 ml, switching, polarity of the electrophoretic cell after each run allows use of the same buffer for up to 4 runs.

Staining solution: 0.25% Coomassie Brilliant Blue R250 (in methanol: Acetic acid: water: 50:10:40)

a. 50 ml methanol

b. 10 ml acetic acid

c. 40 ml distilled water

d. 250 mg Coomassie Brilliant Blue R 250

Methanol, acetic acid and water were mixed in the ratio of 50: 10: 40 and then 250 mg of Coomassie Brilliant Blue R250 was dissolved in it. The stain solution was filtered through a plug of cotton. Note: The stain solution can be used 5-6 times, if stored properly.

Destaining solution:

a. 50 ml methanol

b. 10 ml acetic acid

c. 40 ml distilled water

Methanol: acetic acid: water was mixed in the ratio 50: 10: 40 and the solution were stored at room temperature in a screw cap bottle.

Note: The de-stained solution can be re used once or twice if the solution is passed through activated charcoal powder. (Coomassie Brilliant Blue R250 adsorbs onto the charcoal and the methanol: acetic acid: water becomes clear of the stain).

Fixative: 5% Trichloro acetic acid:

a. 5 g of Trichloro acetic acid

b. 100 ml of distilled water

5 g of Trichloro acetic acid was dissolved in 100 ml of distilled water and the solution was stored at room temperature. Note: This solution can be used only once. Before staining the gels are immersed in the fixative to guard against diffusion of separated components.

Silver staining: (all reagents are to be prepared fresh and in double distilled water)

a. Fixative: Methanol: acetic acid: water:: 50: 12: 38 containing 75 µl of formaldehyde

Methanol, acetic acid and water were mixed in the ratio 50:12:38. To this 75 µl of formaldehyde was added.

b. 50% Ethanol: Ethanol and distilled water were mixed in 1:1 ratio (75 ml+75 ml).

c. 0.02% Sodium thiosulfate: 20 mg of sodium thiosulfate was dissolved in about 80 ml of distilled water and then the volume was made up to 100 ml with distilled water.

d. 0.2% Silver nitrate, containing 0.75% formaldehyde: 200 mg of silver nitrate was dissolved in about 80 ml of distilled water and then the volume was made up to 100 ml with distilled water. To this 75 µl of formaldehyde was added.

e. Developer: 6% sodium carbonate containing 0.05% formaldehyde and 4 ml of 0.02% sodium thiosulfate

6 g of sodium carbonate was dissolved in about 70 ml of distilled water. To this 4 ml of 0.02% sodium thiosulfate was added along with 50 µl of formaldehyde and the volume was made up to 100 ml with distilled water and the solution was stored in dark at room temperature.

f. Stop solution 5% Citric acid: 5 g of Citric acid was dissolved in about 80 ml distilled water and then the volume was made up to l00 ml with distilled water.

Samples

a. Pure human IgG

b. Fraction no 2 (serum samples 1, 2 and 3)

Other requirements

1. Glass beakers 50, l00, 250, 500 and 1000 ml

2. 5 ml, 10 ml glass pipettes

3. 2-20, 20-200 and 200-1000 µl micro pipettes and tips

4. Tissue paper, blotting paper, Whatman No.1 filter paper

5. Dissection needle

6. Hot plate

7. Magnetic stirrer and magnetic bar

8. pH meter

9. Eppendorf tubes (1.5 ml)

10. Gel staining trays & Float

11. Semi log graph sheets, pencil, eraser, scale

12. Adhesive tapes

13. SDS-PAGE electrophoresis unit

14. Power pack

Method

Samples to be analyzed on SDS-PAGE were mixed with sample buffer in a 1:1 ratio. After mixing, the samples were boiled for 10 min. After selecting the appropriate thickness spacers and comb, the glass sandwich was assembled as follows: After placing the glass plate on a leveled surface, silicone grease was applied to the spacers and the spacers were placed on the left and right edge of the plate. After the spacers were fixed on the glass plate the notched plate was placed on the rectangular plate. The glass plate assembly was then clamped in order to keep the assembly tightly together in position. After clamping, the lower end of the glass plate was sealed with 2% agarose in 0.9% saline (Agarose was poured into a boat or gel casting tray and then the assembly was placed in the boat). The whole setup was left undisturbed for the gel to set/solidify. Casting of separating gel and stacking gel was carried out in the gel casting unit (Figure 8).

Preparation of Separating Gel (12%)

After the agarose gel has set, 5-10 ml of the separating gel was prepared by mixing the ingredients given Table 1.

After mixing the solution, it was poured into the chamber between the glass plate sandwich assemblies. Immediately after pouring the separating gel mix, 400 µl of the leveler solution was added on the top of the separating gel to even the surface. The setup was left undisturbed for the gel to polymerize. After the separating gel polymerized, the leveler solution was removed by blotting with filter paper.

Preparation of Stacking Gel

After blotting out the leveler 2.5 ml of the stacking gel was prepared by mixing the ingredients given in Table 2. And then it was poured over the separating gel. Immediately after pouring, glycerol applied comb was inserted into the stacking gel. (Glycerol prevents sticking of the gel to the comb), for the formation of wells into which samples are to be loaded.

Once the stacking gel polymerized the glass plate assembly with the separating and stacking gels was clamped to the electrophoresis apparatus (Care was taken while clamping the plates i.e. the notched plate was facing towards the inner side and the rectangular facing outside). After clamping the plates, electrode buffer was filled into both the upper and lower buffer reservoirs. After the cathodic and anodic buffers were filled, the electrophoresis apparatus was connected to the power pack via connecting cords (Note: black was connected to cathode and red to anode, since proteins are negatively charged they migrate towards anode).

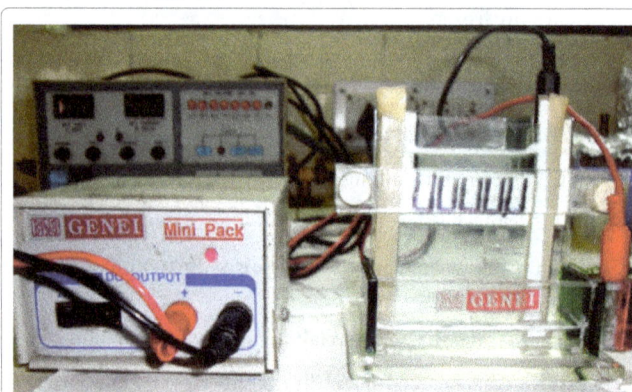

Figure 8: Small vertical gel Electrophoresis system consists of electrophoresis unit.

Relative binding affinity IgG	
Species	Polyclonal
Rabbit	++++ +++
Human IgG	++++ ++++
Dog	+++
IgM	— +
IgD	— +
IgA	— +
Cow	++ ++++
Horse	++ ++++
Goat	- ++
Guinea pig	++++ ++
Sheep	+/- ++
Pig	+++ +++
Rat	+/- ++
Mouse	++ ++
Chicken	— +

Species/ Subclass Protein-A Protein-G		
Monoclonal		
Human	Mouse	Rat
IgG 1 ++++ ++++	IgG 1 + ++++	IgG 1 — +
IgG 2 ++++ ++++	IgG 2a ++++ ++++	IgG 2a — ++++
IgG 3 — ++++	IgG 2b +++ +++	IgG 2b — ++
IgG 4 ++++ ++++	IgG 3 ++ +++	IgG 2c + ++

— (weak or no binding) ++++ (Strong binding)

Table 1: Relative Affinity of Immobilized Protein-A and Protein-G for Various Antibody Species and Subclasses of Polyclonal and Monoclonal IgG.

Fraction No.	Protein Content		
	Serum sample-1 (mg/ml)	Serum sample-2 (mg/ml)	Serum sample-2 (mg/ml)
1	0.49	0.11	0.23
2	2.00	1.05	1.28
3	0.98	0.31	0.78
4	0.70	0.15	0.59

Table 2: Protein Content in Protein-A Fraction.

	Ingredients	7.5%	10%	12%	15%
1	1.5M Tris HCl pH 8.8 (ml)	2.5	2.5	2.5	2.5
2	Acrylamide: Bisacrylamide	2.5	3.33	4.0	5.0
3	Distilled water	4.83	4.0	3.33	2.33
4	10% SDS	0.1	0.1	0.1	0.1
5	TEMED	10	10	10	10
6	10% APS	75	75	75	75

Table 3: Preparation of Separating Gel (10 ml).

After connecting the cords, the protein samples (according to the method of staining (Coomassie Brilliant Blue R250: 10-25 µl/well, or Silver staining: 5-15 µl/well) were loaded into the wells of stacking gel. After loading the samples, the power pack was switched on and the power set to 70 V. After the tracking dye entered the separating gel, the voltage was increased to 100 V and electrophoresis was carried out till the tracking dye reached the other end of the gel. After the completion of the run, the gel was carefully removed from in between the glass plates and subjected to either Coomassie Brilliant Blue R 250 or silver staining.

Staining the gels

Coomassie Brilliant Blue R250 staining: After the completion of the run, the gel was carefully removed from in between the glass plates and it was fixed in 5% TCA for 1 h. (TCA guards against diffusion of separated protein components). After fixing for 1h, the gel was stained in 0.25% Coomassie Brilliant Blue R 250 solution for 3 h. After staining, the gel was destained overnight to remove the background stain. The gel was then photographed for permanent record.

Silver staining: The proteins separated on gels can be visualized by staining either with Coomassie Brilliant Blue R-250 or Amido black 10 B dye. The drawback of the above dyes is that it detects a band containing 1 µg of the protein. In many occasions, the available protein for electrophoresis is so small or some proteins occur in minute amounts

that the detection becomes extremely difficult with these dyes. Under such circumstances a higher sensitive detection system is required. Silver staining is a very useful method in this regard with about 100 fold greater sensitivity over dye staining. It is comparable in sensitivity to autoradiography of labeled polypeptides. There are different methods described by different workers for silver staining. The method given below is very simple and rapid.

Principle: The amino acids particularly aromatic in the protein reduce silver nitrate and form complexes with metallic silver of yellowish-brown to brown color.

Method: After electrophoresis, the gel was fixed in methanol: acetic acid: water (50: 12:38) containing 75 µl of 37% formaldehyde for 1 h or overnight with uniform shaking. Once the gel is removed from the fixer further steps are to be carried out without any delay. After fixing, the gel was treated with 50% ethanol 3 times (3x10 min each). After ethanol treatment, the gel was impregnated with 0.02% sodium thiosulfate exactly for 1 min (This step is very crucial as the band and background staining depends on this step). After sodium thiosulfate treatment, the gel was washed 3 times with distilled water (3×1 min). After washing, the gel was agitated in 0.2% silver nitrate containing 0.075% formaldehyde for 20 min on a shaker. After silver nitrate treatment, the gel was washed 3 times with distilled water (3×1 min each). After washing, the gel was agitated in the developer till the bands developed (Care was taken such that the gel did not takes up the background stain). Immediately after the appearance of the bands the gel was washed several times with excess of distilled water or alternatively the gel was placed in the stop solution (5% citric acid). After staining, the gel was photographed for permanent record [13-17].

Native–PAGE (Activity staining of the enzyme HRP): Native-PAGE analysis of HRP was essentially carried out as described for SDS-PAGE methodology section but with slight modification. Separating and stacking gel was prepared by mixing the ingredients given Tables 3 and 4.

The sample (HRP) was diluted with equal volume of sample buffer and then 30-40 µg of the sample was loaded in to the well. Electrophoresis was carried out at 100 V. After completion of electrophoresis, the gel was stained with substrate solution (DAB system) till the development of colored band.

Conjugation of horseradish peroxidase to antibodies: The present study that describes the conjugation of horseradish peroxidase to anti-

Ingredients		
1	0.5M Tris Hcl pH 6.8	0.75 ml
2	Acrylamide-Bisacrylamide	1.0 ml
3	Distilled water	3.15 ml
4	10% SDS	50 µl
5	10% APS	45 µl
6	TEMED	10 µl

Table 4: Preparation of Stacking Gel (5 ml).

Formulation for Preparation of Different Percentages of Separating Gel (10 ml)					
	Ingredients	7.5 %	10 %	12 %	15 %
1	1.5M Tris HCl pH 8.8 (ml)	2.5 ml	2.5 ml	2.5 ml	2.5 ml
2	Acrylamide: Bisacrylamide	2.5 ml	3.33 ml	4.0 ml	5.0 ml
3	Distilled water	4.93 ml	4.10 ml	3.43 ml	2.43 ml
4	TEMED	10 µl	10 µl	10 µl	10 µl
5	10% APS	75 µl	75 µl	75 µl	75 µl

Table 5: Preparation of Separating Gel and Stacking for Native-Page Analysis.

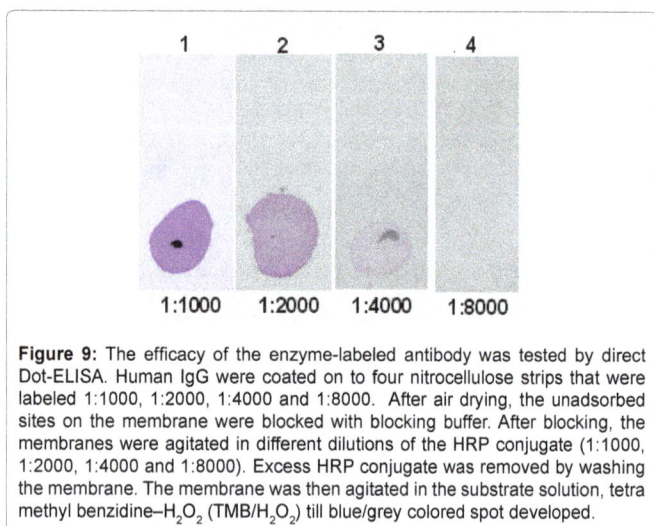

Figure 9: The efficacy of the enzyme-labeled antibody was tested by direct Dot-ELISA. Human IgG were coated on to four nitrocellulose strips that were labeled 1:1000, 1:2000, 1:4000 and 1:8000. After air drying, the unadsorbed sites on the membrane were blocked with blocking buffer. After blocking, the membranes were agitated in different dilutions of the HRP conjugate (1:1000, 1:2000, 1:4000 and 1:8000). Excess HRP conjugate was removed by washing the membrane. The membrane was then agitated in the substrate solution, tetra methyl benzidine–H_2O_2 (TMB/H_2O_2) till blue/grey colored spot developed.

human IgG was initiated because; direct conjugation of enzymes to antibodies has greatly simplified the development and performance of many different types of immunoassays. Horseradish peroxidase (HRPO) conjugates are useful in all types of immunological assays e.g., Horseradish peroxidase–antibody conjugates can be used in ELISA (enzyme-linked immunosorbent assay) and western blotting [18].

Result: Conjugation of horseradish peroxidase (HRPO) to antibody (IgG) was essentially carried out by periodate oxidation method.

The method involves three chemical steps:

1. Sodium periodate ($NaIO_4$)-oxidation of the carbohydrate side chains of HRPO

2. Schiff base formation between activated peroxidase and amino groups of the antibody and

3. Sodium borohydride ($NaBH_4$) reduction of the Schiff base to form a stable conjugate.

After preparing the antibody (IgG) horseradish peroxidase conjugate, as a preliminary study, the efficacy of the enzyme-labeled antibody was tested by direct dot-ELISA. As shown in Figure 9, the titer of the anti-human IgG–HRP conjugate was found out to be 1:4000.

The HRP conjugate was used for the following experiments:

- ELISA

- Western blot analysis

To carry out ELISA and western blot experiments a suitable antigen is required. Towards this goal, IgG from three different human serum samples were purified on protein-A affinity chromatography.

Purification of human IgG employing protein-A affinity chromatography

A protein must be purified before its structure and the mechanism of its action can be studied. However, because proteins vary in size, charge, and water solubility, no single method can be used to isolate all proteins. To isolate one particular protein from the estimated 10,000 different proteins in a cell is a daunting task that requires methods both for separating proteins and for detecting the presence of specific proteins. Any molecule, whether protein, carbohydrate, or nucleic acid, can be separated, or resolved, from other molecules on the basis of their differences in one or more physical or chemical characteristics. The larger and more numerous differences between two proteins, the separation of the proteins is easier and more efficient. The two most widely used characteristics for separating proteins are size, defined as either length or mass, and binding affinity for specific ligand.

Affinity chromatography

Affinity chromatography is the process of bio-selective adsorption and subsequent recovery of a compound from an immobilized ligand. The ability of proteins to bind specifically to other molecules is the basis of affinity chromatography. In this technique, ligand molecules that bind to the protein of interest are covalently attached to the beads used to form the column. Ligand can be enzyme substrates or other small molecules that bind to specific proteins. In a widely used form of this technique, antibody-affinity chromatography, the attached ligand is an antibody specific for the desired protein. An affinity column will retain only those proteins that bind the ligand attached to the beads; the remaining proteins, regardless of their charges or masses, will pass through the column without binding to it. The proteins bound to the affinity column are then eluted by adding an excess of ligand or by changing the salt concentration or pH. The ability of this technique to separate particular proteins depends on the selection of appropriate ligand. The affinity chromatography is the most acceptable method for purification of biomolecules. Theoretically affinity chromatography should give a fairly pure sample in a single step.

Protein-A has an affinity to bind to IgG of various species. This property of protein-A was employed for the purification of IgG from serum of different human serum samples in our studies.

Staphylococcal protein-A

Staphylococcal protein A is an IgG binding protein found in the bacterial cell wall of Staphylococcus aureus. Staphylococcal protein-A binds to most mammalian IgG and can be used for detecting or purifying such antibodies. Affinity chromatography on staphylococcal protein A column is extensively used for purifying monoclonal and polyclonal antibodies.

Protein A: Staphylococcal protein A is a 42 Kilo Dalton protein, exists both in secreted and membrane-associated form. There are 4 Fc binding, highly homologous regions each constituting of 58 to 62 amino acids. These regions are consecutively arranged from the N terminal part of protein. The residual C terminal part is approximately 150 residues long differs to a great extent with respect to primary and secondary structures from the 4 active regions. Furthermore, it is suggested that the protein is bound to the bacterial cell-wall structure through this C terminal part. The binding of protein-A has been well documented for immunoglobulin's from a variety of mammalian species and for immunoglobulin's IgM and IgA as well.

Advantages of protein-A column:

• High immunoglobulin binding capacity

• High flow rate with good resolution

• Highly stability to all physical changes

• The dissociation constant Kd=10-4 to 10-8 permits recovery under mild conditions hence called high performance affinity chromatography (HPAC)

Results: IgG from three different human serum samples were purified on protein-A column. The absorbance values for eluted fractions are given in Table 5. Serum sample (0.5 ml serum+0.5 ml equilibration buffer) was loaded on protein A column that was equilibrated with equilibration buffer (PBS pH 7.4). 1ml fractions were collected and A_{280} for each fraction was read in a UV–Vis spectrophotometer and absorbance values are given in the Table 6 and the elution profiles are shown in Figure 10. As evident from Figure 11, IgG from human serum samples 1, 2 and 3 fractionated into single peak. Among the different fractions in the peak, fraction no.2 showed the highest absorbance value in all the 3 samples. Since peak fraction no.2 showed highest absorbance values, further experiments were carried out with this fraction only.

In order to confirm the presence of IgG in fraction no.2, ELISA and western blot experiments were performed. These experiments were carried out with anti-human IgG HRP conjugate that was prepared in our laboratory.

Enzyme Linked Immunosorbent Assay (ELISA): Modern medicine is dependent up on various tools of investigation for arriving at a correct diagnosis. Most of the in-vitro methods rely up on detection and accurate measurement of a particular component in different physiological fluids and tissues of the human body. Immunoassay exploits the antigen-antibody reaction to achieve this end.

Antigen: An "antigen" (Ag) is defined as any foreign substance, which is capable of eliciting an immune response (humoral, cellular, or most commonly, both) in the host. Lately the term "immunogen" is used more frequently instead of antigen. The most potent immunogens are macromolecular proteins, but polysaccharides, nucleic acids, lipids, synthetic polypeptides and other synthetic polymers such as polyvinyl pyrolidone can act as antigens. The antigenicity of a molecule depends on a number of properties listed below:

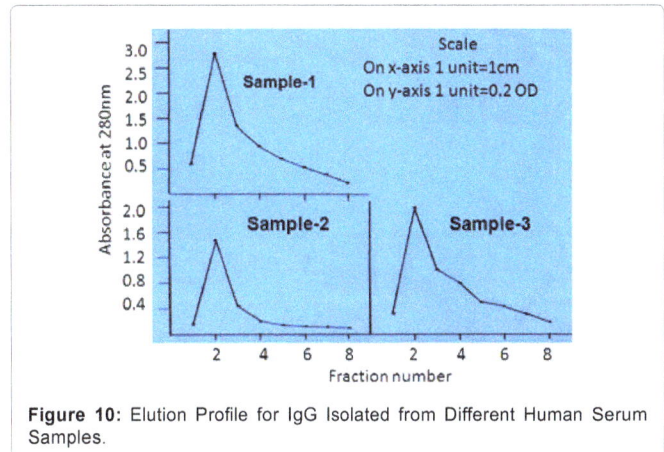

Figure 10: Elution Profile for IgG Isolated from Different Human Serum Samples.

1. Foreignness.

2. Molecular size.

3. Chemical complexity.

4. Genetic constitution of the animal.

5. Method and route of antigen administration.

6. Antigen concentration.

7. Amino acid composition.

Antibody (Immunoglobulin's): Antibodies are heterogeneous group of globulins mainly found in the serum. Antibody is a glycoprotein that is produced in response to introduction of an antigen and which has the ability to combine with the antigen that stimulated its production.

Since the antibodies are glycoproteins, they can also act as antigens when inoculated in to a different mammalian species. There are five types of immunoglobulin's (Igs). The classification of the immunoglobulin's is based on the type of "heavy chain" present in the antibody. Thus, immunoglobulin's containing heavy chains γ, μ, α, δ and ε are called immunoglobulin-G (IgG), immunoglobulin-M (IgM), immunoglobulin-A (IgA), immunoglobulin-D (IgD), and immunoglobulin-E (IgE) respectively. In addition, several sub-classes are also known. When an individual is immunized against single antigen, specific antibodies are produced which usually consist of a mixture of IgM, IgG and IgA. The amount of each immunoglobulin's produced depends on the nature of the antigen and the stage of immunity. Specificity and sensitivity are the two major features of antigen-antibody reactions. Of the various immunoassays was the turning point in the accurate measurement of substances at ultra-low-level. The radioimmunoassay suffers from certain inherent drawbacks such as:

1. Radioimmunoassay requires constant supply of radioisotopes.

2. Short half-life of radiolabel.

3. Requirement of highly skilled technicians.

4. Expensive infrastructure (i.e. Scintillation counters/spectrometers are required.

5. Potential health hazards associated with routine use of radioactive material (i.e. facility for radioactive substances have also to be protected against radiation hazards).

Fraction. No	Serum sample-1 A_{280}	Serum sample-2 A_{280}	Serum sample-3 A_{280}
1.	0.69	0.16	0.32
2.	2.81	1.47	1.80
3.	1.38	0.44	1.10
4.	0.98	0.21	0.83
5.	0.71	0.15	0.50
6.	0.54	0.13	0.43
7.	0.40	0.12	0.31
8.	0.28	0.10	0.19
9.	0.17	0.07	0.11
10.	0.07	0.03	0.06

Table 6: Absorbance Values for Protein-A Eluted Fractions.

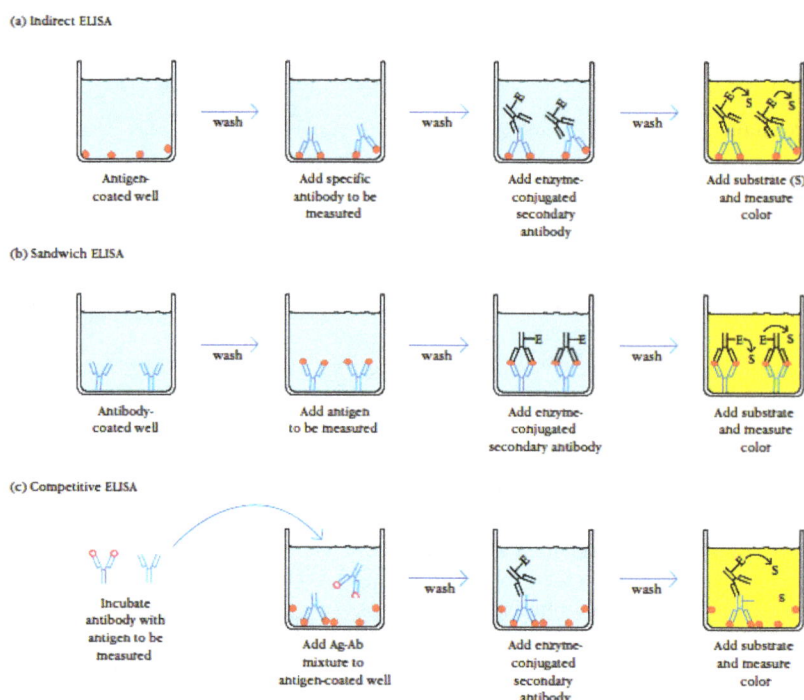

Figure 11: Variations in the Enzyme-Linked Immunosorbent Assay (ELISA) Technique.

6. Problem of disposal of radioactive waste.

These drawbacks of radioimmunoassay have been circumvented by development of an similarly sensitive and versatile technique in which in place of radioactive compound, enzymes have been used as 'labels' or 'markers' in immunoassays. Immunoassays in which either the primary or the secondary antibody is coupled to an enzyme are referred to as enzyme immunoassays. Enzyme immunoassays while incorporating almost all the benefits of radioimmunoassay, Assays, provide additional advantages in terms of ease of operation, safety of personnel and cost effectiveness.

The term "enzyme linked immunosorbent assay" (ELISA) was coined in 1971 by Engvall and Peralman for this enzyme based technique. Over the last two decades, application of enzyme immunoassays has expanded quite dramatically. Enzyme immunoassays are today playing an increasingly important role in diagnostics and research laboratories the world over. The developments in automation of immunoassays and monoclonal antibody technology have enabled introduction of a variety variations in the basic ELISA procedures to fulfill the requirements of different test systems. The application of monoclonal antibodies in ELISA protocols has enhanced specificity of the test. ELISA is presently one of the most commonly used tests for the detection of antigens and antibodies. In addition ELISA can also be employed for the quantitation of antigens and antibodies. There are a variety of modifications of this test and these are collectively known as enzyme immunoassays (EIA).

The basic feature of this group of assays is as follows: One of the immunereagent (usually antigen at times antibody) is immobilized through adsorption on the solid phase support (usually polyvinyl chloride or polystyrene) in such a way that there is no loss to its activity. The second immunoreagent (either primary or secondary antibody) is linked to an enzyme in a way that there is no loss either to immuno reactivity or to the enzyme activity.

All assays are carried out in 96 well flat bottomed polystyrene or polyvinyl chloride plates.

1. Antigen coating: 100-200 µl of the antigen (usually antigen at times antibody) at a concentration of 5-10 µg per ml in 0.1 M sodium carbonate–bicarbonate buffer, pH 9.6 is added into wells of the flat bottomed microtiter plate. After incubation for appropriate time period at 37°C or room temperature or at 4°C, excess antigen or antibody is removed by washing the wells 5 times with rinse buffer.

2. Blocking nonspecific sites: After washing, non-specific sites are blocked by addition of 300 µl serum diluent buffer into each well and the plate is incubated for 1 h at 37°C. After incubation, wells are washed 5 times with rinse buffer to remove excess blocking.

3. Labeled antibody: After washing, 100-200 µl of appropriately diluted antibody enzyme conjugate is added to wells that are coated with antigen. After incubation for appropriate time period at 37°C or room temperature or at 4°C, excess antibody enzyme conjugate is removed by washing the wells 5 times with rinse buffer.

4. Substrate solution: After washing the wells, 100-200 µl of the chromogenic substrate (depending on the enzyme conjugated to the antibody a suitable chromogenic substrate is added to each well) for e.g. if the enzyme Horseradish peroxidase is conjugated to the antibody the chromogenic substrate TMB/H_2O_2 used or if the enzyme Alkaline phosphatase is conjugated to the antibody then substrate paranitrophenol phosphate is used) and the plate is incubated at room temperature in dark till the development of the colour. After colour development, the reaction is stopped by adding 100 µl of either 2N HCl or 2N H_2SO_4 into each well and the colour developed is read at 450 nm for TMB/H_2O_2 or at 405 nm for para nitrophenol phosphate in a micro titter plate auto reader (ELISA reader) or in a colorimeter.

If the two immuno reagents (antigen with its specific antibody) have bound to each other, colour will develop because of the presence of the linked enzyme; If not there will be no colour development due to the lack of the linked enzyme.

Assays for Antibody

Direct method: Salient points of the procedure for direct method are briefed below:

A suitable antigen (antigen in the given sample) is allowed to adsorb on to the wells of micro titer plate. After antigen adsorption, the non-specific sites are blocked by the addition of blocking or serum diluent buffer into the wells. After incubation, the enzyme linked antibodies (Antibody enzyme conjugate) specific for the antigen is added to the wells, and allowed to incubate at 37°C. During this step if the antibody conjugated to the enzyme is specific to the antigen it binds to the adsorbed antigen in the micro titer plate wells. After incubation, a chromogenic enzyme substrate is added to the wells and the plates are incubated till the development of the colour. The reaction is stopped by addition of 2 N HCl and the colour intensity is measured photometrically. (In an ELISA reader).

Indirect method: This method is largely used to measure antibodies in almost all infections (including HIV). The method has become popular because it requires only a single conjugate, for example enzyme labeled anti-human globulins. Different immunoglobulin's classes can be detected differentially by using class specific conjugates (IgG, IgM, and IgA).

Salient points of the procedures for indirect method are briefed below:

A suitable antigen (antigen in the given sample for e.g. antigen Gp120 in the case of HIV detection) is allowed to adsorb on to the wells of micro titer plate. After antigen adsorption, the non-specific sites are blocked by the addition of blocking buffer into the wells. After incubation, in the subsequent step, appropriately diluted serum or any other test sample (serum from HIV patient e.g. which contains antibodies to Gp120 {primary antibody (Ab1)}) is added to an antigen-coated microtiter well and allowed to react with the antigen adsorbed to the well. After incubation, the presence of antibody bound to the antigen is detected by adding an enzyme-conjugated secondary anti-isotype antibody (Ab2). After incubation, enzyme-conjugated secondary anti-isotype antibody (Ab2) is added to the wells and the plates are incubated at 37°C. If the patients sera contains antibody to the specific antigen (e.g. in the case of HIV antibodies to Gp120) the primary antibody specifically binds to the adsorbed antigen in the wells and becomes immobilized. When the enzyme-conjugated secondary anti-isotype antibody (Ab2) is added to these wells it binds to the primary antibody. After incubation, a chromogenic substrate specific for the enzyme coupled to secondary antibody is added to the wells and the plates are incubated in dark till the colour develops. The reaction is stopped after incubation time by the addition of 2 N HCl. The amount of colored reaction product that forms is measured by specialized spectrophotometric plate readers, which can measure the absorbance of all of the wells of a 96-well plate in seconds.

Applications of indirect ELISA:

1. Indirect ELISA is the method of choice to detect the presence of serum antibodies against human immunodeficiency virus (HIV), the causative agent of AIDS. In this assay, recombinant envelope and core proteins of HIV are adsorbed as solid-phase antigens to microtiter wells. Individuals infected with HIV will produce serum antibodies to epitopes on these viral proteins. Generally, serum antibodies to HIV can be detected by indirect ELISA within 6 weeks of infection.

2. The indirect immunosorbent assay currently is being used to test for antibodies to rubella virus (German measles), and to detect certain drugs in serum. For example, antigen-coated latex beads are used in the SUDS HIV-1 test to detect HIV serum antibodies in about 10 minutes.

Sandwich ELISA: Antigen can be detected or measured by a sandwich ELISA. In this technique, the antibody (rather than the antigen) is immobilized on a microtiter well. This method may be the most versatile and sensitive for the detection of antigens in mixtures. No purified antigen is required. However, only multivalent antigens with different or repeating epitopes may be detected in this assay since binding of two antibodies to the antigen is required. This requirement is normally not a limitation for proteins, which are almost always multivalent. This method is also referred to as double antibody method.

Salient points of the procedures for indirect method are briefed below:

In this technique, the antibody (rather than the antigen) is immobilized on a microtiter well. The appropriately diluted unlabeled antibody (Capture antibody) is adsorbed on to the 96 well micro titer plates. After antibody adsorption, the non-specific sites are blocked by the addition of blocking buffer into the wells. After incubation, a sample containing antigen (e.g. culture supernatants from antigens produced by recombinant technology) is added and allowed to react with the immobilized antibody. After incubation, a second enzyme-linked antibody specific for the same epitope (if multiple copies of the same epitope is present) or a second enzyme-linked antibody specific for different epitope on the antigen is added and allowed to react with the bound antigen. After incubation, a chromogenic substrate specific for the enzyme coupled to secondary antibody is added to the wells and the plates are incubated in dark till the colour develops. The enzyme reaction is terminated by the addition of 2N HCl after the incubation period. The amount of colored reaction product that forms is measured by specialized spectrophotometric plate readers, which can measure the absorbance of all of the wells of a 96-well plate in seconds.

In this method, concentrations of the antigen or antibody in a given test sample can be determined by carrying out the ELISA in presence of different concentrations of the standard antigen or antibody and then constructing a standard curve with antigen or antibody concentrations versus absorbance. From the standard curve the concentration of antigen or antibody in the given sample can be determined.

Result

Serum samples 1, 2, 3 and fraction no.2 from serum samples 1, 2, 3 were tested for the presence of IgG using a direct ELISA. Results are shown in Figure 12. It was observed that anti-human IgG HRP conjugate reacted with wells that were coated with serum samples 1, 2, 3 fraction no.2 from serum samples1, 2, 3 and pure IgG (Figure 13) indicating the presence of IgG in these samples. Anti-human IgG-HRP conjugate failed to react with wells that were coated with rabbit IgG (Figure 13). These results not only confirm for the presence of IgG but also indicated the specificities of anti-human IgG-HRP conjugate for human IgG.

Western Blotting

Highly specific enzyme and antibody assays can detect individual proteins

The purification of a protein, or any other molecule, requires a specific assay that can detect the molecule of interest in column

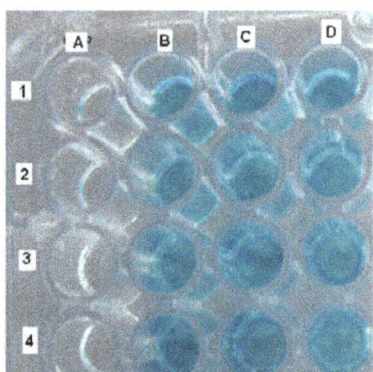

A₁A₂-Antibody Control; A₃A₄-Rabbit IgG (Negative control)
B₁B₂-Serum sample-1; B₃B₄-Serum sample -2;
C₁C₂-Serum sample-3; C₃C₄-Fraction no. 2 (Serum sample 1);
D₁D₂-Fraction no. 2 (Serum sample 2); D₃D₄-Fraction no 2 (Serum sample 3)
Figure 12: Enzyme Linked Immunosorbent Assay.

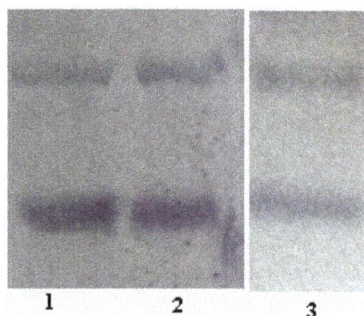

1: Fraction no. 2 (serum sample-1)
2: Fraction no. 2 (serum sample-2)
3: Fraction no. 2 (serum sample-3)
Figure 13: Western Blot Analysis.

luciferase, an enzyme present in fireflies and some bacteria, can be linked to an antibody. In the presence of ATP and luciferin, luciferase catalyzes a light-emitting reaction. In either case, after the antibody binds to the protein of interest, substrates of the linked enzyme are added and the appearance of color or emitted light is monitored. A variation of this technique, particularly useful in detecting specific proteins within living cells, makes use of green fluorescent protein (GFP), a naturally fluorescent protein found in jellyfish.

Western Blotting or immune blotting, a powerful method for detecting a particular protein in a complex mixture combines the superior resolving power of gel electrophoresis, the specificity of antibodies, and the sensitivity of enzyme assays. It is so named in view of the previously used nomenclature for nucleic acid blotting procedure i.e. the transfer of DNA fragments on to nitrocellulose paper or nylon membrane is called southern blotting which has been named after its inventor southern. In a similar way RNA molecules can be transferred to nitrocellulose membrane and the technique is known as northern blotting by analogy to southern blotting.

This multistep procedure is commonly used to separate proteins and then identify a specific protein of interest. Two different antibodies are used in this method, one specific for the desired protein and the other linked to a reporter enzyme. Blotting of protein bands allows them to become more accessible for detection and identification using a variety of methods such as immunological detection, autoradiography, analysis of DNA binding proteins, glycoprotein's, etc. and also the membranes can be stored for much longer period. A number of supporting matrices such as nitrocellulose, diazobenzyl oxymethyl cellulose sheets, nylon membrane and Hydrogen bond membrane etc., are used for this purpose. The use of membrane as a support for protein enables the ease of manipulation, efficient washing and faster reaction during the immuno detection, as proteins are more accessible for reaction.

Blotting is the transfer of resolved proteins from the gel to the surface of a suitable membrane. The transfer of proteins from gels can be achieved by any one of the methods such as simple capillary action, application of vaccum or electrophoretically. Electro blotting is the most efficient and has become a standard method. In this method, the transfer buffer (0.02 M tris-HCl, 0.15 M glycine and 20% methanol) has low ionic strength, which allows electro transfer of proteins. Methanol in the buffer increases the binding of proteins to nitrocellulose and reduces swelling of the gel during transfer. The protein is transferred to the corresponding position on the membrane as on the gel. A mirror image of the gel is formed. However, the protein location and detection can only be assessed after immunodetection.

Result

Fraction no.2 from serum samples 1, 2, 3 was analyzed by western blotting. Results are shown in Figure 13.

Fraction no 2 from serum samples 1, 2, and 3 were analyzed by western blotting. Western blotting is essentially a combination of three techniques viz., electrophoresis (SDS-PAGE), Electrotransfer (protein blotting) and Immunochemical detection (blot development).

Stage I: Separation of proteins on SDS-PAGE: The protein (Fraction no 2 from serum samples 1, 2, and 3) to be analyzed by western blotting is first subjected to separation on a 10-12% separating gel and 6%stacking gel.

Stage II: After electrophoresis the proteins separated on SDS-PAGE were transferred to a more stable medium such as nitrocellulose membrane. During transfer the membrane is kept in close contact with

fractions or gel bands. An assay capitalizes on some highly distinctive characteristic of a protein: the ability to bind a particular ligand, to catalyze a particular reaction, or to be recognized by a specific antibody. An assay must also be simple and fast to minimize errors and the possibility that the protein of interest becomes denatured or degraded while the assay is performed. The goal of any purification scheme is to isolate sufficient amounts of a given protein for study; thus a useful assay must also be sensitive enough that only a small proportion of the available material is consumed. Many common protein assays require just from 10^{-9} to 10^{-12} g of material.

Chromogenic and light-emitting enzyme reactions

Many assays are tailored to detect some functional aspect of a protein. For example, enzyme assays are based on the ability to detect the loss of substrate or the formation of product. Some enzyme assays utilize chromogenic substrates, which change color in the course of the reaction. (Some substrates are naturally chromogenic; if they are not, they can be linked to a chromogenic molecule). Because of the specificity of an enzyme for its substrate, only samples that contain the enzyme will change color in the presence of a chromogenic substrate and other required reaction components; the rate of the reaction provides a measure of the quantity of enzyme present. Such chromogenic enzymes can also be fused or chemically linked to an antibody and used to "report" the presence or location of the antigen. Alternatively,

the gel containing the separated proteins. At the end of electro transfer, proteins from the gel migrate to the nitrocellulose membrane. The protein is transferred to the corresponding position on the membrane as on the gel. A mirror image of the gel is formed. However, the protein location and detection can only be assessed after immunodetection.

Stage III: Immunodetection: The transferred proteins are bound to the surface of the nitrocellulose membrane and are accessible for reaction with immunochemical reagents. Location of protein bands are detected by immunodetection using a enzyme-labeled (anti human IgG-HRP conjugate) and its suitable substrate (TMB/H2O2).

It was observed that anti-human IgG-HRP conjugate reacted with both the heavy and light chain in fraction no 2 (serum samples 1, 2 and 3) resulting in the lighting up of two bands. These results clearly suggest that the proteins present in fraction no 2 (serum samples 1, 2 and 3) were indeed IgG.

Immunofluorescence (dot blot method)

Fluorescent molecules

• Fluorescein, an organic dye that is the most widely used label for immunofluorescence procedures, absorbs blue light (490 nm) and emits an intense yellow-green fluorescence (517 nm).

• Rhodamine, another organic dye, absorbs in the yellow-green range (515 nm) and emits a deep red fluorescence (546 nm). Because it emits fluorescence at a longer wavelength than fluorescein, it can be used in two-color immunofluorescence assays. An antibody specific to one determinant is labeled with fluorescein, and an antibody recognizing a different antigen is labeled with rhodamine. The location of the fluorescein-tagged antibody will be visible by its yellow-green color, easy to distinguish from the red color emitted where the rhodamine-tagged antibody has bound. By conjugating fluorescein to one antibody and rhodamine to another antibody, one can, for example, visualize simultaneously two different cell-membrane antigens on the same cell.

• Phycoerythrinis an efficient absorber of light (~30-fold greater than fluorescein) and a brilliant emitter of red fluorescence, stimulating its wide use as a label for immunofluorescence.

Fluorescent-antibody staining of cell membrane molecules or tissue sections can be direct or indirect.

Direct and indirect immunofluorescence staining of membrane antigen

In direct staining, the specific antibody (the primary antibody) is directly conjugated with fluorescein; in indirect staining, the primary antibody is unlabeled and is detected with an additional fluorochrome-labeled reagent. A number of reagents have been developed for indirect staining. The most common is a fluorochrome-labeled secondary antibody raised in one species against antibodies of another species, such as fluorescein-labeled goat anti-mouse immunoglobulin. Indirect immunofluorescence staining has two advantages over direct staining. First, the primary antibody does not need to be conjugated with a fluorochrome. Because the supply of primary antibody is often a limiting factor, indirect methods avoid the loss of antibody that usually occurs during the conjugation reaction. Second, indirect methods increase the sensitivity of staining because multiple molecules of the fluorochrome reagent bind to each primary antibody molecule, increasing the amount of light emitted at the location of each primary antibody molecule. Immunofluorescence has been applied to identify a number of subpopulations of lymphocytes, notably the CD4+ and CD8+ T-cell subpopulations.

Result

The presence of IgG in fraction no 2 (serum samples 1, 2 and 3) was further confirmed by immunofluorescence (dot blot method) by using goat anti-human fluorescein isothiocyanate (FITC). It was observed that goat anti-human IgG-FITC reacted with fraction no. 2, from serum samples 1, 2 and 3 to produce a fluorescence spot.

Antigen (fraction no 2 from serum samples 1, 2 and 3) were spotted on nitrocellulose membrane strip. Unadsorbed (non-specific) sites on the membrane were blocked by agitating the membrane in 10 ml blocking buffer. After washing, the membrane was agitated in 1:200 diluted goat anti human IgG-FITC. Excess goat anti human IgG-FITC conjugate was removed by washing. After air drying, the membrane was placed in a UV cabinet to observe the fluorescence spot developed. Results of all the three experiments Viz. ELISA, western blotting and immunofluorescence clearly suggested that the protein present in fraction no.2 (serum samples 1, 2 and 3) was IgG. After confirming the prescience of IgG, the next goal was to determine the purity and molecular weight of purified IgG by SDS-PAGE.

Determination of Purity and Relative Molecular Weight of Purified Human IgG

Sodium dodecyl sulphate-polyacrylamide gel electrophoresis (SDS-PAGE)

An important technique for the separation of proteins is based on the migration of charged proteins in an electric field, a process called electrophoresis. These procedures are not generally used to purify proteins in large amounts, because simpler alternatives are usually available and electrophoretic methods often adversely affect the structure and thus the function of proteins. Electrophoresis is, however, especially useful as an analytical method. Its advantage is that proteins can be visualized as well as separated, permitting a researcher to estimate quickly the number of different proteins in a mixture or the degree of purity of a particular protein preparation. Also, electrophoresis allows determination of crucial properties of a protein such as its isoelectric point and approximate molecular weight. The polyacrylamide gel acts as a molecular sieve, slowing the migration of proteins approximately in proportion to their charge-to-mass ratio. Migration may also be affected by protein shape. In electrophoresis, the force moving the macromolecule is the electrical potential, E. The electrophoretic mobility of the molecule, μ, is the ratio of the velocity of the particle molecule, V, to the electrical potential. Electrophoretic mobility is also equal to the net charge of the molecule, Z, divided by the frictional coefficient, f, which reflects in part a protein's shape.

Thus:

$$\mu = \frac{V}{E} = \frac{Z}{f}$$

The migration of a protein in a gel during electrophoresis is therefore a function of its size and its shape.

An electrophoretic method commonly employed for estimation of purity and molecular weight makes use of the detergent sodium dodecyl sulfate (SDS).

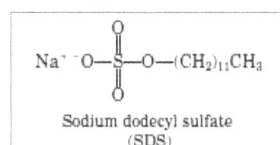

Sodium dodecyl sulfate
(SDS)

SDS binds to most proteins in amounts roughly proportional to the molecular weight of the protein, about one molecule of SDS for every two amino acid residues. The bound SDS contributes a large net negative charge, rendering the intrinsic charge of the protein insignificant and conferring on each protein a similar charge-to-mass ratio. In addition, the native conformation of a protein is altered when SDS is bound, and most proteins assume a similar shape. Electrophoresis in the presence of SDS therefore separates proteins almost exclusively on the basis of mass (molecular weight), with smaller polypeptides migrating more rapidly. After electrophoresis, the proteins are visualized by adding a dye such as Coomassie blue, which binds to proteins but not to the gel itself. Thus, a researcher can monitor the progress of a protein purification procedure as the number of protein bands visible on the gel decreases after each new fractionation step. When compared with the positions to which proteins of known molecular weight migrate in the gel, the position of an unidentified protein can provide an excellent measure of its molecular weight. If the protein has two or more different subunits, the subunits will generally be separated by the SDS treatment and a separate band will appear for each (Table 7).

$$R_f = \frac{\text{Distance travelled by the individual protein!}}{\text{Distance travelled by bromophenol dye}\left(\text{Dye front}\right)}$$

The pioneering work on electrophoresis by A. Tiselius and co-workers was performed in free solution. However, it was soon realized that many of the problems associated with this approach, particularly the adverse effects of diffusion and convection-currents, could be minimized by stabilizing the medium. This was achieved by carrying out electrophoresis on a porous mechanical support, which was wetted in electrophoresis buffer and in which electrophoresis of buffer ions and samples could occur. The support medium cuts down convection currents and diffusion so that the separated components remain as sharp zone. The earliest supports used were filter paper or cellulose acetate strip, wetted in electrophoresis buffer. Nowadays these media are infrequently used; although cellulose acetate still has its use in clinical laboratories for the separation of serum proteins. In particular, for many years' small molecules such as amino acids, peptides and carbohydrates were routinely separated and analyzed by electrophoresis on supports such as paper or thin-layer plates of cellulose, silica or alumina. Although occasionally still used, such molecules are nowadays more likely to be analyzed by more modern and sensitive techniques. While paper and or thin-layer supports are fine for resolving small molecules, the separation of macromolecules such as proteins and nucleic acids on such supports is poor. However, the introduction of the use of gels as support medium led to a rapid improvement in methods for analyzing macromolecules. The earliest

gel system to be used was the starch gel and although this still has some uses, the vast majority of electrophoretic techniques used nowadays involve either polyacrylamide gels or agarose gels. In the case of paper and cellulose acetate electrophoresis, charge on the molecule is the major determinant for its electrophoretic mobility and ultimate separation. The gels in common use, polyacrylamide and agarose, have pores of molecular dimensions whose sizes can be specified. The molecular separations are therefore based on gel filtration as well as the electrophoretic mobilities of the molecules being separated, which shift the equilibrium towards formation of zwitter ions. As zwitter ions do not possess a net charge, they are immobile and carry the current in this region and migrate rapidly in this strong local electric field the pH of the buffer used in the sample is the same buffer that is used in the stacking gel. The Polymerization of Acrylamide and N, N'-Methylene Bis Acrylamide to Form a Cross-Linked Polyacrylamide Gel. Polymerization of acrylamide and N,N'-methylene bisacrylamide to form a cross-linked polyacrylamide gel. The polymerization is induced by free radicals resulting from the chemical decomposition of ammonium persulphate ($S_2O_8^{-2}$ $2SO_4^{-}$) or photodecomposition of riboflavin in the presence of traces of O_2. In either case TEMED, a free radical stabilizer is usually added to the gel mixture. The physical properties of gel and its pore size are controlled by the proportion of polyacrylamide in the gel and its degree of cross-linking. The most commonly used polyacrylamide concentrations are in the range 3 to 15%. Two different porosity gels commonly used in SDS-PAGE are the stacking gel (high porosity gel) and separating or resolving gel (low porosity gel).

When the current is switched on, all the ionic species have to migrate at the same speed otherwise there would be a break in the electrical circuit. Glycine in the upper buffer reservoir exists in two forms; as zwitterions, which does not have a net charge, and as a glycinate anion with a charge of minus one.

When the power is switched on, chloride, protein and glycinate ions begin to migrate towards the anode. Upon entering the stacking gel, the glycinate ion encounters a condition of low pH, which shifts the equilibrium towards formation of zwitter ions. As zwitter ions do not possess a net charge, they are immobile. This immobility of glycine zwitter ions to migrate in to the stacking gel coupled with high mobility of the chloride ions creates a very high localized voltage gradient between the leading chloride and the trailing glycinate ions. Since proteins have their mobility intermediate between the trailing and the leading ions, they carry the current in this region and migrate rapidly in this strong local electric field. The proteins, however, cannot overtake the chloride ions, as the strong local field exists only between the chloride and the glycinate ions. As a result the proteins migrate quickly until they reach the region rich in chloride ions and then drastically slow down. The result is that the three species of interest adjust their concentrations so that [Cl]>[protein-SDS]>[glycinate]. There is only a small quantity of protein-SDS complexes, so they concentrate in a very tight, sharp band between glycinate and Cl (Chloride ions) (Chlorideions) boundaries. (i.e., the faster migration of proteins results in piling or stacking up of the protein samples into a tight sharp disc). It is in this form that the macromolecules enter the running gel. The smaller pores of the separating gel retards the movement of the sharp band of the macromolecules for a long enough time for the glycinate anions to catch up. (The larger pores in the stacking gel do not have any sieving effect therefore the macromolecules migrate faster without any hindrance however, when they reach the separating gel the pores are numerous and of a smaller diameter imparting molecular sieving property to the gel therefore the macromolecules cannot migrate with

Lane	Molecular Weight marker		Relative mobility (cm)	Rf values	Molecular weight (Daltons)
1	Fraction no.2	Band-1	1.2	0.38	50,000
		Band-2	1.8	0.56	24,000
2	BSA		1.0	0.31	66'000
3	Lysozyme		2.2	0.68	14,000
4	Phosphorylase-b	Band-1	0.7	0.22	97,400
	BSA	Band-2	1.0	0.31	66,000
	Ovalbumin	Band-3	1.6	0.50	43,000
	Carbonic anhydrase	Band-4	2.0	0.63	29,000
	β lactoglobulin	Band-5	2.1	0.65	19,400
	Lysozyme	Band-6	2.2	0.69	14,300
5	Fraction No.2	Band-1	1.2	0.38	50,000
	Fraction No.2	Band-2	1.8	0.56	24,000

Table 7: Molecular Weight Determination by SDS-Page.

the same speed as they did in the stacking gel as a result the proteins pile or stack up into a tight sharp disc).

Once the glycinate ions reach the separating gel it becomes more fully ionized in the higher pH environment its mobility increases. (The pH of the stacking gel is 6.8 that of the separating gel is 8.8). Thus the interface between glycinate and Cl⁻ (Chloride ions) leaves behind the protein-SDS complexes which are left to electrophoresis at their own rates. (Upon entering the separating gel the glycinate ions encounter a condition of high pH [pH of the separating gel buffer is about 2 pH units higher than that of the stacking gel] which shifts the equilibrium towards formation of glycinate anion). The negatively charged protein-SDS complexes now continue to move towards the anode under the applied field with the same mobility. However as they pass through the separating gel the proteins separate, owing to the molecular sieving properties of the gel. Quite simply, the smaller the proteins the more easily it can pass through the pores of the gel, whereas larger proteins are successively retarded by frictional resistance due to the sieving effect of the gels. Being a small molecule, the bromophenol blue dye is totally unretarted and therefore indicates the electrophoresis front, i.e. the extent of migration of the dye gives an index of electrophoretic process. The dye migrates faster than all macromolecules. When the dye reaches the bottom of the gel, the current is turned off and the gel is removed from in-between the glass plates and shaken in an appropriate stain solution. A typical gel would take 1 to 1.5 h to prepare and set, 3 h to run at 30 mA. Vertical slab gels are invariably run, since this allows up to 20 different samples to be loaded on to a single gel.

Detection, estimation and recovery of proteins in gels

Gels from a column are removed by forcing water from a hypodermic syringe around the walls of the column, allowing the gels to be extracted under gentle pressure. Slab gels are removed by introducing a thin metal plate (spatula) between the two gel plates and coaxing the plates apart. Before staining, the gels may be immersed in a fixative (5% TCA) to guard against diffusion of separated component (prevent diffusion). The most commonly used general protein stain for detecting proteins on gels is the trimethylamine sulfate dye Coomassie Brilliant Blue R-250 (CBB).

Coomassie Brilliant Blue R 250 (CBB)

Proteins are often visualized by staining with the dye Coomassie Brilliant Blue R 250. Staining is usually carried out using 0.25% (w/v) CBB in methanol: acetic acid: water (50:10:40, by volume). This acid-methanol mixture acts as a denaturant to precipitate or fix the protein in the gel, which prevents the protein from being washed out while it is being stained. Staining of most gels is accomplished in about 2 h and destaining is achieved by gentle agitation in the same methanol-acid-water solution but in the absence of the dye (usually overnight). The Coomassie stain is highly sensitive; a very weakly staining band on a polyacrylamide gel would correspond to about 1 μg of protein. The CBB stain is not used for staining cellulose acetate (or protein blots) because it binds to the paper. In this case, proteins are first denatured by brief immersion of the strip in 10% (v/v) trichloroacetic acid, and then immersed in a solution of a dye that does not stain the support material, for example: Procion blue, Amido black or Procion S. Although the Coomassie stain is highly sensitive, many workers require greater sensitivity and use a silver stain. Silver stains are based either on techniques developed for histology or on methods based on the photographic process. In either case, silver ions (Ag+) are reduced to metallic silver on the protein where silver is deposited to give a black or brown band. Silver stains can be used immediately after electrophoresis or alternatively, after staining with CBB. With the latter approach, the major bands on the gel can be identified with CBB and then minor bands not detected with CBB, identified using the silver stain. The silver stain is at least 100 times more sensitive than Coomassie Brilliant Blue, detecting protein down to 0.l μg amounts.

Results

Determination of purity and molecular weight of purified human IgG: To determine the purity of isolated IgG, fraction no.2 (serum samples 1, 2 and 3), were analyzed on a 12% separating gel and 6% stacking gel. Coomassie blue stained gel is shown in Figure 13. Proteins from fraction no.2 serum samples 1, 2 and 3 resolved in to 2 bands. After determining the purity of isolated IgG, the molecular weight of isolated IgG was determined by running standard protein molecular weight markers along with fraction no.2 serum samples 1 and pure human IgG. Coomassie brilliant blue R 250 stained pattern is shown in the Figure 13.

As shown in Figure 13 (lane-1), two protein bands with molecular weights of and 50 kD and 25kD corresponding to heavy (50 kD) and light chain (25 kD) of immunoglobulin molecule were observed. Similar results were observed in the case of pure human IgG. These results clearly suggest that the IgG purified from human and serum was fairly pure.

Native–PAGE (Activity staining of the enzyme HRP): While SDS–PAGE is the most frequently used gel system for studying proteins, the method is of no use if one is aiming to detect a particular protein (often an enzyme) on the basis of its biological activity, because the protein (enzyme) is denatured by the SDS–PAGE procedure. In this case it is necessary to use non-denaturing conditions. In native or buffer gels, polyacrylamide gels are again used (normally a 7.5% gel) but the SDS is absent and the proteins are not denatured prior to loading. Since all the proteins in the sample being analyzed carries their native charge at the pH of the gel (normally pH 8.7), proteins separate according to their different electrophoretic mobilities and the sieving effects of the gel. It is therefore not possible to predict the behavior of a given protein in a buffer gel but, because of the range of different charges and sizes of proteins in a given protein mixture, good resolution is achieved. The enzyme of interest can be identified by incubating the gel in an appropriate substrate solution such that a coloured product is produced at the site of the enzyme [18]. An alternative method for enzyme detection is to include the substrate in an agarose gel that is poured over the acrylamide gel and allowed to set. Diffusion and interaction of enzyme and substrate between the two gels results in colour formation at the site of the enzyme. Often, duplicate samples will be run on a gel, the gel cut in half and one half stained for activity, the other for total protein. In this way the total protein content of the sample can be analyzed and the particular band corresponding to the enzyme identified by reference to the activity stain gel.

Result: The enzyme HRP was analyzed on a slab gels consisting of 12% separating gel and 6% stacking gel. After electrophoresis, the enzyme was identified by activity staining.

Conclusion: Experiment studies shows binding of enzymes to antibodies. This involves the formation of a stable, covalent linkage between an enzyme & Antibody. During these studies we have done many trials and studies on different Proteins, Preparation of anti-Human IgG is a big task for us which we used in this experiment, Prepared anti-human IgG HRP conjugate we have used for the following experiments like Qualitative ELISA for detection of human IgG and Identification of IgG by western blotting, For this conjugation studies we carried out by Periodate oxidation method and results are noted.

References

1. Healey K, Chandler HM, Cox JC, Hurrell JG (1983) A rapid semi-quantitative capillary enzyme immunoassay for digoxin. Clin Chim Acta 134: 51-58.

2. Jeanson A, Cloes JM, Bouchet M, Rentier B (1988) Comparison of conjugation procedures for the preparation of monoclonal antibody-enzyme conjugates. J Immunol Methods 111: 261-270.

3. Nakane PK, Kawaoi A (1974) Peroxidase-labeled antibody. A new method of conjugation. J Histochem Cytochem 22: 1084-1091.

4. Tijssen P, Kurstak E (1984) Highly efficient and simple methods for the preparation of peroxidase and active peroxidase-antibody conjugates for enzyme immunoassays. Anal Biochem 136: 451-457.

5. Voller A, Bidwell DE, Bartlett A (1976) Enzyme immunoassays in diagnostic medicine. Theory and practice. Bull World Health Organ 53: 55-65.

6. Coligan, JE, Kruisberk AM, Margulies DH, Shevach EM, Strober W (1997) Current protocols in Immunology Wiley, Newyork. Deutscher, M. (Ed.).1997. Guide to protein purification. Academic press.

7. Domon B, Aebersold R (2006) Mass spectrometry and protein analysis. Science 312: 212-217.

8. Herzenberg LA, Weir DM (1996) Weir's handbook of experimental Immunology, Immunochemistry and Molecular Immunology, 5th edition, Oxford, Blackwell Scientific publications, USA.

9. Porath J, Belew M (1987) 'Thiophilic' interaction and the selective adsorption of proteins. Trends Biotechnology 5: 225-229.

10. Matejtschuk P (1997) Affinity separations: A practical approach. Oxford University Press, USA.

11. Porath J, Maisano F, Belew M (1985) Thiophilic adsorption--a new method for protein fractionation. FEBS Lett 185: 306-310.

12. Scopes RK, Cantor C (1994) Protein purification: principles and practice (3rd Edition). Springer Verlag, USA.

13. Subba Rao PV, McCartney-Francis NL, Metcalfe DD (1983) An avidin--biotin microELISA for rapid measurement of total and allergen-specific human IgE. J Immunol Methods 57: 71-85.

14. Law B (1996) Immuno Assay: A practical guide. Taylor and Francis Ltd, London, U.K.

15. Stites DP, Rodgers C, Folds JD, Schmitz J (1997) Clinical laboratory detection of antigens and antibodies. Medical Immunology, 9th Edition, Stites DP, Terry AL, Parslow TG, editors, McGraw-Hill, USA.

16. Stryer L (1995) Biochemistry. Fourth edition, Freeman WH, New York, USA, 62.

17. Burnette WN (1981) "Western blotting": electrophoretic transfer of proteins from sodium dodecyl sulfate--polyacrylamide gels to unmodified nitrocellulose and radiographic detection with antibody and radioiodinated protein A. Anal Biochem 112: 195-203.

18. Laemmli UK (1970) Cleavage of structural proteins during the assembly of the head of bacteriophage T4. Nature 227: 680-685.

12

Gas Chromatography-Mass Spectrometry Analysis of Photosensitive Characteristics in Citrus and Herb Essential Oils

Pei-Shan Wu[1], Yu-Ting Kuo[1], Shen-Ming Chen[2]*, Ying Li[2] and Bih-Show Lou[1]*

[1]Chemistry Division, Center for General Education, Chang Gung University, 259 Wen-Hwa 1st Road, Kwei-Shan, Tao-Yuan 333, Taiwan, ROC
[2]Department of Chemical Engineering and Biotechnology, National Taipei University of Technology, No. 1, Section 3, Chung-Hsiao East Road, Taipei 106, Taiwan, ROC

Abstract

Essential oils (EOs) are commonly used in aromatherapy and offer a number of health benefits. However, the photosensitivity of citrus EO family limits their applications. It is important to characterize the compositional changes of EOs upon possible factors affecting their stability, such as light and water content. In this study, we used gas chromatography equipped with mass spectrometry detector to investigate the constituents of commercial citrus EOs (lemon, orange) and herb EOs (clary sage, lavender). The result indicated that limonene was the most abundant compound in citrus EOs and followed by β-pinene or β-myrcene. Linalyl acetate and β-linalool were the major constituents in herb EOs. It is surprised to find that almost no change in chemical composition under sunlight exposure for 2hr. In contrast, the amount of terpene hydrocarbons decreased greatly in citrus OEs with H_2O addition and under sunlight exposure, which might be converted to oxidative compounds, such as carveol, ρ-cymene and limonene oxide. However, herb EOs was much less photosensitive, which are more potential to become a stable material for daily used products application.

Keywords: Essential oil; Citrus; Herb; Photosensitive; Gas Chromatography-Mass Spectrometry (GC-MS)

Introduction

Essential oils (EOs) are complex mixture and highly concentrated hydrophobic liquid containing volatile plant secondary metabolites belong to terpenoids and aromatic groups. They are usually extracted from various parts of plants (flower, leaf or fruit) by different methods such as steam/water distillation, solvent extraction or cold expression, etc. The natural source and pleasant flavor characters make EOs widely used in the medicine, cosmetic, household products and food industry. EOs was also reported to contain many bioactive compounds, such as terpenoids, alkaloids, flavonoids and carotenes. These components make EOs are extensively represent a green alternative in the pharmaceutical, neutritionanl argticultural field due to their antimicrobial, antiviral, insecticidal, antioxidant, antiinflamatory and stress-repellent properies [1-3].

Citrus fruits are the most common subtropical crops in the world, such as lemon and orange. We usually eat citrus fruits' sweet and juicy fleshes, though recently, people have started to pay attention to reusing of fruits peels. The main reasons are that citrus peels are easily obtained and naturally high in pectin, their antioxidant activity, and relative benefits from vitamin C and flavonoids [1]. Furthermore, when they are made into EOs, the main product, limonene, has been proved to inhibit cancer cells initiation, promotion and progression [4]. Another widespread EOs is extracted from spices and their constituents mainly depend on the plant species such as sage and lavender. Sage is a small evergreen shrub with grayish leaves, blue to purplish flowers and has therapeutic properties of antiinflammatory, antibacterial, antioxidant and stimulant for medicinal purposes [5]. Lavender belongs to mint family and used for centruies as an herbal remedy due to its' sweet overtones. It is also believed to be benefit for stress, exhaustion, headaches, depression and digestion problems, and even have application of food preservation [2].

Gas chromatography equipped with mass spectrometry (GC-MS) as detector is a widely used platform for analyzing volatile complex compounds [6]. A tool with good selectivity and high sensitivity is necessary for natural EOs because they are usually composed of many different ingredients or flavonoids. GC-MS can offer a quick qualitative function based on the integrity of a compound database (ex: NIST), and the quantification can be more precise when isotope standards and selected ion mode (SIM) are used together.

Meanwhile, due to the various components in oil extracts, some of them are light-, oxygen-, temperature- or moisture-sensitive. Most of commercial EOs is highly concentrated and need to be diluted before used to avoid skin or respiratory damage [7]. Proper storage and safe usage become an important issue for EOs to ensure the effectiveness and quality for future development in the medicinal field [8,9]. The effect of sunlight exposure on the hydrolysis or oxidation reaction in the citrus and herb EOs is not revealed yet. In this work, a photosensitivity experiment was designed to understand the composition changes when EOs coexisted with water under sunlight treatment.

Materials and Methods

Chemicals and reagents

Alkane standard solution (contains C8-C20, ~40 mg/L each, in hexane) and LC/MS-grade methanol were purchased from Sigma-Aldrich (St. Louis, MO, USA).

*Corresponding author: Bih-Show Lou, Chemistry Division, Center for General Education, Chang Gung University, 259 Wen-Hwa 1st Road, Kwei-Shan, Tao-Yuan 333, Taiwan, ROC, E-mail: blou@mail.cgu.edu.tw

Shen-Ming Chen, Department of Chemical Engineering and Biotechnology, National Taipei University of Technology, No. 1, Section 3, Chung-Hsiao East Road, Taipei 106, Taiwan, ROC, E-mail: smchen78@ms15.hinet.net

EO samples

Commercial citrus and herb essential oils were provided from an aroma products company (EASECOX, Germany), and the original material plant source was Italy. Before experiment, EOs were kept in dry environment and prevented from light at room temperature. All samples were diluted 1:15 (v/v) in methanol prior to GC-MS analysis.

Photosensitive experiment

EOs were added H_2O (group C) with the ratio 5:1 (EO: H_2O, v/v) and with a non-H_2O added sample (group B) as positive control. The sample vials were sealed well and irradiated with UVA (320-400nm) and UVB (290-320nm) of 100 mW/cm^2 from sunlamps (Xe lamp, 1.5AM) for 2 hours. The irradiated samples and stock without treatment (group A) were collected and prevented from light before GC-MS analysis.

GC-MS analysis

GC-MS analysis was performed on Agilent Technology (Little Falls, California, USA) 6890 series gas chromatography (GC) system, equipped with 5973 mass spectrometry (MS) detector and a 7683 series auto-injector was used. Compounds were separated on Rtx®-Wax capillary column (30 m × 0.25 mm, film thickness 0.25 μm; RESTEK, Pennsylvania, USA). Helium (5N5 grade) was used as carrier gas, with a flow rate of 0.8 mL/min, and the split ratio was 60:1. Sample injection volume was 1 μl and the injector temperature was 230°C. The column oven temperature was held at 70°C for 2 min, and then programmed to 130°C at 30°C/min and change the gradient to 230°C with 10°C/min. Finally, held at 230°C for 6 min and the total run time was 20 min. An electron ionization (EI) system with ionization energy 70 eV was used for detection. The ion source temperature was set at 230°C, the interface temperature was 250°C, detector voltage was 2 kV. The mass spectrum was acquired in scan mode at a scan rate 0.98 scan/sec within a mass range of 20-800 amu. The measurement was performed in duplicate for each sample with solvent delay for 2 min.

Data process and compound identification

The data was processed by software provided by Agilent Technology (MSD ChemStation D.03.00.611). The compounds identification were using their MS data compared to the on-site NIST98 mass spectral library and on-line NIST Chemistry WebBook (http://webbook.nist.gov/chemistry/), and the retention indices (RIs) relative to C8-C20 *n*-alkanes obtained on a nonpolar Rtx®-Wax column.

Results and Discussion

A high throughput CG-MS methodology for essential oil analysis

In this study, we demonstrated that GC-MS is a quick and reliable platform for EOs analysis in both qualification and quantification. Natural plant essential oils were prepared simply by dilution with methanol and a short separation gradient of GC coupled with auto-sampler made this platform more efficient. MS detector provided high sensitivity and good selectivity in samples analysis. In addition, rich GC-MS database, such as NIST library, coupled with retention index supplied a credible method to identify compounds in a complex mixture [10].

Chemical composition of citrus and herb essential oils by GC-MS analysis

GC-MS chromatography for lemon and orange EOs prior to treatment are shown GC-MS chromatograms of lemon and orange EOs prior to treatment as control are shown in Figure 1 I-A and II-A, respectively. The most predominant compound of lemon OE was the limonene shown at the peak 6 of Figure 1 I-A accounting for the % area of 57.71, which is consistent with other reports [8,13-17], and followed compounds were β-pinene and 3-carene at the peak 3 with 13.57% and the peak 7 with 10.54%, respectively. Table 1 summarized peak number, retention time (RI), compound name, formula and % area for all the identified compounds of lemon OE found in this study. Similarly, limonene at the peak 4 of Figure 1 II-A was the highest contained compound with % area of 86.05 shown in Table 2 for orange EO, however, the next two abundance compounds were β-myrcene and L-carvone at the peak 3 with 2.22% and the peak 22 with 1.12 %, respectively. These compounds are common monoterpenoid hydrocarbons found in EO through the addition, cabocation and cyclication reaction of gernyl pyrophosphate under specitfic enzymic control in plants. The structures of major components for lemon and orange EOs were showed in Figure 3. Limonene is a single-cyclic terpenenoid with a strong citrus odor and bitter taste. α-pinene, β-pinene and 3-carene are bi-cyclic terpenenoids found major constituents of turpentine in nature. β-myrcene is a very widespread monoterpenoid in nature, especially in herbs and spices. Carvone is a common monoterpenoid ketone found in EOs and its enantiomers provides the characteristic odor of spearmint [11]. According to both Table 1 and 2, the contents of other type terpenoids, such as sesquiterpenoids (bergamotene and bisabolene), alcohols (β-linalool and α-terpieol), aldehydes (cis- and trans-citral), ketones (sulcatone), acids (octanoic acid) and oxides (cis- and trans-limonene oxide) were occupied low concentration in original citrus EOs before treatment (total amount is lass than % area of 10).

In herb EOs, the chromatograms of untreated clary sage and lavender EOs as control were showed in Figure 2 III-A and IV-A, respectively. Linalyl acetate and β-linalool were the two predominant components in both clary sage and lavender EOs with similar concentration ratio. In general, linalyl acetate found in many flower and spice plants is the acetate ester of linalool, and they both often exist as in accompany and possess antifungal activity [12]. Linalyl acetate was almost taken up half content in both clary sage and lavender shown at the peak 14 of Figure 2 III-A with % area of 46.43 and the peak 16 of Figure 2 IV-A with 33.81%, respectively. On the other hand, β-linalool was found at the peak 13 of Figure 2 III-A with 22.68% and the peak 15 of Figure 2 IV-A with 24.89% for clay sage and lavender, respectively. α-terpineol and 4-terpeneol are the two of four isomers of terpineol, which is a common monoterpene alcohol in herb EOs. Trans-geranyl acetate, a monoterpene with a pleasant floral aroma, was the third highest component in calry sage (peak 24, 4.80%). A bi-cyclic sesquiterpene and FDA approved food additive [9], caryophyllene was the fourth abundant compound (peak 18, 5.83%) in lavender EO. As compared with citrus EOs, the amounts of hydrocarbon terpenoids in clary sage and lavender existed relative low, such as β-myrcene, β-ocimene and germacrene D. Detail chemical compositions and their relative intensities were listed in Table 3 and 4 for clary sage and lavender, respectively. Grand viriability dependens on several factors including plant species, season, location, the part for extraction and the preparation method, and could result in different chemical constituents of citrus and herb Eos found in this study from other literatures [13-17].

Photosensitive investigation

Compared with the chromatograms between A and B of Figures 1 and 2 for both citrus and herb EOs, we found surprisingly that they were almost identical to each other and revealed no effect under sunlight

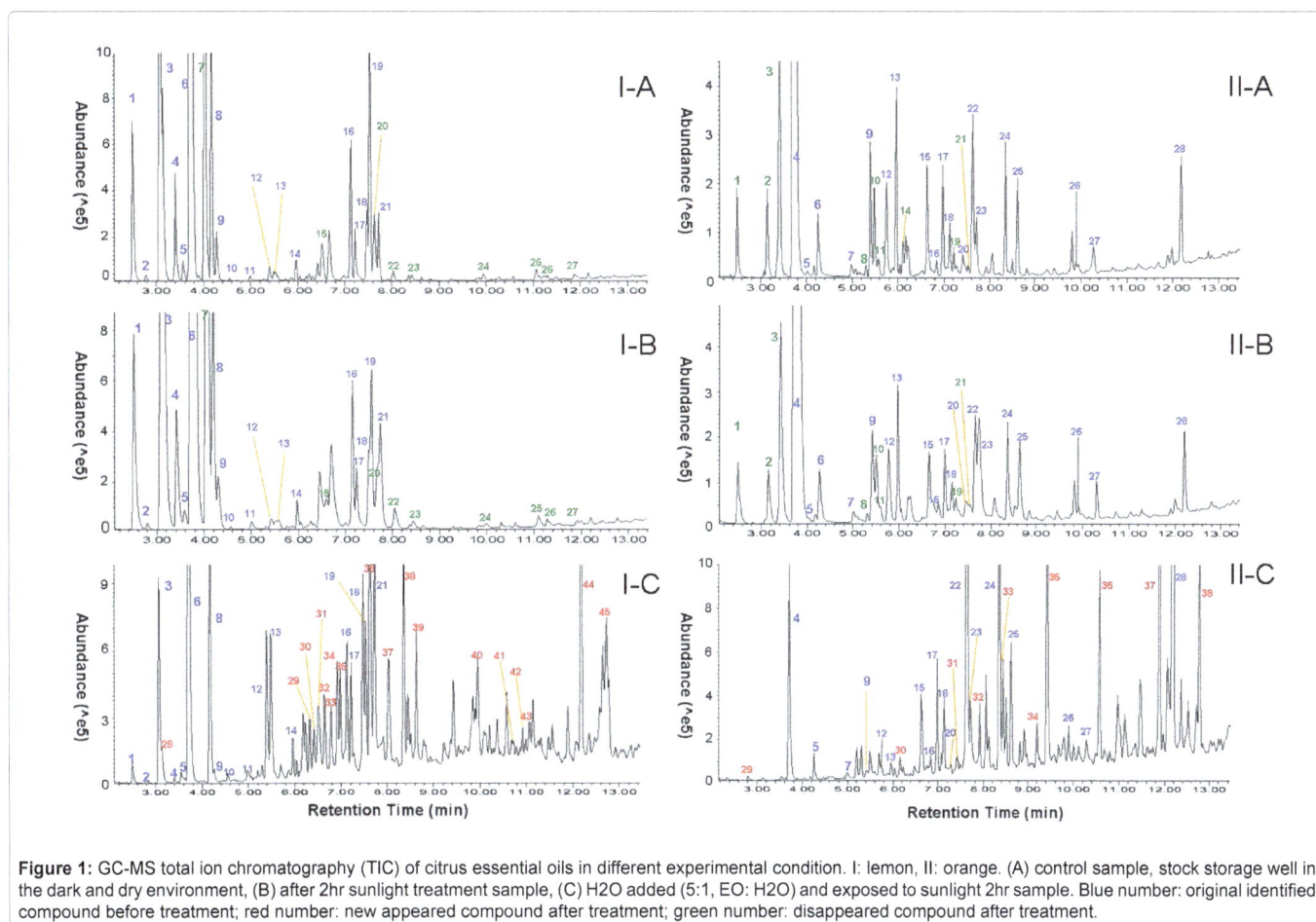

Figure 1: GC-MS total ion chromatography (TIC) of citrus essential oils in different experimental condition. I: lemon, II: orange. (A) control sample, stock storage well in the dark and dry environment, (B) after 2hr sunlight treatment sample, (C) H2O added (5:1, EO: H2O) and exposed to sunlight 2hr sample. Blue number: original identified compound before treatment; red number: new appeared compound after treatment; green number: disappeared compound after treatment.

exposure for 2hr. On the other way, a great effect was observed with H$_2$O addition and under sunlight exposure for 2hr in I-C and II-C of Figure 1 for both citrus EOs. The major component, limonene, decreased dramatically from 57.71% and 86.05% to 19.02% and 4.55% for lemon and orange EO, respectively, and considered that it was conversed into alcohols, aldehyde, ketones and oxides via oxidation, hydration and isomerisation. In lemon oil, many alcohols, ketone oxide compounds were appeared or increased significantly, such as caveol (0% to 5.85%), carvone (0% to 6.23%) and limonene oxie (0.38% to 6.07%). These compounds were the products from limonene hydration and oxidation under water-light combined action and a brief mechanism was proposed in Figure 4A. Limonene-1,2-diol (0% to 7.19%) was limonene involved in further epoxidation and the epoxides had been hydrated to limonene-diol [18]. α-, β-pinene and 3-carene at the peaks 1, 3 and 7 of Figure 1 I-C decreased obviously from 2.27% to 0.36%, 13.57% to 3.36% and 10.54% to 0%, respectively. The oxidative reactions of α-, β-pinene were proposed in Figure 4B [19]. The possible mechanism presented that α-, β-pinene were converted to their hydroperoxides with migration of the double bond under light exposion via myrtenol (peak 37, 0% to 2.44%) and trans-pinocarveol (peak 34, 0% to 1.43%) and finally produced terminal hydroperoxides, myrtenal (peak 33, 0% to 1.10%). The oxidation of 3-carene can be initiated by ozone and OH radical under humidity atmosphere and the ozonide or hydroxyl radical go further to generate different products [20]. The remainder generated monoterpenoid alcohols after water-light treatment, such as α-terpieol, cis- and trans-mentha-2,8-dien-1-ol were also the limonene hydrated and oxided products [21,22]. The aromatic monoterpene p-cymene at peak 8 of Figure 1 I-C increased

from 2.92% to 6.66% after water and sunlight treatment, which has been reported for identification of aged EOs as increase during storage [7,19]. Geranic acid, cis- and trans-geranyl acetate were oxidized from citral shown in Figure 4C [7] and increased in a considerably range from 0% to 3.5 ~ 4%. Citral (mixture of neral and geranial at peaks 16 and 19) is a strong lemon odor aldehyde and is often used as index to estimate the quality of lemon EOs. For orange oil, most compositional changes were similar to lemon oil. An enormously limonene content dropped from 86.05% to 4.55% and the raise of carveol (cis- and trans-form, 1.28% to 12.95%), carvone (1.12% to 17.30%), limonene oxide (cis- and trans-form, 0% to 5.14%) and the major product limonen-1,2-diol (0.68% to 22.31%). Unlike lemon oil, orange oil had fewer hydrocarbon terpenoids, and the limonene hydration and oxidation become the main reaction in total compositional change. There are still aldehyde and ketone of monoterpenoid were observed increased at the peaks 38 and 37 from 0% to 3.36% and 0% to 7.16%, respectively, which might be resulted in the oxidation of limonene and β-linalool under water-light environment [9,23]. The CC profile provides a simple way to understand the compositional change between pre-and post-treatment. According to the color change, it is easier to figure out that the constituent changes of EO from monoterpenes to alcohols and ketones after water-light treatment were quite different between lemon and orange EOs shown in Figure 5 I and II, respectively.

In proportion of herb EOs, there were much less compositional changes in photosensitive experiment and suggested that major compound linalyl acetate was quite stable and not easily hydrated and oxidized under sunlight coupled water treatment. The evidence

Peak Nº. a	RI b	Compounds	Formula	% Area c		
				A	B	C
Hydrocarbons						
1	1031	α-Pinene	$C_{10}H_{16}$	2.27	2.15	0.36
2	1078	Camphene	$C_{10}H_{16}$	0.07	0.05	0.07
3	1121	β-Pinene	$C_{10}H_{16}$	13.57	13.04	3.36
4	1166	β-Myrcene	$C_{10}H_{16}$	1.40	1.24	0.08
5	1189	p-Menthane	$C_{10}H_{20}$	0.28	0.13	0.23
6	1213	Limonene	$C_{10}H_{16}$	57.71	61.22	19.02
7	1255	3-Carene	$C_{10}H_{16}$	10.54	10.09	NA
8	1278	p-Cymene	$C_{10}H_{14}$	2.92	2.26	6.66
9	1292	Terpinolene	$C_{10}H_{16}$	0.59	0.59	0.27
15	1599	α-Bergamotene	$C_{15}H_{24}$	0.72	0.42	NA
20	1746	α-Bisabolene	$C_{15}H_{24}$	0.81	0.31	NA
22	1786	β-Bisabolene	$C_{15}H_{24}$	0.16	0.29	NA
28	1146	β-Terpinene	$C_{10}H_{16}$	NA	NA	0.66
31	1607	β-Bergamotene	$C_{15}H_{24}$	NA	NA	1.54
Oxides						
12	1459	cis-Limonene oxide	$C_{10}H_{16}O$	0.18	0.10	2.84
13	1471	trans-Limonene oxide	$C_{10}H_{16}O$	0.20	0.18	3.23
24	2003	cis-Caryophyllene oxide	$C_{15}H_{24}O$	0.13	0.07	NA
40	2007	trans-Caryophyllene oxide	$C_{15}H_{24}O$	NA	NA	2.06
Alcohols						
14	1538	β-Linalool	$C_{10}H_{18}O$	0.23	0.21	0.64
17	1698	α-Terpieol	$C_{10}H_{18}O$	0.52	0.43	1.39
23	1839	trans-Geraniol	$C_{10}H_{18}O$	0.07	0.08	NA
25	2182	(-)-Spathulenol	$C_{15}H_{24}O$	0.18	0.15	NA
27	2284	α-Bisabolol	$C_{15}H_{26}O$	0.11	0.06	NA
32	1624	cis-Mentha-2,8-dien-1-ol	$C_{10}H_{16}O$	NA	NA	1.32
34	1657	trans-Pinocarveol	$C_{10}H_{16}O$	NA	NA	1.43
35	1665	trans-Mentha-2,8-dien-1-ol	$C_{10}H_{16}O$	NA	NA	1.38
37	1788	Myrtenol	$C_{10}H_{16}O$	NA	NA	2.44
38	1832	trans-Carveol	$C_{10}H_{16}O$	NA	NA	3.81
39	1863	cis-Carveol	$C_{10}H_{16}O$	NA	NA	2.04
41	2140	Cuminol	$C_{10}H_{14}O$	NA	NA	0.26
42	2166	trans-p-Mentha-2,8-dienol	$C_{10}H_{16}O$	NA	NA	0.27
43	2182	(+)-Spathulenol	$C_{15}H_{24}O$	NA	NA	0.68
44	2333	Limonen-1,2-diol	$C_{10}H_{18}O_2$	NA	NA	7.19
Aldehydes						
11	1398	Nonanal	$C_9H_{18}O$	0.08	0.11	0.24
16	1687	cis-Citral (Neral)	$C_{10}H_{16}O$	1.72	1.20	1.56
19	1735	trans-Citral (Gernial)	$C_{10}H_{16}O$	2.91	1.64	1.94
33	1641	Myrtenal	$C_{10}H_{14}O$	NA	NA	1.10
Ketones						
10	1339	Sulcatone	$C_8H_{14}O$	0.03	0.02	0.12
29	1583	Camphenilone	$C_9H_{14}O$	NA	NA	0.94
30	1595	(+)-Nopinone	$C_9H_{14}O$	NA	NA	0.57
36	1747	Carvone	$C_{10}H_{14}O$	NA	NA	6.23
Acids						
26	2218	Nonanoic acid	$C_9H_{18}O_2$	0.05	0.06	NA
45	2412	Geranic acid	$C_{10}H_{16}O_2$	NA	NA	3.91
Esters						
18	1726	cis-Geranyl acetate	$C_{12}H_{20}O_2$	0.64	0.69	3.71
21	1756	trans-Geranyl acetate	$C_{12}H_{20}O_2$	0.72	0.98	3.40
Unknowns				1.20	2.25	13.09

a Peak number labeled on TIC
b Retention index relative to C8-C20 n-alkanes on Rtx®-Wax column
c A: control sample; B: 2hr sunlight treated sample; C: H₂O added and sunlight treated 2hr sample

Table 1: Chemical compositional changes and relative intensity (%) of lemon EO in photosensitive experiment analysis by GC-MS.

Peak No. a	RI b	Compounds	Formula	% Area c		
				A	B	C
Hydrocarbons						
1	1031	α-Pinene	$C_{10}H_{16}$	0.58	0.47	NA
2	1130	α-Phellandrene	$C_{10}H_{16}$	0.57	0.46	NA
3	1166	β-Myrcene	$C_{10}H_{16}$	2.22	1.71	NA
4	1213	Limonene	$C_{10}H_{16}$	86.05	87.82	4.55
5	1278	p-Cymene	$C_{10}H_{14}$	0.09	0.05	0.07
Oxides						
9	1459	cis-Limonene oxide	$C_{10}H_{16}O$	0.76	0.74	0.23
10	1471	trans-Limonene oxide	$C_{10}H_{16}O$	0.48	0.40	NA
33	1838	cis-Carvone oxide	$C_{10}H_{14}O_2$	NA	NA	1.52
35	1950	Limonene dioxide	$C_{10}H_{16}O_2$	NA	NA	5.14
Alcohols						
8	1447	1-Heptanol	$C_7H_{16}O$	0.08	0.06	NA
13	1538	β-Linalool	$C_{10}H_{18}O$	1.07	0.73	0.32
15	1629	trans-p-Mentha-2,8-dienol	$C_{10}H_{16}O$	0.89	0.43	1.82
16	1652	1-Nonanol	$C_9H_{20}O$	0.14	0.12	0.14
17	1671	cis-p-Menth-2,8-dienol	$C_{10}H_{16}O$	0.64	0.41	1.85
19	1698	α-Terpineol	$C_{10}H_{18}O$	0.11	0.21	NA
24	1832	Trans-carveol	$C_{10}H_{16}O$	0.72	0.65	10.46
25	1863	cis-carveol	$C_{10}H_{16}O$	0.56	0.91	2.49
26	2005	Perilla alcohol	$C_{10}H_{16}O$	0.04	0.04	0.84
28	2333	Limonen-1,2-diol	$C_{10}H_{18}O_2$	0.68	0.53	22.31
Aldehydes						
6	1291	Octanal	$C_8H_{16}O$	0.54	0.44	NA
7	1398	Nonanal	$C_9H_{18}O$	0.13	0.10	0.18
11	1482	Citronellal	$C_{10}H_{18}O$	0.04	0.12	NA
12	1503	Decanal	$C_{10}H_{20}O$	0.80	0.63	0.33
18	1687	cis-Citral (Neral)	$C_{10}H_{16}O$	0.19	0.34	0.81
20	1713	Dodecanal	$C_{12}H_{24}O$	0.19	0.17	0.15
21	1735	trans-Citral (Gernial)	$C_{10}H_{16}O$	0.03	0.15	NA
23	1753	β-Methylcrotonaldehyde	C_5H_8O	0.12	0.92	0.77
29	1074	Hexanal	$C_6H_{12}O$	NA	NA	0.10
38	2421	1,3,4-Trimethyl-3-cyclohexenyl-1-carboxaldehyde	$C_{10}H_{16}O$	NA	NA	3.36
Ketones						
22	1747	Carvone	$C_{10}H_{14}O$	1.12	0.27	17.30
30	1563	Limona ketone	$C_9H_{14}O$	NA	NA	0.35
31	1719	Umbellulone	$C_{10}H_{14}O$	NA	NA	0.26
32	1777	p-Acetyltoluene	$C_9H_{10}O$	NA	NA	0.99
34	1925	4-Hydroxy-3-methylacetophenone	$C_9H_{10}O_2$	NA	NA	0.85
36	2128	Cyclooctanone	$C_8H_{14}O$	NA	NA	3.15
37	2302	3-Isopropylidene-5-methyl-hex-4-en-2-one	$C_{10}H_{16}O$	NA	NA	7.16
Acids						
27	2089	Octanoic Acid	$C_8H_{16}O_2$	0.19	0.28	0.62
Esters						
14	1538	Linalyl acetate	$C_{12}H_{20}O_2$	0.13	NA	NA
Unknowns				0.81	0.85	11.87

a Peak number labeled on TIC
b Retention index relative to C8-C20 n-alkanes on Rtx®-Wax column
c A: control sample; B: 2hr sunlight treated sample; C: H₂O added and sunlight treated 2hr sample

Table 2: Chemical compositional changes and relative intensity (%) of orange EO in photosensitive experiment analysis by GC-MS.

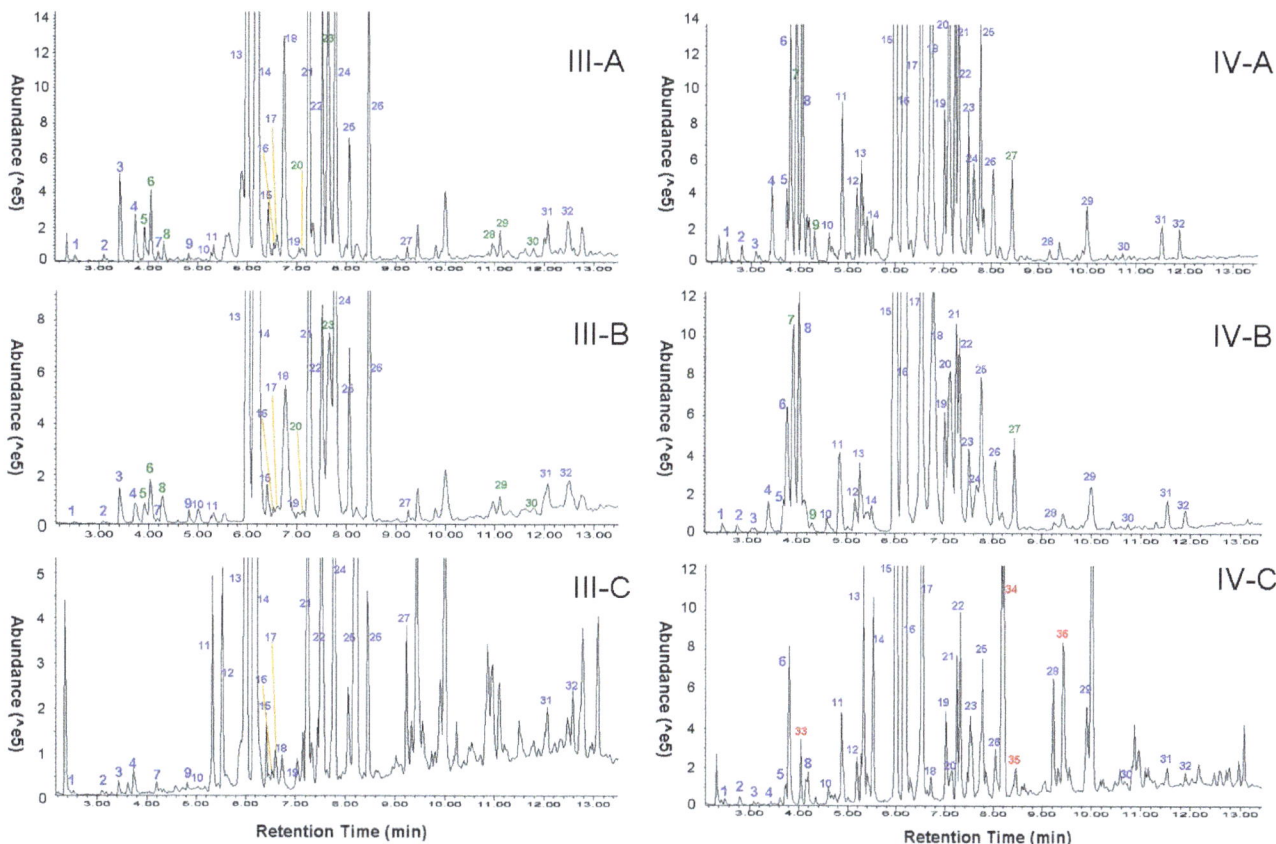

Figure 2: GC-MS total ion chromatography (TIC) of herb essential oils in different experimental condition. III: clary sage, IV: lavender. (A) control sample, stock storage well in the dark and dry environment, (B) after 2hr sunlight treatment sample, (C) H2O added (5:1, EO: H2O) and exposed to sunlight 2hr sample. Blue number: original identified compound before treatment; red number: new appeared compound after treatment; green number: disappeared compound after treatment.

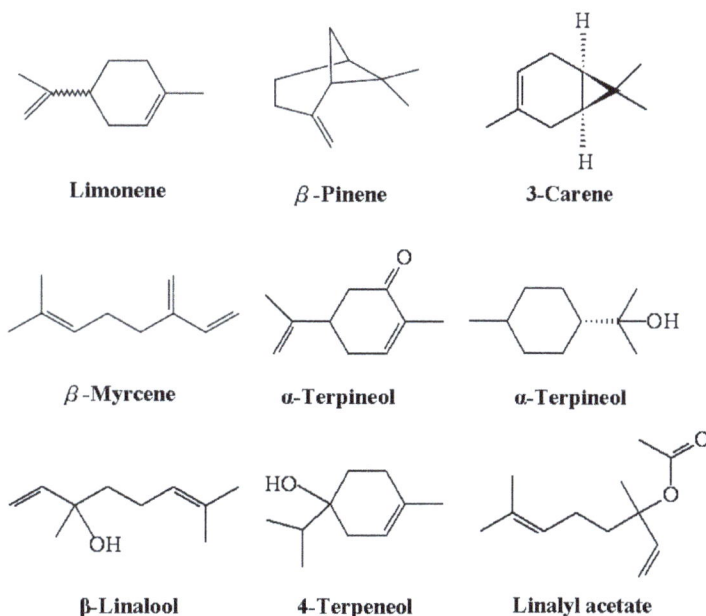

Figure 3: Structures of the major three compositional compounds in citrus and herb essential oils.

Table 3 (left)

PeakNo.[a]	RI[b]	Compounds	Formula	% Area[c] A	B	C
Hydrocarbons						
1	1031	α-Pinene	$C_{10}H_{16}$	0.07	0.02	0.01
2	1121	β-Pinene	$C_{10}H_{16}$	0.06	0.02	0.02
3	1166	β-Myrcene	$C_{10}H_{16}$	0.69	0.32	0.07
4	1213	Limonene	$C_{10}H_{16}$	0.48	0.23	0.16
5	1237	trans-β-Ocimene	$C_{10}H_{16}$	0.32	0.20	NA
6	1254	cis-β-Ocimene	$C_{10}H_{16}$	0.61	0.40	NA
7	1278	p-Cymene	$C_{10}H_{14}$	0.08	0.05	0.05
8	1292	4-Carene	$C_{10}H_{16}$	0.17	0.00	NA
17	1617	β-Elemene	$C_{15}H_{24}$	0.24	0.23	0.14
18	1624	Caryophyllene	$C_{15}H_{24}$	2.71	2.22	0.18
20	1687	cis-β-Farnesene	$C_{15}H_{24}$	0.05	0.09	NA
23	1730	Germacrene D	$C_{15}H_{24}$	2.79	2.18	NA
28	2169	Ledane	$C_{15}H_{26}$	0.18	NA	NA
30	2276	Guaiene	$C_{15}H_{24}$	0.11	0.30	NA
Oxides						
11	1447	Linalool, epoxydihydro-	$C_{10}H_{18}O_2$	0.11	0.09	0.90
12	1476	Linalool oxide	$C_{10}H_{18}O_2$	NA	NA	0.99
32	2368	Sclareoloxide	$C_{18}H_{30}O$	0.52	0.97	0.23
Alcohols						
9	1380	3-Hexen-1-ol	$C_6H_{12}O$	0.06	0.00	0.03
10	1440	1-Octen-3-ol	$C_8H_{16}O$	0.04	0.05	0.07
13	1538	β-Linalool	$C_{10}H_{18}O$	22.68	20.98	21.05
16	1608	4-Terpeneol	$C_{10}H_{18}O$	0.08	0.09	0.04
21	1699	α-Terpineol	$C_{10}H_{18}O$	6.72	5.68	5.18
25	1795	cis-Geraniol	$C_{10}H_{18}O$	0.99	1.07	0.48
26	1839	trans-Geraniol	$C_{10}H_{18}O$	2.02	2.47	0.91
27	1927	2,6-Dimethyl-3,7-octadiene-2,6-diol	$C_{10}H_{18}O_2$	0.09	0.11	0.65
29	2181	β-Spathulenol	$C_{10}H_{18}O$	0.22	0.33	NA
31	2306	β-Eudesmol	$C_{15}H_{26}O$	0.15	0.44	0.09
Esters						
14	1563	Linalyl acetate	$C_{12}H_{20}O_2$	46.43	53.85	51.22
15	1595	Linalyl formate	$C_{11}H_{18}O_2$	0.46	0.35	0.08
19	1672	Geranyl formate	$C_{11}H_{18}O_2$	0.06	0.10	0.18
22	1726	cis-Geranyl acetate	$C_{12}H_{20}O_2$	2.11	1.78	2.17
24	1756	trans-Geranyl acetate	$C_{12}H_{20}O_2$	4.80	4.00	3.95
Unknowns				3.89	1.40	11.16

[a] Peak number labeled on TIC
[b] Retention index relative to C8-C20 n-alkanes on Rtx®-Wax column
[c] A: control sample; B: 2hr sunlight treated sample; C: H₂O added and sunlight treated 2hr sample

Table 3: Chemical compositional changes and relative intensity (%) of clary sage EO in photosensitive experiment analysis by GC-MS.

Table 4 (right)

Peak No.[a]	RI[b]	Compounds	Formula	% Area[c] A	B	C
Hydrocarbons						
1	1031	α-Pinene	$C_{10}H_{16}$	0.12	0.09	0.08
2	1078	Camphene	$C_{10}H_{16}$	0.11	0.08	0.11
3	1121	β-Pinene	$C_{10}H_{16}$	0.08	0.03	0.06
4	1166	β-Myrcene	$C_{10}H_{16}$	0.47	0.35	0.04
5	1213	Limonene	$C_{10}H_{16}$	0.54	0.22	0.18
7	1237	trans-β-Ocimene	$C_{10}H_{16}$	3.31	1.98	NA
8	1254	cis-β-Ocimene	$C_{10}H_{16}$	3.67	2.20	0.49
9	1284	Terpinolene	$C_{10}H_{16}$	0.22	0.12	NA
18	1624	Caryophyllene	$C_{15}H_{24}$	5.83	4.88	0.28
20	1673	trans-β-Famesene	$C_{15}H_{24}$	2.95	2.62	0.52
24	1730	Germacrene D	$C_{15}H_{24}$	0.75	0.64	0.87
26	1779	γ-Muurolene	$C_{15}H_{24}$	0.76	0.93	0.96
31	2238	2-Isopropyl-5-methyl-9-methylenebicyclo[4.4.0]dec-1-ene	$C_{15}H_{24}$	0.24	0.39	0.44
Oxides						
14	1476	Linalool oxide	$C_{10}H_{18}O_2$	0.18	0.31	1.73
29	1991	Caryophyllene oxide	$C_{15}H_{24}O$	0.51	0.84	3.78
Alcohols						
6	1219	Eucalyptol	$C_{10}H_{18}O$	1.86	1.30	1.52
10	1345	1-Hexanol	$C_6H_{14}O$	0.18	0.11	0.13
13	1440	1-Octen-3-ol	$C_8H_{16}O$	0.35	0.26	0.05
15	1538	β-Linalool	$C_{10}H_{18}O$	24.89	26.83	24.80
17	1608	4-Terpeneol	$C_{10}H_{18}O$	8.59	7.22	6.39
19	1668	Lavandulol	$C_{10}H_{18}O$	0.99	0.94	1.01
21	1699	α-Terpineol	$C_{10}H_{18}O$	1.75	1.71	1.06
22	1708	Borneol	$C_{10}H_{18}O$	1.95	1.40	1.60
27	1839	trans-Geraniol	$C_{10}H_{18}O$	0.70	0.88	NA
28	1927	2,6-Dimethyl-3,7-octadiene-2,6-diol	$C_8H_{14}O_2$	0.06	0.18	1.22
30	2149	p-Cymen-7-ol	$C_{10}H_{14}O$	0.04	0.10	0.07
32	2281	α-Bisabolol	$C_{15}H_{26}O$	0.23	0.26	0.28
34	1801	7-Norbornanol	$C_7H_{12}O$	NA	NA	4.61
35	1830	Thymol	$C_{10}H_{14}O$	NA	NA	0.41
Ketones						
33	1256	3-Octanone	$C_8H_{16}O$	NA	NA	0.50
Esters						
11	1359	Octen-1-ol, acetate	$C_{10}H_{18}O_2$	1.23	0.88	1.16
12	1418	n-Hexyl butyrate	$C_{10}H_{20}O_2$	0.46	0.36	0.53
16	1563	Linalyl acetate	$C_{12}H_{20}O_2$	33.81	39.08	36.34
23	1726	cis-Geranyl acetate	$C_{12}H_{20}O_2$	0.92	0.88	1.69
25	1756	trans-Geranyl acetate	$C_{12}H_{20}O_2$	1.65	1.63	1.88
36	1939	cis-3-Hexenyl butyrate	$C_{10}H_{18}O_2$	NA	NA	1.71
Unknowns				0.59	0.28	3.47

[a] Peak number labeled on TIC
[b] Retention index relative to C8-C20 n-alkanes on Rtx®-Wax column
[c] A: control sample; B: 2hr sunlight treated sample; C: H₂O added and sunlight treated 2hr sample

Table 4: Chemical compositional changes and relative intensity (%) of lavender EO in photosensitive experiment analysis by GC-MS.

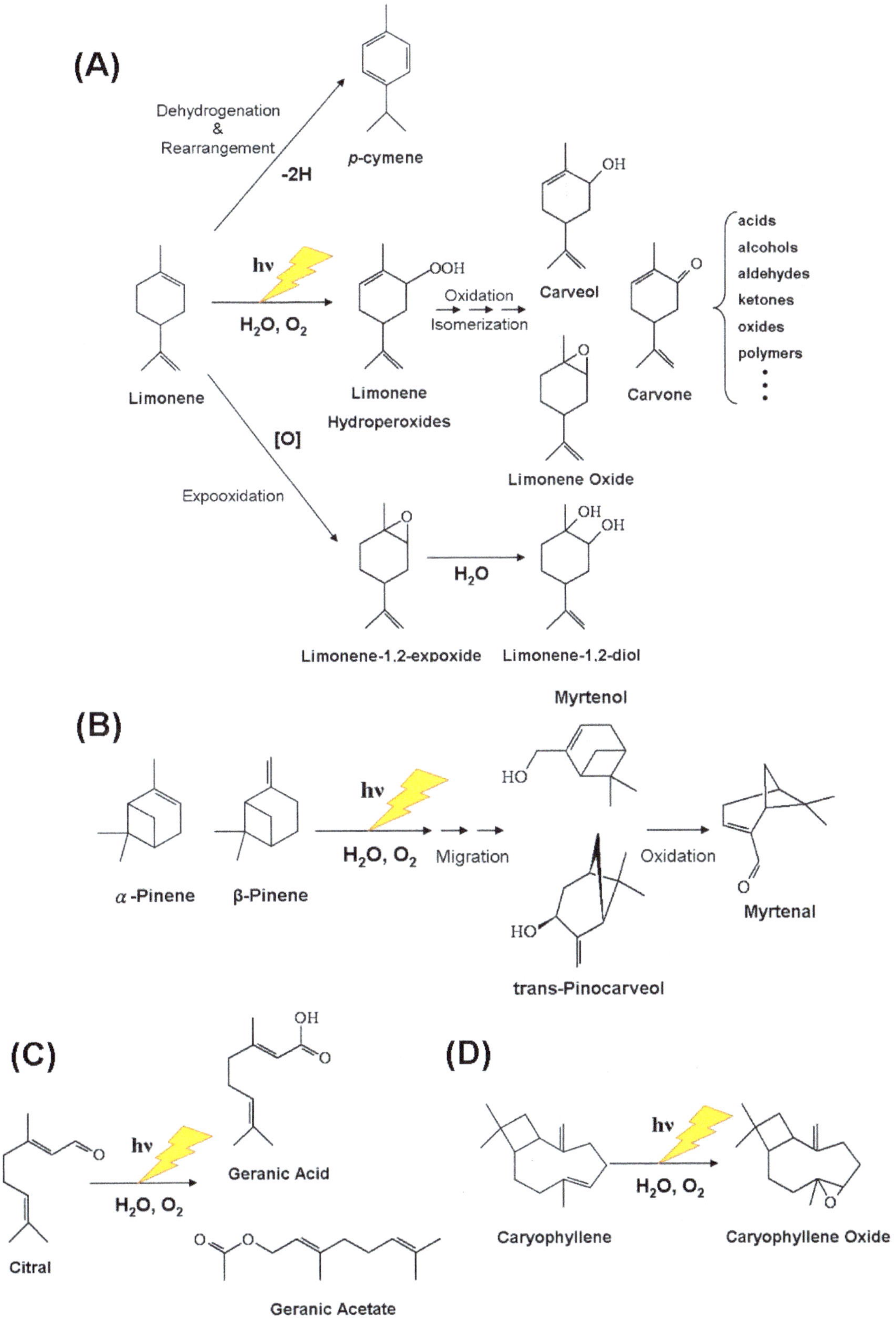

Figure 4: The reaction pathways of terpenes. (A) oxidation of limonene into various oxidative products, (B) the hydroperoxidation pathway of α-pinene and β-pinene , (C) oxidation of caryophyllene into its stable product, caryophyllene oxide, (D) citral oxidative and esterified products.

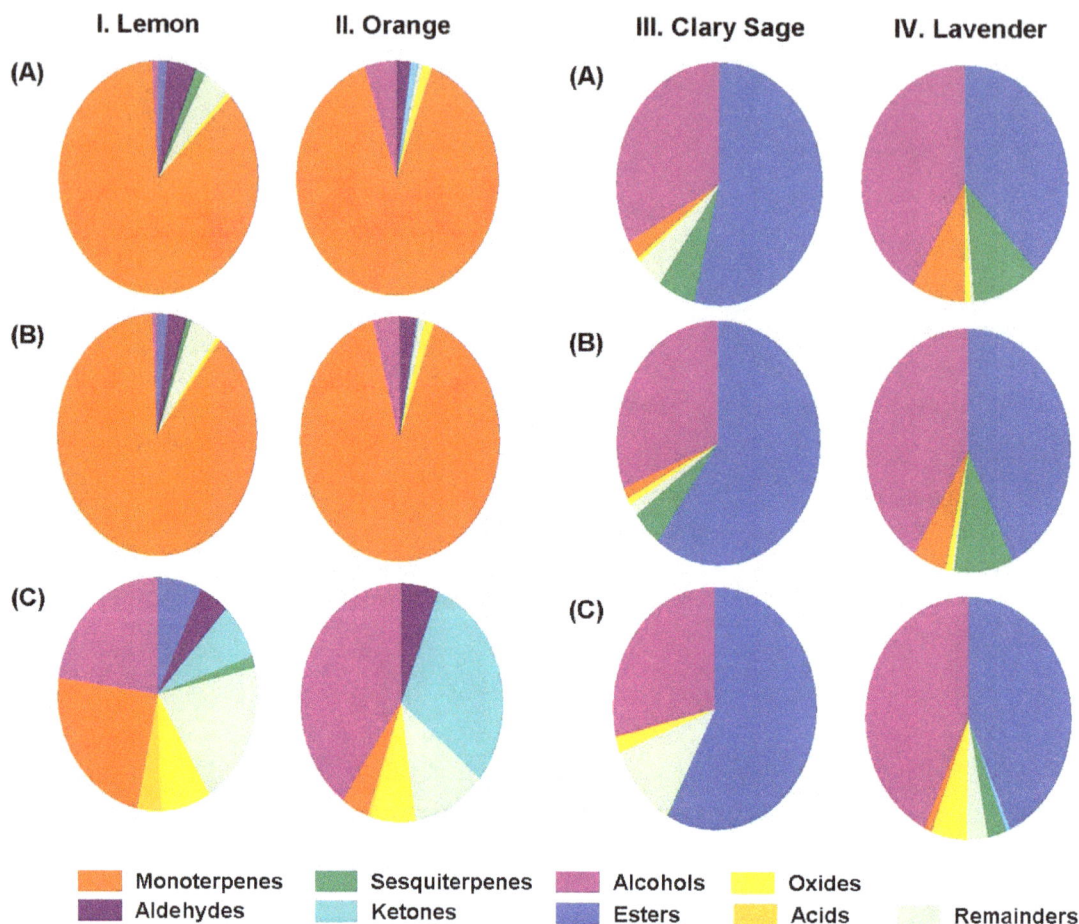

Figure 5: The citrus EO compositional changes in photosensitivity experiment expressed by CC profiles. I: lemon, II: orange, III: clary sage, IV: lavender. (A) control sample, stock storage well in the dark and dry environment, (B) after 2hr sunlight treatment sample, (C) H_2O added (5:1, EO: H_2O) and exposed to sunlight 2hr sample.

of this study suggested herb EOs become a better choice for skin care products. However, there was still some chemical reactions occurred for herb EOs under water-light environment. For instance, β-ocimene decreased 0.93% to 0% and 6.98% to 0.49% in clary sage and lavender, respectively, owing to the unstable properties in air [24]. Caryophyllene, β-famesene and germacrene D were sesquiterpenoids [11] and might converse into oxides or alcohols when H_2O co-exist with sunlight. Compare to lavender oil in group A and C experiment, caryophyllene (peak 18) decreased from 5.83% to 0.28% and caryophyllene oxide (peak 29) increased from 0.51% to 3.78% (Figure 4D). The sub-major compound, β-linalool sometime carries on esterification into linalyl acetate and increases the anticandidal ability [12]. β-linalool revealed a slight decreased in relative abundance and conversed into linalyl acetate for clary sage oil in both light and water-light experiment. According to the CC profiles of clary sage and lavender shown in Figure 5 III and IV, only few compositions varied for both herbs EOs after light/water treatment compare to citrus oils.

In this study, the present of H_2O had a great influence of citrus EOs in composition changes under solarization. The limonene was hydrated, oxidized and isomerized into its alcohol, ketone and oxide products. Those hydroperoxides may cause skin hypersensitivity, respiratory damage and allergic symptom [25-27]. For those reasons, citrus EOs might not be such a suitable material for skin or hair care products and

their storage should avoid to moisture and sunshine. However, herb EOs were much less photosensitive than citrus due to their ester and alcohol composition. This might make herb EOs have more potential to become a stable material for daily used products application. But all type EOs should store in proper environment to reduce the risk of damage, and kept using quality from compositional change.

Conclusion

Citrus and herb EOs are described to contain many bioactive compounds in literature that make them useful in pharmaceutics as an antioxidant, antimicrobial and in aromatherapy as a stress repellent or stimulant [27]. These applications always involve EOs contacting skin or the respiratory tract directly, and the compositional changes may cause body damage or become an ineffective therapy depending on the storage and preparation of the EOs. The knowledge of EO compositional change as a result of different elements (light, water, oxygen and temperature, etc.) can ensure the effectiveness and quality of EOs, in addition to guaranteeing their safety [28,29]. The results in this study provided a suggestion of handing and storage for EO usage.

Acknowledgement

This work was supported by the Nation Science Council of Republic of China (NSC 101-2410-H-182-028) and Chang Gung University (UMRPD5A0071) to Bih-Show Lou.

References

1. Kalemba D, Kunicka A (2003) Antibacterial and antifungal properties of essential oils. Curr Medici Chem 10: 813-829.

2. Hui L, He L, Huan L, XiaoLan L, AiGuo G (2010) Chemical composition of lavender essential oil and its antioxidant activity and inhibition against rhinitis-relate bacteria. African J Microbio Res 4: 309-313.

3. Shabnam J, Ayesha J, Shaista N, Saeed MK, Zaid M, et al. (2014) Phytochemistry, GC-MS analysis, antioxidant and antimicrobial potential of eessential oil from five citrus species. J Agric Sci 6: 201-207.

4. Marti N, Mena P, Canovas JA, Micol V, Saura D (2009) Vitamin C and the role of citrus juices as functional food. Natur Product Commun 4: 677-700.

5. Genovaite B, Ona B, Danute M (2007) Essential oil composition variability in sage (*Salvia offoconalis L.*) CHEMIJA18: 38-43.

6. Luigi M, Alessandro C, Peter QT, Rosaria C, Paola D, et al. (2004) Fast GC for analysis of citrus oils. J Chrom Sci 42: 410-416.

7. Claudia T, Florian CS (2012) Stability of essential oils: a review. Comprehen Review Food Sci & Food Safety 12: 40-53.

8. Crupi ML, Costa R, Dugo P, Dugo G, Mondello L (2007) A comprehensive study on the chemical composition and aromatic characteristics of lemon liquor. Food Chem 105: 771-783.

9. Claudia T, Florian CS (2012) Impact of different storage conditions on the quality of selected essential oils. Food Res Internat 46: 341-353.

10. Oiao Y, JunXie B, Zhang Y, Zhang Y, Fan G, et al. (2008) Characterization of aroma active compounds in fruit juice and peel oil of *Jinchen* sweet orange fruit (citrus sinensis (L.) osbeck) by GC-MS and GC-O. Molecules 13: 1333-1344.

11. Husnu CB, Gerhard B (2010) Handbook of essential oils: science, technology, and application. Taylor & Francis CRC Press.

12. Yana H, Velizar G, Juergen W, Leopold J, Erich S, et al. (2013) Chemical composition ans antifungal activity of essential oil of *Salvia L.* from *Bulgaria* against clinical isolates of *Candida* species. J BioSci Biotech 2: 39-44.

13. Mauuel VM, Yolanda RN, Juana FL, Jose Angel PA (2007) Chemical composition of the essential oils obtained from some spices widely used in mediterranean region. Acta Chim Slov 54: 921-926.

14. Nimet K, Hasan B (2013) Determination of lavender and lavendin cultivars (Lavandula sp.) containing high quality essential oil in Isparat, Turkey. Turkis J Field Crops 8: 58-65.

15. Bhuiyan MN, Begum J, Sardar PK, Rahman MS (2009) Constituents of peel and leaf essential oils of Citrus Medica L. J Sci Res 1: 387-392.

16. Kamal GM, Anwar F, Hussain AI, Sarri N, Ashraf MY (2011) Yield and chemicalcomposition of *Citrus* essential oils as affected by drying pretreatment of peels. Internat Food Res J 18: 1275-1282.

17. ZhiSheng X, QunDi L, ZhiKun L, MingQian Z, XiaoXue Y, et al. (2013) The GC/MS analysis of volatile compenents extracted by different method from *Exocarpium Citri Grandis*. J Analyt Method in Chem 918406.

18. Earl R, John CL (1966) Reaction of the limonene 1,2-oxides. I. The stereospecific reactions of the (+)-*cis*- and (+)-*trans*-limonene 1,2-oxides. Reac of limonene Oxides 31: 1937-1944.

19. Sawamura M, Son US, Choi HS, Kim MSL, Phi NTL, et al. (2004) Compositional changes in commercial lemon essential oil for aromatherapy. Intern J Aromathera 14: 27-36.

20. Leonardo B, Lilian FF, Jacques FD, Edilson CS, Claudio Vinicius FS, et al. (2014) Theoretical study of Δ-3-(+)-carene oxidation. Phys Chem Chem Phys 16: 19376-19385.

21. Lindomar L, Geciane T, Débora O, Leda R, Cláudio D, et al. (2010) Microorganisms screening for limonene oxidation. Cilinc Tecnol Aliment 30: 399-405.

22. Gloria APS, Janeth APV, Claudia COL (2011) Microbial biotransromation of (R)-(+)-limonene by *Penicillium digitatum* DSM 62840 for producing (R)-(+)-terpineol. VITAE 18: 163-172.

23. Ines E, Manef A (2014) Kinetics of extraction of *Citrus aurantium* essential oil by hydrodistillation influence on the yield and the chemical composition. J Mater Environ Sci 5: 841-848.

24. Tomasz B, Agnieszka L, Elwira S, Krystyna SW, Jaroslaw W, et al. (2013) GC-MS analysis of essential oils from *Salvia Officinalis L.*: comparison of extraction methods of the volatile components. Acta Poloniae Pharmaceutica-Drug Res 70: 35-40.

25. Filipsson AF (1996) Short term inhalation exposure to turpentine: toxicokinetics and acute effects in men. Occupational Environ Med 53: 100-105.

26. Hausen B, Reichling J, Harkenthal M (1999) Degradation products of monoterpenes are the sentisizing agents in tea tree oil. Amer J Contact Dermatitis 10: 68-77.

27. Bertuzzi G, Tirillini B, Angelini P, Venanzoni R (2013) Antioxidative action of *Citrus limonum* essential oil on skin. Euorp J Med Plants 3: 1-9.

28. Nadia AA, Enas ND, Hasnaa SA (2013) The effect of enviromental stress on qualitatve abd quantitatve essential oil of aromatic and medicinal plants. Archives Des Sci 66: 100-178.

29. Djilani A, Dicko A (2012) The Therapeutic Benefits of Essential Oils, Nutrition, Well- Being and Health, Dr. Jaouad Bouayed (Eds) InTech.

13

Comprehensive Analysis of Phytopharmaceutical Formulations – An Emphasis on Two-Dimensional Liquid Chromatography

M Luísa S Silva*

Centre of Chemical Research, Autonomous University of Hidalgo State, Carr. Pachuca-Tulancingo km 4.5, 42184 Pachuca, Hidalgo, México

Abstract

In spite of the efforts in increasing the analytical control on herbal medicines, difficulties in the standardization of their components and the absence of a thorough characterization of the chemical composition still persist and constitute obstacles to the safe therapeutic use of these substances. The public consider these products inoffensive and uses them intensively or extensively, including specific groups of population like children and elder people. Thus, it is important to know as much as possible the chemical composition of these phytopharmaceutical products and quantify their pharmacologically active components. The analysis of plant extracts is a difficult task, as they are usually very complex mixtures. The separation power of traditional one-dimensional methodologies is generally inadequate, so two-dimensional separation is required to provide enhanced separation, since it enables to achieve higher overall peak capacities. Furthermore, comprehensive 2D analysis can provide a fingerprinting analysis of samples, allowing the detection of differences in the presence and/or concentration of compounds in complex mixtures.

Keywords: Phytopharmaceutical formulations; Comprehensive analysis; Fingerprinting; Two-dimensional liquid chromatography

Introduction

One-dimensional (1D) separation techniques, such as liquid and gas chromatography and electrophoresis have emerged in the last century, and their sustained development since then has demonstrated their reliability, robustness and broad field of application in Analytical Chemistry. However, as they became mature separation techniques, new challenges were faced by analytical chemists, in what concerns the analysis of more and more complex samples and the need to obtain complete information about sample composition, along with detection and quantification of small differences between samples.

Nowadays, the requirement of a full characterization of samples is observed in a variety of areas, namely food [1,2] and environmental [3,4] control, clinical monitorization [5], drug and biomarker discovery [6,7].

To face these demands, comprehensive separation approaches have been developed, in which all components of the sample are submitted to the separation process. The need to analyze complex samples with hundreds of constituents called for separation systems with larger peak capacities, that is to say, with the ability to separate, with an adequate resolution, a high number of peaks between the first and the last eluting peaks in a chromatogram. One of the ways of increasing the peak capacity is by increasing the separation space (between the first and the last eluting peaks). In order to achieve this, 1D separation has been replaced by multidimensional separation schemes, where two or more separation mechanisms are combined to enable a more efficient separation of the sample's components.

In a two-dimensional (2D) chromatographic system, the final peak capacity is the product of the individual peak capacities of each dimension [8,9]. Though, to attain this high peak capacity, two conditions must occur. First, the resolution obtained on the first dimension should not be lost on the second dimension, so back-mixing of separated peaks during the transference from the first to the second dimension should be avoided. According to Murphy et al. (1998) [10], this is achieved by performing three to four samplings per peak separated on the first dimension. For that reason, in liquid

chromatography × liquid chromatography (LC × LC) modes, the first separation is performed along the dimension that demands the widest space, thus the longest time. Second, the mechanisms used to separate the compounds should be preferably independent (orthogonal). In practice, perfect orthogonal combinations are very difficult to reach, since retention of analytes depends on their physical and chemical properties, which are often correlated. For example, the molecular size of peptides correlates strongly with their hydrophobicity and net charge, thus their separation by ion exchange chromatography on the first dimension, followed by reverse phase liquid chromatography (RP-LC) on the second dimension, will not be completely orthogonal. Therefore, the most orthogonal combination of separation mechanisms is always dependent on the sample's composition and its dimensionality, which is defined as the number of parameters (dimensions) that are necessary to account for the chromatographic properties of its components [11]. In the limit situation, if the two separation mechanisms are the same, the peak capacity will correspond to the sum of the individual peak capacities of each separation [8,9].

The advantages of 2D separations over 1D were already described in the 1970s, by [12,13], for the analysis of complex samples. A significant increase in the number of compounds was obtained if two sufficient orthogonal separation mechanisms were used in the analysis of proteins or plant extracts, compared to 1D separation.

In recent years, several forms of 2D liquid phase separation methods became the basic support in proteomics research [14-21].

*Corresponding author:** Maria Luísa S. Silva, Centro de Investigaciones Químicas, Universidad Autónoma del Estado de Hidalgo, Carr. Pachuca-Tulancingo km 4.5, 42184 Pachuca, Hidalgo, México
E-mail: mluisasilva@portugalmail.pt

Two-dimensional separations have also found application in the metabolomics field, in areas like clinical diagnosis and monitoring, toxicity assessment and nutrigenomics [22,23]. Samples derived from proteomic and metabolomic research are highly complex, thus requiring multidimensional separations for the successful resolution of a huge number of components, present at widely distinct concentrations in the referred samples. Also in the pharmaceutical field, multidimensional separations are required for the identification and quantification of drugs and their metabolites in biological fluids such as plasma or urine [24-28] and for impurity detection and identification as a quality control in drug substances and final products [29,30].

Because of its high resolving power, two-dimensional liquid chromatography (2D-LC) has been one of the most reported and implemented approaches in the comprehensive analysis of complex samples, and where new methodological approaches were developed.

Comprehensive Analysis of Herbal Medicines

The use of herbal medicines

Traditional herbal medicines (THM) have been used for centuries, by billions of people worldwide, in the prevention and treatment of human diseases. WHO estimates that around 80% of the population in developing countries uses traditional medical practices as a primary approach to their health problems [31]. A high percentage of people living in industrialized countries also choose to do so and, in the recent decades, the use of phytopharmaceutical products has been increasing in the developed countries. WHO estimates that over 100 million Europeans are currently users of traditional and complementary medicine (T&CM), with one fifth regularly using these products or practices, and the same number preferring health care which includes T&CM (European Information Centre for Complementary and Alternative Medicine). There are many more T&CM users in Africa, Asia, Australia and North America [32]. This behavior does not imply a break up with conventional medicine, but frequently people use natural products concomitantly with conventional pharmaceuticals. Supplements for obesity control represent one of the phytotherapeutic products most used by population, though some of the substances included in these products are associated with adverse effects [33]. The use of phytotherapeutic products is also increasing in pregnant women, as food supplements [34] and in children [35].

So, the patterns of use of THM vary among countries, and depend on factors such as culture, historical significance and regulations. In some countries, the availability and/or accessibility of conventional health care products is limited, thus people use THM as primarily source of health treatments. This is observed in Africa and some developing countries [36]. People may also use THM due to culture and historical influences. In some Asian countries, where the conventional health system is well-established, the majority of the population commonly uses THM [37]. Finally, people use THM as complementary therapy. This is what happens typically in developed countries, in Europe and North America.

Besides common drives to use THM, other individual reasons underlay its utilization, namely an increased demand for all health services, a desire for more information leading to an increased awareness of available options, an increasing dissatisfaction with existing healthcare services, and a rekindled interest in "whole person care" and disease prevention which are more often associated with T&CM. In addition, T&CM recognizes the need to focus on quality of life when a cure is not possible [38]. In general, there has been an increase in self-healthcare as consumers choose to be more proactive about their own health and people usually have the perception that "natural means safe" (which is not necessarily true) [39].

Regulation and guidelines for quality control of herbal medicines

Most countries have different ways of defining herbal medicines and adopted diverse approaches for licensing, dispensing, manufacturing and trading these products. The situation is more precarious in many developing countries where, despite the large use of THM and empirical knowledge about them, there are very few legislative guidelines to incorporate these products in national drug policies [40]. In fact, some accidents have been reported, associated with the use of herbal formulations, which reflects a lack of attention in this area and/or insufficient research [41].

Therefore, considering the need to assure the quality, safety and efficacy of herbal medicines both in industrialized and developing countries, and in face of the growing use of these products, several official entities have been emitting legislation to regulate the production and quality control of these products [42-47]. According to the legislation, the presentation dossier of these products must include a description of all the components with known therapeutic activity (with molecular and structural formulas, including stereochemistry) as well as the other components. WHO published a series of monographs whose purpose is to provide accurate scientific information on the safety, efficacy and quality control/quality assurance of widely used medicinal plants, and to assist the Member States to develop their own medicinal plant monographs [46]. The European, United States and China Pharmacopoeias have also published monographs on the quality control of herbal materials.

The majority of natural products used for therapeutic purposes is composed by one plant species or by complex mixtures of plants, presented in a natural or in a pharmaceutical form. The tendency is to use standardized plant extracts in the formulation, obtained by a patented extraction process. In either way, they should comply with the requirements that are normally specified in pharmacopoeias. These include assays for standardization of raw materials, such as organoleptic evaluation, micro and macroscopic examination, and assays to identify or to determine a representative profile of the main active constituents of the product, and also to quantify these compounds. Usually, these assays are performed by HPLC. In addition, elaboration of pharmacological, clinical and toxicological assays is intended to assure the three basic parameters that all pharmaceutical products must follow: quality, efficacy and safety.

In spite of the efforts in increasing the control on herbal medicines, their safety remains a problem. First of all, a difficulty in the standardization of their components constitutes an obstacle to the safe therapeutic use of these substances [48]. The chemical composition of herbal formulations depends on several factors, such as climate parameters, botanical species, cultivation conditions, harvest time, anatomical part of the plant used, storage conditions and extraction procedure [41,49]. Additionally, many plants used in phytotherapy don't have their constituents thoroughly characterized under the chemical and pharmaceutical point of view [50]. The majority of plants contain several substances with potential pharmacological activity, which difficult the exact identification of the constituent responsible for the therapeutic or adverse effect [51]. Not rarely, the active compound indicated in the product container is absent in the product itself [52]. By comprising pharmacologically active substances in their composition, they have the potential to induce adverse effects.

Quantitative discrepancies between lots and between similar products can be a source of adverse effects, since the difference between safety and toxicity is often dose-dependent [53]. Also, as they are frequently used along with conventional medicines, interactions may occur, enhancing or diminishing the therapeutic effect of the synthetic drugs. During their production, herbal medicines may suffer adulterations, for example, by addition of potentially toxic, unrelated substances [54,55]. Furthermore, the public consider these products inoffensive, as they are produced from a natural source and are easily reachable without prescription, and uses them intensively or extensively. Their use by specific groups of population, like children and elder people, with sub-optimum metabolism and excretion mechanisms, increases the occurrence of adverse effects and drug interactions. Lastly, the massive use of herbal medicines by population expands these concerns. Hence, for the referred reasons, it is important to know as much as possible the chemical composition of these therapeutic products and quantify their active compounds.

Fingerprinting techniques

Among all the procedures available to perform the quality control of herbal medicines, WHO recommends the use of chromatographic fingerprinting techniques [56]. Other entities, namely the European Medicines Agency [57], the Food and Drug Administration of United States [58] and the State Food and Drug Administration of China [59,60] have also accepted fingerprints as adequate techniques to evaluate the quality of herbal formulations. A fingerprint is a characteristic profile of a sample, which represents its chemical composition in a qualitative and quantitative way. Generally, fingerprints can be obtained using several techniques (chromatographic, electrophoretic and spectroscopic) [41,49,61-63], although the chromatographic ones are more frequently applied. In such cases, the entire chromatogram is the fingerprint and the analysis must fulfil some requirements: (a) a high peak capacity, since all components in the sample are potentially relevant; (b) retention-time stability and (c) detector stability, as recording the chromatograms may take considerable time; (d) a wide dynamic range, because both major and minor components are important and (e) the use of multivariate-analysis techniques (for example partial least squares or principal components regression), to correlate fingerprints with the product specifications and characteristics. The result of a fingerprint analysis may be a collection of chromatograms, a classification of samples or a set of peaks that correlates with a product property [64].

Chromatographic techniques used to obtain fingerprints include thin-layer chromatography (TLC), high-performance liquid chromatography (HPLC), ultra-high performance liquid chromatography (UHPLC), hydrophilic interaction liquid chromatography (HILIC) and gas chromatography (GC).

TLC has the main advantages of readily available technique, easy to implement in the laboratory and flexibility to optimize working parameters, which makes it a good technique for fast screening analysis. On the other hand, its lack of reproducibility and resolution, high compound concentration requirements and the semi-quantitative nature of the technique disables its widespread use for herbal formulations fingerprinting [49]. Nevertheless, some methods have been developed for herbal analysis. For example, a 2D-TLC was developed to discriminate several varieties of *Heracleum spp.* [65]. A review on 2D-TLC in the analysis of secondary plant metabolites illustrates the advantages and drawbacks of using this type of comprehensive chromatographic separation in the analysis of natural compounds [66].

HPLC is the most used technique for herbal products characterization. Its high sensitivity, resolution, automation and the ability to couple the technique with different detectors explains its wide use. For herbal fingerprinting, several methods have been described and reviewed [49,67-71]. Main disadvantages of HPLC rely on the expensive machinery required and large volumes of environmentally unfriendly solvents used, undetected co-eluted compounds and the vulnerability of conventional silica-based columns to extreme mobile phase pHs and high temperatures [49].

The ability of UHPLC to perform fast and high resolution separations, better than HPLC, due to the use of smaller solid phase particles that allow the application of higher pressures, are making this a valuable and emerging technique for herbal medicines analysis. Comparing the results obtained with HPLC and UHPLC for the same samples, there is a reduction in analysis time and an enhancement in selectivity [72-77].

HILIC has demonstrated good performance in herbal fingerprinting due to its good retention and separation of hydrophilic and polar compounds with aqueous and polar organic mobile phases, which are more environmentally friendly compared to normal-phase liquid chromatography. Some applications for herbal products have been reported, associating HILIC and reverse-phase liquid chromatography to create orthogonality and, therefore, improve the separation capacity of the method [78,79].

GC presents high separation efficiency and sensitivity, adequate for the analysis of complex samples. Wang et al. [80] reported the separation of ephedrin-type alkaloids and their enantiomers in raw herbs and commercial herbal products by comprehensive GC × GC analysis. The proposed method showed improved performance, compared to single column GC analysis, providing adequate resolution in the separation of the alkaloids of interest, as well as from potential interference species in the sample matrix. The direct qualification and quantification of the volatile components of *Teucrium chamaedrys* was described by using a comprehensive 2D-GC – time-of-flight mass spectrometry (GC x GC–TOF/MS) system [81]. The GC x GC separation resolved hundreds of components within the sample, and with the separation coupled with TOF/MS for detection, high probability identifications were made for 68 compounds. In spite of the enhanced performance of 2D-GC, the use of high temperatures limits its application in the analysis of herbal formulations to essential oils, whereas the possible degradation of thermo-labile compounds and the necessary volatility of molecules difficult a wider use of the technique, when derivatization is not possible [82-85].

Recently, and to overcome some of the limitations in the conventional chromatographic techniques, there is a trend for miniaturized separation systems, where solvent consumption can be reduced, lessening the impact in environment. Nonetheless, a common drawback in these systems is the limited sample volume that can be introduced, which calls for additional preconcentration steps, increasing the complexity of the procedure and the analysis time [86,87]. The main technique used in miniaturized systems for the fingerprinting herbal formulations is capillary electrophoresis (CE) [88,89]. However, some drawbacks limit its use, namely the occurrence of overlapping peaks in complex samples and some irreproducibility in migration times due to fluctuations in electro-osmotic flow. In order to improve the performance of CE, capillary electrochromatography has been applied in the analysis of herbal samples, combining attractive features from CE and chromatographic techniques. This results in a high-resolution, selective and reproducible technique, adequate for herbal fingerprinting [90-93]. Some difficulties in the use of miniaturized separation systems still remain, such as high backpressure of the columns aggravated by the viscosity of solvents, additional clean-up steps required for samples and low sensitivity, which impairs its wide use in herbal preparation analysis [49].

The need for a comprehensive analysis

Developments in Analytical Chemistry, namely in modern chromatographic, spectrometric and radioimmunological methods allowed to amplify the knowledge on herbal medicines, regarding their chemical composition and the structures of their active components. As a consequence, there is a better quality control of these products. Nevertheless, in the immense plethora of herbal medicines accessible to the public, not all products are controlled as they should be [48]. Besides, even the phytotherapeutic products that comply with all legal requirements should benefit from a more exhaustive analysis of their chemical content, in order to obtain a more complete characterization of their pharmacological profile.

In this scenario, comprehensive analysis provides a useful tool to perform complete characterization of herbal medicines. Unlike the analysis of conventional pharmaceuticals, where the identification and quantification of the synthetic active constituent is relatively easy, the analysis of plant extracts is a difficult task, as they are usually very complex mixtures. The separation power of traditional 1D methodologies is usually insufficient for separation of more complex samples and, in such cases, two-dimensional chromatography provides enhanced separation, due to their higher overall peak capacity. Furthermore, 2D analysis can provide a fingerprinting analysis of samples, allowing the detection of differences in the presence and/ or the concentration of components in complex mixtures. This way, comparison between products of similar nature but different origins or sources can be performed, as well as the evaluation if a manufactured product is off specifications.

The concept of LC × LC was introduced by Erni and Frei [13] and it is based on the use of two chromatographic columns in series. These columns separate compounds in a sample according to two distinct characteristics. Between the first and the second columns there is a modulator that continuously traps small portions of the effluent from the first column and releases them on the second column (Figure 1).

The result is a very detailed 2D-chromatogram presented in three dimensions (two retention time axes and an intensity axis). This form of operation enables an increase in the peak capacity of the whole separation. Theoretically, the total peak capacity equals the peak capacities of the two individual dimensions [94]. Fingerprints obtained with two-dimensional LC surpassed 1D chromatography fingerprints, multidimensional fingerprints generated by hyphenated techniques and multiple fingerprints, and the main disadvantage of the technique is the time needed to perform a full separation [69]. Comprehensive two-dimensional liquid chromatography has been thoroughly reviewed [67-69,95,96].

Typically, phytopharmaceutical compounds are separated by reverse phase (RP) LC. Furthermore, RP columns are generally used in second dimension separations, due to their compatibility with MS. Nevertheless, combinations of normal phase (NP) × RP, HILIC × RP and RP × RP may be used. NP × RP is the combination with the theoretically highest degree of orthogonality, but solvent incompatibility is a serious limitation. Even so, it can be applied if a first dimension column with a small diameter relative to the second dimension column diameter is used [97], reducing the volume of effluent to be injected on the second

column, but also reducing analytical sensitivity. Alternatively, the addition of water to NP solvents may increase their miscibility with RP solvents. HILIC x RP is also a highly orthogonal combination [98]. HILIC uses a polar stationary phase and a water gradient (the mobile phase is what distinguishes HILIC from NP chromatography). The usefulness of this combination was already demonstrated in proteomic separations [99]. Although its application has some difficulties, namely incompatibility of the solvents used in HILIC and RP, impairing their on-line coupling, several improvements have been reported to surpass this limitation [100,101]. As for RP × RP, it has the broadest application, enables gradient elution on both dimensions with fast speed and high peak capacity, but orthogonality strongly depends on the column or mobile phase choice. RP columns need to be prepared with different functionalities in order to make them more orthogonal among themselves. Changing the pH of the mobile phase can also induce drastic alterations of the separation selectivity of RP columns [102]. To reduce time analysis in second dimension and, thus, faster transferences, monolithic columns may also be used. Other parameters to be optimized in method development are the flow rates used in both separations and temperature. Second dimension separation is the rate controlling step in 2D-LC, thus several approaches were reported to speed it up. The use of supra-ambient temperatures on the second dimension separation is one of the possibilities to speed up gradient elution separations. The eluent viscosity decreases with increasing temperatures, enabling the use of higher flow rates at the same pressure. This results in fast and reproducible gradient elutions and a fast column reconditioning [103], thereby shortening the overall comprehensive separation. There is the fear that the use of higher temperatures may induce analyte decomposition, but some studies have shown that even thermally labile pharmaceuticals and peptides can be analyzed with little evidence of decomposition [104].

The choice of eluents has to follow the requirements of on-line LC × LC separations: the eluent used in first dimension must be a weak eluent in the second dimension, both eluents must be miscible and salt precipitation should not occur. The gradients to be used must be optimized, aiming the best compromise between resolution and analysis time.

In sample preparation, if required, extraction techniques such as solid-phase extraction may be applied. On-line extraction is sometimes used, since it enables higher recoveries, as reported [105-109].

Method optimization must contemplate the automation level, considering the application of the developed methodologies in routine analysis. Therefore, the possibility to perform the method procedures on-line, namely sample preparation and cleaning/regeneration of equipment between samples should be explored.

In order to reduce reagent consumption and waste production, a tendency to miniaturize the manifolds has been followed.

Since the optimization of the chromatographic method is a complex task due to the number of parameters affecting the separation, chemometric methods ought to be used, such as factorial design. This approach enables to obtain the combination of variables which provides the best analytical response with a limited number of experiments [110].

Analytical applications of 2D-liquid chromatography in the analysis of phytotherapeutic products

LC x LC techniques have been applied to the analysis of phytotherapeutic products (Table 1). A 2D-LC system with an

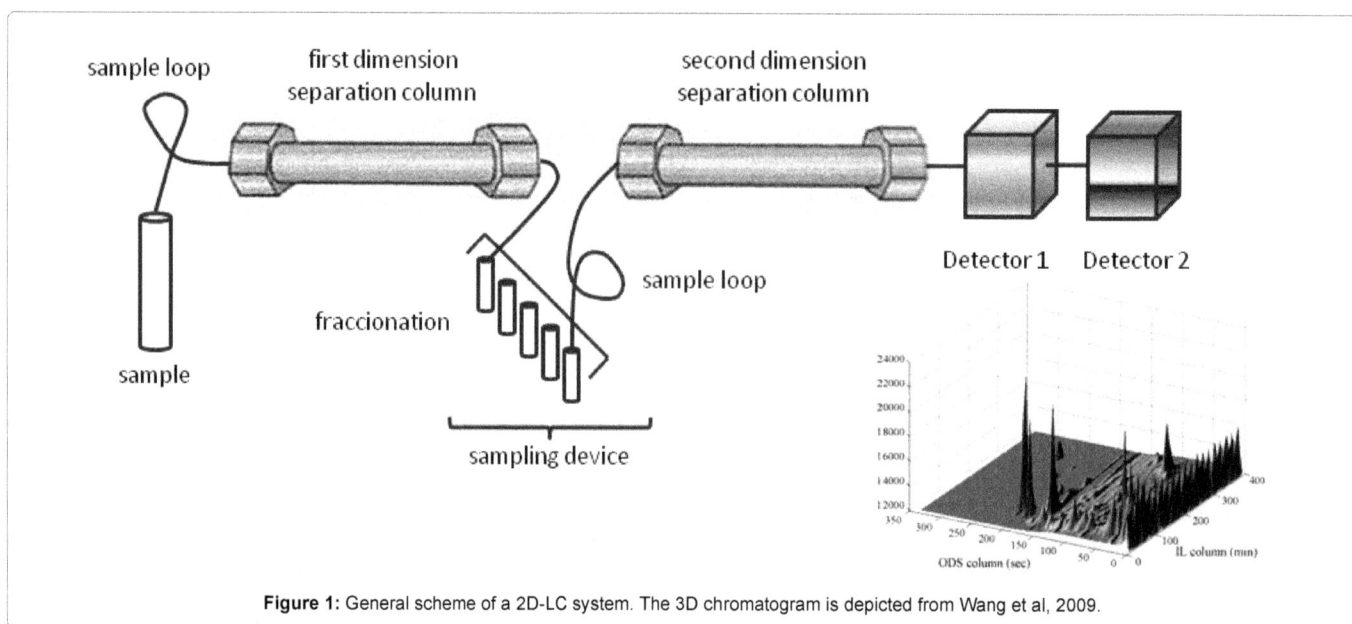

Figure 1: General scheme of a 2D-LC system. The 3D chromatogram is depicted from Wang et al, 2009.

Herbal formulation	Column features	Resolution	Reference
Longdan Xiegan Decoction	1st column: immobilized liposomes 2nd column: RP	more than 50 compounds were separated	Wang et al., 2009
Rhizoma chuanxiong	1st column: cyano 2nd column: ODS	more than 50 compounds were separated	Chen et al., 2004
Saponins in extracts of Panax notoginseng	1st column: HILIC 2nd column: RP	224 saponins were found (peak capacity 10200)	Xing et al., 2012
Extracts of umbelliferae herbs Ligusticum chuanxiong Hort and Angelica sinensis (Oliv.) Diels	1st column: cyano 2nd column: silica monolithic ODS	between 100 and 120 components were efficiently separated in each extract	Hu et al., 2005
Flavonol glycosides in the leaves of Maytenus ilicifolia	1st column: size exclusion 2nd column: RP	better resolution compared to 1D analysis	Souza et al., 2009
Extracts of Psoralea corylifolia	1st column: ion exchange 2nd column: RP	more than 188 components were separated	Chen et al., 2005
Extract of Rheum palmatum L.	1st column: silica-bonded human serum albumin 2nd column: silica monolithic ODS	better resolution compared to 1D analysis	Hu et al., 2006

Table 1: Applications of 2D-LC analysis on the comprehensive characterization of herbal formulations

immobilized liposome chromatography column in conjunction with a RP column was developed for the screening and analysis of the membrane-permeable compounds in a traditional Chinese medicine (Longdan Xiegan Decoction) [111]. The authors reported that more than 50 components in the sample were separated using the developed separation system, which demonstrated to be useful in the identification of membrane permeable natural products in complex matrices such as extracts of traditional Chinese medicines. A different comprehensive 2D-LC separation system based on the combination of a cyano and an ODS column was developed for the separation of components in a traditional Chinese medicine (Rhizoma chuanxiong) [108]. More than 52 components in the extract were resolved and 11 of them were preliminary identified according to their UV and mass spectra. A comprehensive off-line 2D-LC method coupling HILIC and RPLC was developed for the detection of saponins in extracts of Panax notoginseng. The orthogonality of the 2D HILIC × RPLC was up to 81%, and the peak capacity was 10200. In total, 224 saponins were found, and some of them were trace amounts. In addition, a screening table designed by adding molecular weights of possible aglycones and sugars was constructed to help rapidly characterize the saponins using MS information [112]. Hu et al. [113] reported the use of a silica monolithic column in the second dimension separation, for the analysis

of two extracts of umbelliferae herbs. The system performance was superior to that using a conventional column in the second dimension, due to the increase of the elution speed on the second dimensional monolithic column. It was observed that much more peaks for low-abundant components in the herbal extracts could be clearly detected, compared to previous analysis. Between 100 and 120 components were efficiently separated in each extract in shorter analysis time. A 2D (size exclusion – reversed phase) system was developed to analyze flavonol glycosides in the leaves of Maytenus ilicifolia, commonly used in traditional Brazilian medicine [114]. The two-dimensional system showed better performance than 1D analysis and a greater number of compounds could be determined. The strategies used for the analysis of Chinese herbal medicines have been reviewed and include LC × LC, which demonstrated a powerful separation ability, high peak capacity and excellent detectability, compared to single dimension HPLC [86].

Also, there have been improvements in the development of novel 2D-LC systems, by using new interfaces between both columns in order to increase the number of fractions transferred from the first to the second column, reducing the risk to loose resolution of the primary dimension [62] and the application of monolithic columns for phytochemical analysis, which convey some advantages, namely an increase in the flexibility of the technique regarding chemistries and functional compositions of stationary phases, low separation impedance, compatibility with micro and nanoformat separations, low time and labour consumption and cost-efficiency [115]. Monolithic columns are mechanically robust, may be prepared easily in situ and the porous properties can be controlled, its synthesis is flexible (monoliths can be modified with a broad range of functionalities), no void volumes are formed with conventional LC flow rates and the dominance of convection over diffusion in mass-transfer under dynamic conditions allow the use of higher flow rates to carry out the separation without loss of resolution. Since they need short conditioning times, they can be used in successive gradient runs. However, their use is limited by the commercially available column dimensions and they present a lower surface area, which means lower binding capacity [116].

Improvements in instrumentation used for sample preparation in the analysis of plants were also reported [107,109,117-124].

A number of chromatographic applications for the comprehensive analysis of herbal products were reported so far, but the detailed characterization of these formulations could benefit enormously from the implementation of two-dimensional LC since it could provide a sensitive characterization of extracts, in order to allow a chemical fingerprinting which globally addresses their chemical composition. Additionally, it could allow the simultaneous determination of the main compounds associated with therapeutic activity, for quality control, and the differentiation of plant species from different origins.

Conclusions

The small review herein exposed showed that 2D-LC is a very powerful separation technique, but yet not vastly applied in the comprehensive analysis of several complex matrices such as herbal medicines. This may be due to the highly sophisticated technique, the high costs and the requirements for trained personnel to work with this technique, besides the time needed to perform a complete fingerprint analysis. On the other hand, there is no doubt that the comprehensive analysis would benefit this research area and the implementation of two-dimensional-LC methods would greatly expand the analysis possibilities, both qualitative and quantitative, that the phytotherapeutic products, here mentioned, require. In the near future, as the technique becomes more widespread and applied, new applications in the analysis of these samples will surely be developed and reported.

Acknowledgements

The author wishes to thank Professor José Costa Lima and Professor Beatriz Quinaz, from the Faculty of Pharmacy of the University of Porto, Portugal, in recognition for the valuable talks, teaching, supervision of doctoral work and inspiration to continue the research in the field of Pharmaceutical Sciences and Analytical Chemistry.

References

1. Ryan D, Shellie R, Tranchida P, Casilli A, Mondello L, et al. (2004) Analysis of roasted coffee bean volatiles by using comprehensive two-dimensional gas chromatography-time-of-flight mass spectrometry. J Chromatogr A 1054: 57-65.

2. Williams A, Ryan D, Olarte Guasca A, Marriott P, Pang E (2005) Analysis of strawberry volatiles using comprehensive two-dimensional gas chromatography with headspace solid-phase microextraction. J Chromatogr B Analyt Technol Biomed Life Sci 817: 97-107.

3. Focant JF, Sjödin A, Patterson DG Jr (2004) Improved separation of the 209 polychlorinated biphenyl congeners using comprehensive two-dimensional gas chromatography-time-of-flight mass spectrometry. J Chromatogr A 1040: 227-238.

4. Frysinger GS, Gaines RB, Xu L, Reddy CM (2003) Resolving the unresolved complex mixture in petroleum-contaminated sediments. Environ Sci Technol 37: 1653-1662.

5. Song SM, Marriott P, Kotsos A, Drummer OH, Wynne P (2004) Comprehensive two-dimensional gas chromatography with time-of-flight mass spectrometry (GC x GC-TOFMS) for drug screening and confirmation. Forensic Sci Int 143: 87-101.

6. Issaq HJ, Blonder J (2009) Electrophoresis and liquid chromatography/tandem mass spectrometry in disease biomarker discovery. J Chromatogr B Analyt Technol Biomed Life Sci 877: 1222-1228.

7. Wang Y, Zhang J, Liu C, Gu X, Zhang X (2005) Nano-flow multidimensional liquid chromatography with electrspray ionization time-of-flight mass spectrometry for proteome analysis of hepatocellular carcinoma. Anal Chim Acta 530: 227-235.

8. Giddings JC (1984) Two-dimensional separations: concept and promise. Anal Chem 56: 1258A-1260A, 1262A, 1264A passim.

9. Guiochon G, Beaver LA, Gonnord MF, Siouffi AM, Zakaria M (1983) Theoretical investigation of the potentialities of the use of a multidimensional column in chromatography. J Chromatogr A 255: 415-437.

10. Murphy RE, Schure MR, Foley JP (1998) Effect of sampling rate on resolution in comprehensive two-dimensional liquid chromatography. Anal Chem 70: 1585-1594.

11. Giddings JC (1995) Sample dimensionality: a predictor of order-disorder in component peak distribution in multidimensional separation. J Chromatogr A 703: 3-15.

12. O'Farrell PH (1975) High resolution two-dimensional electrophoresis of proteins. J Biol Chem 250: 4007-4021.

13. Erni F, Frei RW (1978) Two-dimensional column liquid chromatographic technique for resolution of complex mixtures. J Chromatogr A 149: 561-569.

14. Bergström SK, Dahlin AP, Ramström M, Andersson M, Markides KE, et al. (2006) A simplified multidimensional approach for analysis of complex biological samples: on-line LC-CE-MS. Analyst 131: 791-798.

15. Cellar NA, Karnoup AS, Albers DR, Langhorst ML, Young SA (2009) Immunodepletion of high abundance proteins coupled on-line with reversed-phase liquid chromatography: a two-dimensional LC sample enrichment and fractionation technique for mammalian proteomics. J Chromatogr B Analyt Technol Biomed Life Sci 877: 79-85.

16. Delahunty C, Yates JR 3rd (2005) Protein identification using 2D-LC-MS/MS. Methods 35: 248-255.

17. Opiteck GJ, Lewis KC, Jorgenson JW, Anderegg RJ (1997) Comprehensive on-line LC/LC/MS of proteins. Anal Chem 69: 1518-1524.

18. Peng J, Elias JE, Thoreen CC, Licklider LJ, Gygi SP (2003) Evaluation of multidimensional chromatography coupled with tandem mass spectrometry (LC/LC-MS/MS) for large-scale protein analysis: the yeast proteome. J Proteome Res 2: 43-50.

19. Regnier F, Amini A, Chakraborty A, Geng M, Li J, et al. (2001) Multidimensional chromatography and the signature peptide approach to proteomics. LC GC 19: 200-213.

20. Shin YK, Lee HJ, Lee JS, Paik YK (2006) Proteomic analysis of mammalian basic proteins by liquid-based two-dimensional column chromatography. Proteomics 6: 1143-1150.

21. Wolters DA, Washburn MP, Yates JR 3rd (2001) An automated multidimensional protein identification technology for shotgun proteomics. Anal Chem 73: 5683-5690.

22. Dunn WB, Ellis DI (2005) Metabolomics: current analytical platforms and methodologies. TrAC 24: 285-294.

23. Dunn WB, Bailey NJ, Johnson HE (2005) Measuring the metabolome: current analytical technologies. Analyst 130: 606-625.

24. Alexander AJ, Ma L (2009) Comprehensive two-dimensional liquid chromatography separations of pharmaceutical samples using dual Fused-Core columns in the 2nd dimension. J Chromatogr A 1216: 1338-1345.

25. Komaba J, Masuda Y, Hashimoto Y, Nago S, Takamoto M, et al. (2007) Ultra sensitive determination of limaprost, a prostaglandin E1 analogue, in human plasma using on-line two-dimensional reversed-phase liquid chromatography-tandem mass spectrometry. J Chromatogr B 852: 590-597.

26. Kueh AJ, Marriott PJ, Wynne PM, Vine JH (2003) Application of comprehensive two-dimensional gas chromatography to drugs analysis in doping control. J Chromatogr A 1000: 109-124.

27. Nishino I, Fujitomo H, Umeda T (2000) Determination of a new oral cephalosporin, cefmatilen hydrochloride hydrate, and its seven metabolites in human and animal plasma and urine by coupled systems of ion-exchange and reversed-phase high-performance liquid chromatography. J Chromatogr B 749: 101-110.

28. Shellie R, Marriott PJ (2002) Comprehensive two-dimensional gas chromatography with fast enantioseparation. Anal Chem 74: 5426-5430.

29. Argentine MD, Owens PK, Olsen BA (2007) Strategies for the investigation and control of process-related impurities in drug substances. Adv Drug Deliv Rev 59: 12-28.

30. Sheldon EM (2003) Development of a LC-LC-MS complete heart-cut approach for the characterization of pharmaceutical compounds using standard instrumentation. J Pharm Biomed Anal 31: 1153-1166.

31. Mahady GB (2001) Global harmonization of herbal health claims. J Nutr 131: 1120S-3S.

32. Barnes PM, Bloom B, Nahin RL (2008) Complementary and alternative medicine use among adults and children: United States, 2007. Natl Health Stat Report : 1-23.

33. Pittler MH, Schmidt K, Ernst E (2005) Adverse events of herbal food supplements for body weight reduction: systematic review. Obes Rev 6: 93-111.

34. Refuerzo JS, Blackwell SC, Sokol RJ, Lajeunesse L, Firchau K, et al. (2005) Use of over-the-counter medications and herbal remedies in pregnancy. Am J Perinatol 22: 321-324.

35. Ernst E (2001) Complementary medicine: its hidden risks. Diabetes Care 24: 1486-1488.

36. Abdullahi AA (2011) Trends and challenges of traditional medicine in Africa. Afr J Tradit Complement Altern Med 8: 115-123.

37. World Health Organization (WHO) (2012) The regional strategy for traditional medicine in the Western Pacific (2011–2020). Manila, WHO Regional Office for the Western Pacific.

38. Roberti di Sarsina P (2007) The Social Demand for a Medicine Focused on the Person: The Contribution of CAM to Healthcare and Healthgenesis. Evid Based Complement Alternat Med 4: 45-51.

39. Mahady GB (1998) Herbal medicine and pharmacy education. J Am Pharm Assoc 38: 274.

40. World Health Organization (WHO) (1998) Traditional medicine. Regulatory situation of herbal medicines. A worldwide review, Genève, Switzerland.

41. Goodarzi M, Russell PJ, Vander Heyden Y (2013) Similarity analyses of chromatographic herbal fingerprints: a review. Anal Chim Acta 804: 16-28.

42. European Medicines Agency (EMEA) (1999) Fixed combination of herbal medicinal products with long term marketing experience – Guidance to facilitate mutual recognition and use of bibliographic data, London, England.

43. European Medicines Agency (EMEA) (2001) Note for guidance on quality of herbal medicinal products, London, England.

44. Poser M (2010) DTCA of prescription medicines in the European Union: is there still a need for a ban? Eur J Health Law 17: 471-484.

45. World Health Organization (WHO) (1992) Division of Drug Management and Policies. Guidelines for good clinical practice (GCP) for trials on pharmaceutical products, WHO Technical Report Series, no 850, Annex 3, Genève, Switzerland.

46. World Health Organization (WHO) (1999) WHO Monographs on selected medicinal plants Volume 1, Genève, Switzerland.

47. World Health Organization (WHO) (2002) Council for International Ethical Guidelines for Biomedical Sciences (CIOMS), International ethical guidelines for biomedical research involving human subjects, Genève, Switzerland.

48. Miller LG, Hume A, Harris IM, Jackson EA, Kanmaz TJ, et al. (2000) White paper on herbal products. American College of Clinical Pharmacy. Pharmacotherapy 20: 877-891.

49. Tistaert C, Dejaegher B, Vander Heyden Y (2011) Chromatographic separation techniques and data handling methods for herbal fingerprints: a review. Anal Chim Acta 690: 148-161.

50. Farah MH, Edwards R, Lindquist M, Leon C, Shaw D (2000) International monitoring of adverse health effects associated with herbal medicines. Pharmacoepidemiol Drug Saf 9: 105-112.

51. Chan TY, Tam HP, Lai CK, Chan AY (2005) A multidisciplinary approach to the toxicologic problems associated with the use of herbal medicines. Ther Drug Monit 27: 53-57.

52. Gilroy CM, Steiner JF, Byers T, Shapiro H, Georgian W (2003) Echinacea and truth in labeling. Arch Intern Med 163: 699-704.

53. Fugh-Berman A (2000) Herb-drug interactions. Lancet 355: 134-138.

54. Ernst E (2002) Adulteration of Chinese herbal medicines with synthetic drugs: a systematic review. J Intern Med 252: 107-113.

55. Saper RB, Kales SN, Paquin J, Burns MJ, Eisenberg DM, et al. (2004) Heavy metal content of ayurvedic herbal medicine products. JAMA 292: 2868-2873.

56. World Health Organization (WHO), 2000, General Guidelines for Methodologies on Research and Evaluation of Traditional Medicine, Geneve, Switzerland.

57. European Medicines Agency (EMEA), 2005, CHMP, Guideline on quality of herbal medicinal medicine products/traditional herbal medicinal products, Committee for medicinal products for human use (CHMP), European Medicines Agency Inspections, July 21, 2005, CPMP/QWP/2819/00 Rev 1, EMEA/CVMP/814/00 Rev 1, 2005.

58. Wu KM, Farrelly J, Birnkrant D, Chen S, Dou J, et al. (2004) Regulatory toxicology perspectives on the development of botanical drug products in the United States. Am J Ther 11: 213-217.

59. State Drug Administration of China (2000) Technical requirements for chromatographic fingerprints of traditional Chinese medicinal injection. Chin Trad Pat Med 22: 671-678.

60. Drug Administration Bureau of China (2002) Requirements for Studying Fingerprints of Traditional Chinese Medicine Injection, Drug Administration Bureau of China, Beijing, China.

61. Alaerts G, Dejaegher B, Smeyers-Verbeke J, Vander Heyden Y (2010) Recent developments in chromatographic fingerprints from herbal products: set-up and data analysis. Comb Chem High Throughput Screen 13: 900-922.

62. François I, de Villiers A, Tienpont B, David F, Sandra P (2008) Comprehensive two-dimensional liquid chromatography applying two parallel columns in the second dimension. J Chromatogr A 1178: 33-42.

63. Liang YZ, Xie P, Chan K (2004) Quality control of herbal medicines. J Chromatogr B Analyt Technol Biomed Life Sci 812: 53-70.

64. van Mispelaar VG, Janssen HG, Tas AC, Schoenmakers PJ (2005) Novel system for classifying chromatographic applications, exemplified by comprehensive two-dimensional gas chromatography and multivariate analysis. J Chromatogr A 1071: 229-237.

65. Cieśla L, Bogucka-Kocka A, Hajnos M, Petruczynik A, Waksmundzka-Hajnos M (2008) Two-dimensional thin-layer chromatography with adsorbent gradient as a method of chromatographic fingerprinting of furanocoumarins for distinguishing selected varieties and forms of Heracleum spp. J Chromatogr A 1207: 160-168.

66. Cieśla L, Waksmundzka-Hajnos M (2009) Two-dimensional thin-layer chromatography in the analysis of secondary plant metabolites. J Chromatogr A 1216: 1035-1052.

67. Dixon SP, Pitfield ID, Perrett D (2006) Comprehensive multi-dimensional liquid chromatographic separation in biomedical and pharmaceutical analysis: a review. Biomed Chromatogr 20: 508-529.

68. Dugo P, Cacciola F, Kumm T, Dugo G, Mondello L (2008) Comprehensive multidimensional liquid chromatography: theory and applications. J Chromatogr A 1184: 353-368.

69. Stoll DR, Li X, Wang X, Carr PW, Porter SE, et al. (2007) Fast, comprehensive two-dimensional liquid chromatography. J Chromatogr A 1168: 3-43.

70. Wu H, Guo J, Chen S, Liu X, Zhou Y, et al. (2013) Recent developments in qualitative and quantitative analysis of phytochemical constituents and their metabolites using liquid chromatography-mass spectrometry. J Pharm Biomed Anal 72: 267-291.

71. Xie P, Chen S, Liang YZ, Wang X, Tian R, et al. (2006) Chromatographic fingerprint analysis--a rational approach for quality assessment of traditional Chinese herbal medicine. J Chromatogr A 1112: 171-180.

72. Kuo CH, Lee CW, Lin SC, Tsai IL, Lee SS, et al. (2010) Rapid determination of aristolochic acids I and II in herbal products and biological samples by ultra-high-pressure liquid chromatography-tandem mass spectrometry. Talanta 80: 1672-1680.

73. Liang X, Zhang L, Zhang X, Dai W, Li H, et al. (2010) Qualitative and quantitative analysis of traditional Chinese medicine Niu Huang Jie Du Pill using ultra performance liquid chromatography coupled with tunable UV detector and rapid resolution liquid chromatography coupled with time-of-flight tandem mass spectrometry. J Pharm Biomed Anal 51: 565-571.

74. Liu H, Du Z, Yuan Q (2009) A novel rapid method for simultaneous determination of eight active compounds in silymarin using a reversed-phase UPLC-UV detector. J Chromatogr B Analyt Technol Biomed Life Sci 877: 4159-4163.

75. Ortega N, Romero MP, Macia A, Reguant J, Angles N, et al. (2010) Comparative study of UPLC-MS/MS and HPLC-MS/MS to determine procyanidins and alkaloids in cocoa samples. J Food Compos Anal 23: 298-305.

76. Qi LW, Wen XD, Cao J, Li CY, Li P (2008) Rapid and sensitive screening and characterization of phenolic acids, phtalides, saponins and isoflavonoids in Danggui Buxue Tang by rapid resolution liquid chromatography-diode array detection coupled with time-of-flight mass spectrometry. Rapid Commun Mass Spectrom 22: 2493-2509.

77. Wang J, Li H, Jin C, Qu Y, Xiao X, et al. (2008) Development and validation of a UPLC method for quality control of rhubarb-based medicine: fast simultaneous determination of five anthraquinone derivatives. J Pharm Biomed Anal 47: 765-770.

78. Chen Y, Bicker W, Wu JY, Xie MY, Lindner W (2010) Ganoderma species discrimination by dual-mode chromatographic fingerprinting: A study on stationary phase effects in hydrophilic interaction chromatography and reduction of sample misclassification rate by additional use of reversed-phase chromatography. J Chromatogr A 1217: 1255-1265.

79. Jin Y, Liang T, Fu Q, Xiao YS, Feng JT, et al. (2009) Fingerprint analysis of Ligusticum chuanxiong using hydrophilic interaction chromatography and reversed-phase liquid chromatography. J Chromatogr A 1216: 2136-2141.

80. Wang M, Marriott PJ, Chan WH, Lee AW, Huie CW (2006) Enantiomeric separation and quantification of ephedrine-type alkaloids in herbal materials by comprehensive two-dimensional gas chromatography. J Chromatogr A 1112: 361-368.

81. Ozel MZ, Göğüş F, Lewis AC (2006) Determination of Teucrium chamaedrys volatiles by using direct thermal desorption-comprehensive two-dimensional gas chromatography-time-of-flight mass spectrometry. J Chromatogr A 1114: 164-169.

82. Bombarda I, Dupuy N, Da JP, Gaydou EM (2008) Comparative chemometric analyses of geographic origins and compositions of lavandin var. Grosso essential oils by mid infrared spectroscopy and gas chromatography. Anal Chim Acta 613: 31-39.

83. Di X, Shellie RA, Marriott PJ, Huie CW (2004) Application of headspace solid-phase microextraction (HS-SPME) and comprehensive two-dimensional gas chromatography (GC × GC) for the chemical profiling of volatile oils in complex herbal mixtures. J Sep Sci 27: 451-458.

84. Qiu Y, Lu X, Pang T, Zhu S, Kong H, et al. (2007) Study of traditional Chinese medicine volatile oils from different geographical origins by comprehensive two-dimensional gas chromatography-time-of-flight mass spectrometry (GC×GC-TOFMS) in combination with multivariate analysis. J Pharm Biomed Anal 43: 1721-1727.

85. Zhu H, Wang Y, Liang H, Chen Q, Zhao P, et al. (2010) Identification of Portulaca oleracea L. from different sources using GC-MS and FT-IR spectroscopy. Talanta 81: 129-135.

86. Li P, Qi LW, Liu EH, Zhou JL, Wen XD (2008) Analysis of Chinese herbal medicines with holistic approaches and integrated evaluation models. Trends Anal Chem 27: 66-77.

87. Marston A (2007) Role of advances in chromatographic techniques in phytochemistry. Phytochemistry 68: 2786-2798.

88. Sun Y, Guo T, Sui Y, Li F (2003) Fingerprint analysis of Flos Carthami by capillary electrophoresis. J Chromatogr B Analyt Technol Biomed Life Sci 792: 147-152.

89. Yu K, Gong Y, Lin Z, Cheng Y (2007) Quantitative analysis and chromatographic fingerprinting for the quality evaluation of Scutellaria baicalensis Georgi using capillary electrophoresis. J Pharm Biomed Anal 43: 540-548.

90. Chen XJ, Zhao J, Wang YT, Huang LQ, Li SP (2012) CE and CEC analysis of phytochemicals in herbal medicines. Electrophoresis 33: 168-179.

91. Han FM, Cheng ZY, Yang X, Chen Y (2000) [Separation and determination of baicalin in prescriptions containing Scutellaria baicalensis Georgi by HPCE]. Se Pu 18: 280-282.

92. Tistaert C, Dejaegher B, Chataigné G, Van Minh C, Quetin-Leclercq J, et al. (2011) Dissimilar chromatographic systems to indicate and identify antioxidants from Mallotus species. Talanta 83: 1198-1208.

93. Xie GX, Qiu MF, Zhao AH, Jia W (2006) Fingerprint analysis of Flos Carthami by pressurized CEC and LC. Chromatographia 64: 739-743.

94. Svec F (1997) Two-Dimensional High-Performance Liquid Chromatography. Chem Educator 2: 1-8.

95. François I, Sandra K, Sandra P (2009) Comprehensive liquid chromatography: fundamental aspects and practical considerations--a review. Anal Chim Acta 641: 14-31.

96. Guiochon G, Marchetti N, Mriziq K, Shalliker RA (2008) Implementations of two-dimensional liquid chromatography. J Chromatogr A 1189: 109-168.

97. Dugo P, Skeriková V, Kumm T, Trozzi A, Jandera P, et al. (2006) Elucidation of carotenoid patterns in citrus products by means of comprehensive normal-phase × reversed-phase liquid chromatography. Anal Chem 78: 7743-7750.

98. Gilar M, Olivova P, Daly AE, Gebler JC (2005) Orthogonality of separation in two-dimensional liquid chromatography. Anal Chem 77: 6426-6434.

99. Boersema PJ, Divecha N, Heck AJ, Mohammed S (2007) Evaluation and optimization of ZIC-HILIC-RP as an alternative MudPIT strategy. J Proteome Res 6: 937-946.

100. Mihailova A, Malerød H, Wilson SR, Karaszewski B, Hauser R, et al. (2008) Improving the resolution of neuropeptides in rat brain with on-line HILIC-RP compared to on-line SCX-RP. J Sep Sci 31: 459-467.

101. Wilson SR, Jankowski M, Pepaj M, Mihailova A, Boix F, et al. (2007) 2D LC separation and determination of bradykinin in rat muscle tissue dialysate with on-line SPE-HILIC-SPE-RP-MS. Chromatographia 66: 469-474.

102. Kovacs JM, Mant CT, Hodges RS (2006) Determination of intrinsic hydrophilicity/hydrophobicity of amino acid side chains in peptides in the absence of nearest-neighbor or conformational effects. Biopolymers 84: 283-297.

103. Antia FD, Horváth C (1988) High-performance liquid chromatography at elevated temperatures: Examination of conditions for the rapid separation of large molecules. J Chromatogr 435: 1-15.

104. Thompson JD1, Carr PW (2002) A study of the critical criteria for analyte stability in high-temperature liquid chromatography. Anal Chem 74: 1017-1023.

105. Ding S, Dudley E, Chen L, Plummer S, Tang J, et al. (2006) Determination of active components of Ginkgo biloba in human urine by capillary high-performance liquid chromatography/mass spectrometry with on-line column-switching purification. Rapid Commun Mass Spectrom 20: 3619-3624

106. Chen L, Song F, Liu Z, Zheng Z, Xing J, et al. (2014) Study of the ESI and APCI interfaces for the UPLC-MS/MS analysis of pesticides in traditional Chinese herbal medicine. Anal Bioanal Chem 406: 1481-1491.

107. Chen X, Kong L, Su X, Fu H, Ni J, et al. (2004) Separation and identification of compounds in Rhizoma chuanxiong by comprehensive two-dimensional liquid chromatography coupled to mass spectrometry. J Chromatogr A 1040: 169-178.

108. Cheng XL, Qi LW, Wang Q, Liu XG, Boubertakh B, et al. (2013) Highly efficient sample preparation and quantification of constituents from traditional Chinese herbal medicines using matrix solid-phase dispersion extraction and UPLC-MS/MS. Analyst 138: 2279-2288.

109. Morgan ED (1991) Chemometrics: Experimental Design, Wiley, Chicester

110. Wang Y, Kong L, Lei X, Hu L, Zou H, et al. (2009) Comprehensive two-dimensional high-performance liquid chromatography system with immobilized liposome chromatography column and reversed-phase column for separation of complex traditional Chinese medicine Longdan Xiegan Decoction. J Chromatogr A 1216: 2185-2191.

111. Xing Q, Liang T, Shen G, Wang X, Jin Y, et al. (2012) Comprehensive HILIC × RPLC with mass spectrometry detection for the analysis of saponins in Panax notoginseng. Analyst 137: 2239-2249.

112. Hu L, Chen X, Kong L, Su X, Ye M, et al. (2005) Improved performance of comprehensive two-dimensional HPLC separation of traditional Chinese medicines by using a silica monolithic column and normalization of peak heights. J Chromatogr A 1092: 191-198.

113. de Souza LM, Cipriani TR, Sant'ana CF, Iacomini M, Gorin PA, et al. (2009) Heart-cutting two-dimensional (size exclusion x reversed phase) liquid chromatography-mass spectrometry analysis of flavonol glycosides from leaves of Maytenus ilicifolia. J Chromatogr A 1216: 99-105.

114. Maruska A, Kornysova O (2006) Application of monolithic (continuous bed) chromatographic columns in phytochemical analysis. J Chromatogr A 1112: 319-330.

115. Saunders KC, Ghanem A, Boon Hon W, Hilder EF, Haddad PR (2009) Separation and sample pre-treatment in bioanalysis using monolithic phases: A review. Anal Chim Acta 652: 22-31.

116. Lanças, F.M., Vilegas, J.H.Y., Queiroz, M.E.C., Marchi, E., 1994. Projeto, construção e aplicação de um sistema "home made" para SFE. I. Aplicações de um sistema em escala analítica no estudo de produtos naturais de origem vegetal. Ciênc. Tecnol. Alim. 14, 45-55.

117. Li DQ, Zhao J, Xie J, Li SP (2014) A novel sample preparation and on-line HPLC-DAD-MS/MS-BCD analysis for rapid screening and characterization of specific enzyme inhibitors in herbal extracts: case study of Î±-glucosidase. J Pharm Biomed Anal 88: 130-135.

118. Liu RL, Zhang J, Mou ZL, Hao SL, Zhang ZQ (2012) Microwave-assisted one-step extraction-derivatization for rapid analysis of fatty acids profile in herbal medicine by gas chromatography-mass spectrometry. Analyst 137: 5135-5143.

119. Sargenti SR, Lanças FM (1994) Design and construction of a simple supercritical fluid extraction system with semi-preparative and preparative capabilities for application to natural products. J Chromatogr A 667: 213-218.

120. Xiao X, Si X, Tong X, Li G (2012) Ultrasonic microwave-assisted extraction coupled with high-speed counter-current chromatography for the preparation of nigakinones from Picrasma quassioides (D.Don) Benn. Phytochem Anal 23: 540-546.

121. Poser M (2010) DTCA of prescription medicines in the European Union: is there still a need for a ban? Eur J Health Law 17: 471-484.

122. European Information Centre for Complementary & Alternative Medicine [web site].

123. Ministério da Saúde do Brasil, Agência Nacional de Vigilância Sanitária. RDC n° 48 de 16 de Março de 2004. Dispõe sobre o registro de medicamentos fitoterápicos. Diário Oficial, Brasília, 18 mar.

124. Opiteck GJ, Ramirez SM, Jorgenson JW, Moseley MA 3rd (1998) Comprehensive two-dimensional high-performance liquid chromatography for the isolation of overexpressed proteins and proteome mapping. Anal Biochem 258: 349-361.

14

Characteristic Fingerprint Analysis of *Mallotus philippinensis* by Ultra Performance Liquid Chromatography Electrospray Ionization Mass Spectrometry

Atul S Rathore, Sathiyanarayanan L and Kakasaheb R Mahadik*

Centre for Advanced Research in Pharmaceutical Sciences, Poona College of Pharmacy, Bharati Vidyapeeth (Deemed University), Pune, Maharashtra, India

Abstract

The developing countries mostly trust on traditional remedies, which include the use of different plant extracts or the bioactive phytoconstituents. For this purpose, analysis such as chemical fingerprinting strongly represents one of the best possibilities in searching of new economic and therapeutically effective plants for medicine. *Mallotus philippinensis* Muell. Arg (Euphorbiaceae) is a large genus of the trees and shrubs mainly distributed in the tropical and subtropical regions and are reported to have widespread range of pharmacological activities. A new, simple and rapid ultraperformance liquid chromatography (UPLC) method with photodiode array (PDA) detector has been developed. Further confirmation was performed by electrospray ionization mass spectrometry (ESI-MS) for the chemical fingerprint analysis in extracts of *Mallotus philippinensis*. The chromatographic separations were obtained on a Waters ACQUITY UPLC BEH C18 (2.1 mm × 50 mm, 1.7 μm) column using a gradient elution with 0.1% (v/v) formic acid/water and acetonitrile as mobile phase at a flow rate of 0.4 mL/min. The UPLC-PDA-ESI-MS characteristic fingerprints were established and 7 characteristic peaks were identified along with 5 unknown peaks within 4.5 min by comparing the retention times, λ max (nm), and MS spectra with the literature data. Therefore, this fingerprint analysis method can be applied for the identification and quality control of *Mallotus philippinensis*.

Keywords: *Mallotus philippinensis*; Euphorbiaceae; UPLC; PDA; Chemical Fingerprinting; ESI-MS

Introduction

Mallotus philippinensis Muell. (commonly called Kamala, Kampillaka, and Kapila) is belonging to the family Euphorbiaceae and is a very common perennial shrub or small tree found in the outer Himalayas ascending to 1500 meters [1,2]. Mature fruits have glandular hairs/trichomes collected as a reddish brown powder by shaking and rubbing the fruits by hand. The collected material is fine, granular powder, dull red, or madder red-colored and easily floats on water. *Mallotus philippinensis (M. philippinensis)* has a widespread natural distribution, from the western Himalayas, Western Ghats through India, Sri Lanka, to southern China, and throughout Malaysia to Australia.

M. philippinensis possess various pharmacological activities such as, antifilarial [3], anticancer, antimicrobial, antiparasitic, anti-inflammatory, immune-regulatory and antioxidant [4-7]. This plant is traditionally used as an anthelmintic, purgative, anti-allergic, carminative and also useful in the treatment of bronchitis, abdominal diseases, spleen enlargement [1,8]. Major phytochemicals present in this genus comprise of different natural compounds, generally phenols, diterpenoids, steroids, flavonoids, cardenolides, triterpenoids, coumarin and isocoumarins [1,9-11]. Current knowledge about this endangered medicinal plant is still inadequate concerning its phytochemistry and biological activity. Although, some researchers have contributed towards the isolation of certain novel constituents and their activity. One of the major anticancer potential phytoconstituent of *M. philippinensis* is Rottlerin (1-[6-[(3-acetyl-2,4,6-trihydroxy-5-methylphenyl)methyl]-5,7-dihydroxy-2,2-dimethyl-2H-1-benzopyran-8-yl]-3-phenyl-2 propen-1-one) [12]. In recent years, chromatographic fingerprinting has been found to be a suitable approach for the quality assessment and control of many natural product/herbal medicines. Few analytical studies were performed on *M. philippinensis* using high-performance liquid chromatography technique [5,7,13-16]. These methods reported chemical fingerprinting limited only for those

peaks which are responsible for particular pharmacological activity and were not extensive. Though, the above methods are very time-consuming with longer run times, low sensitivity and specificity, so they are inappropriate for the analysis of multiple compounds for the authentication of *M. philippinensis*. As a modern-day separation and detection method, ultra-performance liquid chromatography (UPLC) with photodiode array (PDA) coupled with electrospray ionization (ESI) mass spectrometry has attracted increasing attention because of its short analysis time, high throughput, greater resolution, higher peak capacity, lower solvent consumption and extremely high sensitivity. In the present study, UPLC-PDA method have been developed for the chemical fingerprint analysis of chemical constituents in *M. philippinensis*. Also, UPLC coupled with mass spectrometry with electrospray ionization (ESI) interface is defined for the confirmation of phytoconstituents.

Materials and Methods

Reagents and chemicals

Methanol, acetonitrile (LC-MS grade) and formic acid (analytical grade) were of from Fluka (Sigma Aldrich, St. Louis, MO, USA). Ultra high purity water was prepared using a Milli-Q water purification system (Millipore Corporation, Bedford, MA, USA).

*Corresponding author:** Kakasaheb R Mahadik, Centre for Advanced Research in Pharmaceutical Sciences, Poona College of Pharmacy, Bharati Vidyapeeth (Deemed University), Pune-411 038, Maharashtra, India
E-mail: krmahadik@rediffmail.com

M. philippinensis fruits were collected from the plants grown in western ghats of Maharashtra region, India in the month of February. A voucher specimen of *M. philippinensis* (RAAMAP3) was deposited in the Botanical Survey of India, Western Region Centre, Pune, India. The glandular trichomes from fruits were dried at room temperature in shade and ground to powder using a stainless-steel grinder.

Extraction and sample preparation

In the extraction process, 2.5 g powdered glandular trichomes was weighed and extracted with 50 mL methanol in an ultrasonic bath (Equitron, Medica Instrument Mfg. Co., Mumbai, India) at a frequency of 53 KHz and temperature between 40 and 50°C for 60 min. The extract was filtered through a 240 nm pore size filter paper (Whatman, USA). The residues were then re-extracted twice with 25 mL methanol using the same conditions. The extraction solution was collected, pooled and evaporated to dryness under reduced pressure (60 Pa) using rotatory evaporator (Buchi Rotavapor-R2, Flawil, Switzerland) at elevated temperature (50°C) [17]. Dried extract (1 mg) was weighed accurately and dissolved in 1 mL of 100% methanol using ultrasonicator and filtered through a 0.2 μm PTFE syringe filter (Whatman, USA). The filtrate was diluted with methanol to final working solutions and analyzed directly by UPLC-PDA-ESI-MS. Samples were stored in a refrigerator at 4°C until analysis.

Instrumentation

The chromatographic analysis was performed on a Waters ACQUITY UPLC system (Waters, Milford, MA, USA) equipped with autosampler, a binary pump and PDA detector. The chromatographic separation was performed on a Waters ACQUITY UPLC BEH C18 column (2.1 mm × 50 mm, 1.7 μm). The column oven temperature was maintained at 25°C. The mobile phase consisted of 0.1% v/v formic acid in water (A) and acetonitrile (B) with a gradient elution program, i.e., 0–1.5 min, 50-90% B; 1.5–3.5 min, 90% B; 3.5–4.5 min, 90–50% B; and 1 min post-run, 50% B. The flow rate was set at 0.4 mL/min and the injection volume was 5 μL.

For phytoconstituents confirmation, samples were analysed using the same UPLC conditions mentioned above and mass spectrometry detection was conducted on an Applied Biosystem 4000 QTRAP (Applied Biosystems/ MDS Sciex, Concord, ON, Canada) connected to the UPLC system, operating in positive (ESI+) electrospray ionization mode with Q1MS Scan.

The source parameters were: ion spray voltage set at 5500 V; turbo spray source temperature, 450°C; nebulizer gas (gas 1), 50 psi; heater gas (gas 2), 50 psi; collision gas, medium; and the curtain gas (CUR) was set at 20 psi. Q1MS spectra were recorded at unit resolution by scanning in the range of m/z 100–1000 at a cycle time of 9 s with a step size of 0.1 Da. Nitrogen was used as the nebulizer, heater, and curtain gas as well as the collision activation dissociation (CAD) gas. Declustering potential (DP) and entrance potential (EP) were set at 78 and 10 V, respectively as compound dependent parameters. Analyst 1.5.1 software package (AB Sciex) used for instrument control and data processing.

Results and Discussion

Optimization of UPLC conditions

To attain optimum separation in a shorter analysis time, the chromatographic conditions such as the mobile phases, column temperatures, and elution programs were optimized in the initial test. Two different brands of analytical columns, the ACQUITY CSH C18 column (2.1 mm × 100 mm, 1.7 μm) and the ACQUITY BEH C18 column (2.1 mm × 50 mm, 1.7 μm) were compared. The results

showed that the ACQUITY BEH C18 column (2.1 mm × 50 mm, 1.7 μm) produced chromatograms with improve resolution within a shorter run time. In this analysis, different combinations of mobile phases (water-methanol, 0.1% formic acid in water-methanol, water-acetonitrile and 0.1% formic acid in water-acetonitrile) flow rates (0.2, 0.3, and 0.4), column temperatures (25,30 and 40°C) and time were optimized for better chromatographic behavior and appropriate ionization. A suitable chromatographic separation was achieved within 4.5 minutes using gradient elution with 0.1% formic acid in water and acetonitrile at 25°C column temperature with a flow rate of 0.4 mL/min. The PDA detection wavelength was optimized within the range of 190-400 nm, and finally the λ max (nm) with good response for most of the phytoconstituents was selected (Table 1).

Fingerprint analysis of *M. philippinensis*

In order to identify the compounds in the glandular trichomes of *M. philippinensis*, UPLC-PDA and ESI-MS technique displaying the protonated molecular ion ([M+H]⁺, positive ion mode) was used. There are 7 characteristic peaks (peak 2, 3, 4, 7, 8, 10, 11) (Figure 1) were identified along with 5 unknown peaks (peak 1, 5, 6, 9, 12) in the UPLC fingerprint, and their respective retention time, λ max (nm) and protonated molecular ion is shown in Table 1. By comparing the retention times (t_R), λ max (nm) of the UV and ESI-MS spectra with previously reported literature, the compounds were unambiguously recognized. UPLC-UV chromatogram at 290 nm of *M. philippinensis* extract and MS spectra of identified phytoconstituents were shown in Figures 2 and 3 respectively. Unknown peaks were identified due to the lack of appropriate data and literature support.

The major phytoconstituent of *M. philippinensis* is rottlerin (peak 10), which elute at t_R 2.89, exhibit a [M+H]⁺ peak at m/z 517.1 [1,18-20] as shown in Table 1. Two compound (peak 3, 4) were identified at t_R 1.85 i.e., mallotophilippen D (m/z 491.2, [M+H]⁺) [1,21] and 4-hydroxyrottlerin (m/z 532.9, [M+H]⁺) [18,19]. Red compound (peak 2) was identified at t_R 1.67 with [M+H]⁺ m/z 337.2 [20,22]. Similarly at t_R 2.09 (peak 7,8), byakangelicin and mallotus A were exhibit a [M+H]⁺ peak with m/z 335.3 [23] and m/z 339.1 [24,25], respectively. Kamalachalcone C (peak 7) exhibit a [M+H]⁺ peak with m/z 531.4 [18] and elute at t_R 3.36. In the similar way, peak 1 was identified as unknown 1 that exhibit a [M+H]⁺ peak with m/z 322.9 and 581.0 at t_R 1.58. Peak 5,6 (unknown 2 and 3) elute at t_R 1.92 and 2.02 represents [M+H]⁺ peak with m/z 429.1 and 391.1, 407.2, respectively. The other high intensity peaks (peak 9 and 12) identified as unknown 4 and 5 exhibit intense [M+H]⁺ peaks with m/z 461.2 and 338.1, elute at t_R 2.40 and 3.98, respectively.

Peak No.	t_R (min)	m/z [Ion species]	Molecular Formula	Identification	λ max (nm)
1	1.58	322.9, 581.0 [M+H]⁺	-	Unknown 1	276, 384
2	1.67	337.2 [M+H]+	$C_{21}H_{20}O_4$	Red compound	273
3	1.85	491.2 [M+H]+	$C_{30}H_{34}O_6$	Mallotophilippen D	288, 377
4	1.85	532.9 [M+H]+	$C_{30}H_{28}O_9$	4-Hydroxyrottlerin	288, 377
5	1.92	429.1 [M+H]+	-	Unknown 2	273, 290
6	2.02	391.1, 407.2 [M+H]+	-	Unknown 3	292, 351
7	2.09	335.3 [M+H]+	$C_{17}H_{18}O_7$	Byakangelicin	297, 351
8	2.09	339.1 [M+H]+	$C_{20}H_{18}O_5$	Mallotus A	297, 351
9	2.40	461.2 [M+H]+	-	Unknown 4	351
10	2.89	517.1 [M+H]+	$C_{30}H_{29}O_8$	Rottlerin	290, 351
11	3.36	531.4 [M+H]+	$C_{31}H_{30}O_8$	Kamalachalcone C	351
12	3.98	338.1 [M+H]+	-	Unknown 5	339

Table 1: The [M+H]⁺ ions and UV absorption maxima for compounds identified from fruit glandular trichomes extract of *Mallotus philppinensis* by using UPLC/ESI-MS experiment.

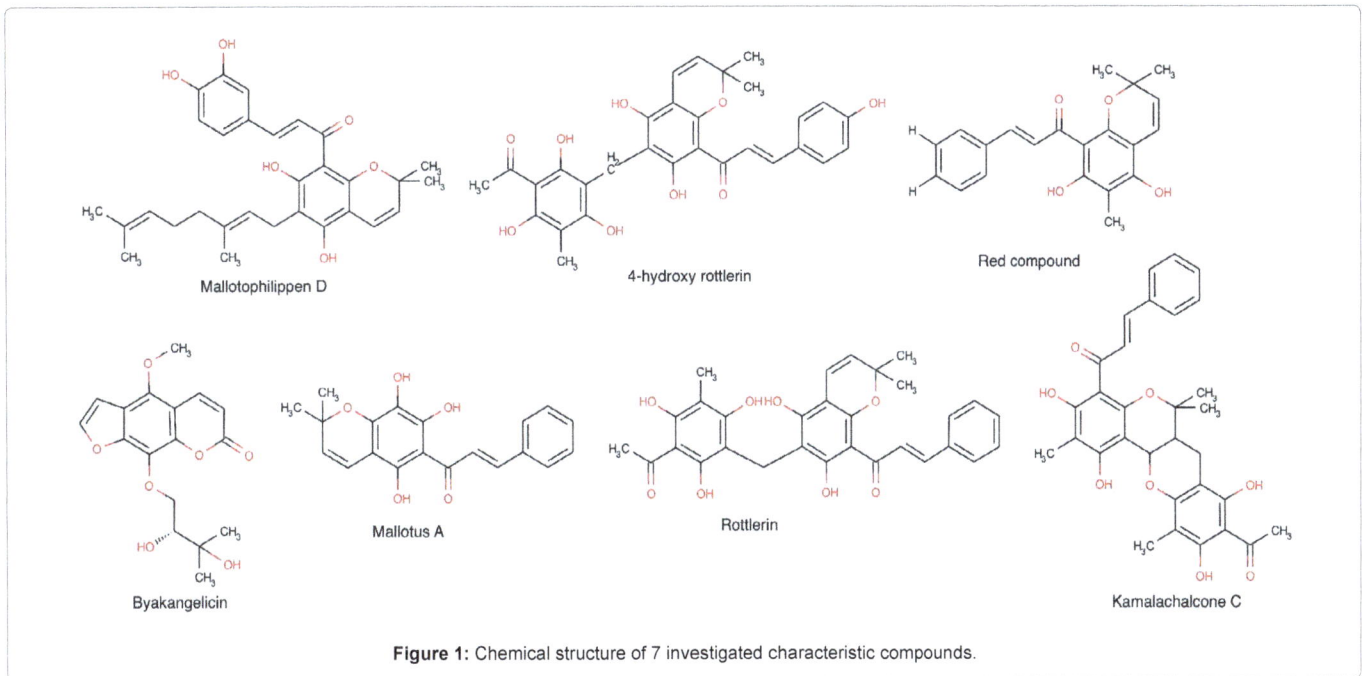

Figure 1: Chemical structure of 7 investigated characteristic compounds.

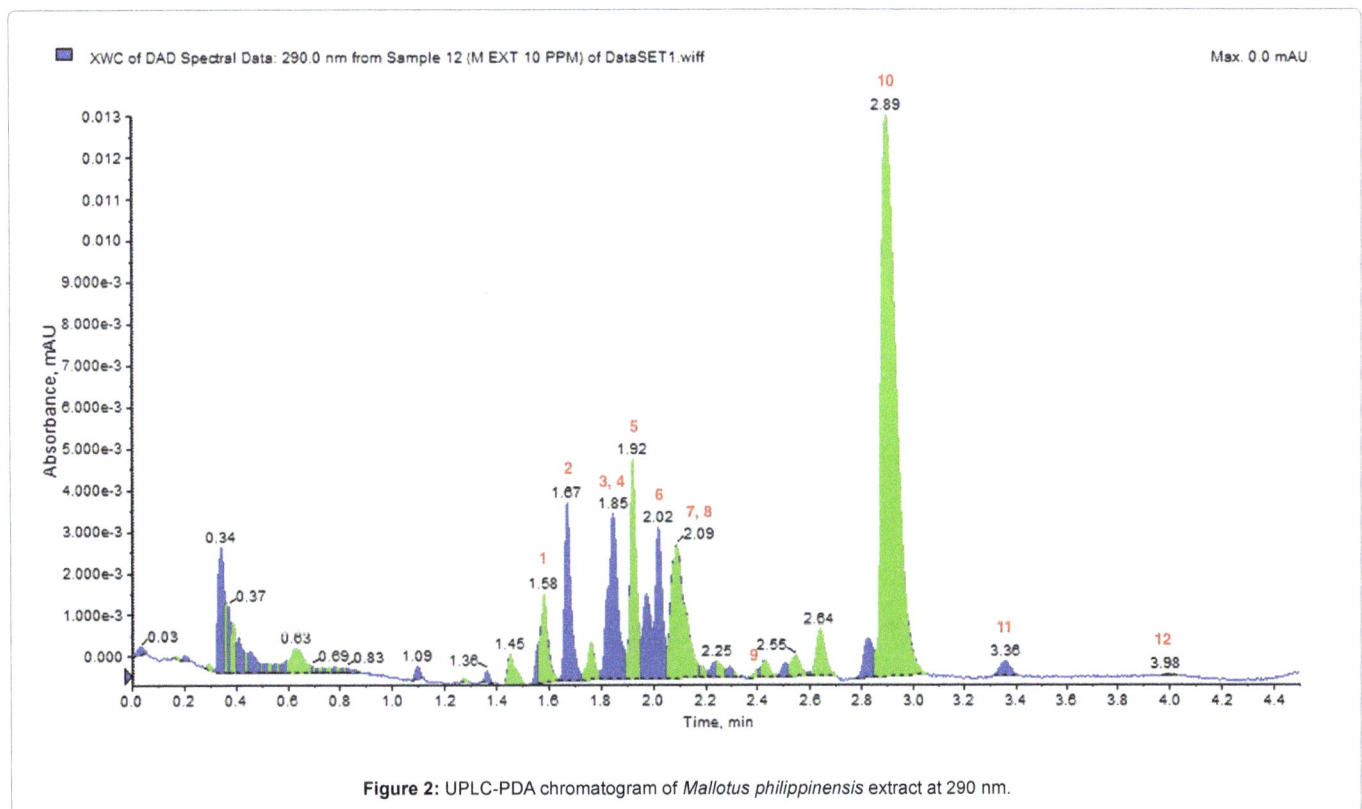

Figure 2: UPLC-PDA chromatogram of *Mallotus philippinensis* extract at 290 nm.

These results suggest that fingerprint analysis method of *M. philippinensis* may be used for authentication and to control adulteration from other species. In addition, this method can also help to standardize the presence of major phytoconstituents in other *M. philippinensis* products.

To summarize, for the first time we have established here a simple and rapid UPLC- PDA and ESI-MS method for the quality control of *M. philippinensis* which plays a significant role in the effectiveness of its clinical purposes. Chromatographic fingerprinting, which has been popular and accepted by experts and scientists universally, is supposed to be a good methodology for quality assessment and control of *M. philippinensis*. The UPLC-PDA combined with ESI-MS offers a powerful instrument for separating and qualifying individual phytoconstituents and creates a characteristic fingerprint profile. The individual peaks are a primary source for quality monitoring of *M. philippinensis* products.

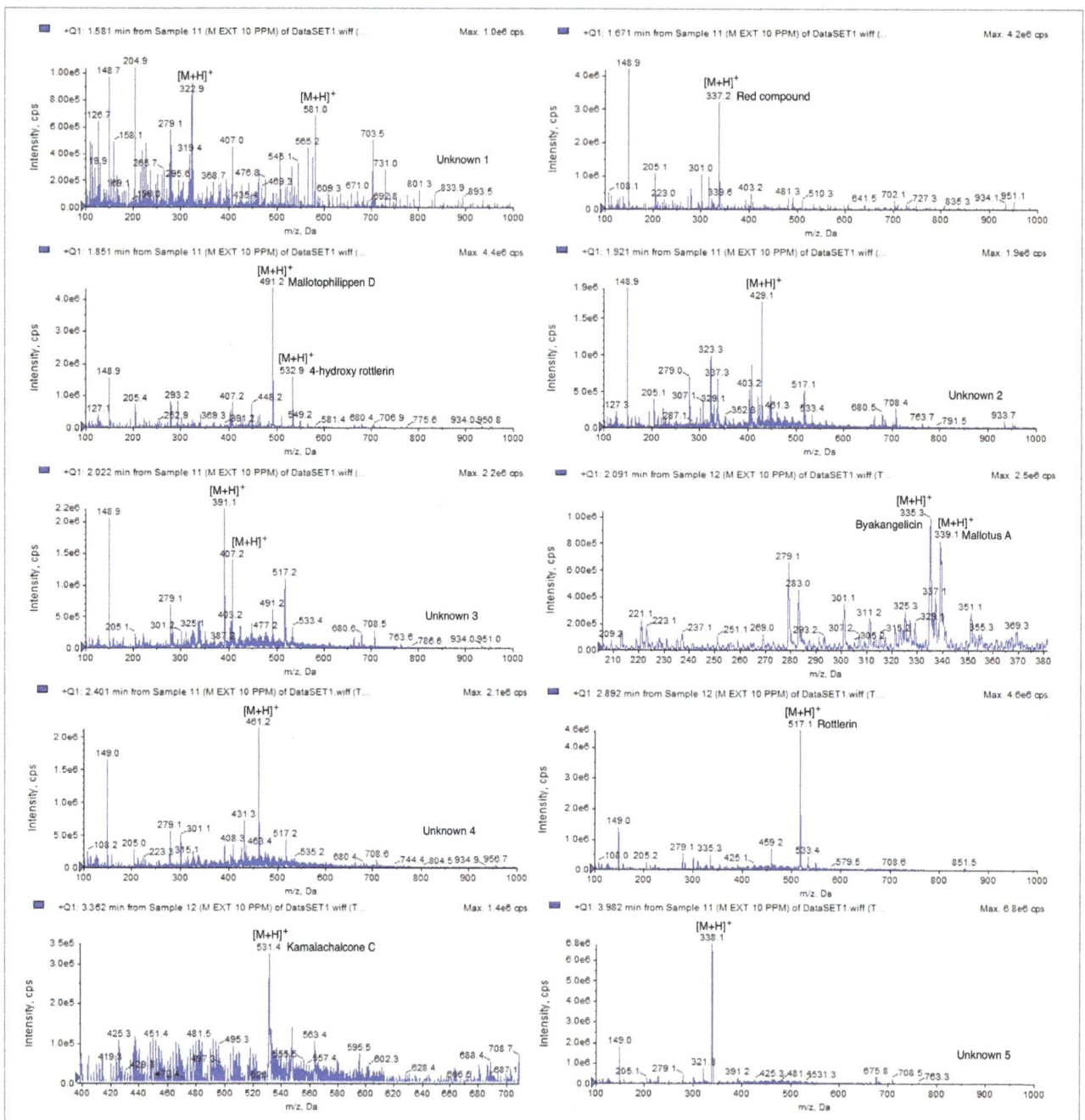

Figure 3: MS spectra of identified compounds.

Acknowledgments

The authors thank Council of Scientific and Industrial Research (CSIR), New Delhi for research fellowship [08/281(0022)/2011-EMR-I] to Atul Singh Rathore. The authors would like to thank Mr. Avinash V. Kapase for his help and support.

References

1. Gangwar M, Goel R, Nath G (2014) Mallotus philippinensis Muell. Arg (Euphorbiaceae): ethnopharmacology and phytochemistry review. BioMed Res Int 2014: 1-13.

2. Widen C, Puri H (1980) Natural occurrence and chemical variability of phloroglucinols in kamala [Mallotus philippensis]. Planta Med 40: 284-287.

3. Singh R, Singhal K, Khan NU (1997) Antifilarial activity of Mallotus philippensis Lam. on Setaria cervi (Nematoda: Filarioidea) in vitro. Indian J Physiol Pharmacol 41: 397-403.

4. Arfan M, Amin H, Karamac M, Kosinska A, Shahidi F, et al. (2007) Antioxidant activity of extracts of Mallotus philippinensis fruit and bark. J Food Lipid 14: 280-297.

5. Bharadwaj R, Chinchansure AA, Kulkarni RR, Arkile M, Sarkar D, et al. (2015) Rottlerin Derivatives and Other Compounds from Mallotus philippinensis Fruits and Their Potential Antimycobactrial Activity. Planta Med Lett 2: e28-e30.

6. Kumar VP, Chauhan NS, Padh H, Rajani M (2006) Search for antibacterial and antifungal agents from selected Indian medicinal plants. J Ethnopharmacol 107: 182-188.

7. Lu QY, Zhang L, Lugea A, Moro A, Edderkaoui M, et al. (2013) Determination of Rottlerin, a Natural Protein Kinases C Inhibitor, in Pancreatic Cancer Cells and Mouse Xenografts by RP-HPLC Method. J Chromatogr Sep Tech 4: 1-4.

8. Chan TK, Ng DS, Cheng C, Guan SP, Koh HM, et al. (2013) Anti-allergic actions of rottlerin from Mallotus philippinensis in experimental mast cell-mediated anaphylactic models. Phytomedicine 20: 853-860.

9. Bandopadhyay M, Dhingra V, Mukerjee S, Pardeshi N, Seshadri T (1972) Triterpenoid and other components of Mallotus philippinensis. Phytochemistry 11: 1511.

10. Nair SP, Rao JM (1993) Kamaladiol-3-acetate from the stem bark of Mallotus philippinensis. Phytochemistry 32: 407-409.

11. Roberts K, Weiss E, Reichstein T (1963) Glycosides and aglycons. CCLII. Cardenolides of the seed of Mallotus philippinensis. Helv Chim Acta 46: 2886-2893.

12. Sharma V (2011) A polyphenolic compound rottlerin demonstrates significant in vitro cytotoxicity against human cancer cell lines: Isolation and characterization from the fruits of Mallotus philippinensis. J Plant Biochem Biotech 20: 190-195.

13. Hoai NN, Dejaegher B, Tistaert C, Hong VNT, Rivière C, et al. (2009) Development of HPLC fingerprints for Mallotus species extracts and evaluation of the peaks responsible for their antioxidant activity. J Pharm Biomed Anal 50: 753-763.

14. Patel Vishal R, Patel Madhavi G, Patel Rakesh K (2009) Development and validation of a RP-HPLC method for quantification of rottlerin in Kamala (Mallotus philippinensis). Drug Invent Today 1: 116-118.

15. Tistaert C, Dejaegher B, Chataigné G, Rivière C, Hoai NN, et al. (2012) Potential antioxidant compounds in Mallotus species fingerprints. Part II: fingerprint alignment, data analysis and peak identification. Anal Chim Acta 721: 35-43.

16. Tistaert C, Dejaegher B, Hoai NN, Chataigné G, Rivière C, et al. (2009) Potential antioxidant compounds in Mallotus species fingerprints. Part I: Indication, using linear multivariate calibration techniques. Anal Chim Acta 652: 189-197.

17. Tistaert C, Chataigné G, Dejaegher B, Rivière C, Hoai NN, et al. (2012) Multivariate data analysis to evaluate the fingerprint peaks responsible for the cytotoxic activity of Mallotus species. J Chromatogr B 910: 103-113.

18. Furusawa M, Ido Y, Tanaka T, Ito T, Nakaya Ki, et al. (2005) Novel, complex flavonoids from Mallotus philippensis (Kamala tree). Helv Chim Acta 88: 1048-1058.

19. Kulkarni RR, Tupe SG, Gample SP, Chandgude MG, Sarkar D, et al. (2014) Antifungal dimeric chalcone derivative kamalachalcone E from Mallotus philippinensis. Nat Prod Res 28: 245-250.

20. Hong Q, Minter DE, Franzblau SG, Arfan M, Amin H, et al. (2010) Anti-tuberculosis compounds from Mallotus philippinensis. Nat Prod Commun 5: 211-217.

21. Daikonya A, Katsuki S, Kitanaka S (2004) Antiallergic agents from natural sources 9. Inhibition of nitric oxide production by novel chalcone derivatives from Mallotus philippinensis (Euphorbiaceae). Chem Pharm Bull 52: 1326-1329.

22. Crombie L, Green C, Tuck B, Whiting D (1968) Constituents of kamala. Isolation and structure of two new components. J Chem Soc C: 2625-2630.

23. Likhitwitayawuid K, Supudompol B, Sritularak B, Lipipun V, Rapp K, et al. (2005) Phenolics with Anti-HSV and Anti-HIV Activities from Artocarpus gomezianus., Mallotus pallidus., and Triphasia trifolia. Pharm Biol 43: 651-657.

24. Harborne JB, Baxter H (1999) The Handbook of Natural Flavonoids. Wiley and Sons.

25. Ahluwalia V, Sharma N, Mittal B, Gupta S (1988) Novel prenylated flavanoids from M. Philippensis Muell Arg. Indian J Chem Sec B 27: 238-241.

15

Determination of Quercetin a Biomarker in Hepatoprotective Polyherbal Formulation through High Performance Thin Layer Chromatography

Arti Gupta[1]*, Navin R Sheth[2], Sonia Pandey[1] and Jitendra Singh Yadav[3]

[1]Maliba Pharmacy College, UkaTarsadia University, Bardoli, Gujarat, India
[2]Department of Pharmacognosy, Saurashtra University, Rajkot, Gujarat, India
[3]Vidyabharati Trust College of Pharmacy, Umrakh, Gujarat, India

Abstract

Background: Quercetin was determined in bioactive fractions of *Ocimum gratissimum*, *Butea monosperma*, *Bauhinia variegate* and polyherbal hepatoprotective formulation by HPTLC method.

Methods: Polyherbal hepatoprotective formulation was developed by using five bioactive fractionated extracts of three plants namely *Butea monosperma*, *Bauhinia variegate* and *O. gratissimum*. All three plants contain quercetin. Chromatographic separation was performed on aluminium foil plates coated with 200 µm silica gel 60F$_{254}$ Linear ascending development with toluene:ethyl acetate:formic acid, 5:4:0.1 (v/v/v) was performed at room temperature (25 ± 2°C) in a twin-trough glass chamber saturated with mobile phase vapor. Compact bands (R$_f$=0.38) were obtained for quercetin. Spectro densitometric scanning was performed in fluorescence mode at 380 nm. The method was validated for precision, recovery, specificity, detection and quantification limits.

Results: Linear regression analysis of the calibration plots showed a good linear relationship (R²=0.9843 ± 0.0001) between peak area and concentration in the range 0.5-2.5 µg/band, respectively. The limits of detection and quantification were 0.089 and 0.26 µ/band. The recovery of the method was 97.33-99.11%.

Conclusion: The above method was a rapid and cost effective quality-control tool for routine analysis of quercetin in herbal extracts and in pharmaceutical dosage form.

Keywords: HPTLC; Quercetin; Liver protective polyherbal formulation

Introduction

Nature still obliges as the man's primary source for the cure of his ailments. Research in preventive medicine showed the importance of functional nutrition in reducing the risk factor of certain chronic diseases. Innate defense system of the human body may be insufficient for the damage caused by continued oxidative stress [1]. Quercetin and other flavonoids, have the structure to act as powerful antioxidants, and have often proven so *in vitro*. Quercetin, being a major constituent of the flavonoid intake, could be a key in fighting several chronic degenerative diseases [2]. Growing scientific evidence has shown adverse side effects, like liver damage and mutagenesis, of synthetic antioxidant [3]. Therefore, recently there has been an upsurge of interest in natural products as antioxidants, as they inhibit the free radical reactions and protect human body from various diseases, such as cancer and diabetes. Recent studies showed that a number of plant products including polyphenolic substances (e.g., gallocatechins, delphinidin, cyanidin, gallic acid, ellagic acid, pelargonidin and sitosterol) and various plants or herbal extracts exert potent antioxidant actions, which are very well known for their healing powers [4].

Stem bark powder is used to apply on injury caused due to axe. Stem juice is applied on goitre of human being. Paste of stem bark is applied in case of body swellings. Bark is acrid, bitter, appetizer, aphrodisiac and laxative, anthelmintic, useful in fractures of the bones, diseases of theanus, dysentery, piles, hydrocele, cures ulcers and tumors. Bark is useful in biliousness, dysmenorrhea, liver disorder, gonorrhea and it also purifies the blood. The ash of young branch is prescribed in combination with other drugs in case of scorpion sting [5].

Bark

Kino-tannic acid, gallic acid, pyrocatechin. The plant also contains palasitrin, and major glycosides as butrin, alanind, allophanic acid, butolic acid, cyanidin, histidine, lupenone, lupeol, (-)-medicarpin, miroestrol, palasimide and shellolic acid [5].

Bauhinia variegata L. was widely used in traditional medicine to treat a wide range of complains. It contained many secondary metabolites which are suitable to be used as medicines. The phytochemical screening revealed that *Bauhinia variegata* contained terpenoids, flavonoids, and tannins, saponins, reducing sugars, steroids and cardiac glycosides. The pharmacological studies showed that *Bauhinia variegate* exertedanticancer, antioxidant, hypolipidemic, antimicrobial, anti-inflammatory, nephroprotective, hepatoprotective, antiulcer, immunomodulating, molluscicidal and wound healing effects [6]. The phytochemical screening of n-hexane chloroform, ethyl acetate and methanolic fractions of *B. variegata* flowers revealed the presence of terpenoids, flavonoids, tannins, saponins, reducing sugars, steroids and cardiac glycosides [7]. Its Constituents isolated from the leaves were included lupeol alkaloids, oil, fat glycoside, phenolics, lignin, saponins, terpinoids, β-sitosterol, tannins, kaempferol-3-glucoside , rutin, quercetin, quercitrin, apigenin, apigenin-7-O-glucoside, amides, carbohydrates, reducing sugars, protein, vitamin C, fibers, calcium and phosphorus [8,9]

*Corresponding author: Arti Gupta, Maliba Pharmacy College, UkaTarsadia University, Bardoli, Gujarat, India, E-mail: arti.gupta@utu.ac.in

O. gratissimum is associated with chemo-preventive, anti-carcinogenic, free radical scavenging, radio protective and numerous others pharmacological use [10]. *O. gratissimum* is used to treat different diseases, e.g., upper respiratory tract infections, diarrhea, headache, ophthalmic, skin diseases, pneumonia, and also as a treatment for cough, fever and conjunctivitis [11,12]. Earlier reports have shown the smooth muscle contracting lipid soluble principles, and antimutagenic activity in organic solvent extracts of *O. gratissimum* leaves [12,13]. This medicinal plant has also potential role as antibacterial [14,15], antifungal [16,17,18], antimicrobial [19,20] and anthelmintic [21]. The aqueous leaf extract and seed oil showed anti-proliferative and chemo-preventive activity on HeLa cells. Nangia-Makker et al. reported that, aqueous extract of *O. gratissimum* leaves inhibits tumor growth and angiogenesis by affecting tumor cell proliferation, migration, morphogenesis, stromal apoptosis and induction of inducible cyclooxygenase (COX-2) [22]. Ursolic acid was determined in dichloromethane and ethyl acetate fractions of methanolic extract of O. gratissimum in previously published report [23].

A limited number of study have been used for the determination of quercetin in *Butea monosperma* bark [24] and *Ocimum gratissimum* leaf and *Bauhinia vareigata* bark [25]. The quercetin concentrations were also determined by UV spectropho-tometry [26], liquid chromatography coupled with different types of detectors [27-31]. Even though these analytical procedures are suitable for the detection of quercetin in samples originating from plants, they have limitations with respect to their applications in the determination of quercetin in plant samples. The reported colorimetric method lacks sensitivity and is tedious and time-consuming. Even though high performance liquid chromatography (HPLC) is a method of choice, it is limited by extensive sample clean-up and requires expensive solvents and longer periods of column stabilization. In comparison to HPLC, HPTLC is a versatile analytical technique that requires less expensive instrumentation and expertise. The present study was carried out to develop a rapid, sensitive and accurate analytical method for estimation of quercetin in bioactive fractions of plant extracts and its pharmaceutical dosage form (hepatoprotective tablet formulation) for the routine analysis of a large number of plant extract samples and their formulations.

Material and Methods

Apparatus

HPTLC system (Linomat 5, Camag, Switzerland) automatic sample applicator, TLC scanner IV (Camag), flat bottom and twin- trough developing chamber (15 × 10 cm), pre-coated silica gel aluminum plate (E. Merck, Darmstadt, Germany), electronic analytical balance, Shimadzu (AUX-220), micro syringe (100 mL) (Hamilton).

Reagents and standards

Quercetin was purchased from Yucca enterprises, Wadala, Mumbai and methanol AR grade from S.d. fine-Chem Ltd., Mumbai.

Plant materials

Polyherbal hepatoprotective tablet was prepared by using fractions obtained from alcoholic extracts of *Butea monosperma*, *Bauhinia variegata* stem bark and *O. gratissimum* leaves (Figure 1). All these ingredients were collected from Maliba Pharmacy College campus and were authenticated by Prof. Minoo H. Parabia, Department of Bioscience, Veer Narmad South Gujarat University, Surat. Voucher specimen (No: MPC/13032010/01, 02 and 03) has been deposited in the Department of Bioscience.

Extraction and fractionation procedures:

The dried and powdered material of each plant (500 g) was extracted with methanol at room temperature for three weeks with shaking and stirring. Combined methanolic extracts were evaporated to dryness under reduced pressure below 40°C and then dissolved in distilled water and subjected to solvent-solvent fractionation.

***Butea monosperma* (Lam.) Taub:** Methanolic extract obtained was fractionated with petroleum ether, benzene, chloroform and acetone (AcO) in the order of increasing polarity to obtain respective fractions [32].

***Bauhinia variegate* L.:** Methanolic extract was fractionated with hexane, ethyl acetate (EtOAc) and n-butanol (n-ButOH) in the order of their increasing polarity to obtain respective fractions [18].

***Ocimum gratissimum* L.:** Alcoholic extract was fractionated with hexane, dichloromethane (DCM) and ethyl acetate (EtOAc) in the order of their increasing polarity to obtain respective fractions [33].

Each fraction was concentrated to dryness under reduced pressure and below (40-50°C) on a rotary evaporator to give Acetone fr. of *Butea monosperma* [yield 9.4% w/w], Ethyl acetate fr. [yield 2.2% w/w] and n-butanol fr. [yield 5.0% w/w] of *Bauhinia variegata* L. and dichloromethane fr. [yield 4.2% w/w] and ethyl acetate fr. [yield 4.8% w/w] of *Ocimum gratissimum* L. respectively.

Establishment of qualitative and quantitative phytoprofile of fractionated extracts

Qualitative phytochemical analysis: Each fraction was subjected to various qualitative chemical tests using reported methods to determine the presence or absence of metabolites viz., alkaloids, tannins, flavonoid, steroid, terpernoids and phenolic compounds, etc. [34].

Chemical test for flavonoids: Chemical tests were performed for flavonoids according to Macdonald et al. [35].

Quantitative phytochemical analysis

Determination of total phenols: Each sample was mixed with 1 mL Folin-Ciocalteu reagent and 0.8 mL of 7.5% Na_2CO_3. The resultant mixture of was measured at 765 nm after 2 hr at room temperature. The mean of three readings was used and the total phenolic content was expressed in milligram of gallic acid equivalents/1 g extract. The coefficient of determination was found to be $r^2=0.992$ [36].

Determination of total flavonoids: Standard quercetin was used to make the calibration curve [0.04, 0.02, 0.0025 and 0.00125 mg/mL in 80% ethanol (v/v)]. The standard solutions and test samples (0.5 mL) of each fraction was mixed with 1.5 mL of 95% ethanol (v/v), 0.1 mL of 10% aluminum chloride (w/v), 0.1 mL of 1 mol/L sodium acetate and 2.8 mL water. The volume of 10% aluminum chloride was substituted by the same volume of distilled water in blank. After incubation at room temperature for 30 min, the absorbance of the reaction mixture of each

Figure 1: A: *Butea monosperma* stem bark, **B:** *Bauhinia variegata* stem bark, **C:** *O. gratissimum* leaves.

sample and standard solution were measured at 415 nm. The mean of three readings was used and the total flavonoid content was expressed in milligram of quercetin equivalents/1 g extract. The coefficient of determination was r²=0.99020 [37].

Sample preparation

Preparation of standard solutions of quercetin: Stock solution of quercetin was prepared by dissolving 50 mg quercetin in 100 mL of methanol (500 µg/mL). Standard solutions of concentration 0.5, 1.0, 1.5, 2.0 and 2.5 in µg/mL were prepared by dilution of the stock solution with methanol.

Samples preparation from each plant extracts fractions: Accurately weighed 100 mg of each, acetone fraction of *Butea monosperma*, ethyl acetate and n-butanol fractions of *Bauhinia variegata* and dichloromethane and ethyl acetate fractions of *Ocimum gratissimum* was transferred to separate 10 mL volumetric flask and dissolved in 10 mL of methanol. These solutions were sonicated for 10 minutes and filtered through Whatman No. 1 filter paper to get solution containing 10 mg/mL each.

Sample preparation from polyherbal tablet: Polyherbal tablets equivalent to about 100 mg of mixture of fractionated extracts of *Butea monosperma*, *Bauhinia variegata* and *Ocimum gratissimum* was weighed and transferred to 10 mL volumetric flask containing 10 mL methanol to get solution containing 10 mg/mL. The resulting solution was centrifuged at 3000 rpm for 5 min and supernatant was analyzed for quercetin content [38].

Instrumentation and chromatographic conditions

HPTLC was performed on 15 cm × 10 cm aluminum backed plates coated with silica gel 60F254 (Merck, Mumbai, India). Standard solution of quercetin and sample solution were applied to the plates as bands 6.0 mm wide, 9.2 mm apart, and 15.0 mm from the bottomedge of the same chromatographic plate by use of a Camag (Muttenz, Switzerland) Linomat V sample applicator equipped with a 100 µL Hamilton (USA) syringe. Ascending development to a distance of 80 mm was performed at room temperature (28 ± 2°C), with toluene: ethyl acetate: formic acid, 5:4:0.2 (v/v/v), as mobile phase, in a Camag glass twin-trough chamber previously saturated with mobile phase vapour for 20 min. After development, the plates were dried with a hair dryer and then scanned at 380 nm with a Camag TLC Scanner with WINCAT software, using the deuterium lamp. The method was validated according to the ICH guidelines [11].

Calibration curve of quercetin

Different volumes of stock solution (500 µg/mL) were spotted on the TLC plate to obtain concentration 0.5, 1.0, 1.5, 2.0 and 2.5 µg /spot of quercetin, respectively. The data of peak areas plotted against the corresponding concentration.

Method Validation

The proposed method was validated as per ICH guidelines [39]. Samples were prepared as per the earlier adopted procedure given in the experiment.

Linearity and range

Linearity is expressed in terms of correlation coefficient of linear regression analysis. The linearity response was determined by analyzing 5 independent levels of calibration curve in the range of 0.5, 1.0, 1.5,

2.0 and 2.5 µg /spot of quercetin respectively. The calibration curve of absorbance vs. concentration was plotted and correlation coefficient and regression line equations were determined.

Precision

Result of precision should be expressed as relative standard deviation (% R.S.D) or coefficient of variance (% C.V.).

Repeatability

Standard solutions were applied by Linomat 5 automatic sample applicator. Sample was spotted seven times for repeatability studies. The peak area obtained with each solution was measured and % C.V. was calculated.

Intraday precision

Mixed solution containing (1.0-2.0 µg/spot) of quercetin was analyzed three times on the same day and % C.V. was calculated.

Interday precision

Mixed solution containing (1.0-2.0 µg/spot) of quercetin was analyzed on three different days and % C.V. was calculated.

Accuracy

It was determined by calculating the recovery of quercetin by standard addition method.

Recovery studies

The accuracy of the method was established by performing recovery experiments at three different levels using the standard addition method. In 1 µl (1 µg/mL) of samples, known amounts of quercetin (0.5, 1.0 and 1.5 µg/spot) standard were added by spiking. The values of percent recovery and average value of percent recovery for quercetin were calculated.

Limits of detection and limit of quantization

The LOD and LOQ were estimated from the set of 5 calibration curves. The LOD and LOQ may be calculated as

LOD=3.3 × (SD/Slope)

LOQ=10 × (SD/Slope)

Where,

SD=Standard deviation of the Y- intercepts of the 5 calibration curves.

Slope=Mean slope of the 5 calibration curves

Specificity

The specificity of the method was ascertained by analyzing the standard drug and extract. The spot for quercetin in the sample was confirmed by comparing the R_f values and spectra of the spot with that of the standard. The peak purity of the quercetin was assessed by comparing the spectra at three different levels, viz. peak start, and peak apex and peak end positions of the spot.

Results and Discussion

Phytochemical screening

Preliminary phytochemical screening of alcoholic extract and its fractions showed the presence of flavonoids, steroids, terpenoids, tannins and phenolic compounds. The chemical tests analysis demonstrated that AcO fraction of *Butea monosperma*, EtOAc fraction and n-BtOH fraction of *Bauhinia variegata*, DCM and EtOAc fractions of *Ocimum gratissimum* were rich in phenolic compounds. The phenolic content in *Butea monosperma* (acetone fraction), *Bauhinia variegata* (Ethyl acetate and n-butanol fractions) and *Ocimum gratissimum* (Dichrolomethane and ethyl acetate fractions) were found to be 452 ± 1.6, 712.4 ± 2.4, 442.5 ± 1.1, 735 ± 2.1 and 1365 ± 1.4 mg gallic acid/1 gm fraction respectively. The flavonoid content in *Butea monosperma* (acetone fraction), *Bauhinia variegate* (Ethyl acetate and n-butanol fractions) and *Ocimum gratissimum* (Dichrolomethane and ethyl acetate fractions) were found to be 251 ± 1.8, 417 ± 2.2, 227 ± 3.2, 394.5 ± 2.4 and 717 ± 5.2 mg quercetin/1 gm fraction respectively. The phenol and flavonoid contents are responsible for hepatoprotective activity; hence these solvent fractions were selected for further study.

Optimization of mobile phase

Various ratios of solvents were tried as a mobile phase and optimum mobile phase was selected was toluene:ethyl acetate:formic acid, (5:4:0.1 v/v/v). This mobile phase allowed good resolution, dense, compact and well-separated spots at R_f value 0.38. Wavelength 380 nm was used for quantification of the drug. Since there is only one peak seen, is shown in Figure 3.

Quantification by HPTLC Method development

In HPTLC chromatogram, all tracks for standard quercetin at wave length 380 nm were shown in Figure 2. The R_f value of standard quercetin was found to be 0.38 and peak area was 9726 (Figure 3).

Method validation

Linearity and range: The linearity was determined for both drugs at five different concentration levels. The linearity of quercetin was in the range of 0.5-2.5 µg/spot and calibration curves are shown in Figure 3. Correlation co-efficient for calibration curve of quercetin was 0.9843.

The regression line equation for quercetin is as follows: y=9076x+6315 (Figure 4).

Precision

Repeatability: The data for repeatability are shown in Table 1. The % C.V for repeatability was found to be 0.5 (24341 ± 125) (Table 1)

Intra-day precision: The data for intra-day precision for quercetin are shown in Table 2. The % C.V of quercetin was found to be in range of 0.69%-0.97% (Table 2).

Inter-day precision: The data for inter-day precision for quercetin are shown in Table 3. The % C.V of quercetin was found to be in range of 0.77%-1.50% (Table 3).

Accuracy: Accuracy of the method was confirmed by recovery at three level of standard addition. Percentage recovery for quercetin was found to be in range of 97.33%-99.11%. The results are shown in (Table 4).

Limits of detection (LOD) and limit of quantitation (LOQ): Limit of detection and quantitation were determined by equation LOD=3.3 × (SD/s) and LOQ=10 × (SD/s) LOD and LOQ results are shown in (Table 5).

Estimation of Quercetin in Fractionated Extracts of *Butea monosperma, Bauhinia variegata* and *Ocimum gratissimum* and Polyherbal Formulation

The peak purity was assessed by comparing the spectra at peak start, peak apex and peak end positions of the spot. Good correlation (R^2=0.9843) was obtained between the standard and the samples in the range of 0.5-2.5 µg/spot.

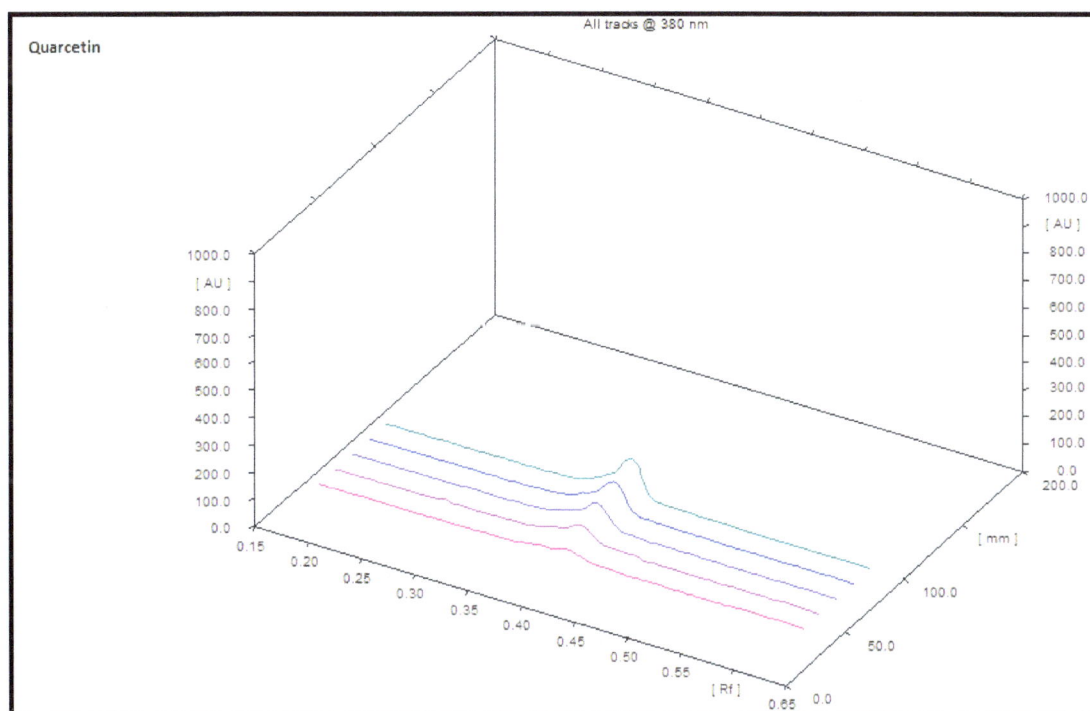

Figure 2: 3D-Chromatogram of quercetin (0.5-2.5 µg/spot).

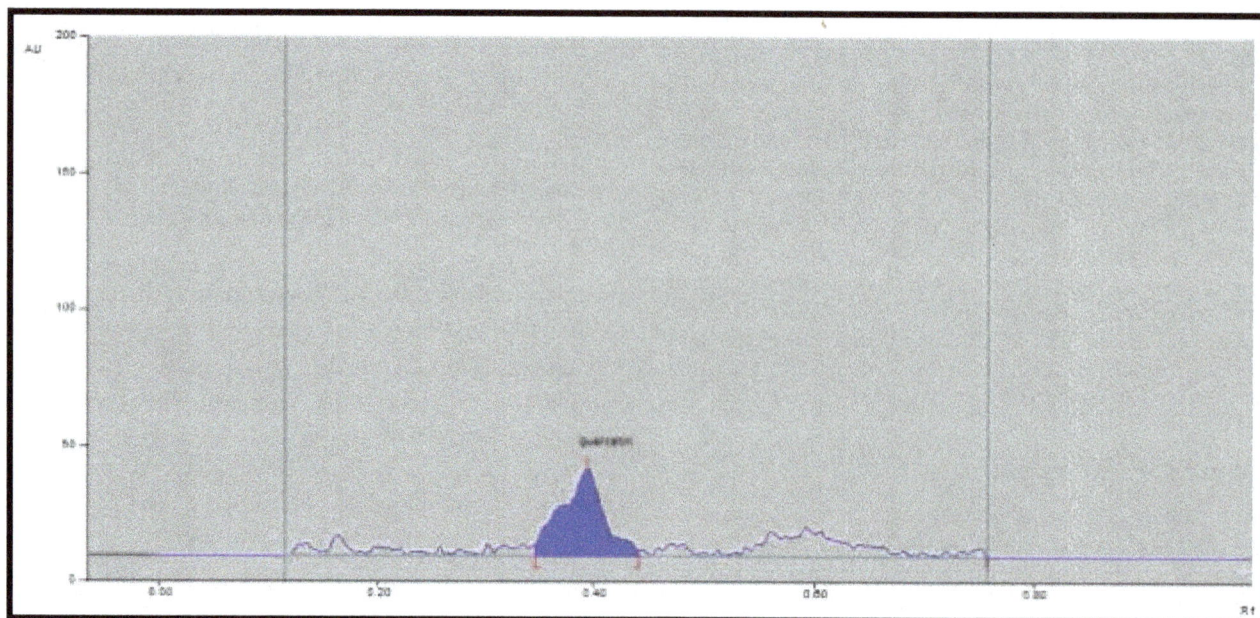

Figure 3: Densitogram of standard quercetin (0.5 μg/spot) at 380 nm.

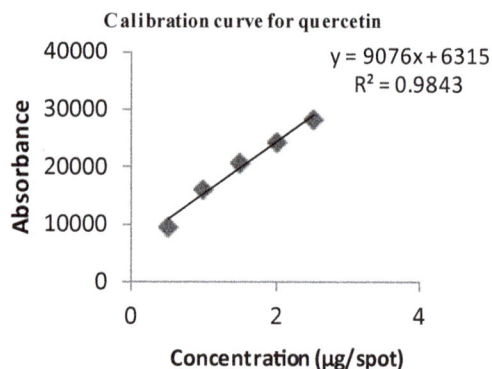

Figure 4: Calibration curve for quercetin (0.5, 1.0 and 1.5 μg/spot).

Concentration (μg/spot)	Peak area
2	24243
2	24558
2	24289
2	24289
2	24332
2	24467
2	24210
Average	24341.14 ± 125.97
% CV	0.51

Table 1: Repeatability data for estimation for quercetin.

Quercetin		
Concentration (μg/spot)	Area Mean ± S.D. (n=3)	% C.V
1.0	16256.91 ± 158.87	0.97
1.5	20777.82 ± 183.9448	0.88
2.0	24363.54 ± 170.40	0.69

Table 2: Intra-day precision data of estimation for quercetin.

Concentration (μg/spot)	Area Mean ± S.D. (n=3)	% CV
1.0	16274.1 ± 245.11	1.50
1.5	20777.82 ± 183.94	0.88
2.0	24364.79 ± 189.13	0.77

Table 3: Inter-day precision data of estimation for quercetin. Range of 0.77%-1.50%.

The identification of quercetin was done on the basis of R_f values. The concentrations of quercetin in acetone fraction of *B. monosperma*, ethyl acetate and n-butanol fractions of *B. variegata*, dichloromethane and ethyl acetate fractions of *O. gratissimum* were found to be 0.395, 0.174 4, 0.1382, 0.3229, 0.6734 (mg/10 mg) respectively. The results are shown in (Table 6). Acetone fraction of *Butea monosperma* showed eight peaks; the fourth peak R_f value (0.39) was coinciding with standard R_f value (Figure 5). The concentration of quercetin was found to be 0.395 (μg/10 mg).

Ethyl acetate fraction of *Bauhinia variegata* showed eight peaks; the fourth peak R_f value (0.38) was coinciding with standard R_f value of quercetin (Figure 6). The concentration of quercetin was found to be 0.174 (μg/10 mg). n- butanol fraction of *Bauhinia variegata* showed six peaks, the third peak R_f value (0.38) was coinciding with standard R_f value (Figure 7). The concentration of quercetin was found to be 0.138 (μg/10 mg). Dichloromethane fraction of *Ocimum gratissimum* showed nine peaks, the third peak R_f value (0.38) was coinciding with standard R_f value (Figure 8). The concentration of quercetin was found to be 0.322 (μg/10 mg). Ethyl acetate fraction of *Ocimum gratissimum* showed seven peaks; the third peak R_f value (0.39) was coinciding with standard R_f value (Figure 9). The concentration of quercetin in ethyl acetate fraction of *Ocimum gratissimum* was found to be 0.673 (μg/10 mg) [39,40]. Polyherbal tablet of formulation showed eighteen peaks, the R_f value (0.38) of seventh peak was coinciding with standard R_f value. The HPTLC densitogram is shown in Figure 10. The concentration of quercetin was found to be 0.113 (μg/10 mg) [38].

Concentration of quercetin in sample (ng/spot)	Amount of quercetin standard added (µg/spot)	Total Concentration (µg/spot)	Mean concentration recovered (µg/spot)	% Recovery	% Recovery mean
1	0.5	1.5	0.49	0.98	
1	0.5	1.5	0.48	0.96	97.33
1	0.5	1.5	0.49	0.98	
1	1	2	0.98	0.98	
1	1	2	0.99	0.99	98.00
1	1	2	0.97	0.97	
1	1.5	2.5	1.49	0.99	
1	1.5	2.5	1.49	0.99	99.11
1	1.5	2.5	1.48	0.98	

Table 4: Recovery data for quercetin.

Quercetin	
Mean slope	9076
SD of intercept	6315
LOD (µg/spot)	0.08
LOQ (µg/spot)	0.26

Table 5: LOD and LOQ data for quercetin.

Tracks	Samples	R_f values	Concentration (mg/spot)
1	Quercetin (std.)	0.38	-
2	Butea monosperma (AcO)	0.39	0.395989423
3	Bauhinia variegata (EtOAc)	0.38	0.174416042
4	Bauhinia variegata (n-BtOH))	0.38	0.138276774
5	Ocimum gratissimum (DCM)	0.39	0.322939621
6	Ocimum gratissimum (EtOAc)	0.38	0.673424416
7	Polyherbal tablet	0.38	0.113155575

Table 6: Estimation of quercetin in fractionated extracts of *Butea monosperma*, *Bauhinia variegata*, *Ocimum gratissimum*.

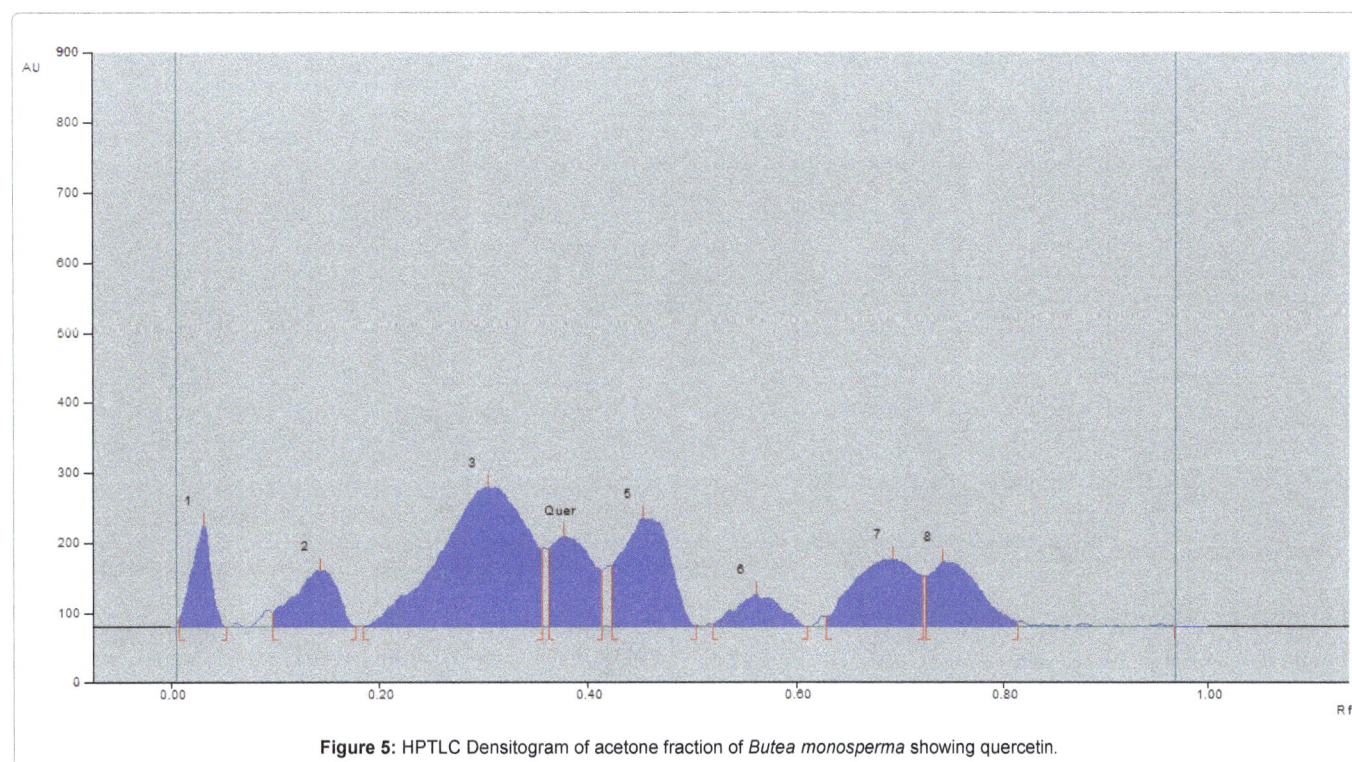

Figure 5: HPTLC Densitogram of acetone fraction of *Butea monosperma* showing quercetin.

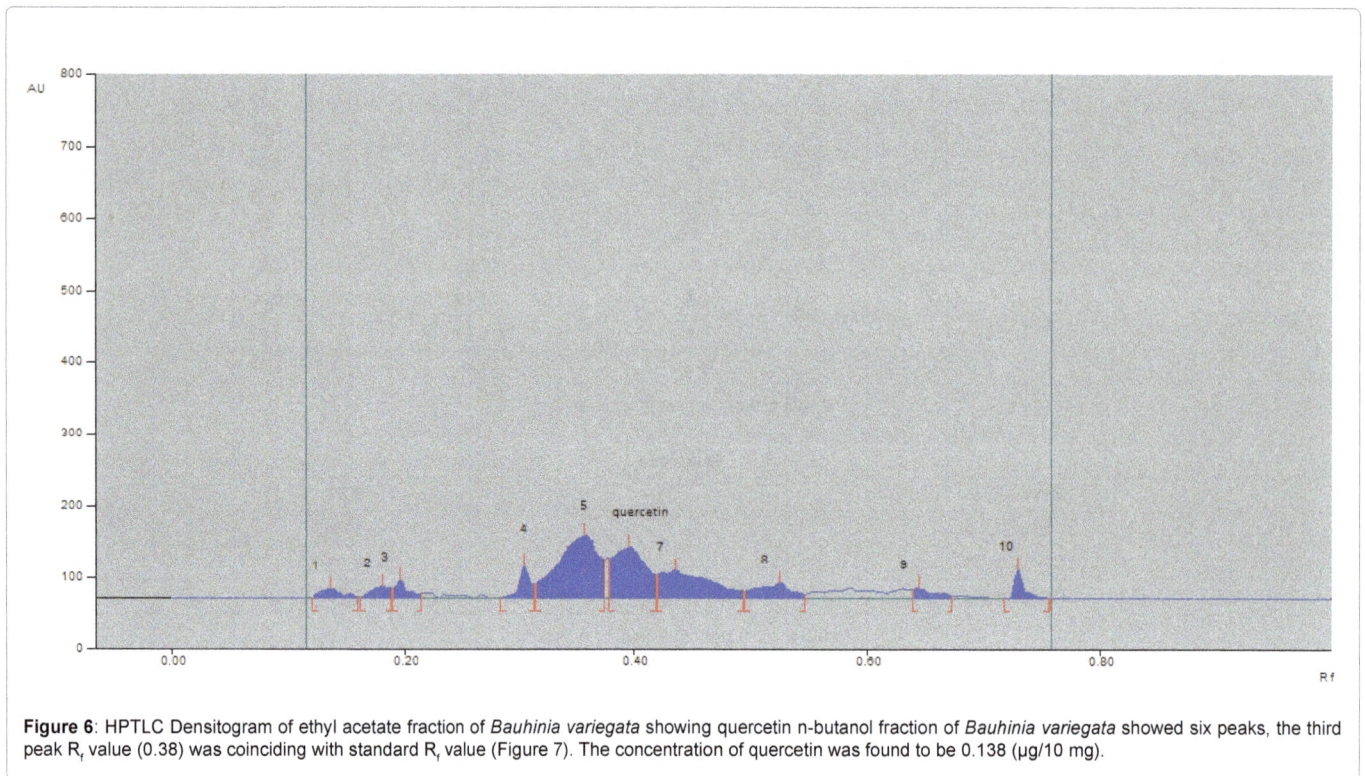

Figure 6: HPTLC Densitogram of ethyl acetate fraction of *Bauhinia variegata* showing quercetin n-butanol fraction of *Bauhinia variegata* showed six peaks, the third peak R_f value (0.38) was coinciding with standard R_f value (Figure 7). The concentration of quercetin was found to be 0.138 (μg/10 mg).

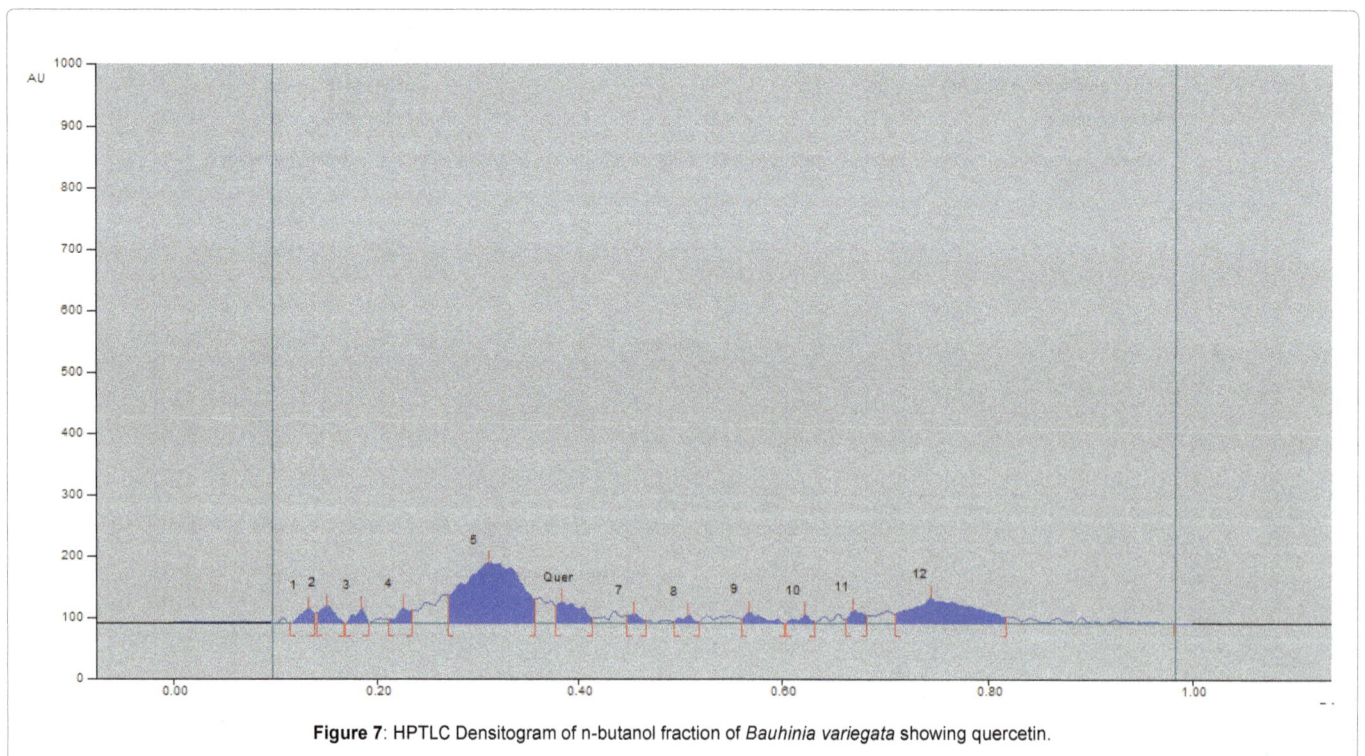

Figure 7: HPTLC Densitogram of n-butanol fraction of *Bauhinia variegata* showing quercetin.

Summary of validation parameters

The detailed summary of validation parameters is described in (Table 7)

Conclusion

A good correlation was obtained among the standard, samples of polyherbal formulation and fractionated extract of plant. An HPTLC method for quantitative estimation of quercetin present in fractionated extract of plants and polyherbal tablet has been developed and validated. The method can be used as a quercetin standard.

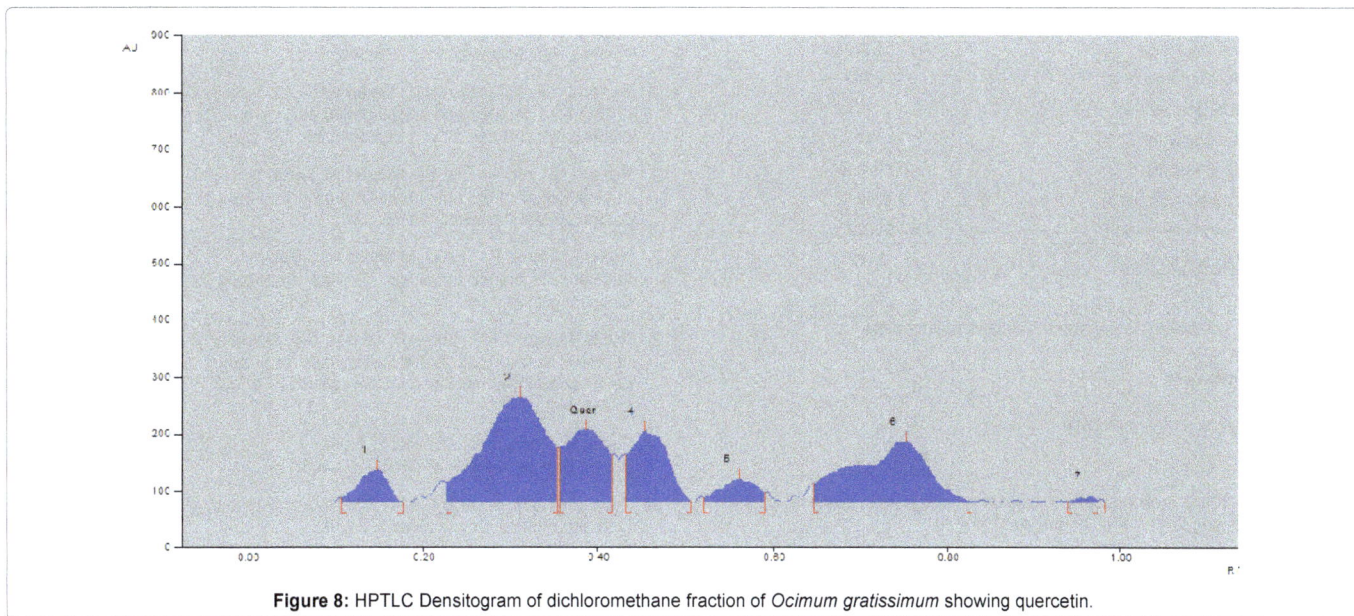

Figure 8: HPTLC Densitogram of dichloromethane fraction of *Ocimum gratissimum* showing quercetin.

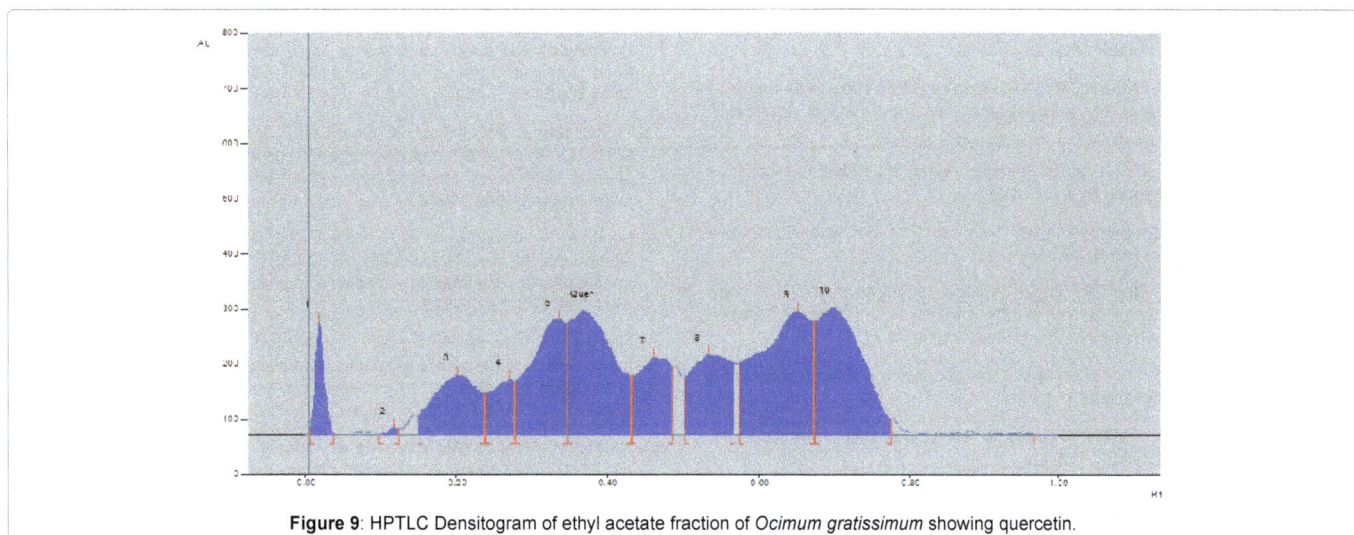

Figure 9: HPTLC Densitogram of ethyl acetate fraction of *Ocimum gratissimum* showing quercetin.

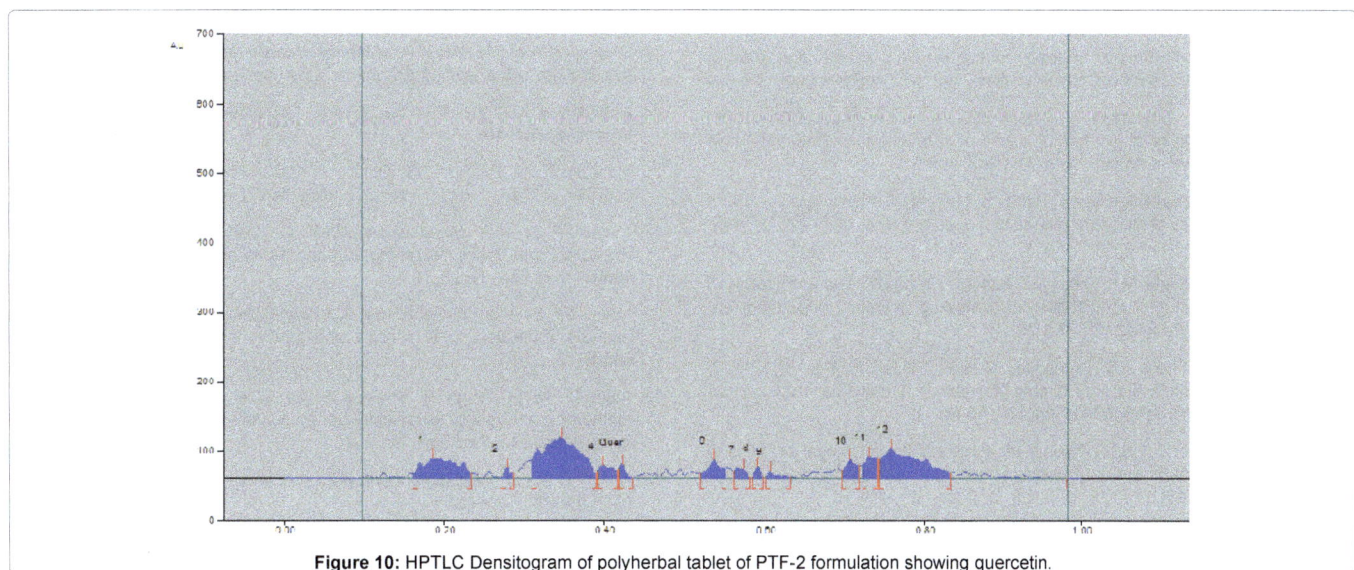

Figure 10: HPTLC Densitogram of polyherbal tablet of PTF-2 formulation showing quercetin.

Parameters	Result for quercetin
Linearity range	0.5-2.5 µg/spot
Correlation coefficient	0.9843
Precision (% CV)	
Repeatability (n=7)	0.5 (24341 ± 125)
Intraday precision (n=3)	0.69-0.97
Interday precision (n=3)	0.77-1.50
Accuracy (% recovery)	97.33-99.11
LOD (µg/spot)	0.08
LOQ (µg/spot)	0.26

Table 7: Summary of validation parameters.

Conflicts of interest

All authors have none to declare.

References

1. Halliwell B (1994) Free radicals, antioxidants, and human disease: curiosity, cause, or consequence? The lancet 344: 721-724.

2. Bouktaib M, Lebrun S, Atmani A, Rolando C (2002) Hemisynthesis of all the O-monomethylated analogues of quercetin including the major metabolites, through selective protection of phenolic functions. Tetrahedron 58: 10001-10009.

3. Singh R, Singh M, Chandra L, Bhat D, Arora M, et al. (2012) In vitro Antioxidant and free radical scavenging activity of Macrotyloma uniflorum (Gahat dal) from Kumauni region. Int J Fundam Appl Sci 1: 9-11.

4. Kiran B, Lalitha V, Raveesha K (2013) Psoralea corylifolia L. a potent medicinal plant with broad spectrum of medicinal properties. Int J Fundam Appl Sci 2: 20-22.

5. Burli D, Khade A (2007) A comprehensive review on Butea monosperma (Lam.) Kuntze. Pharmacogn Rev 1: 333-337.

6. Al-Snafi A (2013) the pharmacological importance of Bauhinia variegata–a review. Int J Pharm Sci Res 4: 160-164.

7. Uddin G, Sattar S, Rauf A (2012) Preliminary Phytochemical, In vitro Pharmacological Study of Bauhinia alba and Bauhinia variegata flowers. Middle-East Journal of Medicinal Plants Research 4: 75-79.

8. Dhale D (2011) Phytochemical screening and antimicrobial activity of Bauhinia variegata Linn. J ecobiotechnol 3: 4-7.

9. Spilková J, Hubik J (1992) Biologische wirkungen von flavonoiden. II. Pharmazie in unserer Zeit 21: 174-182.

10. Gupta SK, Prakash J, Srivastava S (2002) Validation of traditional claim of Tulsi, Ocimum sanctum Linn. as a medicinal plant. Indian J Exp Biol 40: 765-773.

11. Ilori M, Sheteolu AO, Omonigbehin EA, Adeneye AA (1996) Antidiarrhoeal activities of Ocimum gratissimum (Lamiaceae). J Diarrhoeal Dis Res 14: 283-285.

12. Onajobi FD (1986) Smooth muscle contracting lipid-soluble principles in chromatographic fractions of Ocimum gratissimum. J Ethnopharmacol 18: 3-11.

13. Obaseiki-Ebor EE, Odukoya K, Telikepalli H, Mitscher LA, Shankel DM (1993) Antimutagenic activity of extracts of leaves of four common edible vegetable plants in Nigeria (west Africa). Mutat Res 302: 109-117.

14. Nakamura CV, Ueda-Nakamura T, Bando E, Melo AF, Cortez DA, et al. (1999) Antibacterial activity of Ocimum gratissimum L. essential oil. Mem Inst Oswaldo Cruz 94: 675-678.

15. Orafidiya LO, Oyedele AO, Shittu AO, Elujoba AA (2001) The formulation of an effective topical antibacterial product containing Ocimum gratissimum leaf essential oil. Int J Pharm 224: 177-183.

16. Lemos Jde A, Passos XS, Fernandes Ode F, Paula JR, Ferri PH, et al. (2005) Antifungal activity from Ocimum gratissimum L. towards Cryptococcus neoformans. Mem Inst Oswaldo Cruz 100: 55-58.

17. Nwosu MO, Okafor JI (1995) Preliminary studies of the antifungal activities of some medicinal plants against Basidiobolus and some other pathogenic fungi. Mycoses 38: 191-195.

18. Silva M, Oliveira J, Fernandes O, Passos X, Costa C, et al. (2005) Antifungal activity of Ocimum gratissimum towards dermatophytes. Mycoses 48: 172-175.

19. Nakamura CV, Ishida K, Faccin LC, Filho BP, Cortez DA, et al. (2004) In vitro activity of essential oil from Ocimum gratissimum L. against four Candida species. Res Microbiol 155: 579-586.

20. Sartoratto A, Machado ALM, Delarmelina C, Figueira GM, Duarte MCT, et al. (2004) Composition and antimicrobial activity of essential oils from aromatic plants used in Brazil. Braz J Microbiol 35: 275-280.

21. Pessoa LM, Morais SM, Bevilaqua CM, Luciano JH (2002) Anthelmintic activity of essential oil of Ocimum gratissimum Linn. and eugenol against Haemonchus contortus. Vet Parasitol 109: 59-63.

22. Nangia-Makker P, Tait L, Shekhar MP, Palomino E, Hogan V, et al. (2007) Inhibition of breast tumor growth and angiogenesis by a medicinal herb: Ocimum gratissimum. Int J Cancer 121: 884-894.

23. Gupta A, Navin RS, Sonia P, Dinesh RS, Jitendra SY (2013) Determination of ursolic acid in fractionated leaf extracts of Ocimum gratissimum Linn and in developed herbal hepatoprotective tablet by HPTLC. Pharmacognosy Journal 5: 156-162.

24. Geeta R, Prakash R, Navgeet S, Neeru V, Sumit J (2011) Butea monosperma (Lam.) Kuntze: A Review. IJRP 2: 98-108.

25. Jash S, Roy R, Gorai D (2014) Bioactive constituents from Bauhinia variegata Linn. Int J Pharm Biomed Res 5: 51-54.

26. Mahapatra SK, Chakraborty SP, Das S, Roy S (2009) Methanol extract of Ocimum gratissimum protects murine peritoneal macrophages from nicotine toxicity by decreasing free radical generation, lipid and protein damage and enhances antioxidant protection. Oxid Med Cell Longev 2: 222-230.

27. Aaby K, Hvattum E, Skrede G (2004) Analysis of flavonoids and other phenolic compounds using high-performance liquid chromatography with coulometric array detection: relationship to antioxidant activity. J Agric Food Chem 52: 4595-4603.

28. Alonso-Salces RM, Ndjoko K, Queiroz EF, Ioset JR, Hostettmann K, et al. (2004) On-line characterisation of apple polyphenols by liquid chromatography coupled with mass spectrometry and ultraviolet absorbance detection. J Chromatogr A 1046: 89-100.

29. Lommen A, Godejohann M, Venema DP, Hollman PC, Spraul M (2000) Application of directly coupled HPLC-NMR-MS to the identification and confirmation of quercetin glycosides and phloretin glycosides in apple peel. Anal Chem 72: 1793-1797.

30. Nielsen SE, Freese R, Cornett C, Dragsted LO (2000) Identification and quantification of flavonoids in human urine samples by column-switching liquid chromatography coupled to atmospheric pressure chemical ionization mass spectrometry. Anal Chem 72: 1503-1509.

31. Rodríguez-Delgado MA, Malovaná S, Pérez JP, Borges T, García Montelongo FJ (2001) Separation of phenolic compounds by high-performance liquid chromatography with absorbance and fluorimetric detection. J Chromatogr A 912: 249-257.

32. Sharma A, Chakraborti K, Handa S (1991) Antihepatotoxic activity of some Indian herbal formulations as compared to silymarin. Fitoterapia 62: 229-235.

33. Chattopadhyay RR (2003) Possible mechanism of hepatoprotective activity of Azadirachta indica leaf extract: part II. J Ethnopharmacol 89: 217-219.

34. Khandelwal K (2001) Pharmacognosy: Techniques and Experiments. Nirali Prakashan 8: 3-40.

35. Macdonald IO, Oludare AS, Olabiyi A (2010) Phytotoxic and Anti-microbial activities of Flavonoids in Ocimum gratissimum. Life Science Journal 2: 37-40.

36. Yuvaraj P, Louis T, Madhavachadran V, Gopinath N, Rekha S., 2011. Total Phenolic Content And Screening of Antioxidant Activity Of Selected Ayurvedic Medicinal Plants. 91: 25-31.

37. Kosalec I, Bakmaz M, Pepeljnjak S, Vladimir-Knezević S (2004) Quantitative analysis of the flavonoids in raw propolis from northern Croatia. Acta Pharm 54: 65-72.

38. Alam P, Ali M, Singh R, Shakeel F (2011) A new HPTLC densitometric method for analysis of swertiamarin in Enicostemma littorale and commercial formulations. Nat Prod Res 25: 17-25.

39. ICHHT G (2005) Validation of analytical procedures: text and methodology Q2 (R1). IFPMA: Geneva.

40. Prakash J, Gupta SK (2000) Chemopreventive activity of Ocimum sanctum seed oil. J Ethnopharmacol 72: 29-34.

16

Characterization of the Volatile Components of Cannabis Preparations by Solid-Phase Microextraction Coupled to Headspace-Gas Chromatography with Mass Detector (SPME-HSGC/MS)

Sebastiano Arnoldi, Gabriella Roda*, Eleonora Casagni, Lucia Dell'Acqua, Michele Dei Cas, Fiorenza Fare, Chiara Rusconi, Giacomo Luca Visconti and Veniero Gambaro

Department of Pharmaceutical Sciences, University of Milan, Via Mangiagalli 25, 20133, Milan, Italy

Abstract

Solid phase microextraction coupled to headspace sampling and GC/MS technique was applied to the characterization of the volatile components of several Cannabis preparations (hashish). Different parameters of the analytical method (fiber, coating thickness, sampling and exposition temperatures, sample preparation) were evaluated to optimize the characterization of the volatile components. α-Pinene, β-myrcene, limonene, 4-carene, trans-3(10) caren-2-ol, 4,7,7-trimethylbicyclo [4.1.0] heptan-3-ol, caryophyllene, β-humulene, azulene, gurjunene, ledene and caryophyllene oxide were identified among the volatile components of all hashish preparations. Moreover, a suitable internal standard (nonane) was chosen, the reproducibility and linearity of the method were evaluated in order to carry out the quantitative determination of caryphyllene, the most abundant volatile terpene. Its quantity ranged from 800 to 3000 µg/g.

Keywords: Hashish; Cannabis; Volatile components; Caryphyllene; Azulene; HS-SPME-GC

Introduction

Cannabis preparations (marijuana and hashish) are still among the most consumed illicit drugs worldwide, due to their low price and the commonly considered low social impact. Hashish is the sticky resin produced by the Cannabis female flowers and it is particularly rich of psychoactive principles [1,2]. The biologically active compounds are a group of terpenoid secondary metabolites, called cannabinoids. Among them, the main constituents are cannabidiol (CBD), cannabinol (CBN) and µ-9-tetra-hydrocannabinol (µ-9-THC) [3,4]. Hashish looks like a hard yellowish-brownish paste that becomes malleable upon heating and different samples can present a high variety in terms of color, density, content of the active principles and flavor [1]. Besides the biologically active cannabinoids, more than 90 phytocannabinoids have been isolated from Cannabis and its essential oil is a complex mixture containing also monoterpenoids and sesquiterpenoids [5-8], giving it the typical organoleptic properties. The volatiles constituents of Cannabis have been extensively studied [9] because they represent a potential for chemically fingerprinting different cultivars [10]. On the other hand, less attention was dedicated to the volatile components of hashish, despite the fact that they could be very useful for characterizing the different preparations and for establishing the origin and eventual links between different seizures [2]. In this frame, we were interested in analyzing the volatile components of different hashish preparations seized by the judicial authority and delivered to our laboratory for the determination of the content of cannabinoids [11]. For this purpose, we chose solid phase microextraction [12] coupled to headspace sampling, which is based on the adsorption of the volatile analytes by the coating of a suitable fiber and their direct injection into a GC/MS system. This method shows several advantages respect to liquid-liquid extraction (LLE), because it is more selective, less time consuming, it does not require the use of solvents and it is particularly suitable for volatile analytes [13]. Different parameters of the analytical method (fiber, coating thickness, sampling and exposition temperatures, sample preparation) were evaluated to optimize it for the determination of the volatile components of hashish, in particular, the parameters were studied taking into account azulene and caryophyllene, two significant

representatives of the monoterpene and sesquiterpene classes present in hashish (Figure 1). Moreover, the quantitative determination of caryophyllene, the most abundant component was carried out choosing nonane as internal standard (IS).

Materials and Methods

Reagents, chemicals, and standards

All reagents were of analytical grade and were stored as indicated by the supplier. Caryophyllene, nonane (IS), ß-jonone, trans-inane, methyl oleate, α-Pinene, β-myrcene, limonene, 4,7,7-trimethylbicyclo [4.1.0] heptan-3-ol, β-humulene, azulene, gurjunene, caryophyllene oxide and ledene were purchased by Sigma-Aldrich, (Steinheim, Germany). 4-Carene and trans-3(10) caren-2-ol were from ABI Chem (Germany).

Methanol was obtained by Baker (Deventer, The Netherlands).

A IS standard solution was prepared as follows: 100 µL of nonane were brought to 10 mL with methanol (10 µL/mL; 7.2 mg/mL).

Cannabis preparations

Seven different hashish samples were taken into account, seized by the judicial authority and delivered to our laboratory for the determination of the cannabinoid content [11]. The characteristics of the different preparations together with the content of THC, CBD and CBN are reported in Table 1.

*Corresponding author: Gabriella Roda, Department of Pharmaceutical Sciences, University of Milan, Via Mangiagalli 25, 20133, Milan, Italy
E-mail: gabriella.roda@unimi.it

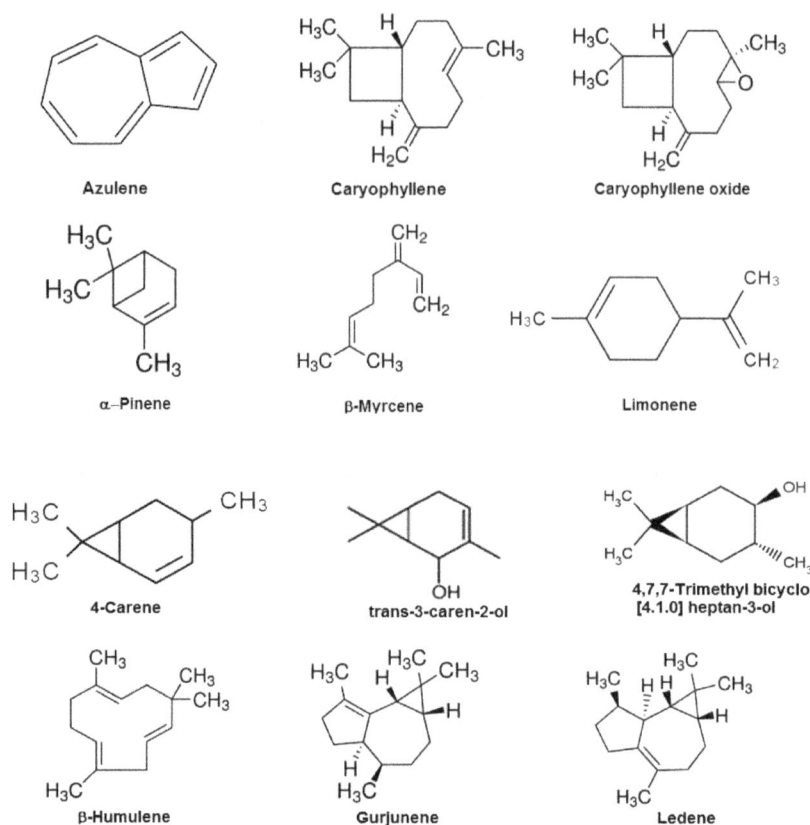

Figure 1: Structures of hashish most abundant terpenes.

Sample	Type	Description	%THC	%CBD	%CBN
1	Hashish	Color: malleable	19.0	5.8	0.5
2	Hashish	Color: dark brown; very hard texture	13.4	6.0	0.6
3	Hashish	Color: black; soft texture, malleable	9.8	5.8	0.6
4	Hashish	Color: light brown; hard texture	11.7	7.9	0.3
5	Hashish	Color: dark brown; compact texture	18.3	7.0	0.4
6	Hashish	Color: brown; soft texture	16.0	6.7	0.9
7	Hashish	Color: greenish-brownish; very hard and compact texture	21.7	9.2	0.6

Table 1: Characteristics of the different preparations and content (%, w/w) of THC, CBD and CBN.

Sample 3 was used to evaluate method repeatability and it was considered as a reference because it was the most abundant sample and it was particularly aromatic, soft and malleable and it had a gold punching as a mark of quality.

GC/MS instrument

Analyses were carried on a 3900 Varian GC system (Agilent, Santa Clara, CA), with a split–splitless injection system operated in a split mode and a Varian Saturn 2100T Detector operated in electron impact mode (70 eV). Data acquisition and analysis were performed using standard software supplied by the manufacturer (Varian Workstation 6.0). The GC was equipped with a capillary column DB-5MS (30 m, 0,25 mm i.d., film thickness 0,25 μm) (Agilent, Santa Clara, CA).

GC/MS conditions

The GC/MS system was operated under following conditions: split ratio, splitless for 2 min then 30:1; solvent delay, 2.0 min; injector temperature, 250°C; interface transfer line, 300°C; ion source, 180°C; oven temperature program, from 200°C to 250°C, at 10°C/min, final isotherm, 15 min. Analysis time 20 min.

Helium was used as the carrier gas at a flow rate of 1.3 mL/min. The MS detector was operated in SCAN mode, mass range: 70 to 500 m/z.

Linearity

For the study of linearity, a sample of hashish (10 mg of sample 3) was exhausted eliminating the volatile components (heating under vacuum at 100°C for 6 h) and then 10 μL of a standard solution of caryophyllene (1000 μg/mL) were added. The chromatographic response was the same as that obtained carrying out the SPME on 10 μL of the standard solution alone. The calibration curve was then built using 10 μL of standard solutions of caryophyllene at suitable concentrations (4.3, 8.6, 17.2, 34.4, 68.8, 137.6, 275, 550, 1100, 2200, 4400, 8800 μg/mL).

Sample preparation for qualitative characterization

A slice was cut from every preparation, frozen and finely chopped. 10.0 mg were weighed, put in a 20 mL headspace vial and immediately sealed using a silicone/PTFA septum and a magnetic cap. Before the analysis 10 μL of the IS standard solution (70 mg) were added by a microsyringe.

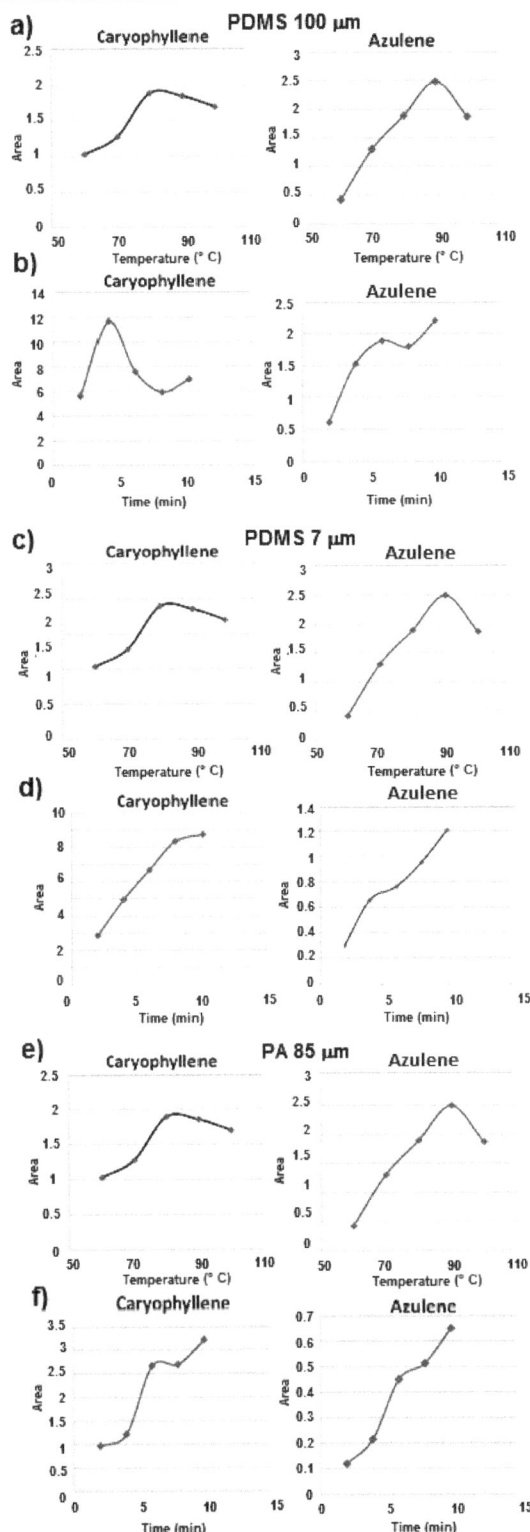

Kinetics of adsorption of azulene and caryophyllene. (a) PDMS fiber 100 μm; extraction time 5 min; extraction temperature: 60°C, 70°C, 80°C, 90°C, 100°C (b) PDMS fiber 100 μm; extraction time 2, 4, 6, 8 and 10 min; extraction temperature: 80°C (c) PDMS fiber 7 μm; extraction time 5 min; extraction temperature: 60°C, 70°C, 80°C, 90°C, 100°C (d) PDMS fiber 7 μm; extraction time 2, 4, 6, 8 and 10 min; extraction temperature: 80°C (e) PA fiber 85 μm; extraction time 5 min; extraction temperature: 60°C, 70°C, 80°C, 90°C, 100°C (f) PA fiber 85 μm; extraction time 2, 4, 6, 8 and 10 min; extraction temperature: 80°C.

Figure 2: Results obtained are depicted.

SPME extraction

Solid phase microextraction coupled to headspace sampling is based on the adsorption of the volatile analytes by the coating of a suitable fiber and their direct injection into a GC/MS system. The sample vial was equilibrated at 80°C for 5 min. For adsorption, the needle of the SPME device containing the extraction fiber (SPME fiber assembly polyacrylate df 85 μm, Supelco, Sigma Aldrich, Steinheim, Germany) was inserted through the septum of the vial and the fiber was exposed to the headspace in the vial for 5 min. Finally, the SPME fiber with the absorbed compounds was introduced into the injection port of the GC/MS for 5 min to accomplish complete desorption of the analytes.

Results and Discussion

We were interested in analyzing the volatile components of different hashish preparations seized by the judicial authority and delivered to our laboratory for the determination of the content of cannabinoids (Table 1), in order to characterize the different preparations and to establish the origin and eventual links between different seizures.

For this purpose we chose solid phase microextraction coupled to headspace sampling, which shows several advantages because it is more selective, less time consuming and it is particularly suitable for volatile analytes. Different parameters of the analytical method were evaluated to optimize it for this kind of determination. Sample preparation was investigated and we decided to carry out the extraction on the vegetable material as it was without adding a solvent. In this way, the aromatic and volatile component of the hashish preparations is preserved and the chromatograms obtained do not show the peak of the solvent but only peaks related to the matrix. Due to the high sensitivity of this technique, we decided to use 10 mg of hashish preparations.

Three different fibers were evaluated: two of them with a polydimethylsiloxane (PDMS) non polar coating but with a different thickness, respectively 100 μm and 7 μm; the third with a polyacrilate (PA) polar coating and a thickness of 85 μm. These three fibers represent a pool recommended by the suppliers for the analysis of a wide range of analytes; in particular, low molecular weight or volatile compounds usually require a 100 μm polydimethylsiloxane (PDMS)-coated fiber. Larger molecular weight or semivolatile compounds are more effectively extracted with a 7 μm PDMS fiber. To extract very polar analytes from polar samples, an 85 μm polyacrylate-coated fiber is required. The thickness of the coating also show an influence on the extraction of the analytes: thicker coatings favor the adsorption of low molecular weight volatiles, while lower thicknesses adsorb preferentially semivolatile analytes.

Five different adsorption times (2, 4, 6, 8 and 10 min) and temperatures (60, 70, 80, 90 and 100°C) were taken into account. The adsorption of two representative terpenes (the monoterpene azulene and the sesquiterpene caryophyllene, Figure 1 in the volatile portion of hashish samples was considered.

Heating favors the evaporation of the volatile analytes from the vegetable material, so the higher the temperature, the higher the vapor pressure of the analyte and consequently the adsorption should be. Temperatures higher than 100°C were not tested because some terpenes are not stable. The different classes of terpenes show different boiling points and consequently adsorption profiles: monoterpenes, such as limonene (b.p.=175,5-176,5°C) are more volatile and an excessive increase of the temperature leads to the desorption from the fibers. On the other hand, sesquiterpenes such as caryophyllene (b.p.=262°C) need a higher temperature to pass to the vapor phase, but above a certain temperature, the desorption process became predominant.

The curves obtained demonstrated that 80°C is the best adsorption temperature for caryophyllene, in the case of azulene, the maximum of adsorption is reached at about 90°C, but it is high also at 80°C, so we decided to set the adsorption temperature at 80°C. As regard as the time of adsorption, 5 min resulted to be the most suitable time. In fact lower times do not allow the establishment of the adsorption equilibrium, higher times likely bring to the desorption of the less adsorbed analytes. So the optimal adsorption conditions were 80°C for 5 min. On the basis of the results obtained reported in Figure 2, the 100 μm PDMS fiber was chosen, because the area of both caryohyllene and azulene resulted higher, indicating a higher adsorption of the analytes.

To evaluate the reproducibility of the method a suitable internal

Analyte	Retention times (min)	KI
$C_{12}H_{26}$	4.332	1200
$C_{14}H_{30}$	6.524	1400
$C_{16}H_{34}$	8.846	1600
$C_{18}H_{38}$	11.034	1800
$C_{20}H_{42}$	13.053	2000
Nonane	2.347	950
trans-Pinane	2.839	995
α-Pinene	2.610	1040
β-Myrcene	2.789	1056
Limonene	3.156	1090
4-Carene	3.571	1127
Trans-3(10)caren-2-ol	4.538	1216
4,7,7- trimethylbicyclo[4.1.0]heptan-3-ol	4.708	1231
Caryophyllene	7.229	1461
β-Humulene	7.690	1503
β-Ionone	8.021	1533
Azulene	8.109	1541
Gurjunene	8.690	1594
Caryophyllene oxide	9.274	1647
Ledene	10.100	1722
Methyl oleate	14.294	2104
CBD	17.316	2380
THC	18.174	2458
CBN	18.721	2508

Table 2: Kovats retention indices (KI).

Withdrawal at the same point	IS Area	Caryophyllene Area
A	0.561	6.584
B	0.511	6.637
C	0.479	7.048
D	0.488	8.005
E	0.507	7.381
F	0.535	6.717
Mean	0.51	7.06
Standard deviation	0.03	0.55
%CV	5.89	7.81
Withdrawal at different points	IS Area	Caryophyllene Area
A	0.500	7.677
B	0.481	6.437
C	0.516	7.410
D	0.483	7.645
E	0.484	5.660
F	0.478	7.248
Mean	0.49	7.01
Standard deviation	0.01	0.80
%CV	3.01	11.44

Table 3: Reproducibility of the method evaluated on sample 3.

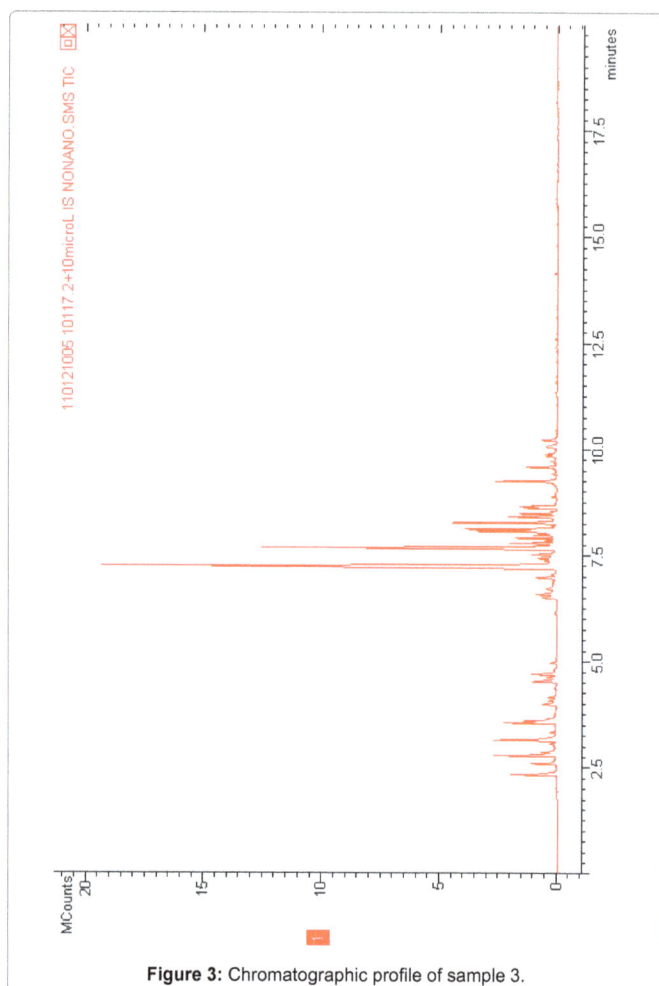

Figure 3: Chromatographic profile of sample 3.

standard was needed. To this end, Kovats retention index (KI) of a series of terpenes evidenced in the analysis of hashish preparations were calculated. As the Kovats index is related to the retention time of the analyte normalized to the retention times of adjacently eluting n-alkanes, decane, dodecane, tetradecane, hexadecane, octadecane and icosane were injected by an autosampler in the same chromatographic conditions used for the analysis of hashish volatile components. THC, CBD and CBN were also analyzed and identified in the hashish volatile components by means of their Kovats retention index. The Kovats retention indices obtained for the most abundant peaks of the volatile components of hashish preparations are reported in Table 2, together with the values obtained for the series of linear alkanes, THC, CBD and CBN. Limonene, β-ionone, trans-pinane, nonane and methyl oleate were taken into account as possible internal standards.

Limonene could not be chosen because it is present in the vegetable material, while β-ionone and trans-pinane are not present in the matrix but they have retention times too close to those of compounds present in the vegetable material. Methyl oleate is the internal standard routinely used for the analysis of cannabinoids by GC/FID technique, but it has a too high boling point (218°C) for this kind of application. On the other hand, nonane that was indicated in the literature [10] as internal standard for the analysis of cannabinoids and terpenes in Cannabis, showed the suitable features to be used as internal standard.

To evaluate the reproducibility of the method different analyses were carried out withdrawing six samples from the same portion of the stick of hashish (sample 3) or from different portions of the same

Sample	A	B	C	D	E	F	I	II	III	IV	V	VI	IS
1	0.24	0.57	0.77	0.40	0.32	0.26	7.073	2.865	0.938	0.353	0.541	0.208	0.457
2	1.09	0.80	0.55	0.51	0.20	0.22	7.164	2.385	0.743	0.157	1.229	0.377	0.539
3	0.18	0.09	0.21	0.15	0.10	0.28	7.248	2.206	1.561	2.264	0.533	0.589	0.478
4	0.04	0.03	0.04	0.09	0.22	0.15	6.204	2.652	1.274	0.173	0.587	0.168	0.450
5	0.30	0.67	1.13	0.54	0.37	0.28	8.997	3.395	1.169	0.210	0.827	0.331	0.484
6	0.03	0.02	0.36	0.76	0.17	0.17	3.589	1.731	0.591	0.317	1.271	0.250	0.522
7	0.21	0.31	0.44	0.34	0.36	0.43	1.780	5.430	1.503	0.585	1.218	0.343	0.473

Table 4: Area of the peaks of the terpenes identified in hashish preparations.

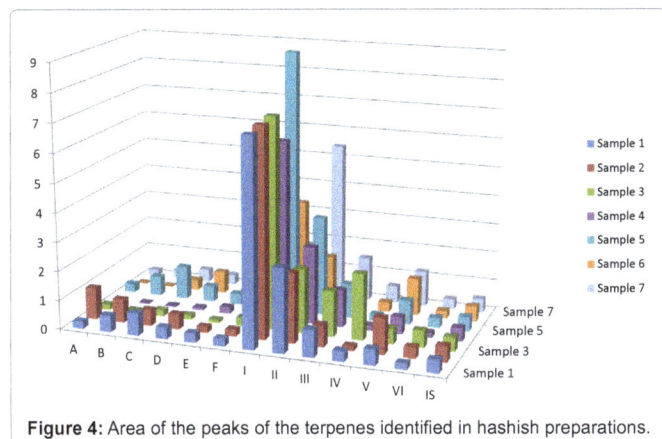

Figure 4: Area of the peaks of the terpenes identified in hashish preparations.

Sample	I (µg/g)
1	1280
2	1688
3	1545
4	1477
5	796
6	2963
7	2945

Table 5: Content of caryophyllene (I, µg/g) in the different Hashish preparations.

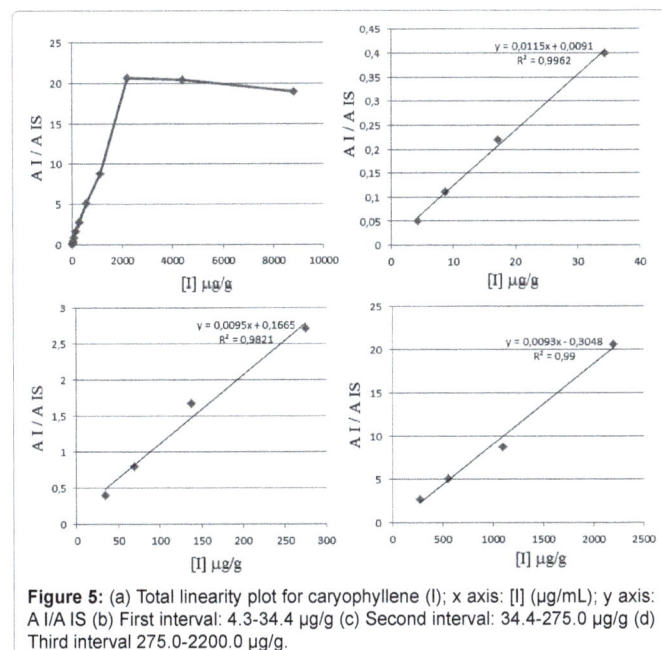

Figure 5: (a) Total linearity plot for caryophyllene (I); x axis: [I] (µg/mL); y axis: A I/A IS (b) First interval: 4.3-34.4 µg/g (c) Second interval: 34.4-275.0 µg/g (d) Third interval 275.0-2200.0 µg/g.

stick. The area of the peaks related to caryophyllene, the most abundant component were compared (Table 3).

Percentage coefficients of variation show that the method has a good reproducibility, this is particularly important for this kind of vegetable material, which has a lower degree of homogeneity, respect to other kind of matrices.

At this point we analyzed all the different preparation of Cannabis (samples 1-7), comparing the chromatographic profile. In Figure 3, the chromatographic profile obtained for sample 3 is reported.

By comparison with the relative standards and the avaluation of Kovats indices, the following terpenes were identified: α-pinene (A), β-Myrcene (B), Limonene (C), 4-Carene (D), *trans*-3(10) caren-2-ol (E), 4,7,7- trimethylbicyclo[4.1.0]heptan-3-ol (F), Caryophyllene (I), β-Humulene (II), Azulene (III), Gurjunene (IV), Caryophyllene oxide (V), Ledene (VI) whose areas are reported in Table 4 and Figure 4.

The most abundand terpenes in all hashish preparations were caryophyllene (I) and β-Humulene (II), two sesquiterpenes, confirming literature data [2]. To carry out quantitative determinations, the linearity of the method was studied for caryophyllene (I), the most abundant component. Twelve standard solution, at different concentrations were analyzed in the range 4.3 µg/mL and 8800.0 µg/mL. The linearity plot obtained plotting the ratio between the area of the peak of caryophyllene (I) and the area of the peak of the internal standard (A I / A IS) against the concentration of caryophyllene ([I]) is reported in Figure 5.

Three different linearity intervals were identified: the first interval from 4.3 to 34.4 µg/g (linearity equation y=0.0115x+0.091; R^2=0,9962); the second from 34.4 to 275.0 µg/g (linearity equation y=0.0095x+0.1665; R^2=0,9821) and the third from 275.0 to 2200.0 µg/g (linearity equation y 0.0093x−0.3048; R^2=0,9900). In the fourth region from 1100.0 to 8800.0 µg/g of the linearity curve a plateau is reached probably due to the saturation of the fiber and the method is not linear any more. The quantitative determination of caryophyllene was carried out using the linearity equation of the third interval. In fact, the concentration of caryophyllene in hashish preparation follows in this range. The results obtained are reported in Table 5. The content of caryophyllene of the last two samples were outside the linearity range, therefore, these samples were determined weighting 5 mg of hashish.

Conclusions

The HS-SPME-GC/MS method studied herein allows the rapid and reproducible determination of the volatile components of Cannabis. This analysis could be particularly useful for the characterization of Cannabis preparations (hashish) and to establish the region of origin and eventual links between different seizures. To this end, the parameters of the SPME and of the analytical method were optimized using azulene and caryophyllene, two representative terpenes of hashish. Ten different hashish preparations were analyzed and the quantitative determination of caryphyllene, the most abundant component was carried out, using nonane as internal standard.

References

1. Ferioli V, Rustichelli C, Pavesi G, Gamberini G (2000) Analytical Characterisation of Hashish Samples. Chromatographia 52: 39-44.

2. Marchinia LM, Charvozb C, Dujourdyb L, Baldovinia N, Filippia JJ (2014) Multidimensional analysis of cannabis volatile constituents: Identification of 5,5-dimethyl-1-vinylbicyclo [2.1.1] hexane as a volatile marker of hashish, the resin of Cannabis sativa. J Chromatogr A 1370: 200-215.

3. Fischedick JT, Hazekamp A, Erkelens T, Choi YH, Verpoorte R (2010) Metabolic fingerprinting of Cannabis sativa L., cannabinoids and terpenoids for chemotaxonomic and drug standardization purposes. Phytochemistry 71: 2058-2073.

4. Pertwee RG (2006) Cannabinoid pharmacology: the first 66 years. Br J Pharmacol 147: 163-171.

5. Hillig KW (2004) A chemotaxonomic analysis of terpenoid variation in Cannabis. Biochem Syst Ecol 32: 875-891.

6. Hendriks H, Malingre ThM, Batterman S, Bos R (1978) The essential oil of Cannabis sativa L. Pharm Weekblad 113: 413-424.

7. Turner CE, ElSohly MA, Boeren EG (1980) Constituents of Cannabis sativa L. XVII. A review of the natural constituents. J Nat Prod 43: 169-234.

8. Ross SA, ElSohly MA (1996) The volatile oil composition of fresh and air-dried buds of Cannabis sativa. J Nat Prod 59: 49-51.

9. Martin L, Smith DM, Farmilo CG (1961) Essential oil from fresh Cannabis sativa andits use in identification. Nature 191: 774-776.

10. Giese MW, Lewis MA, Giese L, Smith KM (2015) Development and Validation of a Reliable and Robust Method for the Analysis of Cannabinoids and Terpenes in Cannabis. J AOAC Int 98: 1503-1522.

11. European Commission (1999) Community method for the quantitative determination of Δ9-THC (Tetrahydrocannabinol) content in hemp varieties. In: Commission Regulation (EC) No 2316/1999 laying down detailed rules for the application of Council Regulation (EC) No 1251/1999 establishing a support system for producers of certain arable crops. J Eur Comm L 280: 43-65.

12. Arthur CL, Pawliszyn J (1990) Solid phase microextraction with thermal desorption using fused silica optical fibers. Anal Chem 62: 2145-2148.

13. Lachenmeier DW, Kroener L, Musshoff F, Madea B (2004) Determination of cannabinoids in hemp food products by use of headspace solid-phase microextraction and gas chromatography–mass spectrometry. Anal Bioanal Chem 378: 183-189.

17

Highly Sensitive Liquid Chromatography-Mass Spectrometry Detection of Microcystins with Molecularly Imprinted Polymer Extraction from Complicated Aqueous Ecosystems

Reddithota J Krupdam*, Darshana Gour and Govind Patel

National Environmental Engineering Research Institute, Nehru Marg, Nagpur 440020, India

Abstract

In the present study, a liquid chromatography-mass spectrometry (LC-MS) method has been developed and validated to monitor traces of microcystins (MCs) in lake and marine waters. The molecularly imprinted polymer (MIP) formulated with itaconic acid as the functional monomer and ethylene glycol dimethacrylate as the cross-linking monomer has been used to selectively enrich MCs from the aqueous solutions. The extraction capacity and selectivity of MIP was higher when comparison with conventionally used resin XAD and powdered activated carbon (PAC). The MIP showed an outstanding selectivity for microcystin-LR (MC-LR) in a mixture of MCs from aqueous solutions in the pH range 6-9. The LC-MS analysis of MCs after MIP extraction showed an excellent linearity in the working range (R^2=0.998) with high repeatability (RSD%, <6.3) and recoveries above 90%. Interference of dissolved ions and solution pH on MCs trace quantification in the lake and marine water samples were quantified. The limits of quantification (LOQ) and lower limit of detection (LOD) for the MC-LR were 10 and 1 ng L^{-1}, respectively, which satisfies the strictest World Health Organization standard for MC-LR in drinking water (1 ng mL^{-1}). The proposed analytical approach is simple, efficient and comparable with the detection limit of the traditional and expensive ELISA method of MCs analysis.

Keywords: Microcystins; Molecularly imprinted polymer; Solid-phase extraction; Liquid chromatography; Environmental trace analysis

Introduction

The production and release of cyanotoxins by cyanobacteria in freshwaters around the world has been well documented [1,2]. Microcystins (MCs) are the most frequently occurring class of cyanobacterial toxins, of which microcystin-LR (MC-LR) is the most toxic and primarily detected congener [3]. The general structure of MCs is cyclo-(-D-Ala-L-R1-D-erythro-β-methyllisoAsp-L-R2-Adda-D-iso-Glu-N-methylde hydroAla), where R1 and R2 represent two variable L-amino acids and Adda stands for 3-amino-9-methoxy-2,6,8-trimethyl-10-phenyldeca-4,6-dienoic acid. There are about 60 MC variants known, and MC-LR, MC-RR and MC-YR are the three most abundant vairants reported in natural algal blooms across the world [4]. Contamination of drinking water by MCs has been linked to cases of primary liver cancer in China and the deaths of 76 patients undergoing dialysis in Brazil [5,6]. Recently, blooms of microcystis have resulted in health alerts in Nebraska and other parts of the Midwestern United States [7]. Due to adverse health effects, the World Health Organization (WHO) established a provisional concentration limit of 1 µg L^{-1} for MC-LR in drinking water [8] and the United States Environmental Protection Agency (USEPA) has placed MCs on the Drinking Water Contaminants List [9]. Reduction of threats to human health and aquatic life involves toxic cyanobacteria blooms and/or MCs to be monitored and removed from water columns, in particular, public water supplies. Because of this, there is a growing demand for the development of more reliable, rapid, cost-effective and sensitive analytical methods for monitoring of MCs in water.

MCs are commonly measured by HPLC with photodiode array (PDA) in full UV spectrum acquisition, over the range 200-300 nm [10]. They developed a SPME-microbore-LC/Q-TOF-MS method for measuring MCs in water [11]. This technique requires small volumes (2 mL) and provides sensitive and information-rich analysis of unknown toxins. Another sensitive method reported for MCs and nodularins analysis in tap water by [12]. Using the LC/ESI-MS/MS method the quantification limits of 0.25-0.90 µg L^{-1} were achieved for MCs, with a short run time of 10 min. [13,14] reported the analysis of polar dimethyl microcystin variants that are common in nature but for which there exist no commercial standards. [15] developed a capillary zone electrophoresis (CZE) and micellar electrokinetic chromatography (MEKC) to the simultaneous separation of cyanobacterial toxins (anatoxin-a, MC-LR, cylindrospermopsin) [14]. Long et al. [16] developed a fast and sensitive detection method for MC-LR with a portable trace organic pollutant analyzer (TOPA) based on the principle of immunoassay and total internal reflection fluorescence [15]. The limit of detection of 0.03 µg L^{-1} and the quantitative detection range of 0.1–10.1 µg L^{-1} was obtained while the cross-reactivity against a few compounds structurally similar to MC-LR was little. The recovery of MC-LR added to water samples at different concentrations ranged from 80 to 110% with RSD values less than 5%. [17] reported eight reversed-phase columns intended for rapid HPLC were assessed for the separation of thirteen MCs and nodularins [16]. Cong et al. [18] presented a novel chromatography electron spray ionization (ESI-MS) tandem triple quadruple mass spectrometry method to determine the trace amounts of major MCs in water. Solid phase extraction consuming 10 mL of water samples was used for sample clean-up and analyte enrichment. Limits of quantification (LOQ) and lower limit of detection were 0.05 and 1.0 ug/L, respectively [19].

*Corresponding author: Reddithota J Krupdam, National Environmental Engineering Research Institute, Nehru Marg, Nagpur 440020, India
E-mail: rj_krupadam@neeri.res.in

Successful extraction of target compounds from complex environmental samples using molecularly imprinted polyerms (MIPs) have been demonstrated for triazine herbicides [20], carcinogenic air pollutants polycyclic aromatic hydrocarbons [21,22], nicotine [23] and other environmentally relevant analytes [24]. Various experimental designs have been used including MIP packed into columns or cartridges, batch-mode where MIP is incubated with samples, and on-line SPE in combination with HPLC or HPLC-MS [25]. Very limited reports are available in literature on MC-LR imprinted polymers for both the purposes - solid phase extraction and sensory material. Using the combination of SPE followed by detection with piezoelectric sensor the minimum detectable amount of MC-LR was 0.35 nM [26]. The use of MIP-SPE provided up to 1000 fold pre-concentration, which was more than sufficient for achieving the required detection limit for MC-LR in drinking water (1 μg L^{-1}) [27]. In the present study, three MIPs specific for MC-LR were prepared with different formulations and then optimized the SPE conditions for MC-LR analysis in lake and marine waters by LC/MS detection. The merit of MIPs as specific SPE material for extraction of MC-LR with regard to sample pH and concentration of dissolved ions was described. The analytical data was treated for the quantitative aspects of the analysis, and accuracy, precision, and limits of detection.

Experimental

Chemical and Reagents

The analyte, MC-LR and other structural analogues of MCs namely MC-RR and MC-YR were purchased from Alexis Biochemicals (San Diego, USA). The functional monomers itaconic acid (IA), methacrylic acid (MAA), and 2-Acrylamido-2-methyl-1-propanesulfonic acid (AMPSA) and cross-linking monomer ethylene glycol dimethacrylate (EGDMA) were purchased from Sigma-Aldrich (Bachs, Switzerland); while the solvents acetonitrile, methanol and trifluoroacetic acid were procured from Merck (Darmstadt, Germany). 2,2'-azobisisobutyronitrile (AIBN) was brought from Acros Organics (Geel, Belgium). The molecular structures of chemicals used in the study are shown in (Figure 1). The monomers were purified prior to use via standard procedures in order to remove stabilizers. The polymerization initiator, AIBN was re-crystallised from acetone and the acetonitrile was purified by passing over over molecular sieves. All reagents used

Microcystins, targeted analytes

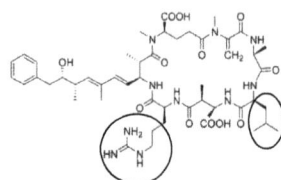

Microcystin-LR

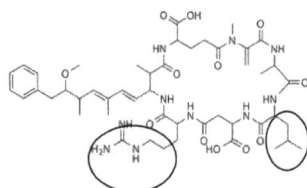

Microcystin-RR

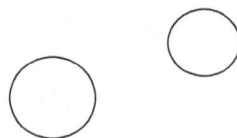

Microcystin-YR

MC variant	Molecular Weight	X	Z
MC-LR	994	Leucine	Arginine
MC-RR	1037	Arginine	Arginine
MC-YR	1044	Tyrosine	Arginine

Polymer Precursors

Functional Monomers

Methacrylic acid

Itaconic acid

2-Acrylamido-2-methyl-1-propanesulfonic acid

Cross-linking Monomer

Ethylene glycol dimethacrylate

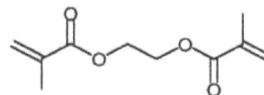

Figure 1: Molecular structures of MCs and other polymer precursors used in the preparation of molecularly imprinted polymers.

Polymer SPE material	Template	Functional monomers (μmol)			Cross-linking Monomer (mmol)	ACN mL	S $m^2 g^{-1}$	V_p $cm^3 g^{-1}$	d_p (A)
	MC-LR (μmol)	APMSA	MAA	IA	EGDMA				
MIP-1	1	-	10	-	1.5	5	380 ± 5	0.632	125
MIP-2	1	10	-	-	1.5	5	393 ± 4	0.645	118
MIP-3	1	-	-	10	1.5	5	416 ± 3	0.653	97

SPE, Solid Phase Extraction; MIP-1, molecularly imprinted polymer formulated with MAA and EGDMA; MIP-2, molecularly imprinted polymer formulated with APMSA and EGDMA and MIP-3, molecularly imprinted polymer formulated with IA and EGDMA.
MC-LR, microcystin-LR; APMSA, 2-Acrylamido-2-methyl-1-propanesulfonic acid; MAA, methacrylic acid; IA, itaconic acid; EGDMA, ethylene glycol dimethacrylate; ACN, acetinitrile.
S, BET surface area; V_p, specific pore volume; and d_p, average pore diameter were calculated from nitrogen absorption measurements.

Table 1: Composition and surface properties of molecularly imprinted polymers used in this study.

were either HPLC grade or analytical grade. Deionized water was obtained by passing tap water through a Milli-Q system and the treated water showed conductivity <15 MΩ cm^{-1} and TOC <5 μg L^{-1}. The stock solution of MCs were prepared at the concentration of 10 μg mL^{-1} in deionized water containing 0.1% trifluoroacetic acid. The reference SPE materials powdered activated carbon (PAC) and polystyrene divinylbenze resin (XAD) were purchased from Merck (Darmstadt, Germany). The lake water sample was collected from Ambazari lake (Nagpur, India) while the marine water sample was collected from Bay of Bengal near East Coast of Kakinada where enriched mangrove vegetation was found; and the salinity of the water was 17%.

Preparation of MIPs

Molecularly imprinted polymer specific to MC-LR was prepared by dissolving (9.93 mg, 1.0 mmol) in 10 mL acetonitrile in a 10 mL glass vial. To this, the functional monomer itaconic acid (210 mg, 2 mmol) was added and then the glass vial was placed in a refrigerator at 0°C for 30 min. Later, 5 mmol (991 mg) cross-linking monomer EGDMA and 10 mg AIBN were added to the solution. The sealed glass vial containing reaction mixture was freeze- thaw-degassed by submerging the tube in liquid nitrogen and holding the frozen tube under a vacuum of 100 mTorr for a period of 15 min. The polymer reaction mixture in the tube was sonicated for 5 min and kept in water bath at 40°C for 16 h. Upon completion of polymerization, the tube was taken out of the water bath and polymer monolith was ground in a ball mill to polymer particles of size 75 μm or less (200 mesh). The template MC-LR was extracted from the polymer matrix washing with methanol. The washing procedure was repeated (10 times) until no MC-LR found in the extraction solvent. Finally, the particles were dried under vacuum for further use. Apart from this MIP, two different MIPs were prepared with using methacrylic acid (MAA) and 2-Acrylamido-2-methyl-1-propanesulfonic acid (AMPSA) as the functional monomers. The composition of polymers was given in Table 1. The corresponding non-imprinted polymers (NIPs) was prepared in parallel in the absence of MC-LR and treated in the same manner.

MC-LR extraction experiments

The extraction capacity of MIP and other SPE materials was determined in batch mode. The dry SPE materials (MIPs formulated with 3 different compositions, XAD, or PAC; 10 mg) were weighed into 5mL glass vessels, and to this were added 0.1, 0.2, 0.3, 0.4, and 0.5 mL of a standard MC-LR solution (10 μg L^{-1}) mixed with a water in separate glass vessels. The samples were then shaken in a water bath at 25°C for 1 h. After sedimentation of the SPE materials, the residual concentration of MC-LR in aqueous phase was measured using LC/MS. The amount of MC-LR adsorbed was calculated by subtraction, using a calibrating curve obtained from the same experiment leaving out the SPE material. The reference SPE materials powdered activated carbon (PAC) and resin XAD were used for comparison with MIP for their adsorption

capacity and selectivity. For each SPE material, these experiments were repeated at least three times.

SPE cartridges and pre-concentration experiments

The SPE protocol was developed in an off-line mode using a SPE manifold supplied by SUPELCO (Bellefonte, PA) connected to vacuum pump. Two hundred milligrams of MIP was transferred into a 10 ml screw-cap scintillation vial and incubated with 3 ml of methanol. The sealed vial was allowed to stand 24 h at ambient temperature with occasional shaking. Then, the slurry was transferred into an 1 ml SPE cartridge equipped with a polyethylene frit. The polymer was allowed to settle for 5 min. The MIP bed was stabilized by careful insertion of a second frit, avoiding any compression of the polymer filling. The pre-conditioned MIP cartridges were sealed with pluggers and stored at 4 °C to prevent drying out by solvent evaporation. Prior to any extraction the polymer was washed with an eluting mixture of methanol/water (4:1) containing 1% trifluoroacetic acid until no more residual analyte (MC-LR) was eluted from the polymer. For SPE experiments, the MIP was conditioned with 10 mL of methanol and 10 mL of acidified milli-Q (pH 5.4). The required sample volume was applied to the conditioned cartridges and the polymer then washed with 10 mL of methanol.

For investigation of the reusability of the MIP , a single MIP cartridge was employed for 10 consecutive SPE clean-up cycles for a given MC-LR spiked deionized water (10 μg L^{-1}). In between the cycles, the MIP packed SPE column was reconditioned by washing with methanol/water (4:1) containing 1% trifluoroacetic acid (10 mL). The same procedure was followed for PAC and XAD filling, conditioning and pre-concentration of MC-LR.

The standard solution (10 μg L^{-1}) of MC-LR was prepared in methanol:water (4:1) solution, diluted with water obtained from the Milli-Q system, and then filtered through an empore disc to prepare a calibration curve ranging from 0.1-100 μg L^{-1}. Environmental water samples (lake water and marine waters) were filtered through a 0.45 um filter prior to any experiments. A standard spiked lake and marine water samples were used. The lake and marine water samples was preliminarily applied to total dissolved ions, and pH measurements to clarify its matrix effect, then filtered through 0.45 μm glass fiber filter prior to use without pH adjustment. The dissolved ions measurement was performed using conductivity meter. To verify the accomplishment of the molecular imprinting, the performance of the MIP was compared to that of non-imprinted polymer. Samples of 1 mL of lake water or marine water were analyzed using identical dimensions of SPE cartridge packed with MIP, XAD and PAC cartridges, separately, and the operational conditions of SPE were optimized. The selectivity of the MIP was determined by comparing the binding capacity of MC-LR on to the NIP and MIP packed SPE cartridges by analyzing 1 mL of samples of lake and marine waters with the optimized operational conditions.

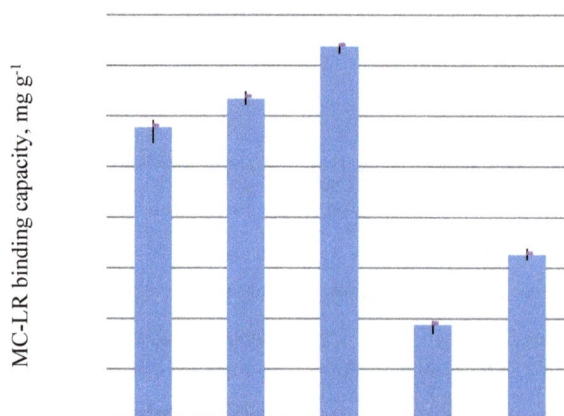

Figure 2: Binding capacity of MC-LR onto molecularly imprinted polymers (MIPs) and other SPE materials used in the present study.

SPE materials	Extraction capacity at different solution pH, mg g^{-1}					
	5.0	6.0	7.0	8.0	9.0	10.0
MIP-1	2.76 ± 0.13	2.83 ± 0.11	2.89 ± 0.13	2.82 ± 0.16	2.85 ± 0.16	2.72 ± 0.11
MIP-2	3.11 ± 0.15	3.15 ± 0.13	3.17 ± 0.16	3.15 ± 0.11	3.17 ± 0.11	2.96 ± 0.13
MIP-3	3.63 ± 0.11	3.65 ± 0.12	3.69 ± 0.19	3.65 ± 0.12	3.68 ± 0.13	3.12 ± 0.09
XAD	0.62 ± 0.08	0.85 ± 0.06	0.94 ± 0.11	0.96 ± 0.11	0.82 ± 0.15	0.59 ± 0.13
PAC	1.47 ± 0.11	1.51 ± 0.13	1.63 ± 0.12	1.68 ± 0.11	1.55 ± 0.09	1.32 ± 0.15

The values reported in the table are mean of three experimental determinations (n=3) and the initial concentration of MC-LR was 10 µg L^{-1}.

Table 2: Effect of solution pH on MC-LR extraction by SPE materials.

Instrumentation

The microcystins were qauntified using LC/MS with a PE series 200 Quaternary Pump and a PE series 200 auto sampler (PerkinElmer, Shelton CT, USA). The chromatographic separation was carried out on a reversed-phase column C18 with dimensions-3.9 mm×150 mm id 5µm (Aqua, Torrance CA, USA). Apart from using C18, the columns packed with different SPE materials such as MIPs, PAC and XAD were also used for chromatographic separation of MCs. The injection volume was 1 mL min^{-1}. The mobile phase was a mixture of deionized water, ammonium acetate (10 mMl pH, 7.0), methanol, and acetonitrile. Mass spectrometry (MS) measurements were performed using an API 165 MS with an atmospheric pressure ionization (API) source operating in turbo-ion spray (TIS) mode (Applied Biosystems, Foster city CA, USA). The elute from the LC column was transferred to the MS device using split ratio of 5:1 (volume to waste/volume transferred) and nitrogen (heated to 450°C; 7 L min^{-1}) was applied to dry ion spray aerosol. Nitrogen was also used as nebulizer gas with a glow of 0.7 L^{-1}. The ionization voltage of the TIS interface was set to 5.5 kV. The MS system was operated in positive multiple ion detection (MID) mode to give highest sensitivity and selectivity [M+H]$^+$ ions cantered at: m/z, 520.5 (MC-RR); 995.7 (MC-YR); and 1045.8 (MC-LR) were monitored. For determination of unknown MCs the following mass ranges were scanned in additional runs: 500-600 Da (doubly charged MCs) and 800-1200 Da (singly charged MCs).

Analytical Performance

MIP- SPE coupled to LC/MS was developed to determine MCs in lake and marine water samples. The linearity of the analytical method was evaluated by a calibration curve in the range of 0.1–100 ng L^{-1} of MC-LR ($n = 5$). The Milli-Q water was spiked with MC-LR to achieve final concentrations of 0.1, 0.5. 1. 5, 10, 20, 50, and 100 ng L^{-1}. The limit of detection (LOD) was defined as three times ratio of signal to noise. The loading volume of aqueous MC-LR standard solution was 100 mL. Finally, quantitative figures of merit for the optimized SPE-LC/MS method were determined through analytical curves estimated using water samples of the MCs in the concentration range from 0.1 to 100 ng L^{-1}. The environmental samples were collected in glass bottles from lake and marine waters; then filtered through 0.45 um glass fibre filter before storing in dark at 4°C. The SPE procedure was accomplished within 24 h to avoid any microbial degradation of MCs.

Results and Discussion

Extraction of MCs

The MCs extraction capacity of MIPs (formulated with different polymer composition) and other SPE materials were determined using batch equilibrium experiments. The extraction capacity of MIP, XAD and PAC were 3.69, 0.94, and 1.63 mg g^{-1} respectively, and also extraction capacity of MIP was 6-folds higher than the corresponding non-imprinted polymer (NIP). The MIP formulated with IA and EGDMA showed significantly higher capacity than the other MIPs and SPE materials (Figure 2). Such high MCs extraction capacity is expected because of existence of specific binding sites in the polymer. The binding capacity of MIPs with the formulation of MAA - EGDMA and AFMSA -EGDMA were less pronounced than the MIP prepared with IA-EGDMA formulation; this could be explained based on formation of the lower affinity binding sites during molecular imprinting of former cases. The extraction capacities of the MIP prepared with IA-EGDMA formulation showed superior MCs binding capacity than those of most MIPs reported in the literature [25,26]. The highest extraction capacity of MIP for MC-LR reported in the literature was 0.98 mg g^{-1} which was about one-third of the MIP prepared in this study. The reason may be formation of clear and high populous binding sites during molecular imprinting process, thereby increase in the surface area of the MIP (formulated with IA-co-EDGMA) compared with NIP suitably justify high MCs extraction capacity. The parameters contributing to the molecular imprinting was formation of the porous structure with an excellent specific surface areas for binding and the increasing number of effective binding sites. The microscopic characteristics of the imprinted polymer, and porous surface could be clearly observed. The specific surface area, pore volume, and pore size distribution obtained from nitrogen adsorption experiments were given in Table 2. The nitrogen porosimeter data showed that the pore size of IA formulated MIP was quite smaller than the MIPs formulated with AFMSA and MAA functional monomers. Earlier studies related to MC-LR were reported primarily in organic media, where higher binding capacity and selectivity were reported [27,28]. In this study, the high affinity of MIP for MC-LR in water was demonstrated, where the MIP bind

kinetically faster to MC-LR (more than 90% of MC-LR within 60 min) and the binding equilibrium was attained within 10 min between MIP and MC-LR (Figure 3). However, the conventional SPE materials, PAC and XAD extracted lower than 25% of the MC-LR when compared with MIP in the same period. The binding equilibrium attained for other SPE materials XAD and PAC was about 2 h. The short contact time needed to reach the binding equilibrium as well as the high binding capacity of the MIP are very much relevant for practical application of sample pretreatment.

The selective extraction of MC-LR among other MCs (MC-YR, and -RR) was studied onto MIP and other SPE materials. From the (Figure 4), it would be clear that only MC-LR was selectively adsorbed onto the MIP from aqueous solution, however, 14% and 19% of MC-RR and -YR variants were adsorbed, respectively. Based on the adsorption equilibrium experimental data MIP formulated with IA-co-EGDMA was used as SPE material for packing column, while the non-MIPs packed SPE columns showed almost equal adsorption capacity for all the MCs and the nature of adsorption would be non-specific.

Chromatographic evaluation of the SPE materials

The extraction capacity and selectivity data showed that the MIP possesses higher selective extraction capacity than the other SPE materials studied for MC-LR extraction from water. The hydrogen bonding between MC-LR and the functional groups of IA, which often occurs in a polar solvents cause specific hydrogen bonds responsible for suppressing the nonspecific hydrophobic interactions. Nevertheless, nonspecific hydrophobic interactions always exist in the adsorption process of MIP in the water samples. To avoid interference of water soluble compounds in the analysis of the target molecule, a washing step is normally included in the analytical protocol.

The washing step was the most crucial procedure to maximize the specific interactions between the MCs and binding sites of MIP, and to simultaneously decrease non-specific interactions to discard matrix components in the polymer [29]. However, the comparative analysis between NIP and MIP was carried out denote useful information about MIP's selective extraction of MC-LR from water samples. Acetonitrile was proved the most effective washing solvent, though MC-LR could not be eluted from NIP completely. Thus, acetonitrile was chosen as the washing solution. When acetonitrile was used at a volume of 1 mL, about 40% of MC-LR loaded on NIP cartridge was washed off while MC-LR bound on MIP was still retained. With an increase of the acetonitrile volume to 2 mL, the amount of MC-LR eluted from the NIP cartridge increased to 70%. Therefore, acetonitrile was mixed with small quantity of water (4:1, v/v) to improve the polarity showed an excellent washing step. 2 mL of MC-LR standard solution (10 ng L^{-1}) in water was applied to the MIP and NIP cartridges. After loading MC-LR solution, both the MIP and NIP cartridges were submitted to washing step, then the cartridges were eluted with 2 mL of acetonitrile/ water (4:1, v/v). Both the washing and elution fractions of the solvent were collected and analyzed. Table 3 shows the recoveries of MC-LR in the washing and elution fractions after pre-concentration on the MIP and NIP cartridges by using 2 mL of each of the washing solvents. For the elution, five aliquots of methanol: trifluoroacetic acid (4:1, v/v), each of 1 mL in volume, were used to elute MC-LR from the MIP cartridge after washing step. The recovery for every 1 mL aliquot of methanol:trofluoroacetic acid (4:1, v/v) was calculated separately. The results showed that 2 mL of the eluting solution was sufficient to elute MC-LR from MIP cartridge completely.

The pre-concentration capacity of a SPE column was evaluated using breakthrough volume experiments. The breakthrough volume

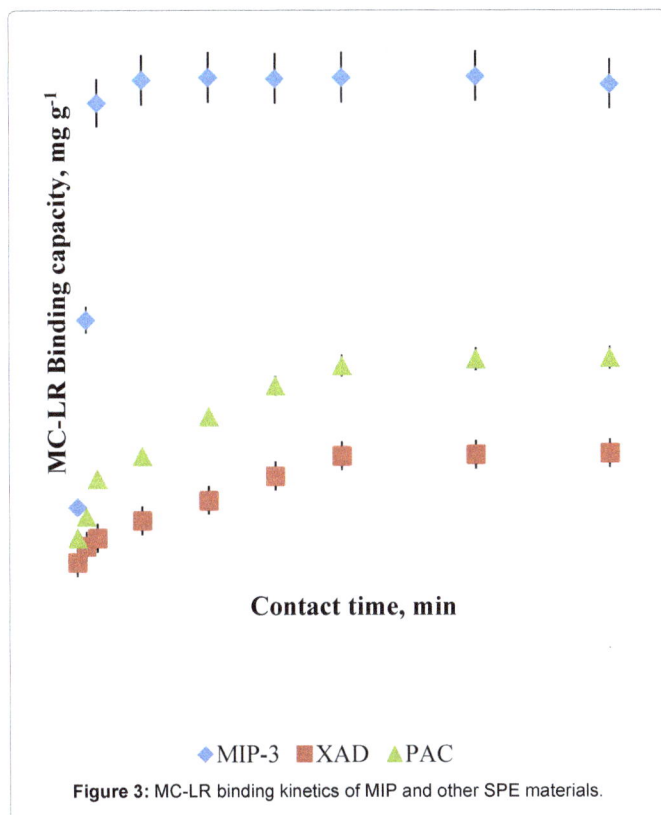

Figure 3: MC-LR binding kinetics of MIP and other SPE materials.

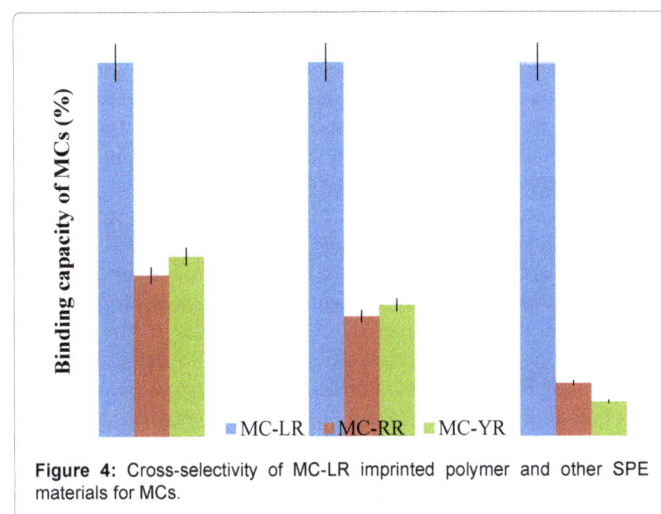

Figure 4: Cross-selectivity of MC-LR imprinted polymer and other SPE materials for MCs.

Compound	Linearity			DL[a] (µg L^{-1})	RSD (%); n=5
	Range (ug/L)	Equation	r		
Microcystin-LR	0.1-10	Y=264+4272X	0.996	0.25	4.2
Microcystin-RR	0.5-10	Y=186+2164X	0.991	0.39	5.8
Microcystin-YR	0.3-10	Y=490+1550X	0.993	0.43	4.7

[a]Detection limits were estimated on the basis of 3:1 signal to noise ratios.
RSD, relative standard deviation (concentration range tested: 01 - 10 µg MC-LR on MIP packed column.

Table 3: The linear range, detection limit (DL) and RSD of MIP-SPE method coupled with HPLC for the determination of microcystins (microcystin-LR, RR, YR).

	Recovery parameter	MIP			XAD			PAC		
		MC-LR	MC-RR	MC-YR	MC-LR	MC-RR	MC-YR	MC-LR	MC-RR	MC-YR
Lake water	Recovery (%)	96	91	93	82	86	77	78	75	77
	RSD (%)	3.2	3.7	4.1	5.8	4.9	6.2	6.5	7.2	6.2
Marine water	Recovery (%)	89	76	82	73	61	77	68	65	61
	RSD (%)	3.5	8.9	5.3	6.4	8.9	7.3	8.3	7.9	8.1

Recovery is based on comparison of peak areas obtained by MC-LR and mixture of MCs by injecting the same amount of cyanotoxin at a concentration of 10 µg L^{-1}.

Table 4: Recoveries of microcytins from the spiked lake and marine waters (n=5).

of the MIP packed SPE column was thrice as much as PAC and XAD columns; which indicates that MIP used in the present study could be used as SPE material for extracting MC-LR from aqueous samples, where high volumes of water are usually loaded. As shown in Table 4, there is no MC-LR loss even if 100 mL of MC-LR standard solution was loaded. Moreover, the MIP packed SPE column had two advantages than the reported MC-LR imprinted polymers; one was the MIP prepared in this study was high extraction capacity for the MC-LR extraction from aqueous solutions than the earlier reported. The other one was that the MIP packed SPE could selectively retain the MC-LR from lake and marine samples and avoided the bleeding of traces of retained MC-LR from the MIP packed SPE column.

Matrix effect on MC-LR extraction

The level of total dissolved solids (TDS) in water and wastewater detrimentally interfere in extraction of MC-LR onto the SPE materials. The interference of TDS was investigated using different strengths of sodium chloride solution. In deionized water, the TDS solutions of concentration from 100 to 5000 mg/L were prepared. The extraction of MC-LR from different strengths of TDS was investigated in batch mode and the data obtained is shown in Figure 5. As the increase in TDS strength, it was found that, MC-LR extraction also increased exponentially. The possible reason would be the addition of NaCl enhanced the mass transfer in the extraction process, and also water molecules solvate around ions leads to poor solubility [30]. This enhanced the extraction of MC-LR on MIP. Since, the diffusion of MC-LR onto the MIP from water is the rate-limiting step, it suggested that the addition of salt speeds up diffusion of MC-LR. Another important matrix parameter is the solution pH. The effect of pH on MC-LR binding onto MIP was investigated by varying the pH of the MC-LR solution from pH 2 to pH 9. Interestingly, as the pH of MC-LR solution increases, MIP displayed much higher retention on MIP than the corresponding NIP. This binding could be due to the formation of ionic interactions between MC-LR and the MIP. The optimum pH for MC-LR binding was 7.3. It would be possible that at the optimum pH (7.3), the MIP lose its charge and the non-specific binding caused by the ionic bonds decreases. This statement would be supported by the dominance of H-bonding arising from functional monomer, IA during molecular recognition of MIP.

LC/MS analysis of MCs

20 mL of standard aqueous MC-LR solution containing 10 µg was passed through the SPE column packed with MIP was passing with the

sample flow rate 2 ml min^{-1}. The columns packed with NIP and other SPE materials (XAD and PAC) were also used for extraction of MC-LR from standard solution. The chromatogram obtained using the MIP packed SPE column showed clear and distinct peaks for MCs; while the chromatograms of cartridges packed with XAD and PAC were quite small and almost similar to the chromatogram of non-imprinted polymer. The sample preparation using MIP packed columns is very effective for MC-LR at trace quantities. These results indicated that the MC-LR was selectively extracted onto the MIP column, and the selectivity of MIPs was quite high for MC-LR, and the recognition sites formed in the MIP originated from the molecular imprinting.

The analytical figures of merit for the proposed sample preparation with MIP packed SPE extraction followed by LC/MS analysis for trace quantification of MC-LR was evaluated under optimal experimental conditions. The mass spectra of [M+H]$^+$ ions of MCs revealed that the site of protonation for MCs was, the methoxy group in the Adda. While the differences in mass spectra between MC-LR and MC-YR was explained based on structural variations. Besides, detecting typical fragment ions of MCs in the mass spectra of MC-YR (m/z, 135 and 213), the other fragment ions, such as m/z 141, 271, 602 and 952, were characteristic for MC-YR. The fragmented ions observed for the three variants of MCs (MC-LR, -RR, and -YR) arise predominantly from the loss of water and then consecutive cleavages of the amide bonds that provide information concerning the amino acid sequences. With a sample loading flow rate of 2 ml min^{-1} for a 10 min extraction, the enrichment factor obtained by the slopes of the linear portion in comparison with the direct injection of 10 µL standard sample solution was 1.045. The detection limit (S/N=3) of 1.0 ng L^{-1} was based on three times of the signal-to-noise (S/N) ratio of baseline near the analyte peak obtained from the LC/MS analysis. The chromatograms of standard mixture of MCs are depicted in Figure 6a. Reproducibility was evaluated by pre-concentrating nine replicate runs of water samples spiked at 10 µg L^{-1}, and the results were satisfactory with relative standard deviation (RSD) of <7%. Calibration curve was obtained for MC-LR in this range by five-point calibration with concentration coefficients for the linear regression curve of 0.998, and the linear range of the calibration graph was 0.05-10 µg L^{-1}. The reliability of an analytical method is proven when it is applied to the real samples. For this purpose, two different type of samples were analyzed with the methodology developed in this work. One sample was from the Lake Ambazari (Nagpur, India) and the another sample was collected from Godavari estuary (Bay of Bengal Sea, India). The chromatograms obtained after pre-concentration of

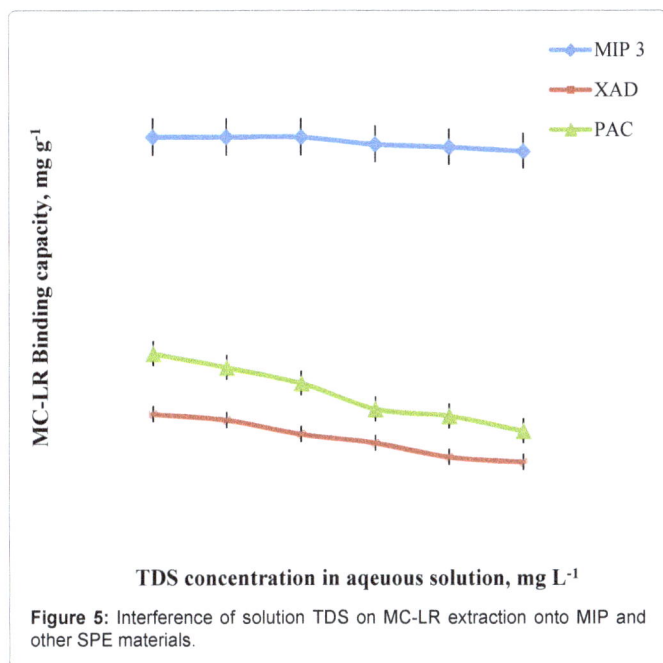

Figure 5: Interference of solution TDS on MC-LR extraction onto MIP and other SPE materials.

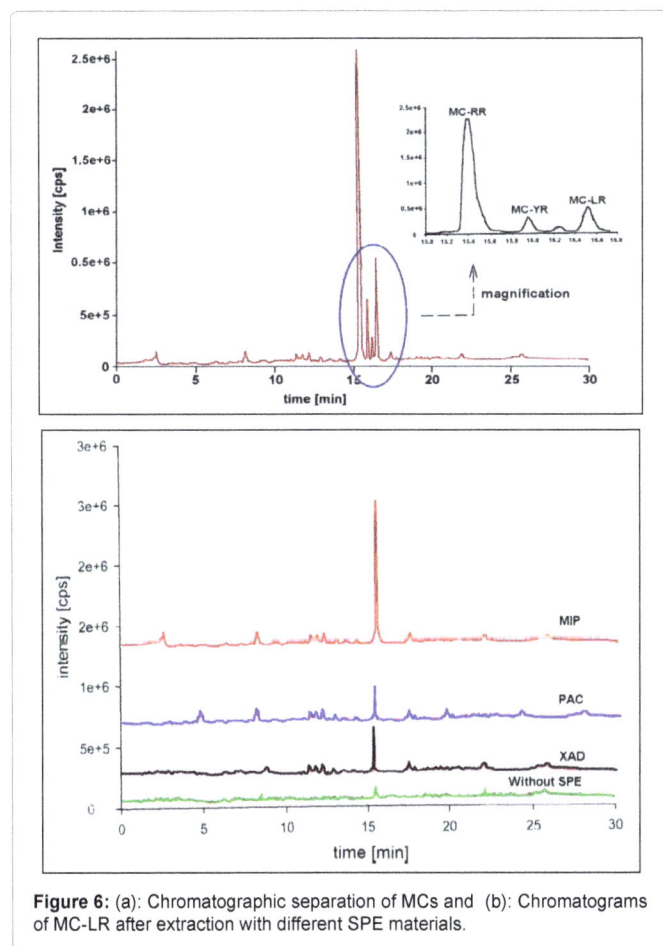

Figure 6: (a): Chromatographic separation of MCs and (b): Chromatograms of MC-LR after extraction with different SPE materials.

aqueous samples through the cartridges of (i) the MIP (ii) the XAD and (iii) the PAC, the MIP chromatogram presents a clearer baseline, and the MCs were identified and detected (Figure 6b). The concentration of MC-LR in aqueous samples were given in Table 4. Recovery of

MCs from spiked lake water and marine water was determined by the comparison of peak areas obtained by standard MC-LR (10 µg L⁻¹) and the mixture of MCs (each MC concentration was 10µg L⁻¹) injecting into HPLC of sample volume 2 mL. The reproducibility of this method was observed because of the high recoveries of MIP. To our surprise, the observed reproducibility of the method was satisfactory in three sequential cycles of extraction. Certainly, the extractions were repeated on a single MIP packed SPE column, and thus results also demonstrated the regeneration of the column. MIP imprinted with MC-LR showed highest recovery (96%) for MC-LR demonstrates that the imprinting is based on the synergistic effect of shape and size complementarily and the interaction of functional groups of the analyte, MC-LR with those of binding cavities. The lower recoveries for MC-RR and MC-YR are due to mismatch of size and shape of MIP's binding cavity. The other SPE materials XAD and PAC showed lower recoveries for all MCs, however, the recoveries are non-selective in nature. The amount of MC-LR in the lake water and marine water were 0.87 and 1.13 mg L⁻¹ respectively, and these values are higher than the WHO standard for MC-LR. The concentration of MC-LR was sufficiently high and could be easily assayed using SPE coupled with LC/MS detection. Normally, an additional clean-up step before LC/MS detection is very much essential to remove matrix interfering parameters (dissolved solids and organic matter). However, since selective interactions between the MC-LR and the binding sites of MIP are achieved using imprinted polymers allows lower coefficients of variation in the measurements, where there is no need of sample clean-up step.

Conclusions

The new analytical method proposed using LC/MS analysis with MIP extraction is highly efficient method for trace quantification of MC-LR in lake and marine waters. The experimental variables were optimized for MIP packed column for selective separation of MCs from aqueous samples. The MC-LR recovery data suggested that the MIP packed column provided a reliable and effective recovery of MC-LR (i.e., about 95%) in the concentration range 0.1-10 µg L⁻¹ from aqueous samples. The results obtained for calibration linearity, precision, accuracy and mass ratio stability of MC-LR represents that the proposed method is quite efficient for the trace quantification of MC-LR in environmental samples. With a 100-fold selective pre-concentration step, the LOD is 1.0 µg L⁻¹ which represent one of the most sensitive detection methods of MC-LR in environmental analysis.

Acknowledgment

Financial support from the Council of Scientific & Industrial Research (CSIR) and Planning Commission of India under Molecular Environmental Science and Engineering Research (ME3ER) are gratefully acknowledged.

References

1. Carmichael WW (1992) Cyanobacteria secondary metabolites--the cyanotoxins. J Appl Bacteriol 72: 445-459.

2. Carmichael WW, Azevedo SM, An JS, Molica RJ, Jochimsen EM, et al. (2001) Human fatalities from cyanobacteria: chemical and biological evidence for cyanotoxins. Environ Health Perspect 109: 663-668.

3. Kaebernick M, Neilan BA (2001) Ecological and molecular investigations of cyanotoxin production. FEMS Microbiol Ecol 35: 1-9.

4. Welker M, Fastner J, Erhard M, von Döhren H (2002) Applications of MALDI-TOF MS analysis in cyanotoxin research. Environ Toxicol 17: 367-374.

5. Lindon M, Heiskary S (2009) Blue-green algal toxin (microcystin) levels in Minnesota lakes. Lake and Reservior Manag 25: 240-252.

6. Dodds WK, Bouska WW, Eitzmann JL, Pilger TJ, Pitts KL, et al. (2009) Eutrophication of U.S. freshwaters: analysis of potential economic damages. Environ Sci Technol 43: 12-19.

7. Pouria S, de Andrade A, Barbosa J, Cavalcanti RL, Barreto VT, et al. (1998) Fatal microcystin intoxication in haemodialysis unit in Caruaru, Brazil. Lancet 352: 21-26.

8. Martínez Hernández J, López-Rodas V, Costas E (2009) Microcystins from tap water could be a risk factor for liver and colorectal cancer: a risk intensified by global change. Med Hypotheses 72: 539-540.

9. Chen J, Xie P (2005) Seasonal dynamics of the hepatotoxic microcystins in various organs of four freshwater bivalves from the large eutrophic lake Taihu of subtropical China and the risk to human consumption. Environ Toxicol 20: 572-584.

10. Fastner J, Codd GA, Metcalf JS, Woitke P, Wiedner C, et al. (2002) An international intercomparison exercise for the determination of purified microcystin-LR and microcystins in cyanobacterial field material. Anal Bioanal Chem 374: 437-444.

11. Hyenstrand P, Metcalf JS, Beattie KA, Codd GA (2001) Effects of adsorption to plastics and solvent conditions in the analysis of the cyanobacterial toxin microcystin-LR by high performance liquid chromatography. Water Res 35: 3508-3511.

12. Allis O, Dauphard J, Hamilton B, Shuilleabhain AN, Lehane M, et al. (2007) Liquid chromatography-tandem mass spectrometry application, for the determination of extracellular hepatotoxins in Irish lake and drinking waters. Anal Chem 79: 3436-3447.

13. Wang J, Pang X, Ge F, Ma Z (2007) An ultra-performance liquid chromatography-tandem mass spectrometry method for determination of microcystins occurrence in surface water in Zhejiang Province, China. Toxicon 49: 1120-1128.

14. Rapala J, Erkomaa K, Kukkonen J, Sivonen K, Lahti K (2002) Detection of microcystins with protein phosphatase inhibition assay, high performance liquid-chromatography-UV detection with enzyme linked immunosorbent assay: Comparison of methods. Anal Chim Acta 466: 213-231.

15. Vasas G, Szydlowska D, Gáspár A, Welker M, Trojanowicz M, et al. (2006) Determination of microcystins in environmental samples using capillary electrophoresis. J Biochem Biophys Methods 66: 87-97.

16. Long F, He M, Zhu AN, Shi HC (2009) Portable optical immunosensor for highly sensitive detection of microcystin-LR in water samples. Biosens Bioelectron 24: 2346-2351.

17. Neffling MR, Spoof L, Meriluoto J (2009) Rapid LC-MS detection of cyanobacterial hepatotoxins microcystins and nodularins--comparison of columns. Anal Chim Acta 653: 234-241.

18. Cong L, Haung B, Chen Q, Lu B, Zhang J, et al. (2006) Determination of trace amount of microcystins in water samples using liquid chromatography coupled with triple quadruple mass spectrometry. Anal Chim Acta 569: 157-168.

19. Fontanals N, Marce RM, Borrull F (2005) New hydrophilic materials for solid-phase extraction. Trends Anal Chem 24: 394-406.

20. Krupadam RJ, Khan MS, Wate SR (2010) Removal of probable human carcinogenic polycyclic aromatic hydrocarbons from contaminated water using molecularly imprinted polymer. Water Res 44: 681-688.

21. Krupadam RJ, Bhagat B, Wate SR, Bodhe GL, Sellergren B, et al. (2009) Fluorescence spectrophotometer analysis of polycyclic aromatic hydrocarbons in environmental samples based on solid phase extraction using molecularly imprinted polymer. Environ Sci Technol 43: 2871-2877.

22. Krupadam RJ, Venkatesh A, Piletsky SA (2013) Molecularly imprinted polymer receptors for nicotine recognition in biological systems. Mol Imprint 1: 27-34.

23. Pichon V, Chapuis-Hugon F (2008) Role of molecularly imprinted polymers for selective determination of environmental pollutants--a review. Anal Chim Acta 622: 48-61.

24. He C, Long Y, Pan J, Li K, Liu F (2007) Application of molecularly imprinted polymers to solid-phase extraction of analytes from real samples. J Biochem Biophys Methods 70: 133-150.

25. Oehrle SA, Southwell B, Westrick J (2010) Detection of various freshwater cyanobacterial toxins using ultra-performance liquid chromatography tandem mass spectrometry. Toxicon 55: 965-972.

26. Mekebri A, Blondina GJ, Crane DB (2009) Method validation of microcystins in water and tissue by enhanced liquid chromatography tandem mass spectrometry. J Chromatogr A 1216: 3147-3155.

27. Chianella I, Lotierzo M, Piletsky SA, Tothill IE, Chen B, et al. (2002) Rational design of a polymer specific for microcystin-LR using a computational approach. Anal Chem 74: 1288-1293.

28. Chianella I, Piletsky SA, Tothill IE, Chen B, Turner AP (2003) MIP-based solid phase extraction cartridges combined with MIP-based sensors for the detection of microcystin-LR. Biosens Bioelectron 18: 119-127.

29. Pichon V (2007) Selective sample treatment using molecularly imprinted polymers. J Chromatogr A 1152: 41-53.

30. Turner NW, Piletska EV, Karim K, Whitcombe M, Malecha M, et al. (2004) Effect of the solvent on recognition properties of molecularly imprinted polymer specific for ochratoxin A. Biosens Bioelectron 20: 1060-1067.

18

Simple Method for Standardization and Quantification of Linoleic Acid in *Solanum nigrum* Berries by HPTLC

Arpan Chakraborty[1], Arka Bhattacharjee[2], Pallab Dasgupta[3], Debasmita Manna[4], Won Chun OH[5] and Goutam Mukhopadhyay[3]*

[1]*Department of Pharmaceutical Technology, Jadavpur University, Kolkata, West Bengal, India*
[2]*Department of Pharmaceutical Sciences and Technology, Birla Institute of Technology, Mesra, Ranchi, India*
[3]*BCDA College of Pharmacy and Technology, Kolkata, West Bengal, India*
[4]*NSHM College of Pharmaceutical Sciences, Kolkata, West Bengal, India*
[5]*Department of Material Science and Engineering, Hanseo University, South Korea*

Abstract

Linoleic acid is a polyunsaturated omega-6 fatty acid found as phytoconstituent in many plant species. It is reported to have activity such as 5-alpha-reductase-inhibitor, antianaphylactic, antiarthritic, antiatherosclerotic, anticancer, anticoronary, antieczemic, antifibrinolytic and many others. In this research work we have developed a method by HPTLC to identify and standardize linoleic acid in methanolic extract of *Solanum nigrum* berries. Calibration curve of Linoleic acid was plotted and was found to have r^2 value of 0.99451. The percentage content of Linoleic acid in *Solanum nigrum* berries methanolic extract was found to be 9.32% w/w.

Keywords: Linoleic acid; *Solanum nigrum*; Finger print; HPTLC

Abbreviations: HPTLC: High Performance Thin Layer Chromatography; μg: Microgram; mL: Mililitre; R_f: Retention Factor; nm: Nanometer; μL: Microlitre; RSD: Relative Standard Deviation; TLC: Thin Layer Chromatography; °C: Degree Centigrade.

Introduction

Solanum nigrum (black nightshade) is a medicinal plant member belonging to the family Solanaceae. *S. nigrum* has been extensively used traditionally to treat various ailments. The juice of the berries used as an antidiarrhoea, opthalmopathy and hydrophobia. It is also used in anasarca and heart disease. Berries are reported to posses tonic, diuretic and cathartic properties. Seeds are useful in dipsia and giddiness [1].

Linoleic acid is a polyunsaturated omega-6 fatty acid. It is reported to have a number of useful physiological activities including 5-alpha-reductase-inhibitor, antianaphylactic, antiarthritic, antiatherosclerotic, anticoronary, anticancer, antieczemic, antifibrinolytic, antigranular, antihistaminic, anti-inflammatory, antimenorrhagic, antiprostatitic, hepatoprotective, hypocholesterolemic, immunomodulatory, insectifuge, metastatic, nematicide [2]. *Solanum nigrum* oil is reported to have linoleic acid as the most abundant unsaturated fatty acid found in it [3,4].

For the establishment of a consistent biological activity of any herbal item, standardization is an important step. It provides a consistent chemical profile, or simply a quality assurance program for production and manufacturing of an herbal drug. Standardization thus, is a tool in the quality control process. HPTLC is a modern adaptation of Thin Layer Chromatography with better and advanced separation efficiency and detection limits. HPTLC provides a number of advantages including easy separation process for with colored compounds. HPTLC can be used for different modes of evaluation, allowing identification of compounds having different light-absorption [5]. HPTLC is a method to standardize and identify the chemical ingredients which is expected to be present in a medicinal plant. This is done from regulatory perspective to ensure the efficacy, quality as well safety of the herbal drugs present in a plant. Thus it provides a very reliable way of determining the purity and percentage content of the active biomarker in the plant extracts.

Standardization of herbal products is a current issue of interest.

For quality control of these herbal materials or their extracts one needs to proceed by selecting one of the different phytoconstituents of the product, preferably the one showing maximum desired bioactivity and subsequent method of quantification of that specific constituent is required to be developed. The method so accepted should be simple and cost effective. Previously mentioned standardization methods of *Solanum nigrum* berries states use of gas chromatography, high performance liquid chromatography which by quantifying linoleic acid or other phytoconstituents [6,7]. But these methods are expensive. Standardization and quantification of linoleic acid in *Solanum nigrum* berries by HPTLC for quality control of this herbal material is an effective and simple method and may be used with necessary modifications for estimation of the a fore said phytoconstituent in other plant materials containing the same.

Materials and Methods

Instrumentation and reagents

The CAMAG HPTLC system used comprised of WINCATS software, LINOMAT V automatic sample applicator, and automatic development chamber, scanning densitometer CAMAG scanner 3 and photo documentation apparatus CAMAG reprostar-3. Stationary phase was used as aluminum based silica gel plate 60 F_{254} (Merck, Mumbai) with 10 cm × 10 cm in a particle size of 5-10 μm. All the solvents were used of analytical grade.100 μL syringe (Hamilton, Switzerland) was used for sample application on HPTLC plates. Linoleic acid was purchased from Sisco Research Laboratories (SRL). Methanol, n-hexane and ethyl acetate (analytical grade) were procured from Merck (Mumbai, India).

*Corresponding author: Goutam Mukhopadhyay, BCDA College of Pharmacy and Technology, Kolkata, West Bengal-700 127, India
E-mail: gmukhopadhyay8@gmail.com

All the samples were filtered through Whatman's syringe filter (NYL 0.45 μm). The plant material was collected from local area of Kolkata, West Bengal, India and the identity was confirmed by taxonomist. The shade dried powdered berries of *Solanum nigrum* were extracted with methanol by cold maceration. The extract solution was dried under reduced pressure with the help of rotary evaporator.

Methods

Preparation of standard solution: About 1 mg of linoleic acid standard was taken in 1 mL eppendorf tube 1.0 mL of methanol was added to it and mixed in vortex mixture till the material got completely dissolved. It was then filtered through 0.45 μ syringe filter and kept for further study.

Preparation of calibration curve of linoleic acid: HPTLC analysis was performed using isocratic technique. The mobile phase was optimized with n-hexane and ethyl acetate in a ratio of 5:4 v/v. The temperature was kept at 25°C and mobile phase was developed in a twin trough glass chamber. The standard solution was applied 2, 4, 6, 8, 10 μL in a band wise fashion. After development the plate was dried. Then the dry plate was treated by spraying sulphuric acid-anisaldehyde reagent. The plate was kept at 110°C for 5 minute in hot air oven and evaluation was carried out at 366 and 540 nm. Calibration curve of linoleic acid was obtained by plotting peak areas versus concentrations of linoleic acid applied.

Identification and quantification of linoleic acid in *Solanum nigrum* berries

Preparation of sample solution: About 5 mg of *Solanum nigrum* berries methanolic extract was dissolved in 1 mL methanol in an eppendorf tubes. It was then mixed in vortex mixture and subjected to ultrasonication bath till the material completely dissolved. Then it was filtered through 0.45 μ syringe filter. Sample solution was applied consequently in the range of 4, 8 and 12 μL and was subjected to a fore mentioned chromatography.

Evaluation of method by some parameters

Instrumental precision: Instrumental precision was checked by repeated scanning (n=5) of the same spot of linoleic acid (2 μg/spot) and expressed as relative standard deviation (%RSD).

Repeatability: The repeatability of the method was affirmed by analyzing 2 μg/spot of linoleic acid TLC plate (n=5) and expressed as %RSD.

Recovery: The accuracy of the method was assessed by performing recovery study at three different levels (50%, 100% and 125% addition of linoleic acid). The percent recoveries and the average percent recoveries were calculated.

Results

Calibration curve of linoleic acid

The calibration curve was found to be linear with the equation of Y=3141.508 × X+1366.840 (correlation coefficient=0.9954), where X represents amount of linoleic acid and Y represents area under the curve. R_f value of standard linoleic acid was found to be 0.65.

Quantification of linoleic acid in *Solanum nigrum* berries

The percentage content of linoleic acid in *Solanum nigrum* berries methanolic extract was found to be 9.32% w/w. R_f value of standard linoleic acid was found to be 0.65. Specificity was confirmed by comparing the R_f of standard and sample. Figures 1 and 2 represents HPTLC Chromatogram of standard Linoleic acid and that of extract of *Solanum nigrum* berries. Photodocumentaton of *Solanum nigrum* berries methanolic extract at 366 nm and 540 nm are shown in Figures 3 and 4. Figure 5 represents all tracks at 540 nm having 5 standard and three samples respectively from left hand side.

Evaluation of method by some parameters

Instrumental precision was within range of 1.97-2.08 with% RSD of 1.65% whereas repeatability was found to be in range of 1.98-2.08 with % RSD of 1.95% percentage recovery data is reported in Table 1. with standard deviation of the amount of marker found.

Discussion

Chromatographic fingerprint analysis has proven to be a rational and feasible approach for the assessment of quality and authentication of species of traditional medicine [8-10]. It effectively uses chromatographic techniques to form specific patterns of recognition for phytochemicals. The developed fingerprint pattern of components can thereafter be utilized to check the presence of markers of interest as well as the ratio of all detectable analytes. Though there are some shortcomings of high performance thin layer chromatography, such as

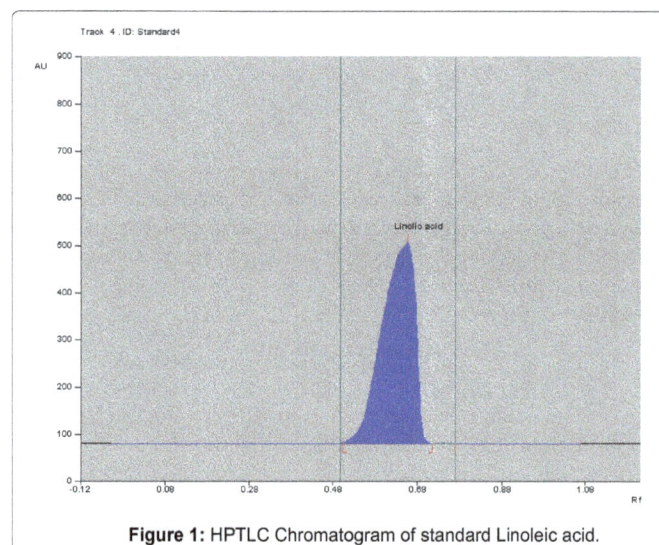

Figure 1: HPTLC Chromatogram of standard Linoleic acid.

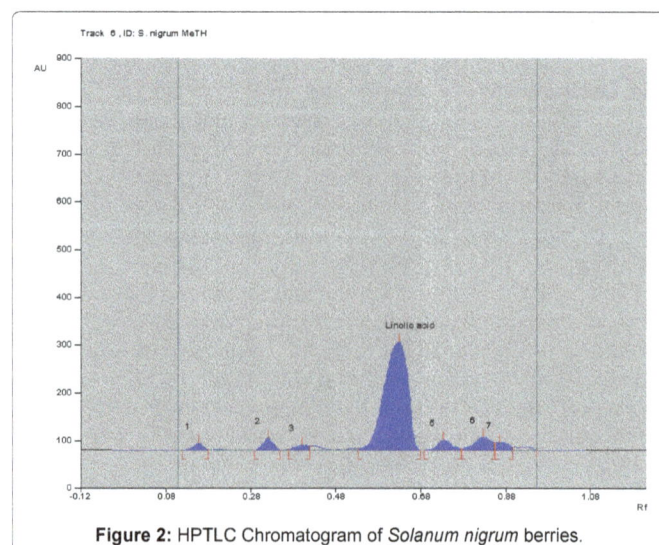

Figure 2: HPTLC Chromatogram of *Solanum nigrum* berries.

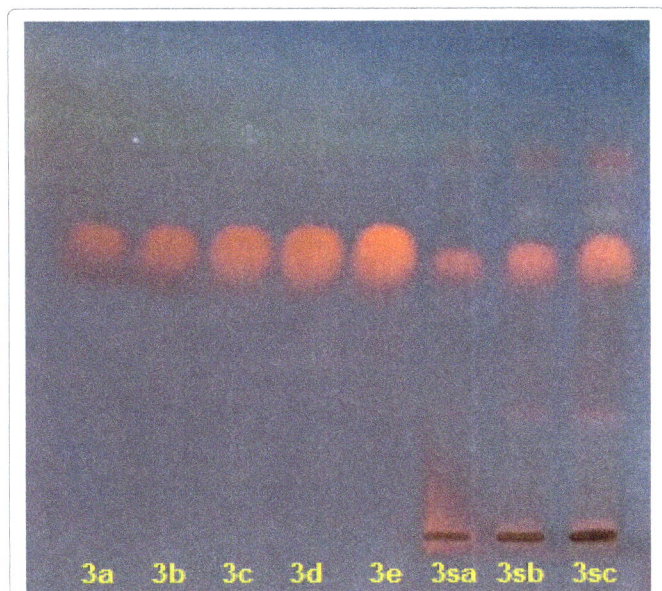

Figure 3: HPTLC Fingerprint analysis of *Solanum nigrum* berries methanolic extract in UV 366 nm. Track 3a, 3b, 3c, 3d and 3e: Linoleic acid standard solution at 2, 4, 6, 8 and 10 µL concentration respectively. Track 3sa, 3sb and 3sc: Sample solution at 4, 8 and 12 µL concentration respectively.

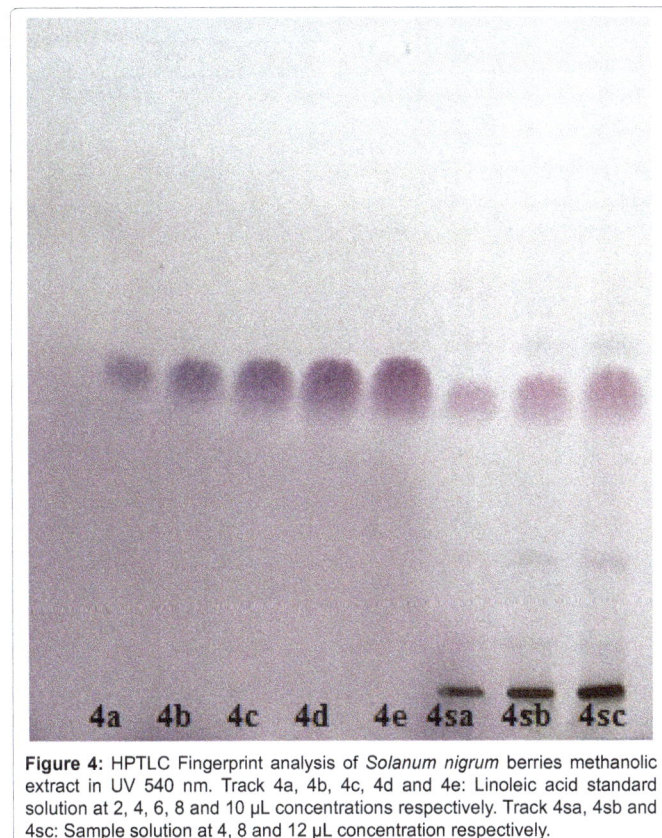

Figure 4: HPTLC Fingerprint analysis of *Solanum nigrum* berries methanolic extract in UV 540 nm. Track 4a, 4b, 4c, 4d and 4e: Linoleic acid standard solution at 2, 4, 6, 8 and 10 µL concentrations respectively. Track 4sa, 4sb and 4sc: Sample solution at 4, 8 and 12 µL concentration respectively.

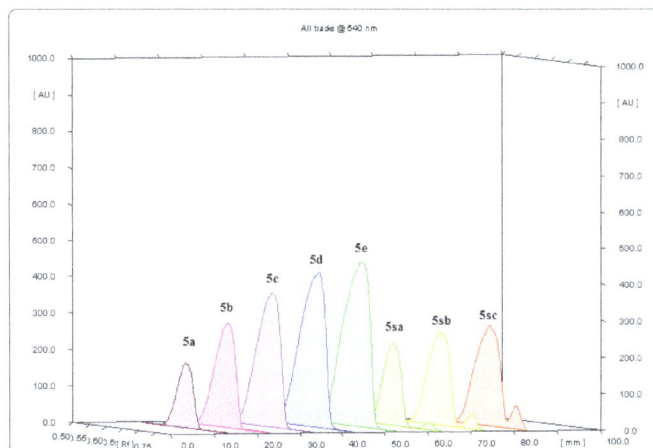

Figure 5: HPTLC 3D Chromatogram of *Solanum nigrum* berries methanolic extract in UV 540 nm. Track 5a, 5b, 5c, 5d and 5e: Linoleic acid standard solution at 2, 4, 6, 8 and 10 µL concentrations respectively. Track 5sa, 5sb and 5sc: Sample solution at 4, 8 and 12 µL concentration respectively.

Marker	Amount of marker present (µg)	Amount of marker added (µg)	Amount of marker found (µg)	Recovery (%)	Average recovery (%)
Linoleic acid	5	2.5	7.56 ± 1.11	100.13	100.63
	5	5	10.08 ± 1.05	100.86	
	5	6.25	11.35 ± 1.04	100.91	

Table 1: Recovery data of linoleic acid.

coupled with the digital scanning profile is an attractive and useful tool for construction of herbal chromatographic fingerprint. The linoleic acid in sample of extract was identified and its presence was confirmed by comparing the R_f value of standard linoleic acid to that of the extract. This method is reproducible and has shown satisfactory results on precision, accuracy and recovery study data. There is no report of detection and quantification of linoleic acid in *Solanum nigrum* berries by HPTLC. Hence, we developed a simple and precise method for quantification of this marker.

Conclusion

The developed method provides a simple, precise and accurate analytical method for the identification and quantification of linoleic acid in berries of *Solanum nigrum*. Quick and easier process for sample preparation, high method sensitivity and reproducible results by the mobile phase are the important features of the developed method. It can be expected that this HPTLC technique may be applied successfully for evaluation of linoleic acid as marker in different herbal extracts.

References

1. Khattak JZK, Anwar Z, Aftab S, Afzal M, Islam M, et al. (2012) Solenum nigrum as potent therapy: A Review. British Journal of Pharmacology and Toxicology 3: 185-189.

2. Duke JA, Bogenschutz-Godwin MJ, Du-Cellier J, Duke PK (2000) CRC handbook of medicinal spices. CRC Press LLC, NW.

3. Sarma H, Sarma A (2011) Solanum nigrum L., a nutraceutical enriched herb or invasive weed?. International Conference on Environment and BioScience IPCBEE 21: 105-109.

4. Saleem TMS, Chetty CM, Ramkanth S, Alagusundaram M, Gnanaprakash K, et al. (2009) Solanum nigrum Linn.- A Review. Pharmacognosy review 3: 342-345.

5. Shivatare RS, Nagore DH, Nipanika SU (2013) HPTLC' an important tool in standardization of herbal medical product: A Review. Journal of Scientific and Innovative Research 2: 1086-1096.

the limited developing distance and lower plate efficiency as compared to high performance liquid chromatography and gas chromatography, still it is an effective tool for evaluation of herbal drugs due to its simplicity. Moreover, the formerly mentioned limitations can be curbed by separately developing fractions of different polarity on two or several thin layer plates. The unique feature of the picture like image of HPTLC

6. Bai Y, Lin CH, Luo WM, Wang LY, Sun C, et al. (2012) Simultaneous quantification of four saponins, three alkaloids and three fatty acids in Solanum nigrum Linn. by HPLC-ELSD. Journal of Medicinal Plants Research 6: 3632-3639.

7. Xie P, Chen S, Liang Y, Wang X, Tian R, et al. (2006) Chromatographic fingerprint analysis-a rational approach for quality assessment of traditional Chinese herbal medicine. Journal of Chromatography A 1112: 171-180.

8. Qiao C, Han Q, Song J, Mo S, Kong L, et al. (2007) Chemical fingerprint and quantitative analysis of Fructus Psoraleae by high-performance liquid chromatography. Journal of Separation Science 30: 813-818.

9. Lu H, Liang Y, Chen S (2006) Identification and quality assessment of Houttuynia cordatainjection using GC–MS fingerprint: a standardization approach. Journal of Ethnopharmacology 105: 436-440.

10. Li K, Wang S (2005) Fingerprint chromatogram analysis of extracts from the leaves of Tripterygium wilfordii Hook. F. by high performance liquid chromatography. Journal of Seperation Science 28: 653-657.

19

Liquid Chromatography MS/MS Responses on Lentinan for Structure Characterization of Mushroom Polysaccharide β-D-Glucan

Nor Azreen Mohd Jamil*, Noraswati Mohd Nor Rashid and Norasfaliza Rahmad

Agro-Biotechnology Institute, Malaysia, National Institute of Biotechnology, C/O MARDI Headquarters, 43400 Serdang, Selangor, Malaysia

Abstract

Lentinan, a complex polysaccharides β-D-glucan found in Shiitake Mushroom has been known to have medicinal effect such as anticancer and immune-modulatory properties. Analysis on lentinan through application of high-end biotechnology instrument such as mass spectrometry is still limited. In this study, Liquid Chromatography Mass Spectrometry-Quadropole Time of Flight (LC/MS Q-TOF) has been used to optimize the MS/MS response in elucidating the lentinan characteristics. Analysis was done by using lentinan standard at different concentrations through different combinations of four types of columns and four types of buffers. Four levels of MS/MS collision energies or known as Collision Induced Dissociation (CID) were applied. Data presented shown that the optimum MS/MS CID that generated all six product ion of lentinan at the most combinations of column and buffer was at 20.0 V. Combination of Altima Hydrophilic Interaction Liquid Chromatography (HILIC) column without any buffer in mobile phase (AH; NB) and Hypercarb column with the present of 0.1% acetic acid in mobile phase (TH; 0.1% AA) respectively were found to be the best combinations among other combinations as these combinations were able to generate product ions of lentinan at lentinan concentration as low as 0.156 mg/mL. This suggested that these column and buffer combinations have given good sensitivity to enhance peak efficiency of lentinan at low concentration which was required to meet MS/MS abundance threshold. As the conclusion, suitable combinations of column and buffer are crucial in order to enhance peak efficiency of lentinan at optimum ionization intensity to allow optimum MS/MS responses for elucidation of their structure characteristics. Determination of appropriate collision energy also plays an important role to generate MS/MS spectra of all product ions for the lentinan.

Keywords: Mushroom β-D-glucan; Lentinan; LC/MS Q-TOF; MS/MS spectra; Structure characterization

Introduction

Lentinan, a polysaccharide β-D-glucan, is widely used as an alternative medicine and also as dietary supplement nowadays as it claimed to have a lot of health benefits such as antimicrobial, antitumor, hypocholesterolemic actions and antioxidants. It was detected for years in a well-known edible mushroom namely Shiitake Mushroom (*Lentinus edodes*) [1- 4]. This compound has been known as Biology Response Modifier (BRM) as it is actively mediates anticancer activity through activation of human immune system [1,5-7]. Lentinan has been reported to have ability in treating many types of cancers including liver, stomach, lung, ovarian and colon cancers through regression of tumour formation [2,6,8]. Daily consume of lentinan on rats has also shown the ability of the compound on weight gains and increased of white blood cells such as monocytes [5].

Lentinan is a type of 1,3/1,6-β-D-glucan with molecular mass of 400000 Dalton [1,3,4]. It is comprised of repeated units of glucose with seven glucose molecules in each repeating unit (Figure 1) that gives the molecular formula of $C_{42}H_{72}O_{36}$ and molecular mass of 1152 Dalton [1,3]. The solubility of lentinan reported was varies. Chihara et al. has reported that lentinan is mostly soluble in alkali solution and formic acid, slightly soluble in hot water and not soluble in cold water, acid solution and organic solvents such as alcohol, ether, chloroform, pyridine and hexamethylphosphoramide [9]. However, latter, this compound has been reported as water soluble β-D-glucan which is readily soluble in water [4,7] but not soluble in 50% (v/v) ethanol [4]. There was a study shown that no peak able to be detected by LC/MS/MS Q-TOF from lentinan extracted by alkaline solution as compared to lentinan from hot water extract [10].

Several methods have been applied by researchers to determine the presence of lentinan in mushroom such as GC/MS (for detection of monosaccharide), methylation analysis (for determination of the positions of glycosidic linkages) and NMR spectroscopy (for determination of degree of branching and the degree of polymerization) [7]. Nowadays, Liquid Chromatography-Mass Spectrometry (LC-MS) also has been used on determination of lentinan [10]. LC/MS/MS Quadropole-Time of Flight is one of very powerful instruments that can provides exact mass information (molecular weight) and fragmentation patterns (product ions) for elucidation on the compound structure [11].

The use of this high end instrument in determining the optimum MS/MS response for Lentinan is still not reported. Selection of right columns, types of buffers and level of MS/MS collision energy need to be emphasized in order to obtain better response [12]. Therefore objective of this study was to gain better LC/MS/MS response in elucidating lentinan characteristics through application of different types of columns, mobile phase's buffers and level of collision energies.

Materials and Methods

Chemicals

All solvents were LC/MS grade. Water was Milli-Q grade with the

*Corresponding author: Nor Azreen Mohd Jamil, Agro-Biotechnology Institute, Malaysia, National Institute of Biotechnology, C/O MARDI Headquarters, 43400 Serdang, Selangor, Malaysia, E-mail: arin2505@yahoo.com.my

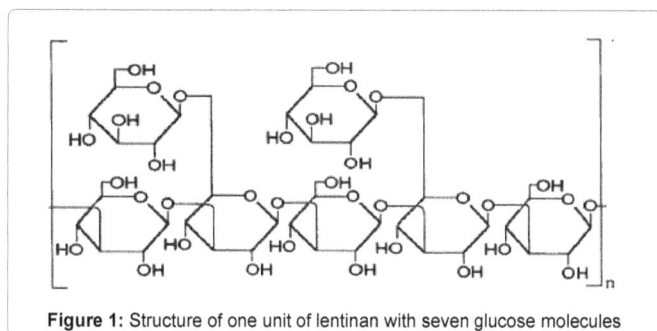

Figure 1: Structure of one unit of lentinan with seven glucose molecules

total organic compound less than 3 ppb.

Sample material

Lentinan Standard isolated from *Lentinus edodes* was purchased from CarboMer Incorporation (San Diego, USA).

Sample preparation

Lentinan standard was prepared at 2.5 mg/mL with clean Milli-Q water. The standard was dissolved in Milli-Q water and mixed well. The solution was heated at 80°C in sonicator water-bath for complete mixing. Serial dilution of the standard was carried out up to 0.078 mg/mL.

LC/MS Q-TOF analysis

Analysis was performed by using an Agilent 1290 RRLC (Rapid Resolution Liquid Chromatography) series equipped with 6550 iFunnel Q-TOF LC/MS System. Lentinan standard was run from the lowest concentration to the highest concentration. Clean Milli-Q water was used as blank. Four types of columns were used for the analysis (Table 1). Mobile phases used were 100% water (A) and 100% acetonitrile (B) with addition of different types of buffers; i) No buffer; ii) 0.1% formic acid; iii) 0.1% acetic acid; and iv) 10 mM ammonium formate respectively. Columns parameters were set as per Table 1.

Gradient for normal phase was 90%-75% B (0-5 min); 75%-50% B (5-10 min); 50%-5% B (10 –12 min); 5% B (12-13 min) and 5%-90% B (13-15 min). Reverse phase gradient was set at 5%-60% B (0-7 min); 60%-90% B (7-11 min); 90%B (11-12 min) and 90%-5% B (12-15 min). The total run time was 15 minutes for each sample.

Analysis was performed in MS/MS run mode at negative ion polarity with the following settings:- Gas Temperature: 250°C; Gas Flow: 13 L/min; Nebulizer pressure (N_2): 35 psig; Sheath gas: 11 L/min at 350°C; Capillary voltage: 3500 V; Nozzle voltage: 1000 V; Fragmentor voltage: 175 V; and Drying gas: 5 L/min at 350°C. MS/MS collision energies were set at 0.0 V, 10.0 V, 20.0 V and 40.0 V.

The profile patterns of total ion chromatogram (TIC), retention time, generation of product ions (fragmentation patterns) and MS/MS response sensitivity on lentinan were analysed.

Results and Discussion

The MS/MS chromatogram for total ion chromatogram (TIC) of 2.5 mg/mL lentinan was showed in Figure 2. Lentinan peak was in circled area. Results showed that different combinations of columns and buffers used were led to different TIC patterns. The same type of column with the presence of different types of buffers in mobile phases has produced different TIC patterns of the lentinan standard. The TIC pattern was also totally different with the use of different types of columns. As shown, there was no lentinan peak at the MSMS TIC

showed by the Prevail Carbohydrate (PC) column with the presence of 10 mM Ammonium Formate (10 mM AF) in mobile phases. Absence of the MS/MS peak determined the failure generation of MS/MS spectra. This was probably due to insufficient abundance threshold of the lentinan for MS/MS generation. Small changes in pH are able to cause extreme sensitivity of some compounds [13]. Therefore this can suggested that combination PH: 10 mM AF was unable to increase sensitivity of lentinan to be ionized at particular abundance threshold. TIC pattern of lentinan standard that passed through the combination of Poroshell120 HILIC (PH) column with 0.1% Formic Acid (FA) was resulted to a lot of ion suppression. The details of causes of ion suppression are not clear. The use of common buffers also might lead to formation of ion suppression. Proper sample preparation needs to be emphasized in order to minimize or eliminate the ion suppression [14]. This suggested that extra precaution is required when analysing lentinan using this PH: 0.1% FA combination.

Figure 3 showed the Extracted Ion Chromatogram (EIC) for retention time of lentinan. As shown, the retention time was changed with different combination of columns and buffers. However some combination was not affecting the retention time of lentinan. As shown in the figure, lentinan that passed through Altima HILIC (AH) column showed the same retention time at 9.3 min with the use of mobile phases that contained of No Buffer (NB), 0.1% Formic Acid (0.1% FA) and 0.1% Acetic Acid (0.1% AA). The retention time was shifted to 7.5 min with the use of mobile phases that contained 10 mM Ammonium Formate (10mM AF). Retention time of the lentinan was different at all with the presence of these four types of buffers which were 6.7 min (NB), 2.6 min (0.1% FA), 4.7 min (10mM AF) and 6.8 min (0.1% AA) through Poroshell120 HILIC (PH) column. Use of Prevail Carbohydrate (PC) column with mobile phases containing 0.1% FA and 0.1% AA remained the same retention time for lentinan (12.4 min), slightly changed to 12.3 min with the mobile phases without any buffer and fasten to 10.9 min with the use of mobile phases containing 10mM AF. While Hypercarb (TH) column has retained the lentinan at almost the same time with the use of the mobile phases containing NB (4.7 min), 0.1% FA (4.8 min), 10 mM AF (5.0 min) and 0.1% AA (4.8 min). In general, retention time of a compound will be different with the use different types of columns [13]. Therefore different characteristics of columns have affected the time of lentinan to be eluted. However, application of different types of buffer in mobile phases through the same types of column has also led to the changes of retention time of lentinan. Combination of column and buffer could contribute to the modification of the column characteristics. The buffer used has adjusted the pH in mobile phases and caused an effect on the stability of columns. It then affects indirectly the peak efficiency and retention of analytes [15]. Furthermore, the adjustment of pH could also influence the column characteristics which contribute to the polarity of analytes that eluting analytes at different times [16]. Alteration of pH

Column types	Parameters			
	Altima HP HILIC (AH)	Poroshell120 HILIC (PH)	Prevail Carbohydrate (PC)	Hypercarb (TH)
Particle size (μ)	3	2.7	5	3
Column Size (mm)	2.1 x150	2.1 x 150	4.6 x 150	2.1 x 100
Injection volume (μl)	1	1	1	0.5
Flow rate (mL/min)	0.2	0.3	0.7	0.15
Gradient type	Normal phase	Normal phase	Normal phase	Reverse phase
Column temperature (°C)	25	25	30	25

Table 1: Parameters setup for different types of columns (as reported by column packing note)

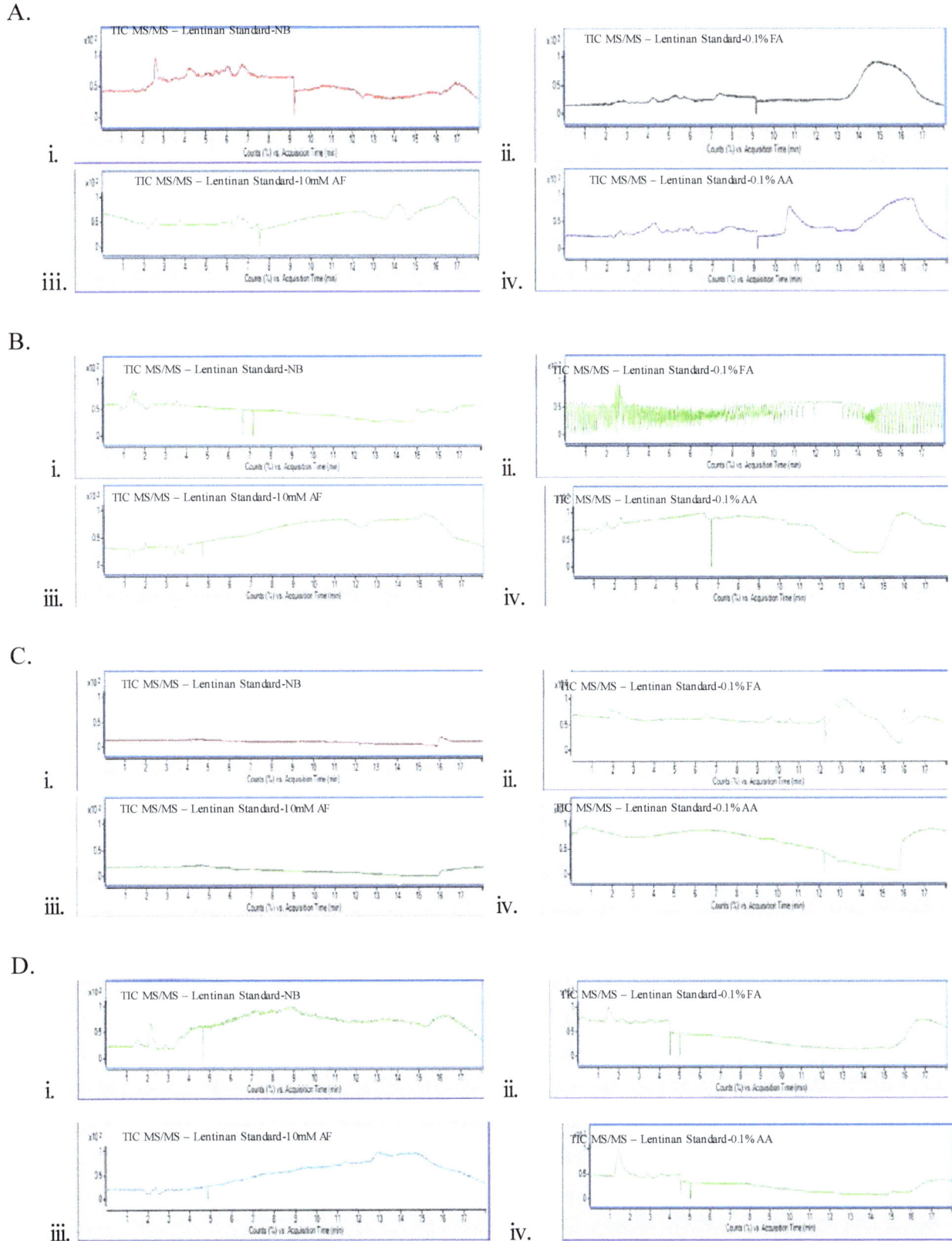

A. Altima HILIC (AH) column; B. Poroshell120 HILIC (PH) column; C. Prevail Carbohydrate (PC) column; and D. Hypercarb (TH) column with the presence of (i) No buffer (NB), (ii) 0.1% formic acid (FA), (iii) 0.1% acetic acid (AA) and (iv) 10mM ammonium formate (AF) in mobile phases.

Figure 2: Total Ion Chromatogram (TIC) of lentinan standard at 2.5 mg/mL

A.

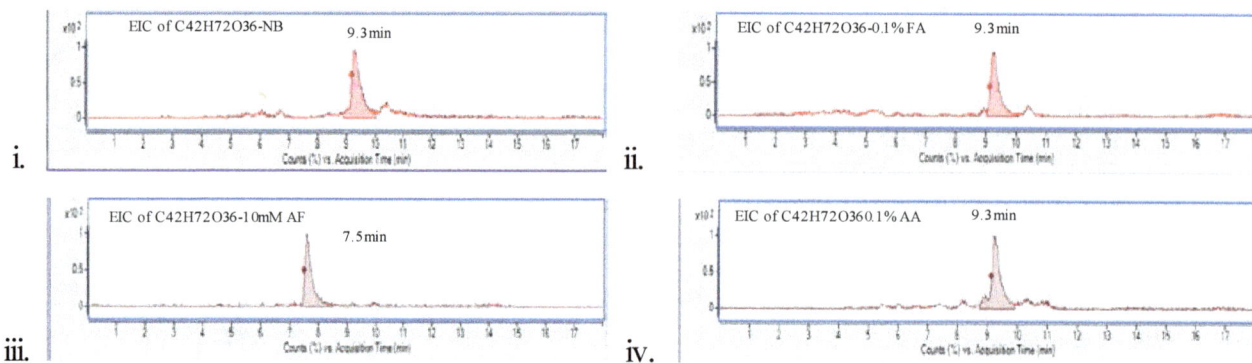

B.

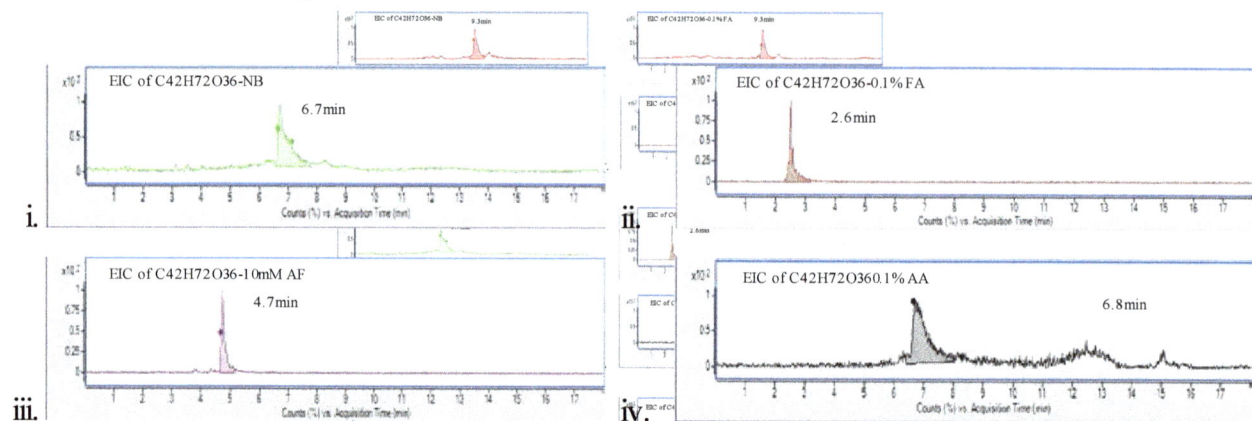

C.

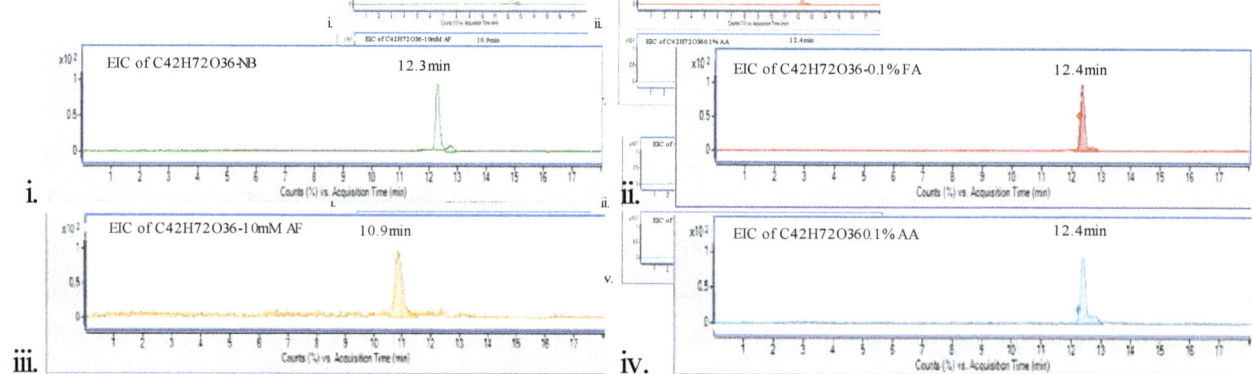

D.

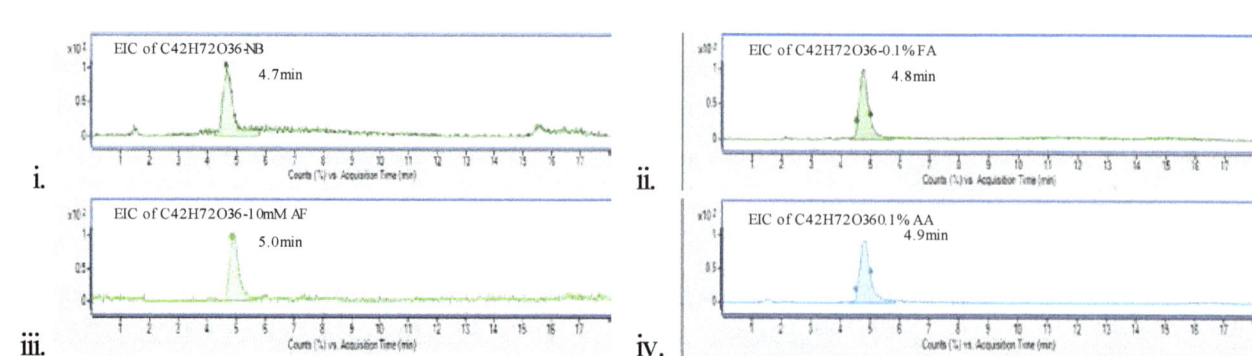

A. Altima HILIC (AH) column; B. Poroshell120 HILIC (PH) column; C. Prevail Carbohydrate (PC) column; and D. Hypercarb (TH) column with the presence of (i) No buffer (NB), (ii) 0.1% formic acid (FA), (iii) 0.1% acetic acid (AA) and (iv) 10mM ammonium formate (AF) in mobile phases.

Figure 3: Retention times of lentinan at 2.5 mg/mL standard concentration

is one of the most important parts in the control of retention time of compounds [13,15]. These suggested that the different pH level of the mobiles phases has resulted to different reaction to the lentinan which then affected it retention time.

Development of optimum MS/MS spectra of lentinan is necessary to study the fragmentation pattern as well as structural characteristic of this compound. Lentinan is a β-D-glucan polysaccharide of 1151 m/z that consists mainly and only repeating units of glucose molecules. The cleavage of the lentinan at specific chains releases glucose units (molecular mass of 162) and resulted to generation of product ions of 989 m/z, 827 m/z, 665 m/z, 503 m/z, 341 m/z and 161 m/z (341 m/z minus 179 [glucose with OH group from water]). Further optimization to fragment the lentinan was done by applying different collision energies or Collision Induced Dissociation (CID) to the compound. Figure 4 showed the MS/MS spectra (fragmentation patterns) of 2.5 mg/mL lentinan that obtained from CID of 0.0 V, 10.0 V, 20.0 V and 40.0 V. As summarized in Table 2, the numbers of product ions of the lentinan generated were different at some CID levels. Optimum MS/MS spectra with all the six targeted products ions (989 m/z, 827 m/z, 665 m/z, 503 m/z, 341 m/z and 161 m/z) was successfully obtained at CID of 20.0V for lentinan that passed through Altima HILIC column with No Buffer (AH; NB), 0.1% Formic Acid (AH;0.1% FA) and with 0.1% Acetic Acid (AH; 0.1% AA) respectively, Poroshell120 HILIC without any buffer (PH; NB) and with 10 mM Ammonium Formate (PH;10 mM AF) respectively, Prevail Carbohydrate with No Buffer (PC; NB), with 0.1% Formic Acid (PC; 0.1% FA) and 0.1% Acetic Acid (PC; 0.1% AA) respectively and also through Hypercarb column with No Buffer (TH; NB), 0.1% Formic Acid (TH; 0.1% FA) and 0.1% Acetic Acid (TH; 0.1% AA) respectively. Collision energies of 0.0 V, 10.0 V and 40.0 V under these column and buffer combinations were able to generate one to five product ions only. However, some columns and buffer were also able to generate the optimum fragmentation at CID of 10.0 V such as at AH; 0.1% AA, PH; 10 mM AF, PC; NB and PC; 0.1% AA. This shown that the optimum MS/MS CID that generated all six product ion of lentinan at most of combinations of column and buffer was at 20.0 V. There were certain column and buffer combinations that unable to generate optimum MS/MS spectra for lentinan at any CID level. This can be seen at lentinan standard that passed through AH; 10mM AF, PH; 0.1% FA, PH; 0.1% AA, PC; 10mM AF and TH; 10 mM AF. No product ion of lentinan was generated at 10.0 V for the standard that passed through PH; 0.1% FA. The presence of 10mM ammonium formate in mobile phases through Prevail Carbohydrate (PC) column was totally failed to generate MS/MS spectra at any collision energy even at high concentration of the standard. These results suggested that the presence of buffers which changed the pH condition of mobile phases have affected the ionic level or conformation shift of the lentinan. It then has reflected the sensitivity of certain column to retain the lentinan and probably led to induction or reduction on peak efficiency and ion transmission level of the lentinan that affected the ion abundance threshold required for MS/MS responses. Therefore it suggested that only combinations of AH; NB, AH; 0.1% FA, AH; 0.1% AA, PH; NB, PH; 10 mM AF, PC; NB, PC; 0.1% FA, PC; 0.1% AA, TH; NB, TH; 0.1% FA and TH; 0.1% AA were able to meet the ion abundance threshold for MS/MS generation at optimum fragmentation pattern at 20.0 V CID.

Six different concentrations of lentinan standard (0.078 mg/mL, 0.156 mg/mL, 0.312 mg/mL, 0.625 mg/mL, 1.25 mg/mL and 2.50 mg/mL) were evaluated at CID 20.0 V to determine the lowest level of the lentinan concentration can be fragmented to produce all the six product ions of lentinan through different combinations of column and buffer.

Supplemenatry Files (1-4) showed the MS/MS spectra of different lentinan concentrations at the column and buffer combinations that generated optimum MS/MS spectra in previous analysis. Each lentinan concentration showed a peak of 1151.3731. However, the lowest concentration of lentinan that successfully generated MS/MS spectra was different for different combinations of buffer and column.

As shown in Table 3, there was no MS/MS spectra generated for lentinan standard at concentration of 0.078 mg/mL. The lowest concentration of lentinan that generated MS/MS spectra with all the six product ions was at 0.156 mg/mL which only showed by AH; NB and TH; 0.1% AA combination respectively. MS/MS spectra for lentinan standard that used PC; NB, PC; 0.1% FA, PC: 0.1% AA, TH; 0.1% FA combinations were started to be generated at 0.312 mg/ml standard concentration. Generation of MS/MS spectra for lentinan that passed through Altima HILIC with the presence of 0.1% FA and 0.1% AA respectively was only be obtained for the highest three concentrations which were at 0.625 mg/mL, 1.25 mg/mL and 2.5 mg/mL. This was same for the standard that passed through Poroshell120 HILIC without buffer and with 10 mM AF. Lentinan that passed through TH: NB has shown the generation of MS/MS spectra at only 2.5 mg/mL of the standard. The results shown that the combinations of column and buffer were probably had influence the differences on MS/MS responses through alteration of column and lentinan characteristics at different pH condition which might be reflected the sensitiveness of the columns to retain the lentinan and hence the peak efficiency. These presented data have suggested that the best combinations found in this study was Altima HILIC column without any buffer in mobile phase (AH; NB) and Hypercarb column with the present of 0.1% acetic acid in mobile phase (TH; 0.1% AA) respectively as the combinations have given good sensitivity to enhance peak intensity of lentinan at low concentration that required for MS/MS spectra generation.

Conclusion

Compound analysis by Liquid-Chromatography Mass Spectrophotometry is based on mass to charge ratio (m/z). β-D-glucan such as lentinan, schizophyllan and krestin is a highly polar compound that consists of long chain polysaccharides. The number of charges for this big compound (such as Lentinan with 400000 Da, 1151 m/z) could be more than one. Analysis of lentinan through LC/MS that based on m/z ratio could produce the same m/z ratio as other types of polysaccharides such as oligosaccharides which are also polar compound and consist of long chain polysaccharides. Optimization of the MS/MS response is so important in order to determine the actual characteristics of lentinan in term of molecular weight, retention time and also the product ions for structure confirmation purpose. Selection of suitable collision energy or CID during fragmentation of lentinan for generation of all the related product ions should be made. However, the successfulness of the fragmentation process through MS/MS mode is also depending on the level of ionization intensity of the particular compound. Enhancement of peak efficiency through suitable combination of column and buffer is so crucial in order to enhance the detection sensitivity of the LC/MS to meet certain MS/MS requirement. It could be done by strengthening the ionic condition of the compound and increasing the columns competency to retain the compound through modification of pH condition by certain buffers. As the conclusion, this study has found that the optimum performance of MS/MS response for lentinan fragmentation was at 20.0 V CID. Combinations of Altima HILIC with no buffer and Hypercarb column with 0.1% acetic acid were found to be the best combinations in this study as the MS/MS spectra able to generate all the six lentinan product ions at the lentinan concentration as low as 0.156 mg/mL. Use of

Column & Buffer Combination	Collision energy (CID), v	Generation of product ions (m/z) at 2.5 mg/mL lentinan						
		1151	989	827	665	503	341	161
AH;NB	0	√	√	√	√	√	x	x
	10	√	√	√	√	x	√	x
	20	√	√	√	√	√	√	√
	40t	x	x	√	√	√	√	√
AH;0.1% FA	0	√	√	√	√	√	x	√
	10	√	√	√	√	√	x	x
	20	√	√	√	√	√	√	√
	40	√	x	√	√	√	√	√
AH;10 mM AF	0	√	√	√	√	x	x	x
	10	√	√	√	√	√	√	x
	20	√	√	√	√	√	x	√
	40	x	x	√	√	√	√	√
AH;0.1% AA	0	√	√	√	√	x	x	x
	10	√	√	√	√	√	√	√
	20	√	√	√	√	√	√	√
	40	x	x	√	√	√	√	√
PH;NB	0	√	√	√	x	x	√	x
	10	√	√	√	√	√	x	√
	20	√	√	√	√	√	√	√
	40	√	x	√	√	√	√	√
PH;0.1% FA	0	√	x	x	x	x	x	x
	10	x	x	x	x	x	x	x
	20	√	x	x	√	x	x	x
	40	x	x	√	√	x	x	x
PH;10 mM AF	0	√	√	√	√	√	√	x
	10	√	√	√	√	√	√	√
	20	√	√	√	√	√	√	√
	40	√	x	√	√	√	√	√
PH;0.1% AA	0	√	√	x	x	x	x	√
	10	√	√	√	√	x	x	x
	20	√	√	√	√	x	√	x
	40	x	x	√	√	x	√	√
PC;NB	0	√	√	√	√	√	√	x
	10	√	√	√	√	√	√	√
	20	√	√	√	√	√	√	√
	40	√	x	√	√	x	√	√
PC;0.1% FA	0	√	√	√	x	x	x	x
	10	√	√	√	√	√	x	x
	20	√	√	√	√	√	√	√
	40	x	x	√	√	√	√	√
PC;10 mM AF	0	x	x	x	x	x	x	x
	10	x	x	x	x	x	x	x
	20	x	x	x	x	x	x	x
	40	x	x	x	x	x	x	x
PC;0.1% AA	0	√	√	√	√	x	x	√
	10	√	√	√	√	√	x	√
	20	√	√	√	√	√	√	√
	40	x	x	√	√	√	√	√
TH;NB	0	√	√	√	√	x	x	x
	10	√	√	√	√	√	x	x
	20	√	√	√	√	√	√	√
	40	x	x	√	√	√	√	√
TH;0.1% FA	0	√	√	√	x	x	√	x
	10	√	√	√	√	x	x	x
	20	√	√	√	√	√	√	√
	40	x	x	√	√	√	√	√
TH;10 mM AF	0	√	√	√	x	x	x	x
	10	√	√	√	√	x	√	√
	20	√	√	√	√	√	x	x
	40	x	x	√	√	√	√	√
TH;0.1% AA	0	√	√	√	x	x	x	x
	10	√	√	√	√	x	√	√
	20	√	√	√	√	√	√	√
	40	√	x	√	√	√	√	√

Table 2: Summary of product ions of 2.5 mg/mL lentinan generated at different CID levels for various combinations of column and buffer

Column & Buffer Combination	MS/MS spectra generation at different concentration (mg/mL) of lentinan					
	0.078	0.156	0.312	0.625	1.25	2.5
AH;NB	x	√	√	√	√	√
AH;0.1% FA	x	x	x	√	√	√
AH;10 mM AF	x	x	x	x	x	x
AH;0.1% AA	x	x	x	√	√	√
PH;NB	x	x	x	√	√	√
PH;0.1% FA	x	x	x	x	x	x
PH;10 mM AF	x	x	x	√	√	√
PH;0.1% AA	x	x	x	x	x	x
PC;NB	x	x	√	√	√	√
PC;0.1% FA	x	x	√	√	√	√
PC;10 mM AF	x	x	x	x	x	x
PC;0.1% AA	x	x	√	√	√	√
TH;NB	x	x	x	x	x	√
TH;0.1% FA	x	x	√	√	√	√
TH;10 mM AF	x	x	x	x	x	x
TH;0.1% AA	x	√	√	√	√	√

Table 3: MS/MS response (spectra generation of six lentinan product ions) of different lentinan concentrations at 20.0V CID

ammonium formate was not suitable as there was no MS/MS spectra generated at any lentinan concentration for any columns used except for Poroshell 120 HILIC column. However the MS/MS spectra obtained from this column were at the highest three concentrations only.

Acknowledgement

The authors acknowledge Agro-Biotechnology Institute, Malaysia (ABI) and Ministry of Science, Technology and Innovation (MOSTI) for providing research grant (Science fund : 02-05-19-SF0004) to support this study.

References

1. Tomassen MM, Hendrix EAHJ, Sonnenberg ASM, Wichers HJ, Mes JJ (2011) Variation Of Bioactive Lentinan-Containing Preparations in *Lentinula Edodes* Strains And Stored Products. Section of Mycosourced molecules and nutritional quality, Proceedings of the 7th International Conference on Mushroom Biology and Mushroom Products (ICMBMP7) 2011, France.

2. Kuppusamy UR, Chong YL, Mahmood AA, Noorlidah A, Vikineswary S (2009) Lentinula edodes (Shiitake) Mushroom Extract Protects Against Hydrogen Peroxide Induced Cytotoxicity In Peripheral Blood Mononuclear Cells. Indian J Biochem Biophys 46: 161-165.

3. Wang Y, Wang H, Yan D, Wang L, Sun Z, et al. (2013) Lentinan Extracted From Shiitake Mushrooms (*Lentinus edodes*) Improves The Non-Specific Immunity of Sea Cucumber (*Apostichopus japonicas*). Aquaculture International 21:1261-1277.

4. Brauer D, Kimmons TE, Phillips M, Brauer DE (2010) Potential for manipulating the polysaccharide content of shiitake mushrooms. Current Research, Technology and Education Topics in Applied Microbiology and Microbial Biotechnology 2: 1136-1142.

5. McCormack E, Skavland J, Mujic M, Bruserud Ø, Gjertsen BT (2010) Lentinan: hematopoietic, immunological, and efficacy studies in a syngeneic model of acute myeloid leukemia. Journal of Nutraceutical Cancer 62(5):574-83.

6. Oba K, Kobayashi M, Matsui T, Kodera Y, Sakamoto J (2009) Individual Patient Based Meta-analysis of Lentinan for Unresectable/Recurrent Gastric Cancer. Journal of Anticancer Research 29:2739-2746.

7. Zhang Y, Li S, Wang X, Zhang L, Cheung PCK (2011) Advances in lentinan: Isolation, structure, chain conformation and bioactivities. Food Hydrocolloids 25(2):196–206.

8. Ng ML, Yap AT (2002) Inhibition of Human Colon Carcinoma Development By Lentinan From Shiitake Mushrooms (Lentinus Edodes). J Altern Complement Med 8(5):581–589.

9. Chihara G, Hamuro J, Maeda Y, Arai Y, Fukuoka F (1970) Fractionation and Purification of the Polysaccharides with Marked Antitumor Activity, Especially Lentinan, from *Lentinus edodes* (Berk.) Sing. (an Edible Mushroom). Cancer Res 30:2776-2781.

10. Jamil NAM, Rahmad N, Rashid NMN, Yusoff MHYM, Shaharuddin NS, et al. (2013) LCMS-QTOF Determination of Lentinan-Like β-D-Glucan Content Isolated by Hot Water and Alkaline Solution from Tiger's Milk Mushroom, Termite Mushroom, and Selected LocalMarket Mushrooms. Journal of Mycology 2013:8 pp.

11. Lim CK, Lord GA (2002) Current developments in LC-MS for pharmaceutical analysis. Biological and Pharmaceutical Bulletin 25(5):547-557.

12. Peng L, Rustamov I, Loo L, Farkas T (2007) Improved Results for LC/MS of Basic Compounds Using High pH Mobile Phase on a Gemini® C18 Column. Application note TN-1031.

13. Dolan J. A Guide to HPLC and LC/MS Buffer Selection ACE HPLC Columns Application Notes: 16 pp.

14. Waters Corporation (2005) Topics in Solid Phase Extraction Part 1- Ion Suppression in LC/MS Analysis: A review. Waters Application Notes: 6pp.

15. Kaushal CK, Srivastava B (2010) A process of method development: A chromatographic approach. Journal of Chemistry Pharmaseutical Research 2(2): 519-545.

16. Buszewski B, Noga S (2012) Hydrophilic interaction liquid chromatography (HILIC)—a powerful separation technique. Analytical Bioanalysis Chemistry 402:231–247.

20

Premnine HCl Estimation in Selected Formulations of 'Dashmul' and in Chloroform Extract of Premna integrifolia L. by a Selective, Validated and Developed HPTLC Fingerprint Method

Attarde DL[1]*, Pal SC[2] and Bhambar RS[1]

[1]*Department of Pharmacognosy, Mahatma Gandhi Vidyamandir's Pharmacy College, Panchavati, Nashik, Maharashtra, India*
[2]*Department of Pharmacognosy, RG Sapakal College of pharmacy, Kalyani Hills, Trimbakeshwar, Nashik, Maharashtra, India*

Abstract

HPTLC technique developed as validated method for estimation of Premnine HCl in *P. integrifolia* chloroform extract (PI-AK) and in selected marketed formulations 'Dashmul arishta','Dashmul kadha' in 3 × 3 batches as per ICH guidelines. Premnine HCl (Pr-s) was isolated as per literature from *P. integrifolia* and focused first time as standard bioactive marker for quantification. Developed mobile phase Toluene: Acetone: Diethylamine (7:2:1) for Pr-S gave Rf 0.59 at λ max 283 nm in densitometric scan, focused for specificity, fingerprint and estimation study in PI-AK and selected formulation successively. Linearity assessed in range of 8 to 16 µg /band with regression coefficient of 0.9983, LOD 0.742 µg/Band, LOQ 2.225, also robust for Pr-S. Accuracy for % recovery performed on extract as well on formulation, further subjected for precision study with application one way ANOVA for finding F value, found within limit therefore no significance of variance. A rapid and selective HPTLC method shows good linearity, recovery and high precision, useful method for analysis of Pr-S and as quality control parameter for raw material as well formulation as per foremost need of WHO, FDA and Pharmacopoeia.

Keywords: *Premna integrifolia;* HPTLC; *Dashmul arishtha; Dashmul kadha;* Premnine HCl

Introduction

Premna integrfolia Linn. (syn. *P. corymbosa* auct., *P. obtusifolia* R., *P. spinosa* Roxb., *P. serratifolia*) is thorny deciduous shrub belonging to Verbenaceae, common along India, Andaman costs, tropical and subtropical Asia, Africa Africa and the specific Islands, known as Agnimanth, Arni, Girikarnika in Sanskrit, Malbau in Malay language useful for inflammation, brochitis, dyspepsia, piles, constipation, fever and root forms an ingredients of *'Dashmul'* which is renowned traditional Ayurvedic remedy tonic for liver, uterus, kidney, also detoxifies and strengthens body. Decoction of root useful for convulsion, rheumatism, and neuralgia [1-3].

Literature survey shows three novel diterpenoids isolated from root bark and simultaneous quantification of it by HPTLC method [4,5] a verbascoside iridoid glycosides from leaves [6], two alkaloid premnine and ganiarine isolated from root [7], volatile constituents isolated from flower buds [8], flavonoids Luteolin 7-O-methyl ether and Apigenin 5,7-O-dimethyl ether isolated from leaves [9], identification of volatile constituents of leaves and roots analysed by GC-MS [10], stem bark and wood alcoholic extracts shows cardiotonic effects and β adrengic effects [11] and also antioxidant effects [12], antimicrobial activity of root extracts [13], wood extract for antiarthritic activity [14], methanolic extract of root evaluated for immunomodulatory effects [15].

Premnine alkaloid shows raise in blood pressure by contracting blood vessels but decreases force of contraction of heart, and dilates pupil [7].

In this original research Premnine alkaloid isolated as per literature and its salt as Premnine HCl first time here focused as unique characteristic bio-active marker for qualitative and quantitative analysis in chloroform extract of *P. integrifolia* and in its marketed formulation as *'Dashmul arishtha'* and *'Dashmul kadha'* with view to develop standardise parameter that will serve as crucial quality control parameter for drug.

Experiment

Equipment

HPTLC Instrument (CAMAG, Switzerland): Linomet Syringe V, TLC scanner V, Digistore- Reprostar 3, Win CATS version 1.4.2. Software, Twin trough Chamber, Pre-coated silica gel 60 F_{254} aluminium plates (0.2 mm thick, Merck, Germany).

Chemicals and solvents

Premnine HCl- isolated as per literature [7].

Analytical grade solvents: Ethanol, Conc. HCl, Pet. Ether (60-80°C), chloroform, Toluene, Acetone, Diethyl amine (Merck), Dragendorff's reagent (Freshly prepared).

Plant material

P. integrifolia shrubs are located in Trimbakeshwar forest area of Nashik District, herbarium was prepared of flowering branch. It was deposited to Botanical Survey of India, Pune for identification purpose. Certificates were issued as i.e., Ref.: BSI /WRC/Tech./2012/DVR-1 dt. 02/1/2012 for *Premna integrifolia L.(Verbenaceae)*. Stem branches were sliced into small pieces, dried under shade for about 20 days, powdered, passed through sieve 25/30 no.-600 µ. Dried powder packed in air tight container and stored to cool, dry condition for further use.

---

*Corresponding author: Attarde DL, Department of Pharmacognosy, Mahatma Gandhi Vidyamandir's Pharmacy College, Panchavati, Nashik, Maharashtra, India
E-mail: daksha511@rediffmail.com

Isolation of premnine HCl alkaloid

As per literature [7] stem powder (2 kg) charged in soxhlet assembly in several batches and exhaustively extracted with ethanol. Alcoholic extract concentrated and poured syrupy mass with mechanical stirring into 1% warm HCl and kept for 2 hr on magenetic stirrer. Filtered precipitated black resinous mass. And extracted with chloroform, each chloroform extract fraction washed with 1% HCl and washing returned to acidic alcoholic fraction. pH of acidic alcoholic fraction now adjusted to 9 with dilute ammonia solution and exhaustively extracted with chloroform. Chloroform extract concentrated, dried and yield noted. It was dissolved in few ml of chloroform and gasoline added dropwise to precipitate out blackish resinous mass and filtered. This is repeated several times for purification. This yellowish semi purified alkaloid fraction obtained was divided into two fraction. One fraction subjected for isolation as follow, dissolve into dry ether and few drops of absolute alcohol saturated with HCl gas were added to it drop by drop and stirred continuously. Yellowish precipitate obtained was filtered, dissolved in alcohol and recrystallized this isolated *Premnine* HCl melting point taken, recorded as 212-214°C, matched with reference value of 211-213°C, as per literature no [7]. It was further confirmed as alkaloid with Dragendorff's test. It was designated as Pr-s.

Preparation of purified chloroform extract of *P. integrifolia*

From above procedure second portion of purified chloroform extract was dissolved in few ml of 1% alcoholic HCl, allow to stand for 2 hr and then dried to solid mass, yield noted and designated as PI-AK.

Selected Marketed Formulation and Treatment

'*Dashmul arishtha*': Manufacture I: selected 3 batches, Manufacture II: selected 3 batches,

'*Dashmul kadha*': Manufacture III: selected 3 batches, batch no., volume, label claim noted and designated with codes. Coded batches were concentrated using rotary evaporator, made hydro alcoholic. pH adjusted to 9 with dilute ammonia solution and extracted with chloroform in several batches. All chloroform extracts were concentrated and dried separately and further dissolved in 1% alcoholic HCl batch wise and allow to stand for 2 hr and then dried to solid mass, yield noted and designated as for Manufacturer I, II, III batches as: DA1, DA2, DA3, SA4, SA5, SA6, BA7, BA8 and BA9 respectively.

Sample preparation for HPTLC

Standard Pr-S solution: 2 mg of standard Pr-S was dissolved in 1 ml of alcohol in volumetric flask (2000 ppm), sonicated for 15 min.

PI-AK solution: 20 mg of purified chloroform extract PI AK was dissolved in 1 ml of methanol (20000 ppm), sonicated for 15 minutes, filtered through whatman filter paper No.1.

Marketed formulations solution: Coded batches as mentioned and treated above-prepared as 20 mg/ml each in methanol separately and designated as DA1, DA2, DA3, SA4, SA5, Sa6, BA7, BA8 and BA9.

HPTLC condition

Stationary phase: Pre-coated Silica Gel G 60 F_{254} HPTLC Plates

Mobile Phase: Toluene: Acetone: Diethylamine (7:2:1)

Derivatizing Reagent (Visualising agent): Dragendorff's Spray Reagent

Wavelength: 283 nm (λ max for standard Premnine HCl),

Band application point -X-8 mm, Y-15 mm, Sample Band length-8 mm,

Chamber saturation Time: 10 min, Solvent Run -80 mm, Lamp: Deuterium,

Scanning slit width 6 × 0.45 mm, Scanning speed 20 mm/s, Area Temperature-22+/-2°C Applicator Syringe-100 µl, Sample Application speed 0.2 µl/s.

Spray gas- Nitrogen Inert gas, Scanner- TLC scanner 5 (version 1.14.26),

Photo documentation-Digistore-Reprostar 3.

HPTLC method

The HPTLC analysis was performed using above chromatographic condition, 5, 10, 15 µl of Pr-S applied on pre-coated TLC Silica gel 60 F_{254} plates as band length of 8 mm using Linomat 5 syringe, spots bands were air dried, mobile phase Toluene: Acetone: Diethylamine (7:2:1) poured to CAMAG twin trough chamber and allowed to saturate for 10 min. than developed till 80 mm, dried with dryer, scanned over TLC scanner 5 (1.14.26) with absorption/remission mode at scan speed 20 mm/s at 254 nm initially. Spectral scan done in between 200-400 nm with 100 nm/s speed, spectra of Pr-S wavelength noted, again plates were rescanned at 283 nm in detection scanner mode. Rf (Retention factor), AUC (Area Under Curve) for standard Pr-S noted, used the data for further detection.

Specificity and fingerprinting of Pr-S as standard in PI-AK and in DA1, SA4, BA7 formulations

For specificity and fingerprinting study sample applied as PI-AK (track 1 to 3: 3 µl each), Pr-S (track 4 ,5:5 ,10 µl), DA1(track 6,7:5, 7 µl), SA4 (track 8,9:5, 7 µl), BA7 (track 10, 11:5, 7 µl), respectively sequentially and plate developed as above and scanned in detection at 283 nm λ max for Standard Pr-S. Rf obtained for it is noted, identified and marked in all samples track. The identified and selected bands subjected for spectral scan between 200-400 nm and over laid spectra for confirmation of specificity for standard. The purity of the bands was confirmed at start, middle and end position of chromatogram. Then finally plate derivatized by dipping into freshly prepared Dragendorff's spray reagent, dried, colour visualized as identity for alkaloids at Pr-S Rf and photo documented. Specificity and fingerprinting chromatogram are as shown in Figure 1a and 1b.

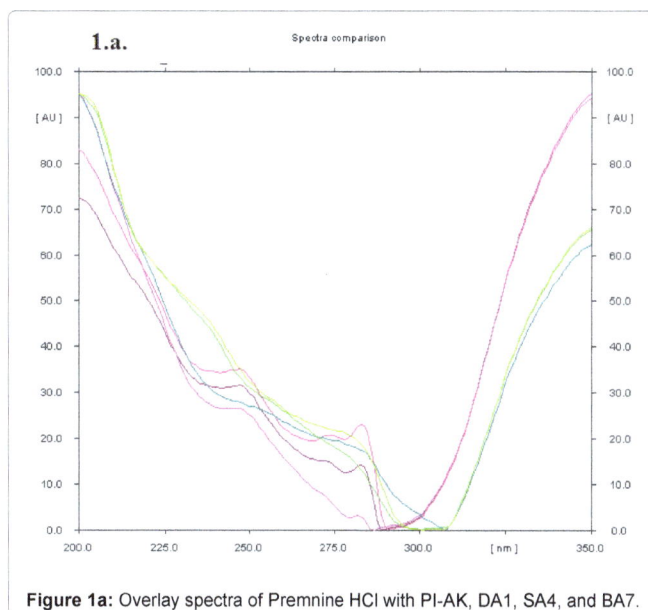

Figure 1a: Overlay spectra of Premnine HCl with PI-AK, DA1, SA4, and BA7.

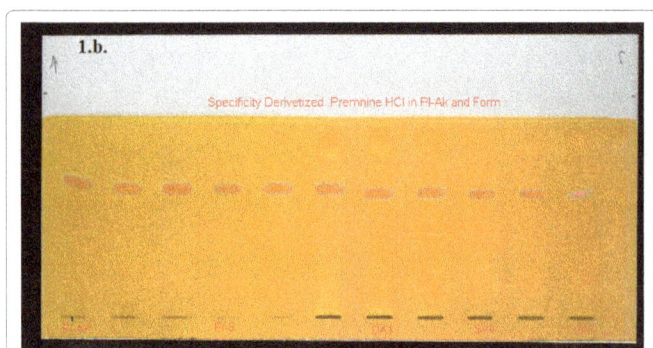

Figure 1b: HPTLC Chromatogram of Standard Premnine HCL with PI-AK, DA1, SA4, BA7 photo documented after derivatization with Dragendorff's reagent.

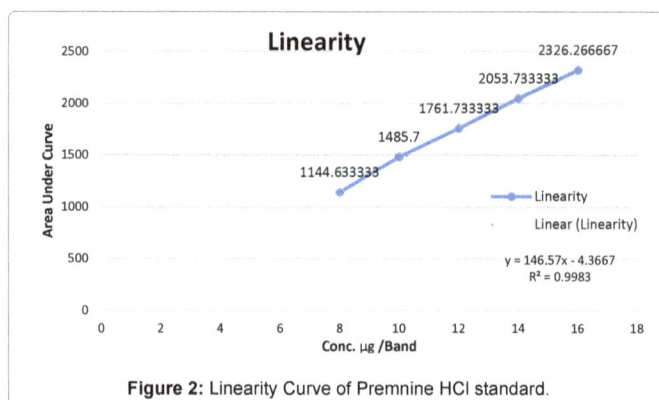

Figure 2: Linearity Curve of Premnine HCl standard.

Parameter	Value
Wavelength, nm	283 nm
Mobile Phase	Toluene:Acetone:Diethylamine (7:2:1)
RF Value	0.59
Linearity Range µg/band	8-16
Regression equation	Y=146.57 × - 4.366
Correlation Coeffiecient	0.9983
Limit Of Detection µg/band LOD	0.472 µg/Band
Limit Of Quantification µg/band LOQ	2.225 µg/Band
Specificity	Specific- Overlay Spectra, RF, Derivatization with Dragendorff's reagent

Table 1: Validation Parameter for Estimation of Premnine HCl.

Analyte	Premnine HCl Concentration Obtained
PI-AK	3.9% w/w
Powder Of *P. integrifolia*	0.3590 mg/gm%
Dashmul arishtha	
Manufacturer I Bottle size : 225 ml, Label Claim: Each 100ml contains 0.52 gm of *'Dashmula'* each.	
DA1*	1.231 mg/ml%
DA2*	0.481 mg/ml%
DA3*	0.497 mg/ml%
Dashmul arishtha	
Manufacturer II Bottle size : 220 ml, Label Claim: Each 10 ml contains 0.50 gm of *'Dashmula kwath'*	
SA4**	2.73 mg/ml %
SA5**	2.52 mg/ml %
SA6**	3.04 mg/ml %
Dashmul kadha	
Manufacturer III Bottle size : 227 ml, Label Claim: Each 100ml contains 418.91 mg of *'Dashmula'* each	
BA7***	1.58 mg/ml %
BA8***	1.046 mg/ml %
BA9***	2.66 mg/ml %

PI-AK- *Premna integrifolia* chloroform extract
*Manufacutre -1: Three Batches Coded As DA1, DA2, DA3
** Manufacutre -2: Three Batches Coded As SA4, SA5, SA6
*** Manufacture-3: Three Batches Coded As BA7, BA8,BA9

Table 2: Quantification of premnine HCl in PI-AK extract and selected marketed dashmula formulations.

Linearity assessment for Pr-S standard and its estimation in PI-AK and in selected marketed formulation

For linearity samples applied as Pr-S: 4, 4, 5, 6, 7, 8 µl (8 to 16 µg/band), for estimation on same plate sample applied as PI-AK- 2 µl, DA1, DA2, DA3- each 8 µl SA4, SA5, SA6- each 6 µl BA7, BA8 and BA9 each 7 µl over track 1 to 16 respectively with band length 8 mm. As per above condition chromatogram developed till 80 mm in above mobile phase, scanned at 283 nm and recorded at Rf 0. 59 for Pr-S, repeated thrice and average AUC noted. Plates were derivatized with Dragendorff's reagent for confirmation of Pr-S in samples. Standard linearity curve prepared by plotting Area Under Curve vs conc. in ng as shown Figure 2 and using calibration curve of concentration of Pr-S in PI-AK and in marketed formulations DA1, DA2, DA3, SA4, SA5, Sa6, BA7, BA8 and BA9 calculated as shown in Tables 1 and 2, Figures 3a-3j.

Validation of HPTLC method

As per International Conference on Harmonization Guidelines-Q2 (R1), 2005 the above estimation method was validated for various parameters as follows.

Linearity for Pr-S standard was assessed in range of 8-16 µg /band. The calibration curve established by plotting Peak area (AUC) *vs.* Conc. in µg/band, regression equation with slope, intercept and coefficient of correlation was calculated.

Using following formula, the Limit of detection (LOD) and limit of quantification (LOQ) were calculated as: LOD=3.3 × σ/S (1)

LOD-10 × σ/S (2)

Where σ=standard deviation of the response; S=slope of the regression line.

Repeatability precision

The repeatability of the method was assessed by three concentration 8, 12, 16 µg/band in triplicate for Standard Pr-S. The percentage relative standard deviation was calculated and expressed as Table 3.

Accuracy (% recovery)

The accuracy of the method was calculated by recovery experiments over known concentration on Extract–PI-AK and formulation SA4 at 80%, 100% and 120% spike of Standard Pr-S in triplicate for each experiment and analyzing it with %RSD. As per Table 4a, 3D graph of accuracy study shown in Figures 4a and 4b.

Intermediate precision

Precision study for Standard Pr-S done for three concentration as 8,

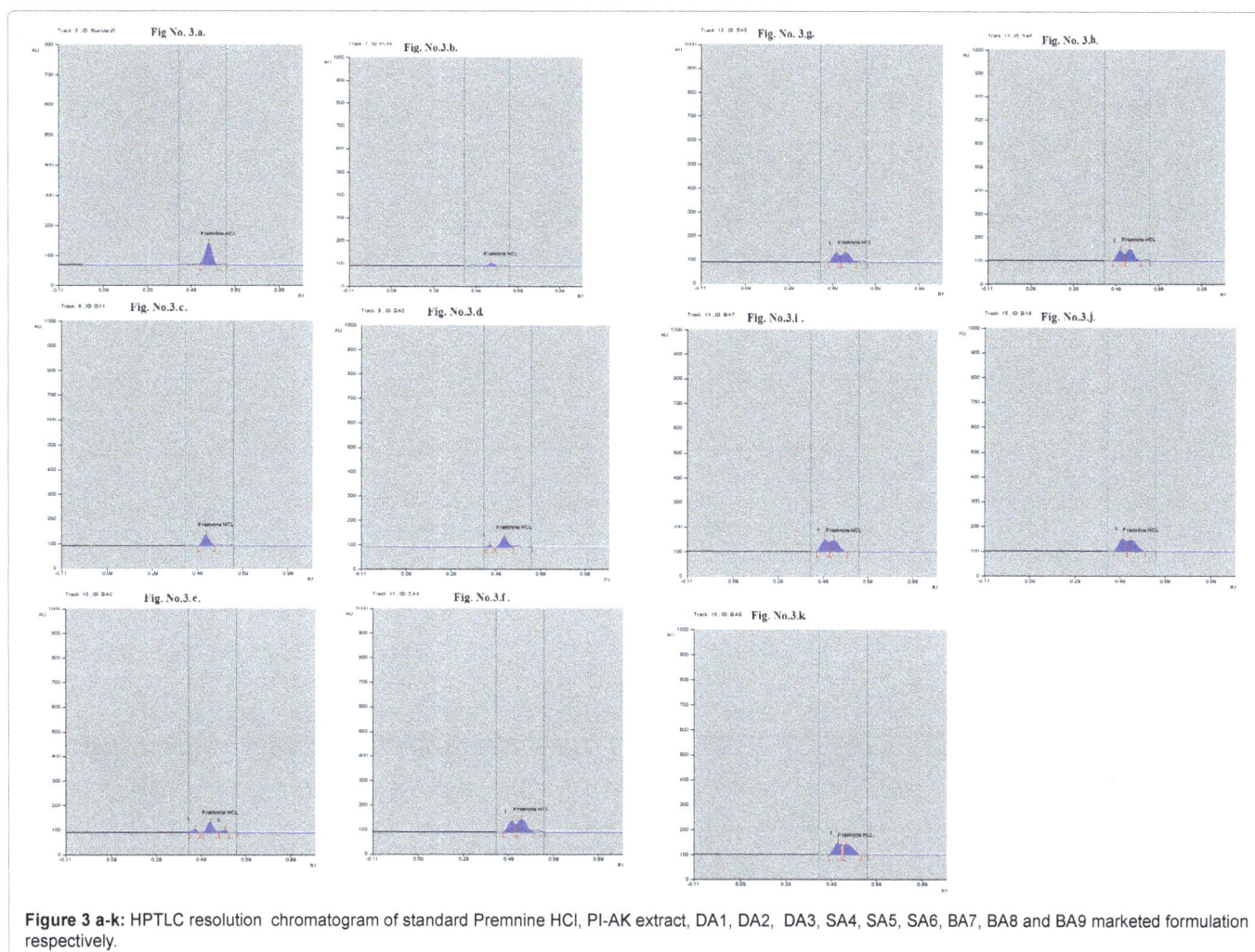

Figure 3 a-k: HPTLC resolution chromatogram of standard Premnine HCl, Pl-AK extract, DA1, DA2, DA3, SA4, SA5, SA6, BA7, BA8 and BA9 marketed formulation respectively.

Concentration[T] (µg/band)	Average Concentration[a] (n=3) Intraday (µg band)	Intra day %RSD	Average Concentration[a] (n=3) Interday (µg /band)	Inter day %RSD
8	7.78	1.9842	7.79	0.6465
12	12.22	0.5664	12.02	1.4634
16	15.79	1.6727	15.88	0.5044

T- Therotical, a- Obtained, % RSD- Relative Standard Deviation.

Table 3: Intermediate Precision Study for Standard Premnine HCl.

	Level of % Recovery	Amount of Pr-S in Pl-AK (µg)	Amount of Pr-S Added (µg)	Total amount of Pr-S taken (µg)	Total amount of Pr-S obtained (µg)	%recovery ± S.D.(n=3)
In Pl-AK + Pr-S						
	80%	3.95	3.16	7.11	7.37	103.65 ± 0.956
	100%	3.95	3.95	7.9	7.74	97.97 ± 0.535
	120%	3.95	4.74	8.69	8.20	94.91 ± 0.563
In SA4						
+ Pr-S	80%	5.14	4.11	9.25	9.23	99.78 ± 0.936
	100%	5.14	5.14	10.28	8.31	103.21 ± 0.526
	120%	5.14	6.16	11.3	11.34	100.35 ± 1.27

(a) Pl-AK- *Premna integrifolia* chloroform extract, (b) SA4-Manufacture-2 -coded Batch, (c) Pr-S- Premnine HCl Standard

Table 4a: Accuracy (%Recovery) for Pl-AK extract and SA4 extract with spike of Standard Premnine HCl (Pr-S) at 80% ,100% and 120% level.

(µg) Amount	(n=3)	Amount of Standard Obtained in µg			WMS	BMS	F value
Pl-AK + Pr-S		Day 1	Day 2	Day 3			
80 %	Mean	7.3	7.26	7.27	0.017189	0.000933	0.054299
(7.11)	%RSD	1.883	1.641	1.86			
100%	Mean	7.75	7.9	7.77	0.009922	0.018433	1.857783

(7.9)	%RSD	0.858	1.74	1.031			
120%	Mean	8.43	8.44	8.47	0.0231	0.001378	0.059644
(8.69)	%RSD	1.44	1.899	1.996			
SA4 + Pr-S							
80%	Mean	9.23	9.25	9.07	0.037567	0.028078	0.747412
(9.25)	%RSD	1.54	1.828	1.877			
100%	Mean	10.57	10.56	10.32	0.031556	0.060233	1.908803
(10.28)	%RSD	1.643	1.640	1.79			
120%	Mean (n=3)	11.14	11.34	11.31	0.029256	0.035233	1.20433
(11.3)	%RSD	1.86	0.759	1.70			

Table 4b: Precision study for PI-AK and SA4 with spike of Standard Premnine HCL at 80% ,100% and 120% level. (One Way ANOVA).

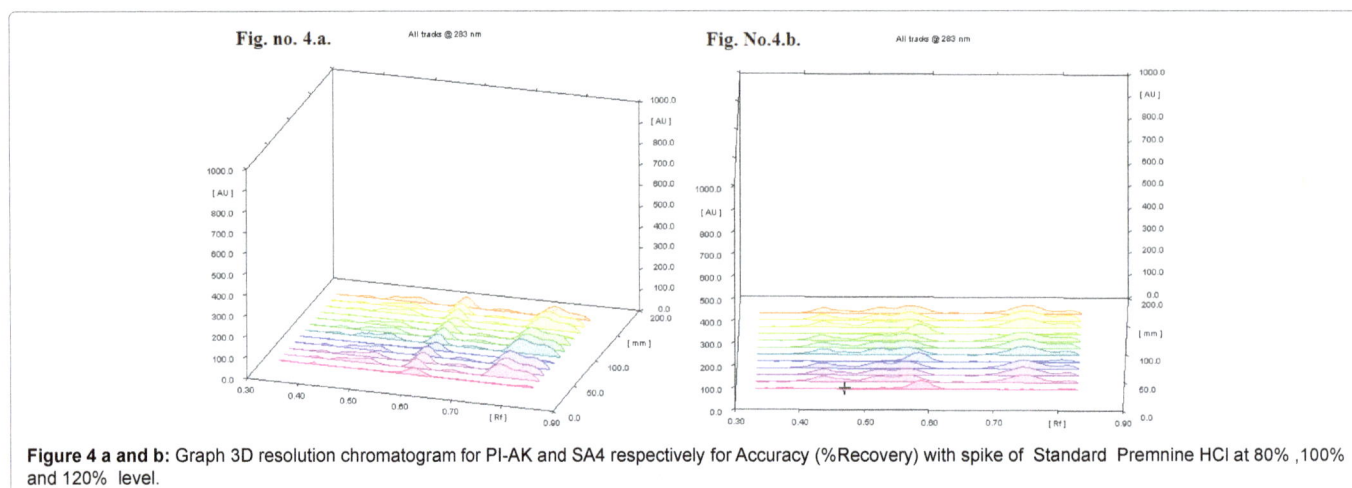

Figure 4 a and b: Graph 3D resolution chromatogram for PI-AK and SA4 respectively for Accuracy (%Recovery) with spike of Standard Premnine HCl at 80% ,100% and 120% level.

12, 16 μg /band, done in triplicate for intraday (Three times a day with 2 hr' interval) and inter day precision

(Three day consecutively).

Precision study also checked for extract and formulation as for PI-AK and SA4 (selected for recovery and precision study) done separately in triplicate for 80%, 100% and 120% triplicate for intraday and inter day precision with 3 × 3 model for each experiment and analyzing it with % RSD.

Further result obtained were subjected for one way analysis of variance and with-day mean square compared to between -day mean square by F test [25] (Table 4b).

Robustness

Mid concentration of 12 μg/band in triplicate of standard Pr-S subjected for robustness study using variability like wavelength 283-5 nm, slit width change as 6 × 0.3 mm, scan speed change as 40 mm/s, mobile phase change of saturation time as 12 min., and mobile phase change in composition as Toluene: Acetone: Diethylamine (7.2:1.8:1) and (6.8:2.2:1).

Results and Discussion

For specificity and fingerprinting study from each manufacturer one batch selected as DA1, SA4 and BA7 and letter on for estimation study all selected nine batches were subjected for study. In specificity study with developed chromatographic condition in mobile phase Toluene: Acetone: Diethylamine (7:2:1), Rf obtained 0.59 for Pr-S HCl standard at λ max 283 nm. There is single spot over track for standard. Track of PI-AK, DA1, SA4, KA7 shows spots with similar Rf for Pr-S. Spectral scan of this selected spots gave specific overlay for of wavelength at 283 nm in all selected batches of different manufacturer. Derivetized plate

with dragendorff's reagent gave orange red spot for Pr-S all sample at Rf 0.59 that confirms presences and helped for fingerprinting of Pr-S bioactive alkaloid in PI-AK and in selected formulations as shown in Figures 1a and1b. The calibration curve of Pr-S was found to be linear in range of 8-16 μg/band with good regression coefficient 0.9983. Table 1 summarize the validation parameter.

Estimation analysis of Pr-S as bio active marker for Extract PI-AK and formulation as shown in Table 2: Variability observed between manufactures and even within batches. Pr-S was selected as standard marker due to its bioactive properties and also specific content of *Premna* species that it can selectively justify presence of raw material in formulation.

Resolution of Pr-S peak in PI-AK extract and formulations while estimating are precise, checked for peak purity of the bands at start, middle and end position of chromatogram as shown in Figures 3a-3j, though co-eluted compound as peak no.1 are observed in resolution chromatogram of Figures 3f-3k, may be having structural similarity with standard Premnine HCl, but it is not alkaloid as do not give Dragendorff's test. And the peak area scanned for these peak and peak corresponds to standard marked as separate peak No. and integrated with red mark line as shown in Figures 3f-3k, noted as different with peak area by Camang -TLC scanner 5 software, therefore that value correspond to Premnine only consider for calculation. Precision study with repeatability performed over low, mid high concentration over Premnine HCl expressed as % RSD inter and intraday as per Table 3. Accuracy study with % recovery at three concentration level performed in PI-AK and in one of selected formulation SA4 shows result as per Table 4a.

Accuracy with % recovery at 80, 100,120% level with spike of Pr-S subjected for intraday and inter day, results at each level subjected to one way analysis of variance and the F value for each level were determined as per Table 4b.

F value as ratio of BMC/WMC, compared with tabulated $F_{(2,6)}$ value which is 5.14 and all calculated values are below it [16,17], therefore there was no significant difference between intra and inter day variability, suggesting good intermediate precision of the method.

Robustness study with change in wavelength by 5 nm, slit width, scan speed for mid concentration, for Pr-S shows %RSD value as 2.003, 1.4321, 1.756 respectively. Changes in mobile phase saturation time for 10 to 12 min and composition change as 7.2: 1.8:1, 6.8:2.2:1 do not show change in Rf from 0.59 and average ng values found as 12.29, 12.34, 12.24 and % RSD 1.74,1.24,1.76 respectively. Therefore, shows no significant changes in values so method is robust.

Conclusion

With present research study of development, validation and estimation conclusion is drawn as, HPTLC method is simple, precise, rapid, robust and selective for estimation of Pr-S alkaloid in Dashmul-*P. integrifolia* chloroform extract and its marketed formulations.

Pr-S selected as standard marker due to its bioactive properties and that has proven fingerprinting and specificity parameter as unique content in *Premna* species that it can etch presence of raw material in formulation.

Although variance is observed in % content of Pr-S alkaloid between manufactures and even in batches but method is precise and accurate even on extract and formulation through validation study therefore can be recommended as method for qualitative and quantitative analysis for quality control parameter for 'dashmul' *P. integrifolia* and its formulation.

Acknowledgement

We are thankful to our Institute and management for providing facility to work over HPTLC instrument.

References

1. The Wealth of India (1972) Dictionary of Indian raw materials and industrial products, New Delhi: Council of Scientific and Industrial Research 8: 239-240.

2. Nadkarnis KM (1976) Dr. Nadkarni's KM 'The Indian Materia Medica' Bombay Popular Prakashan I: 1009.

3. Kirtikar KR, Basu BD (1984) An ICS. Indian Medicinal Plants. Dehradun, India.

4. Deepti Y, Neerja T, Madan G (2010) Diterpenoids from Premna integrifolia. Phytochemistry Letters 3: 143–147.

5. Yadav D, Tiwari N, Gupta MM (2011) Simultaneous quantification of diterpenoids in Premna integrifolia using a validated HPTLC method. Journal of Separation Science 34: 286-291.

6. Otsuka H, Watanabe E, Yusana K, Ogimi CA (1993) A verbascoside iridoid glucoside conjugate form Premna corymbosa Rott. Phytochem 32: 983-986.

7. Basu NK, Dandiya PC (1947) Chemical investigation of Premna integrifolia linn. Journal of the American Pharmaceutical Association 36: 389–391.

8. Taivini T, Bianchini J, Lafontaine AC, Cambon A (1998) Volatile constituents of the flower buds concrete of Premna serratifolia L. Journal of Essential Oil research 10: 307-309.

9. Saidatul HBS (2005) Isolation, identification of flavonoid components and antioxidant activities determination from Morinda citrifolia (Meng kudu) and Premna serratifolia (Bebuas), Abstract of thesis presented to the Senate of Universiti Putra Malaysia in fulfillment of the requirement for the degree of Master of Science.

10. Ravinder SC, Nelson R, Krishnan PM, Pargavi B (2011) Identification of Volatile Constituents from Premna serratifolia L. Through GC-MS. International Journal of Pharm Tech Research 3: 1050-1058.

11. Rajendran R, Suseela L, Meenakshi SR, Saleem BN (2008) Cardiac stimulant activity of bark and wood of Premna serratifolia. Bangladesh J Pharmacol 3: 107-113.

12. Rajendran R, Basha N, Ruby S (2009) Evaluation of In Vitro Antioxidant Activity of stem-bark and stem-wood of Premna serratifolia Lin., (Verbenaceae) Research. J Pharmacognosy and Phytochemistry 1: 11-14.

13. Rajendran R, Basha S (2010) Antimicrobial activity of crude extracts and fractions of Premna serratifolia Lin. Root. International Journal Of Phytomedicines and Related Industries 2: 33-38.

14. Rekha R, Ekambaram K (2010) Antiarthritic activity of premna serratifolia linn. Wood against adjuvant induced arthritis. Avicenna J med Biotech 2: 101-106.

15. Gokani RH, Lahiri SK, Santani DD, Shah MB (2012) Evaluation of Anti-inflammatory and Antioxidant Activity of Premna integrifolia Root. Journal of Complementary and Integrative Medicine 1: 1553-3840.

16. Reich E, Schibli A (2006) High Performance Thin Layer chromatography for the analysis of medicinal plants. Thieme New York.

17. Balton S, Bon CC (2005) Pharmaceuticals Statistics practical and clinical Applications, Marcel dekker Inc, pp: 215-220.

21

Screening of Polyphenol Composition and Antiradical Capacity of Wild Erica arborea: A Native Plant Species from the Portuguese Flora

Alfredo Aires[1]* and Rosa Carvalho[2]

[1]Centre for the Research and Technology for Agro-Environment and Biological Sciences, CITAB, Universidade de Trás-os-Montes e Alto Douro, UTAD, Quinta de Prados, 5000-801 Vila Real, Portugal
[2]Agronomy Department, Universidade de Trás-os-Montes e Alto Douro, UTAD, Quinta de Prados, 5000-801 Vila Real, Portugal

Abstract

With this study, we aimed to determine the polyphenol composition and antiradical activity of activity of *Erica arborea* a native plant species from the Portuguese and Mediterranean flora and often reported as having important bioactivities, but without consistent scientific evidences. The analysis of polyphenol profile and content was performed by HPLC-DAD/UV-Vis and *in vitro* bioassay of 2,2'-azinobis-(3-ethylbenzothiazoline-6-sulfonic acid) (ABTS•+) method was used to evaluate the radical scavenging activity. The HPLC analysis showed a great diversity of compounds. *E. arborea* presented high contents in flavonols (52%) and flavanols (25%) and the average content of antiradical activity higher than 85%. The polyphenols identified are often associated by literature as having preventive properties and protective effect against of degenerative and inflammatory processes, thus, our findings confirm the empirical health properties often associated to this endemic plant. Therefore, based in our study the *E. arborea*, can be used to extract and purify phytochemicals with potentially beneficial effects on health.

Keywords: Endemic flora; Phytochemicals; Bioactivities; Pharmaceutical value

Introduction

The genus Erica is the largest genus of the family Ericaceae with 67 genera with 1,400 species mainly confined to the Europe and Asia [1]. *Erica arborea* (white flowers) are member of this family and are largely present in the mountains of Northern Portugal and Spain, but it is possible to find in other areas of Mediterranean basin. Despite this predominance in these regions no systematic studies have been made, at least to our knowledge, about their phytochemical composition and biological potential. Although, it is well known that Ericaceae species are used for centuries in folk medicine to treat urinary tract infections and other inflammatory diseases [2-4], none of their claimed properties were consistently proven. Different studies have shown that Ericaceae plants contain a larger number of biologically active secondary metabolites, including terpenoids, tannins and polyphenols [5-7]. Recently, investigations have shown that Ericaceae plants such as *Arbutus unedo* and *Erica australis*, largely present in the same region, are used in folk medicine as infusions and teas to treat diabetes, hypertension, prostate inflammation, bladder, kidneys diseases, and they are claimed to have anti-inflammatory and sedative properties [8-11]. It was also found that *E. arborea* displayed remarkable anti-inflammatory and antinociceptive activities due to its richness in phytochemicals, particularly phenolic acids, flavonoids and tannins which seems to be responsible for the high antiradical activities observed [12]. Another studies with other Ericaceae plant species, *Gaultheria procumbens* and *Gauhheria leucocarpa*, showed that high content of salicylic acid derivatives, especially methyl salicylate, might be responsible for its claimed strong anti-inflammatory and anti-rheumatoid properties [13,14]. Despite the large presence of these shrubs in Portugal and Iberian landscapes, no comprehensive profile of these species has been reported. Moreover, their putative beneficial effects are highly speculative, and no systematic and consistent study have been made to correlate their bioactivities with full phytochemical characterization, particularly polyphenols. In fact, polyphenols have been associated with important beneficial effects in the human health. In the past recent decades, epidemiological studies have shown that polyphenols have several biological benefits, such as reducing the risk of appearance of chronic diseases, scavenging oxidative and toxic free radicals, antiaging

and antimicrobial properties. The objective of the present study was to characterize the phenolic composition of *E. arborea* by HPLC-DAD/UV-Vis and to investigate their biological value by an *in vitro* radical scavenging activity. This information will help to clarify if both plants can be used as natural source of phytochemicals by pharmaceutical and agro-food industry.

Materials and Methods

Plant material and sampling

One kilogram of fresh weight of *Erica arborea* (leaves and flowers) were collected during the spring season at April 2015 in the natural open fields in Northern Portugal, Vila Real Region (400 meters of altitude), Natural Park of Alvão (N 41°17'35.538", W 7°44' 29.6268"). After harvest, the fresh samples were properly and botanically identified and dried in a freeze-drier system (UltraDrySystemsTM, USA), milled and reduced to a fine powder, and stored in dark flasks at 4 °C in a dark environment until extraction. Fresh and dry weight were registered and the amount of dry matter was determined. Three replications were taken from independent biological samples.

Extraction procedure

One hundred mg of powdered dry samples were placed in a screw cap tubes (10 mL) and mixed with 10 mL of 70% aqueous methanol (methanol:water). Each mixture was vigorously agitated in a vortex (Genie 2, Fisher Scientific, UK), heated at 70 °C (1083, GFL-Gesells

*Corresponding author: Alfredo Aires, Integrated Member, Centre for the Research and Technology for Agro-Environment and Biological Sciences, CITAB, Universidade de Trás-os-Montes e Alto Douro, UTAD, Quinta de Prados, 5000-801 Vila Real, Portugal, E-mail: alfredoa@utad.pt

chaft ffur Labortechnik mbH, Germany) during 20 minutes and agitated every 5 minutes. Then, the mixtures were centrifuged at 4000 rpm and 15 min (Kubota, Japan), and filtered through a Whatman No. 1 paper. The supernatants were then filtered through PTFE 0.2 μm, Ø 13 mm (Teknokroma, Spain) to amber HPLC vials (Chromabond 2-SVW(A) ST-CPK, Sigma-Aldrich, Tauferkichen, Germany). Finnaly, the obtained extracts were stored (no more than one week) under refrigeration (-20 °C) prior to the HPLC and antiradical colorimetric bioassay analysis.

Polyphenol composition by HPLC-DAD/UV-Vis

The separation and quantification of polyphenols present in *E. cinerea* and *E. arborea* hydro-alcoholic extracts were performed by HPLC-DAD-UV/VIS [15]. After extraction, each extract was injected in HPLC-DAD/VIS-UV system (Gilson Inc., Middleton, WI, USA) equipped with an eluent composed by water with 1% of trifluoroacetic acid (TFA) (solvent A) and acetonitrile with 1% TFA (solvent B). The elution was performed at a flow rate of solvent of 1 mL min^{-1}, with a gradient starting with 100% of water, with an injection volume of 10 μL. The chromatograms were recorded at 280, 320, 370 and 520 nm with a C18 column (250 × 46 mm, 5 μm) (ACE® HPLC columns, Advanced Chromatography Technologies, Ltd., Aberdeen, Scotland). The polyphenols were identified using peak retention time, UV spectra and UV max absorbance bands and trough comparison with external commercial standards (Extrasynthese, Cedex, France, and Sigma-Aldrich, Tauferkichen, Germany), as well as by comparing with published literature [16-19] The external standards were freshly prepared in 70% methanol (methanol:water) at concentration of 1.0 mg mL^{-1} and running in HPLC-DAD-UV/VIS before the samples. The quantification was done using response factor for each compound detected compared to each similar pure standard compound. The results were expressed as μg g^{-1} dry weight (dw). Methanol and acetonitrile were HPLC gradient and purchased from Panreac chemistry (Lisbon, Portugal) and Sigma-Aldrich (Taufkirchen, Germany), respectively. The aqueous solution were prepared using ultra-pure water (Milli-Q, Millipore, Massachusetts, USA).

ABTS radical scavenging activity

The ABTS radical scavenging activity of *E. arborea* extracts was evaluated using the (ABTS●+) radical-scavenging activity colorimetric method [20] conducted in a 96-well microplate. For that different concentrations of both hydro-alcoholic extracts were prepared, ranging from 0.195 to 10.0 mg mL^{-1} and used in the *in vitro* AA assay. A radical ABTS solution was freshly prepared by mixing a 7 mM of ABTS at pH 7.4 (5 mM NaH$_2$PO$_4$, 5 mM Na$_2$HPO$_4$ and 154 mM NaCl) with 2.5 mM potassium persulfate (final concentration) followed by storage in the dark at room temperature for 16 h before use. After, the mixture was diluted with ethanol to give an absorbance of 0.70 ± 0.02 units at 734 nm using a multiscan microplate reader (Multiskan™ FC Microplate Photometer, USA). Then, to each well, an aliquot of 15 μL of each extract or standard were added followed by addition of 285 μL fresh ABTS solution. The microplates were then incubated at room temperature and dark during 10 minutes. At the end, the absorbance values were measured in a multiscan microplate reader at 734 nm. The results were expressed as percentage (%) of ABTS radical scavenging activity, using the following formula:

ABTS radical scavenging (%)=[((absorbance solvent-absorbance sample)/absorbance solvent) × 100]

The concentration of antioxidants which scavenge the free radical ABTS● about 50% (IC$_{50}$) for both plant extracts was also determined.

A positive control with high recognized antioxidant and antiradical activity, the ascorbic acid, was included in the AA assay.

Statistical analysis

All determinations were carried out in triplicate and the results were expressed as mean values ± standard deviation (SD). The Software SPSS v.17 (SPSS-IBM, Orchard Road-Armonk, New York, USA) was used to carry out these analysis.

Results and Discussion

The analysis by HPLC-DAD/UV-Vis of the methanolic extracts of *E. arborea* revealed the presence of different classes of polyphenols as illustrated in the Figure 1, in which is presented a typical example of chromatogram find in the current study. The quantity of each polyphenol identified in each plant extract is presented in Table 1, in which is also included the data of retention time and λ max in the visible region for each polyphenol, and in Table 2 is presented the chemical structure of each polyphenol identified. According our results, the major individual polyphenols were quercetin-3-*O*-rutinoside (13% of total polyphenols identified), 5-*O*-p-coumaroylquinic acid (12%), kaempferol-3-*O*-rutinoside (12%), epicatechin (11%) and myricetin-3-*O*-rhamnoside (8%), which are reported by literature as having important bioactive properties, such antiradical and antioxidative capacity. However, our results showed a lower antiradical activity when compared to a positive control with recognized antioxidant or antiradical activity, the ascorbic acid (Figures 2 and 3), but when compared to other medicinal plants [21,22] the *E. arborea* showed a moderate to higher antiradical activity.

Polyphenols, antioxidant and antiradical activities have been considered parameters to measure the biological property of foods.

Polyphenols	R$_t$ (min)	UV detection (nm)	UV λ max (nm)	Quantity (μg g^{-1} dw)
trans-cinnamic acid	15.72	320	317	3.5 ± 0.2
5-*O*-caffeoylquinic acid	16.45	320	280, 320	8.6 ± 0.9
Gallocatechin	17.36	280	271	25.5 ± 0.9
Myricetin-3-*O*-glucoside	18.21	370	271, 347	56.8 ± 1.0
Myricetin-3-*O*-galactoside	18.32	370	276, 346	20.4 ± 0.8
Epigallocatechin	18.78	280	276	15.2 ± 0.2
Myricetin-3-*O*-rhamnoside	19.16	370	276, 342	64.0 ± 1.1
Epicatechin	19.67	280	277	91.0 ± 0.4
Epigallocatechin gallate	19.93	280	279	38.3 ± 0.6
p-coumaroylquinic acid	20.02	320	282, 311	11.9 ± 1.0
Gallocatechin gallate	20.38	280	277	15.7 ± 2.3
Kaempferol-3-*O*-rutinoside	20.67	370	265, 358	97.5 ± 2.7
Kaempferol-3-*O*-galactoside	20.79	370	265, 356	49.4 ± 1.6
Myricetin	21.03	370	275, 341	5.8 ± 0.2
Kaempferol-3-*O*-glucoside	21.44	370	267, 355	11.2 ± 0.7
Epicatechin gallate	22.71	280	278	23.1 ± 0.5
Quercetin-3-*O*-rutinoside	22.08	370	253, 354	109.7 ± 0.2
Quercetin-3-*O*-galactoside	22.21	370	256, 355	8.7 ± 1.9
5-*O*-p-coumaroylquinic acid	22.30	320	285, 312	99.1 ± 0.1
Ellagic acid	24.56	370	354	56.4 ± 0.2
Chrysin	26.04	370	256, 313	11.3 ± 1.0
Kaempferol	26. 39	370	266, 357	4.1 ± 0.4

Values expressed as mean ± standard deviation (SD) of three replicates

Table 1: Profile and contents of polyphenols, respective retention time (Rt), maximum absorption (λ max) in methanolic extracts of *Erica arborea* (by elution order).

Polyphenols and chemical structures		
5-O-caffeoylquinic acid	Gallocatechin	Kaempferol-3-O-rutinoside
5-O-p-coumaroylquinic acid	Gallocatechin gallate	Myricetin-3-O-galactoside
Chrysin	Epigallocatechin gallate	Myricetin-3-O-glucoside
Ellagic acid	Kaempferol	Myricetin-3-O-rhamnoside
Epicatechin	Kaempferol-3-O-galactoside	Quercetin-3-O-galactoside
Epicatechin gallate	Kaempferol-3-O-glucoside	Quercetin-3-O-rutinoside

Table 2: Chemical structures of the main polyphenols identified *Erica arborea* extracts, by alphabetical order.

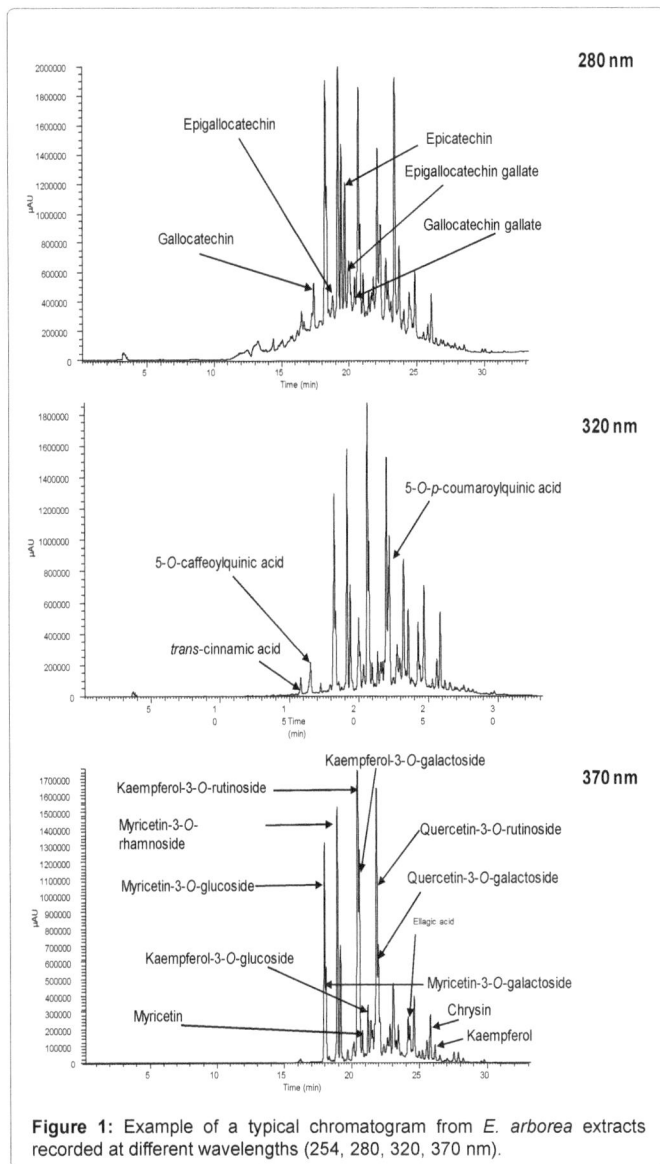

Figure 1: Example of a typical chromatogram from *E. arborea* extracts recorded at different wavelengths (254, 280, 320, 370 nm).

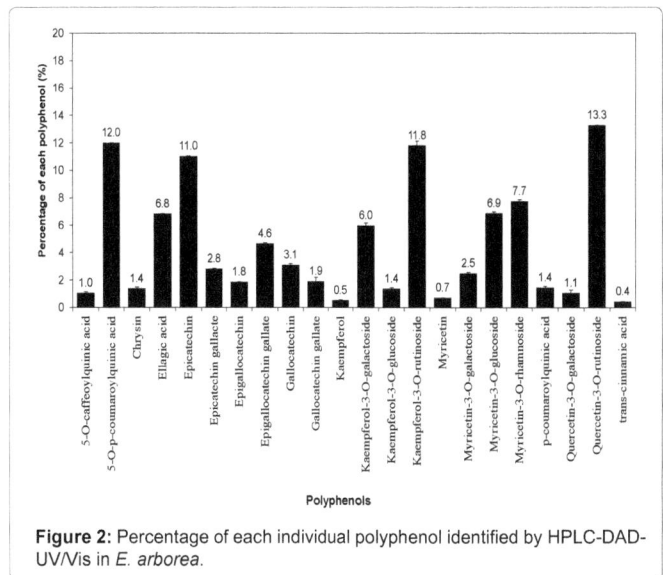

Figure 2: Percentage of each individual polyphenol identified by HPLC-DAD-UV/Vis in *E. arborea*.

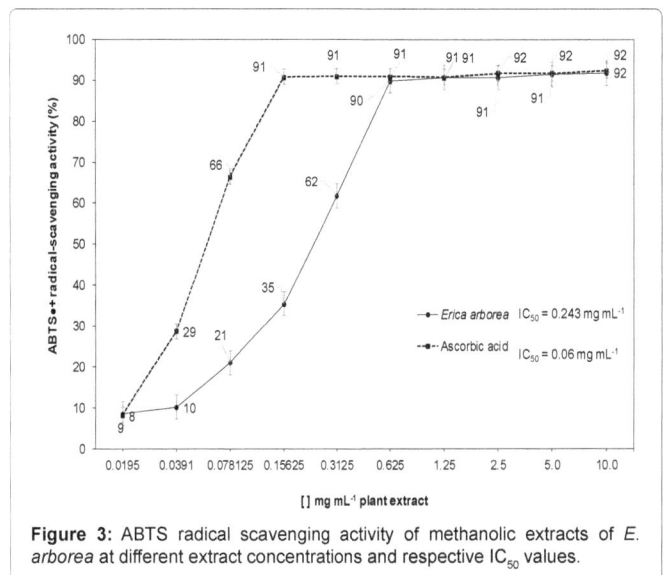

Figure 3: ABTS radical scavenging activity of methanolic extracts of *E. arborea* at different extract concentrations and respective IC_{50} values.

In our study, extracts of *E. arborea* revealed the presence of different classes of polyphenols, often associated with anti-inflammatory activities in other plant matrices. Our results also shown that flavonoids (kaempferol, myricetin, quercetin and luteolin glycosides) and phenolic acids (chlorogenic, neochlorogenic and caffeic acids) are associated with the antiradical activity. These compounds act as scavengers of various oxidizing species such as superoxide anion, hydroxyl radical or peroxy radicals and as quenchers of singlet oxygen [23] and their antiradical capacity arise from their ability to inhibit lipid peroxidation, chelate redox-active metals, and reduce all processes involving reactive oxygen species [24]. They may exert protection of cell structures through different mechanisms, and one in particular is their ability to increase the levels of the powerful intracellular antioxidant enzymes [25,26]. Also, recent works have demonstrate that phenolic acids like chlorogenic and neochlorogenic acids can have a strong anti-inflammatory activity [27,28]. Similar observations were made recently to caffeic acid by Song et al. [27] who reported that caffeic acid is able to inhibit the activity of acetylcholinesterase (AChE) and butyrylcholinesterase (BChE) in a dose-dependent manner, protecting the brain against neuronal inflammatory processes and diseases. Therefore, the high content of

such important compounds in *E. arborea* may be used as an argument to explain their high biological potential, and may enhance the interest to study this plant species as natural source to extract and purify bioactive compounds for pharmaceutical and agro-food industry.

Acknowledgements

The authors acknowledge the support of European Investment Funds by FEDER/COMPETE/POCI–Operational Competitiveness and Internationalization Programme, under Project POCI-01-0145-FEDER-006958 and National Funds by FCT, Portuguese Foundation for Science and Technology, under the project UID/AGR/04033/2013.

Conflicts of Interest

The authors declare no conflict of interest.

References

1. Christenhusz MJM, Byng JW (2016) The number of known plants species in the world and its annual increase. Phytotaxa 261: 201-217.

2. Akkol EK, Yeşilada E, Güvenç A (2008) Valuation of anti-inflammatory and antinociceptive activities of Erica species native to Turkey. J Ethnopharmacol 116: 251-257.

3. Pavlović RD, Lakušić B, Došlov-Kokoruš Z, Kovačević N (2009) Arbutin content and antioxidant activity of some Ericaceae species. Pharmazie 64: 656-659.

4. Nunes R, Anastácio A, Carvalho IS (2012) Antioxidant and free radical scavenging activities of different plant parts from two Erica Species. J Food Quality 35: 307-314.

5. Luís Â, Domingues F, Duarte AP (2011) Bioactive compounds, RP-HPLC analysis of polyphenols, and antioxidant activity of some Portuguese shrub species extracts. Nat Prod Comm 6: 1863-1872.

6. Maleš Ž, Šarić D, Bojić M (2013) Quantitative determination of flavonoids and chlorogenic acid in the leaves of Arbutus unedo L. using thin layer chromatography. J Anal Method Chem 2013: ID 385473.

7. de Jesus NZT, de Souza Falcão H, Gomes IF, de Almeida Leite TJ, de Morais Lima GR, et al. (2012) Tannins, peptic ulcers and related mechanisms. Int J Mol Sci 13: 3203-3228.

8. Neves JM, Matos C, Moutinho C, Queiroz G, Gomes LR (2009) Ethnopharmacological notes about ancient uses of medicinal plants in Trás-os-Montes (northern of Portugal). J Ethnopharmacol 124: 270-283.

9. Barros L, Carvalho AM, Morais JS, Ferreira ICFR (2010) Strawberry-tree, blackthorn and rose fruits: Detailed characterization in nutrients and phytochemicals with antioxidant properties. Food Chem 120: 247-254.

10. Nunes R, Carvalho IS (2013) Antioxidant activities, distribution of polyphenols and free amino acids of Erica australis L. leaves and flowers collected in Algarve, Portugal. Nat Prod Res 27: 1664-1667.

11. Nunes R, Rodrigues S, Pasko P, Tyszka-Czochara M, Grenha A, et al. (2014) Effect of Erica australis extract on Caco-2 cells, fibroblasts and selected pathogenic bacteria responsible for wound infection. Ind Crop Prod 52: 99-104.

12. Amezouar F, Badri W, Hsaine M, Bourhim N, Fougrach H (2013) Antioxidant and anti-inflammatory activities of Moroccan Erica arborea L. Pathol Biol 61: 254-258.

13. Zhang D, Liu R, Sun L, Huang C, Wang C, et al. (2011) Anti-inflammatory activity of methyl salicylate glycosides isolated from Gaultheria yunnanensis (Franch.) Rehder. Molecules 16: 3875-3884.

14. Liu WR, Qiao WL, Liu ZZ, Wang XH, Jiang R, et al. (2013) Gaultheria: phytochemical and pharmacological characteristics. Molecules 18: 12071-12108.

15. Aires A, Carvalho R, Rosa EAS, Saavedra MJ (2013) Phytochemical characterization and antioxidant properties of organic baby-leaf watercress produced under organic production system. CyTA - Journal of Food 11: 343-351.

16. Sakakibara H, Honda Y, Nakagawa S, Ashida H, Kanazawa K (2003) Simultaneously determination of all polyphenols in vegetables, fruits and teas. J Agric Food Chem 51: 571-581.

17. Wang H, Provan, GJ, Helliwell K (2003) HPLC determination of catechins in tea leaves and tea extracts using relative response factors. Food Chem 81: 307-312.

18. Lin LZ, Harnly JM (2007) A screening method for the identification of glycosylated flavonoids and other phenolic compounds using standard analytical approach for all plant materials. J Agric Food Chem 55: 1084-1096.

19. Spáčil Z, Nováková L, Solich P (2008) Analysis of phenolic compounds by high performance liquid chromatography and ultra-performance liquid chromatography. Talanta 76: 189-199.

20. Sratil P, Klejdus B, Kubán V (2006) Determination of total content of phenolic compounds and their antioxidant activity in vegetables-evaluation of spectrophotometric methods. J Agric Food Chem 54: 607-616.

21. Ozkan G, Kamiloglu S, Ozdal T, Boyacioglu D, Capanoglu E (2016) Potential use of Turkish medicinal plants in the treatment of various diseases. Molecules 21: 257.

22. Singh G, Passsari AK, Leo VV, Mishra VK, Subbarayan S, et al. (2016) Evaluation of phenolic content variability along with antioxidant, antimicrobial, and cytotoxic potential of selected traditional medicinal plants from India. Front Plant Sci 407: 1-12.

23. Treml S, Šmejkal K (2016) Flavonoids as potent scavengers of hydroxyl radicals. Compr Rev Food Sci F 15: 720-738.

24. Grassi D, Desideri G, Ferri C (2010) Flavonoids: antioxidants against atherosclerosis. Nutrients 2: 889-902.

25. Myhrstad MCW, Carlsen H, Nordström O, Blomhoff R, Moskaug JØ (2002) Flavonoids increase the intracellular glutathione level by transactivation of the γ-glutamylcysteine synthetase catalytical subunit promoter. Free Radical Bio Med 32: 386-393.

26. Durgo K, Vukovi L, Rusak G, Osmak M, Colic JF (2007) Effect of flavonoids on glutathione level, lipid peroxidation and cytochrome P450 CYP1A1 expression in human laryngeal carcinoma cell lines. Food Technol Biotech 45: 69-79.

27. Song F, Lia H, Sun J, Wang S (2013) Protective effects of hydroxycinnamic acid and cinnamic aldehyde on isoproterenol-induced acute myocardial ischemia in rats. J Ethnopharmacol 150: 125-130.

28. Ruan Z, Shumei M, Zhou L, Zhou Y, Jing L, et al. (2016) Chlorogenic acid enhances intestinal barrier by decreasing MLCK expression and promoting dynamic distribution of tight junction proteins in colitic rats. J Funct Foods 26: 698-708.

22

Isolation of Plastoquinone from Spinach by HPLC

Marco Malferrari* and Francesco Francia

Laboratorio di Biochimica e Biofisica, Dipartimento di Farmacia e Biotecnologie, FaBiT, Università di Bologna, 40126 Bologna, Italy

Abstract

We report a method for the purification of plastoquinone-9 (PQ), a prenylquinone cofactor involved in the photosynthetic electron transport chain. The described procedures relies on spinach-chloroplast isolation followed by PQ extraction and chromatographic fractionation. Extraction of PQ was achieved using partition of chloroplast suspension with methanol:petroleum ether. This procedure removed large amounts of green pigments from the extract and thus facilitates the subsequent chromatographic isolation of PQ. To obtain pure PQ, the developed extraction was combined with a two-step chromatographic approach using orthogonal stationary phases, i.e. alumina and octadecylsilane (C18). A small scale protocol for analytical reverse-phase high performance liquid chromatography (RP-HPLC), which may be implemented in most laboratories equipped with conventional systems, is described. The reported methodology represents a valuable tool for the fast production of small amounts of PQ, for which there are no commercial standards available.

Keywords: *Spinacia oleracea*, plastoquinone-9, HPLC.

Introduction

In plants, algae and cyanobacteria the light-driven redox reactions responsible for the primary energy transduction by photosynthesis involve three major membrane protein complexes named photosystem II (PSII), cytochrome b_6f, and photosystem I (PSI). Active centers of these complexes communicate via two redox carriers: membrane-soluble plastoquinone-9 (PQ) connects the reducing activity of PSII to the oxidizing activity of the b_6f complex, while the soluble protein plastocyanin, is reduced by the b_6f activity and oxidized by PSI [1,2].

Besides its main role as proton and electron carrier, PQ is an active player in the short and long term light adaptation responses of photosynthetic cells [3-6]. Recently the involvement of PQ as singlet oxygen scavenger has been reported in high light stressed *Chlamydomonas reinhardtii* cultures [7].

In spite of the reborn interest for the different PQ functions, which are at present intensively investigated, PQ reference standards are no longer commercially available and the purification methods reported in the literature are often fragmentary and dated, being based on protocols published in the 60-80's [8-11]. In the present paper we describe a procedure in which PQ is directly extracted from water suspensions of chloroplasts by partition with methanol:petroleum ether. The enriched extract is then chromatographed on two orthogonal phases for maximum selectivity using alumina and reverse phase columns. The described protocol, which allows straightforward isolation of pure PQ from spinach chloroplasts, is applied on a small scale requiring HPLC instrumentation normally present in a laboratory.

Materials and Methods

Chloroplast preparation and solvents

Chloroplasts were prepared from spinach leaves (*Spinacia oleracea* L.) following standard protocols described in Barr and Crane [8]. The total chlorophyll (Chl) content was determined spectrophotometrically upon extraction of a few μL of chloroplast suspension with acetone, accordingly to Arnon [12]. Acetone, methanol, petroleum ether (bp 40-60°C), diethyl-ether, ethanol and acetonitrile of analytical or HPLC grade, used as solvent for PQ extraction and isolation, were from Sigma-Aldrich.

Extraction and fractionation of plastoquinone

As an alternative to the Barr and Crane method [8], we tested for PQ extraction from chloroplasts a protocol set up by Venturoli et al. [13] for the extraction of ubiquinone-10 from membranes of photosynthetic bacteria. To this aim, 100 mL of sugar free chloroplasts were diluted in water to a final Chl concentration (a+b) of 1 mg/mL. A 20 time volume excess of a methanol:petroleum ether 3:2 mixture was directly added to the liquid chloroplast sample and vigorously shaked for 1 min. After phase separation, the petroleum ether was collected, and the extraction repeated twice by re-adding fresh petroleum ether to the methanol-water lower phase. The petroleum ether fractions were combined, dried by rotary evaporation and resuspended in 10 mL of petroleum ether.

The extract obtained was fractionated by open column chromatography on acid washed alumina (SIGMA-Aldrich) following the scheme of Barr and Crane [8]. 100 g of alumina were deactivated by adding 6 ml of double distilled (d.d.) water, resuspended in petroleum ether and placed into a glass column with diameter and length of 2.5 cm and 40 cm, respectively. The petroleum ether extract, containing PQ, was loaded into the column and 7 fractions (1-7) were obtained by eluting in sequence with 100 mL of 0, 0.2, 2, 4, 8, 16 and 20% of diethyl-ether in petroleum ether (v/v).

HPLC plastoquinone isolation

Analytical HPLC analysis was performed at 40°C by using a Jasco Pu-1580 pump, a C-18 reverse phase column (Waters Spherisorb 5 μm ODS2, 4.6 x 250 mm), and a Jasco UV 970 detector operating at 255 or 290 nm. The mobile phase (flow rate 1.5 mL/min) was a mixture of acetonitrile:ethanol 3:1, as described in Yoshida et al. [14].

---

*Corresponding author: Marco Malferrari, Laboratorio di Biochimica e Biofisica, Dipartimento di Farmacia e Biotecnologie, FaBiT, Università di Bologna, 40126 Bologna, Italy, E-mail: marco.malferrari2@unibo.it

UV-visible absorption spectra of isolated PQ were recorded using a Jasco V-550 UV/VIS spectrophotometer.

Results and Discussion

The method of PQ purification described in the present work is discussed in the following by considering separately: (i) the extraction with organic solvents; (ii) the fractionation by aluminium oxide; (iii) the isolation by HPLC and recovery of pure components.

Extraction with organic solvent

From 500 g of spinach leaves, 95 mL of chloroplasts containing chlorophyll (a + b) at a concentration of 1.34 mg/mL were obtained. After washing chloroplasts in water to get a sugar free chloroplast suspension, at variance to established protocols [8], extraction of PQ was achieved by partition between a methanol:water phase and a petroleum ether phase (see Materials and Methods for details). This procedure, extensively used for ubiquinone extraction form chromatophores of photosynthetic bacteria [13], was found to be effective also in the case of PQ. Although a small portion of PQ remained in the methanol:water phase, resulting in a slightly lower extraction yield as compared to the established method [8], this extraction procedure provided significant advantages. First time-consuming chloroplast lyophilization could be avoided, which speeded up the procedure and prevented possible unwanted reactions during lyophilization. Second, since large amounts of green pigments remained trapped in the methanol-water phase, the subsequent purification steps were considerably facilitated. In order to further separate PQ from chlorophyll and other pigments, the obtained extracts were then treated on an aluminium oxide column as described below.

Fractionation by aluminium oxide

The use of an acidic alumina column, besides removing to a large extent green pigments and carotenoids, has the advantage that plastoquinone oxidation takes place during this procedure. Indeed it has been shown [15,16] that a substantial fraction (30-40%) of the extracted plastoquinone is in the doubly reduced and protonated form, i.e. plastoquinol PQH_2. The conversion to the more stable oxidized form facilitates the separation and the collection of the purified PQ.

The residue from the methanol:petroleum ether extraction was loaded on the top of a 100 g aluminium oxide column and eluted with fractions of 100 mL petroleum ether with increasing v/v percentages of diethyl ether (see Extraction and fractionation of plastoquinone). At variance to what reported in established protocols [8], PQ eluted in fraction 5. This difference in the elution profile was likely due to the much lower concentration of green pigments in the methanol:petroleum ether extract. To maximize the final yield of PQ, an HPLC confirmatory analysis of the fractions (from 1 to 7) should be, however, always done to ascertain the presence of PQ in the respective fractions.

Isolation and recovery of plastoquinone

The dried residue of the eluted fraction 5 from the aluminium oxide column was dissolved in ethanol and subjected to HPLC analysis; the obtained chromatograms are shown in Figure 1. Oxidized PQ is known to present an absorption maximum at 255 nm (ε_{255} = 17.94 $mM^{-1}cm^{-1}$) that shifts to 290 nm upon reduction (ε_{290} = 3.39 $mM^{-1}cm^{-1}$) [17]; this implies that the detection of the oxidized or the reduced PQ peaks are maximized at 255 nm and 290 nm, respectively. As shown in the upper panel of Figure 1, a large peak with a retention time of 19.2 minutes is present in the chromatogram at 255 nm (black trace), characteristic of the oxidized PQ [14]. Essentially no peak appeared in

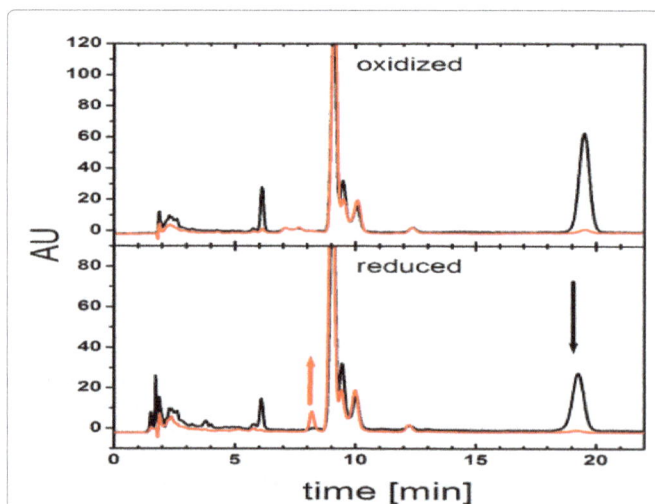

Figure 1: HPLC elution profiles after fractionation with aluminium oxide. Upper panel: chromatogram of the alumina column fraction 5 from the methanol : petroleum ether extract of water suspended chloroplasts. Red and black chromatograms refer to elution profiles detected at 290 or 255 nm, respectively. In the lower panel, black downward arrow indicates the elution of the PQ oxidized form (retention time of 19.2 min); the peak of the oxidized form at 255 nm is decreased upon reduction. Red upward arrow indicates the elution of the PQ reduced form (retention time of 8.1 min) whose peak at 290 nm becomes visible upon reduction.

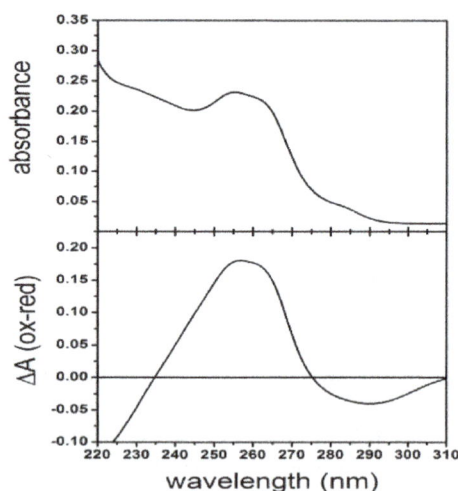

Figure 2: UV spectra of HPLC purified PQ.
Upper panel: spectrum of PQ collected from the HPLC. Lower panel: oxidized-reduced difference spectrum. The oxidized PQ eluting at 19.2 min (see Figure 1) has been collected from analytical HPLC runs, dried under nitrogen flow, and resuspended in ethanol.

the chromatogram at 290 nm (red trace) at the retention time expected for reduced plastoquinone (8.1 min, see below and [14]). This confirms that during the fractionation through the aluminium oxide column, reduced PQH_2 gets oxidized [19]. After addition of $NaBH_4$ to reduce PQ [10,14], the elution peak at 19.2 minutes in the 255 nm chromatogram (lower panel of Figure 1, black trace) decreased, and the reduced PQ eluted earlier with a retention time of 8.1 min, better detected at 290 nm (lower, left panel of Figure 1, red trace). Although reduction of the oxidized PQ form upon $NaBH_4$ addition was incomplete, the elution profiles showed peaks with the same retention times and the same qualitative response to reducing conditions as those reported in [14].

UV absorption spectrum of purified PQ, obtained after collection of the 19.2 minutes retention time HPLC peak, is presented in Figure 2, upper panel. As better evidenced in the oxidized-reduced difference spectrum (lower panel), the absorption profile has a maximum at 255 nm which is typical of PQ. The oxidized-reduced difference spectrum exhibits isosbestic points at 235, 275, and 310 nm, in good agreement with values reported in the literature [18]. Variations of these values in a limited range (1-2 nm) are usually ascribed to small changes in the background absorption of the sample, following the addition of $NaBH_4$ before acquiring the spectrum of the reduced form [18].

The concentration of the HPLC purified PQ was estimated from the oxidized-minus-reduced difference spectrum at 255 nm (ox-red ΔA_{255nm}), assuming a differential extinction coefficient $\Delta\varepsilon_{255nm} = 15$ mM^{-1}cm^{-1} [14] for the difference band. Injection of a known amount of PQ in the analytical HPLC system allowed us to estimate the extraction yield. The yield was found to be 0.023 molesPQ/mg Chl, slightly less than what reported for established method [8]. In our working conditions with analytical HPLC, we obtained about 30 µg of plastoquinone for each 20 µL of PQ-containing-fraction injected in the column.

Conclusions

The involvement of PQ in photosynthetic electron transport, redox signaling pathways and the recent evidence that PQ can act as singlet oxygen scavenger under high light stress are renewing the interest of researchers for this cofactor.

In the present work we have shown that PQ isolation can be easily done using partition of chloroplast suspension with methanol:petroleum ether followed by orthogonal chromatographic steps. Since PQ is not commercially available, our procedure is expected to facilitate the production and usage of this molecule for bioassay or quantification purposes in a normally equipped laboratory.

Acknowledgment

We thank Fabiana Antonioni and Carlotta Pia Cristalli for help in analytical HPLC plastoquinone isolation, and Giovanni Venturoli for critical reading of the manuscript.

F.F. and M.M. acknowledge financial support by University of Bologna, Italy (Grant FARB2012-FFBO122037).

References

1. Crofts AR, Wraight CA (1983) The electrochemical domain of photosynthesis. Biochim Biophys Acta 726: 149-185.

2. Cramer WA, Knaff DB (1990) Energy transduction in biological membranes. A textbook of bioenergetics. Springer.

3. Allen JF, Bennett J, Steinback KE, Arntzen CJ (1981) Chloroplast protein phosphorylation couples plastoquinone redox state to distribution of excitation-energy between photosystems. Nature 291: 25–29.

4. Zito F, Finazzi G, Delosme R, Nitschke W, Picot D, et al. (1999) The Qo site of cytochrome b_6f complexes controls the activation of the LHCII kinase. EMBO J 18: 2961–2969.

5. Pesaresi P, Hertle A, Pribil M, Kleine T, Wagner R, et al. (2009) Arabidopsis STN7 kinase provides a link between short- and long-term photosynthetic acclimation. Plant Cell 21: 2402-2423.

6. Pfalz J, Liebers M, Hirth M, Grübler B, Holtzegel U, et al. (2012) Environmental control of plant nuclear gene expression by chloroplast redox signals. Front Plant Sci 3: 1-9.

7. Nowicka B, Kruk J (2012) Plastoquinol is more active than α-tocopherol in singlet oxygen scavenging during high light stress of Chlamydomonas reinhardtii. Biochim Biophys Acta 1817: 389-394.

8. Barr R, Crane FL (1971) Quinones in Algae and Higher Plants. Meth Enzimol 23: 372-408.

9. Bucke C, Leech RM, Hallaway M, Morton RA (1966) The taxonomic distribution of plastoquinone and tocopherolquinone and their intracellular distribution in leaves of Vicia faba L. Biochim Biophys Acta 112: 19-34.

10. Kruk J (1988) Charge-transfer complexes of plastoquinone and alpha-tocopherol quinone in vitro. Biophys Chem 30: 143-149.

11. Okayama S (1984) Reversed-phase high-performance liquid chromatography of prenylquinones in green leaves using an electrochemical detector. Plant Cell Physiol 25: 1445-1449.

12. Arnon DI (1949) Copper Enzymes in Isolated Chloroplasts. Polyphenoloxidase In Beta Vulgaris. Plant Physiol 24: 1-15.

13. Venturoli G, Fernandez-Velasco JG, Crofts AR, Melandri BA (1986) Demonstration of a collisional interaction of ubiquinol with the ubiquinol-cythocrome c2 oxidoreductase complex in chromatophores from Rhodobacter sphaeroides. Biochim Biophys Acta 851: 340-352.

14. Yoshida K, Shibata M, Terashima I, Noguchi K (2010) Simultaneous determination of in vivo plastoquinone and ubiquinone redox states by HPLC-based analysis. Plant Cell Physiol 51: 836-841.

15. Martinis J, Kessler F, Glauser G (2011) A novel method for prenylquinone profiling in plant tissues by ultra-high pressure liquid chromatography-mass spectrometry. Plant Methods 7: 23.

16. Kruk J, Karpinski S (2006) An HPLC-based method of estimation of the total redox state of plastoquinone in chloroplasts, the size of the photochemically active plastoquinone-pool and its redox state in thylakoids of Arabidopsis. Biochim Biophys Acta 1757: 1669-1675.

17. Kruk J, Strzalka K, Leblanc RM (1992) Monolayer study of plastoquinones,?-tocopherolquinone, their hydroquinone forms and their interaction with monogalactosyldiacylglycerol. Charge-transfer complexes in a mixed monolayer Biochim Biophys Acta 1112: 19-26.

18. Henninger MD, Barr R, Crane FL (1966) Plastoquinone B. Plant Physiol 41: 696-700.

19. Barr R, Henninger MD, Crane FL (1967) Comparative Studies on Plastoquinone II. Analysis for Plastoquinones A, B, C, and D. Plant Physiol 42: 1246-1254.

23

Simultaneous Determination of Six Insecticides in Okra (*Abelmoschus esculentus L. Moench*) by UHPLC-MS/MS

Husham NM Hussan[1], Xingang Liu[2], Fengshou Dong[2], Jun Xu[2] and Yongquan Zheng[2]*

[1] Pesticides Residue Analysis Laboratory, Crop Protection Research Centre, Agricultural Research Cooperation, Wad Medani, Sudan
[2] State Key Laboratory for Biology of Plant Diseases and Insect Pests, Institute of Plant Protection, Chinese Academy of Agricultural Sciences, Beijing 100193, P.R. China

Abstract

The sensitive analytical method using quick, easy, cheap, effective, rugged and safe (QuEChERS) method for simultaneous determination of six insecticides in Okra was developed by UHPLC -MS-MS. The six insecticides were extracted from Okra matrices using acetonitrile and subsequently clean up using only octadecylsilane as sorbent prior to LC-MS-MS analysis. The determination of the target compounds were achieved in less than 2.0 min using an electrospray ionization source in positive mode (ESI+) for imidacloprid, acetamiprid, thiacloprid, thiamethoxam, clothianidin and negative mode (ESI-) for flonicamid. The method showed that the limits quantification (LOQ) ranged from 0.13 to 5.9 µgkg^{-1} in all matrices. The matrix-matched standard gave satisfactory recoveries (72.4%-105.1%) and relative standard deviation (2.2 - 20.0%) values in different matrices at three spiked levels (0.01, 0.1, and 1 mgkg^{-1}) for Okra.

Keywords: Residues analysis; Insecticides; Okra; Crop

Introduction

Pesticides, such as insecticides, herbicides, fungicides and acaricides, have been widely applied during the cultivation and the post-harvest storage of crops. These pesticides were used to prevent the destruction of edible crops by controlling agricultural pests or unwanted plants and thereby increases and improved food production [1].

Okra (*Abelmoschusesculentus L. Moench*) originated in Africa [2], is one of important vegetable crops and place in under Malvaceae family. Now it is growing in tropical and sub-tropical regions in the world. Okra is a nutritional vegetable, which it can promote gastrointestinal peristalsis and prevent constipation efficacy.

Acetamiprid a new generation from neonicotinoid insecticides and highly active to protect the various vegetable crops, by controlling mites and insect pests [3]. Imidacloprid is a systemic chloronicotinyl insecticide that enters the target pest via direct contact or ingestion. It is applied to seeds, soil, crops, and structures for controlling some insects. In addition, it can be used as a topical flea control treatment on domestic pets. Thiacloprid, a neurotoxic insecticide, belongs to family of the neonicotinoids [4]. Thiamethoxam is a second-generation neonicotinoid; it belongs to the thianicotinyl subclass that interferes with the nicotinic acetylcholine receptors in the insect's nervous system, it has systemic and residual activity in several crop plants against broad range pests [5]. Clothianidin, the newest member of the chloronicotinyl insecticide family, it has a high activity against a broad range of insects, including sucking insects, chewing insects, and some lepidopterans [6]. Flonicamid can be against to the aphids and other sucking insects and whiteflies, and other pests [7]. The Chemical structures of imidacloprid, acetamiprid, thiacloprid, thiamethoxam, clothianidin and flonicamid are represented in Figure 1.

The determination of these insecticides in the crops is very important to ensure food safety, and evaluate the risks posed by these insecticides to human, animals and environment. Some methods for determination of imidacloprid, individually in food and environmental matrices have been published using gas chromatography (GC andGC/MS) [8], Liquid chromatography-atmospheric pressure chemical ionization-mass spectrometry (LC–APCI–MS) [9], liquid chromatography–mass spectrometry (LC-MS) [10-13]. Imidacloprid and acetamiprid were analyzed by high performance liquid chromatography(HPLC) [14-16]. Others some reports for determination analysis of acetamiprid, Thiacloprid, Thiamethoxam, flonicamid and clothianidin in different matrices can be found by the HPLC [17] to analysis acetamiprid by liquid and gas chromatography coupled to mass spectrometry for determination for determination acetamiprid, clothianidin, thiacloprid [18]. Some also reported for analysis of thiacloprid, the ion chromatography(IC) [19]. There are some reports have been made for the analysis of thiamethoxam by HPLC with an electrochemical detector and post-column photochemical reactor [20]. Thiamethoxam and Clothianidin by Using GC with-µECD or HPLC-UVD [21,22]. A few reports have made for analysis in food, such as analysis flonicamid in dried hops by liquid chromatography (LC) tandem mass spectrometry (MS/MS) [23], simultaneous determination of flonicamid and its metabolites in vegetables using QuEChERS and reverse-phase liquid chromatography-tandem mass spectrometry [24]. There are simultaneous determinations of five insecticides in bees (acetamiprid, thiacloprid, thiamethoxam and clothianidin) by LC-MS/MS QuEChERS [16-25].

The LC-MS/MS has already been proved powerful and widely used technique. Thanks to progress of the chromatography technique, ultra-high-performance chromatography (UHLPC) was developed by using columns containing particles with a diameter of <2 µm than conventional LC and fluidic systems that operate at higher pressures, resulting in shorter analysis time and increase the peak resolution, capacity, and sensitivity [26].

*Corresponding author: Yongquan Zheng, Institute of Plant Protection, Chinese Academy of Agricultural Sciences, Key Laboratory of Pesticide Chemistry and Application, Ministry of Agriculture; Beijing 100193, P.R China
E-mail: zhengyongquan@ippcaas.cn

This paper described a simultaneous determinations analysis of six insecticides (acetamiprid, imidacloprid, thiacloprid, thiamethoxam, flonicamid and clothianidin) were extracted from Okra matrices using acetonitrile and subsequently clean up using only octadecylsilane (C_{18}) as sorbent prior to LC-MS-MS analysis. To our knowledge, this is the first report to establish an analytical method for determination of six insecticides (acetamiprid, imidacloprid, thiacloprid, thiamethoxam, flonicamid and clothianidin) in Okra using UHPLC-MS/MS. The developed method was also validated by application to the analysis sample.

Materials and Methods

Reagents and materials

Analytical standards of imidacloprid (99.9% purity), clothianidin (99.8%) and thiamethoxam (99% purity) were bought from Agro-Environmental Protection Institute, Ministry of Agriculture (Tianjin, China). Acetamiprid (97.2% purity) and thicacloprid (99% purity) were supplied from Bayer Crop Science, (Frankfurt, Germany), Floncamid

(100% purity) was obtained from Ishihara sangyoKaisha. ISK (Beijing, China). Analytical grade methanol, acetonitrile for pesticides residue analysis were purchased from Beihua fine-Chemical Co. (Beijing-China). Acetonitrile and formic acid (chromatography grade) were obtained from Fisher Scientific (New Jersey, USA). Anhydrous Magnesium Sulfate ($MgSO_4$) and (NaCl) was purchased from Sinopharm Chemical Reagent Co. Ltd (Beijing-China). The octadecylsilane (C18, 40 μm) sorbets purchased from Angela Technologies Int. (Tianjin, China). Purified water was prepared by using Milli-Q water purification system (Millipore Purification Systems).

Preparation of standard solution

The stock standard solution of imidacloprid (100 mg L^{-1}), acetamiprid (100 mg L^{-1}), thiacloprid (100 mg L^{-1}), clothianidin (100 mg L^{-1}), thiamethoxam (100 mg L^{-1}) and flonicamid (100 mg L^{-1}) were prepared in acetonitrile. This solution was diluted to obtain 10.0, 1.0, 0.5, 0.1, 0.05, 0.01 mg L^{-1} in acetonitrile. All solution was stored in a refrigerator in the dark at 4°C and the working standard solutions underwent

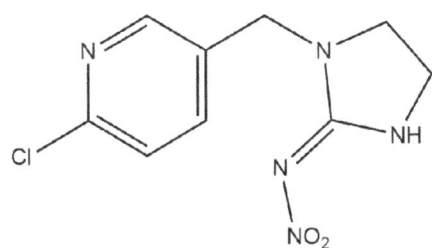

Imidacloprid

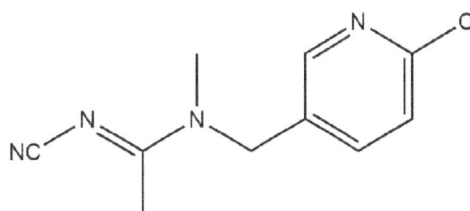

Acetamiprid

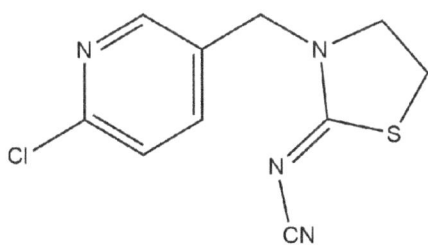

Thiacloprid

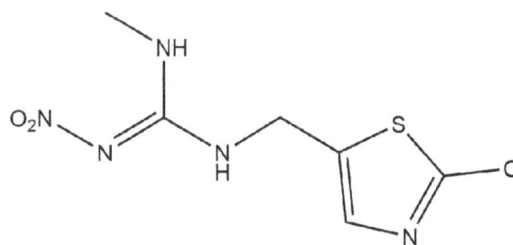

Clothianidin

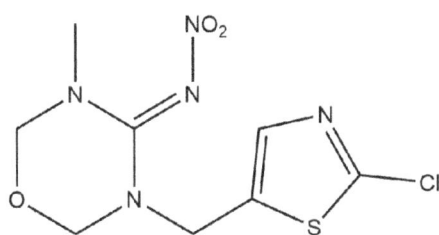

Thiamethoxam

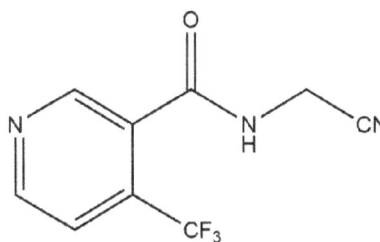

Flonicamid

Figure 1: The Chemical structures of imidacloprid, acetamiprid, thiacloprid, thiamethoxam, clothianidin and flonicamid.

no degradation for 6 months. The Okra (*Abelmoschusesculentus L. Moench*) was obtained from market in Beijing. The matrices were not applied or contaminated by six insecticides, and they were put into polyethylene bags. They were transported to the laboratory and stored in the dark at less than -18°C until analysis.

Instrumentation and chromatography conditions

Extraction equipment: The mobile phase solvent were distilled and passed through a 0.22 μm prose size filter before use. Purified water obtained by purifying demineralized water in Milli-Q Integral 3 water (Millipore, Bedford, MA, USA), a shaker (model HZO-CA, Jintan, China) and centrifuge (model Excelsius II, Fanen, Brazil) were used for sample preparation. An octadecylsilance (C_{18}) column and a 0.22 μm obtained from Agela Technologies (Tianjin, China).

Apparatus and chromatography: Chromatographic separation for six insecticides was performed on a Waters (Milford, MA, USA). ACQUITY UPLC system equipped with a Waters ACQUITY UPLC BEH bridged ethylene hybrid C_{18} column (50 mm × 2.1 mm, 1.7 μm particle size) (Milford, MA, USA). The mobile phase of acetonitrile (solvent A) and 0.1% (v/v) formic acid in water (solvent B), were pumped at a flow rate of 0.3 mL min⁻¹. Simultaneous separations were completed using a gradient profile of 0.0 min/10% A, 1.5 min/70% A, 2.5 min/90% A, 2.6 min/10% A, and 4.5 min/10% A, respectively. The injection volume was 3 μL. The column was kept at 40°C to decrease the viscosity, and the temperature in the sample manager was set at 5°C. All six Insecticides were eluted within 2.0 min.

Analysis of insecticides (Imidacloprid, acetamiprid, thiacloprid, thiamethoxam, clothianidin and flonicamid) compounds was conducted on a triple-quadrupole mass spectrometer (TQD, Waters Crop.) equipped with electrospray ionization source (ESI). The nebulizer gas was 99.95% nitrogen, and the collision was 99.999% argon with a pressure of 2×10^{-3} mbar in the T-wave cell. MS/MS detection was performed in positive and negative ion mode and the monitoring conditions were optimized for target compounds. The conditions were typically as follows: The capillary voltage was set at 3.0 kV, while the source temperature and desolvation temperature were held at 120°C and 350°C, respectively. A 50 Lh⁻¹ cone flow and 500 Lh⁻¹ desolvation gas flow were used respectively. Multi-reaction monitoring (MRM) was

used for the detection of all compounds with a dwell time of 0.05 ms. Infusion experiments of each compound were conducted to optimize the intensity in both positive and negative ionization modes. All other ESI and MS parameters were optimized individually for each target insecticide and were listed in Table 1. The Masslynx NTV.4.1 (Waters, USA) software was used to collect and analyze the data obtained.

Sample preparation: Approximately 500 g of Okra samples were chopped and homogenized by Ultra-Turrax T25 Mixer at 9500 rpm. In total 5 g aliquots homogenized samples were weighed into 50 mL centrifuge tube, 5 mL acetonitrile and 3 mL of water were added on okra. The samples in tubes were shake vigorously for 3 min by using vortex mixer to ensure that the solvent interacted well with the entire samples. The samples tubes were then stored in a refrigerator at -20°C for 30 min, after that 3 g NaCl and 2 g MgSO₄ were added to the sample in the tube and vortexed immediately for 1 min and then the extracts were centrifuged for 5 min at 5000 rpm. A volume of 2 mL prepared aliquot sampled from upper layer into another 5 ml centrifuge tube containing 50 mg only C_{18} use it as sorbent. All the samples vortexed again for 1 min and then centrifuged for 5 min at 5000 rpm. After that 1.5 mL of upper layer was filtered using 0.22 μm nylon syringe filter and transferred into an auto-sampler vial for chromatography injection at LC-MS/MS.

Method validation: Blank samples of Okra were analyzed to verify the absence of interfering species at about the retention time of the analytes. The linearity of the method was studied by analyzing different matrix – matched standard solutions in triplicate at five concentrations ranging from 0.01 to 1 mg L⁻¹. The parameter of linear regression equations, the standard deviation and the correlation coefficient (R^2) were calculated in Table 2. Precision (when repeated independent analysis was performed), accuracy (when recovery assays were performed), sensitivity.

The matrix-dependent limits of quantitation (LOQ) and limit of detection (LOD) were calculated for the analytical methodology using the blank and calibration standards of Okra. The LODs of six insecticides are the concentrations that produce a signal to noise (peak to peak) ratio of 3. The LOQs are defined based on signal-to-noise ratio of 10. The LOQs are estimated from the chromatogram corresponding

!Compound	Molecular formula	MW	t_R (min)	Ion Source	CV (V)	Quantification ion transition	CE 1 (eV)	Diagnostic ion transition	CE 2 (eV)
Thiamthoxam	$C_8H_{10}ClN_5O_3S$	291.7	1.11	ESI+	35	2 92→211	18	292→133	20
Imidacloprid	$C_9H_{10}ClN_5O_2$	255.6	1.21	ESI+	30	2 56→209	18	256→175	20
Clothianidin	$C_6H_8ClN_5O_2S$	249.6	1.19	ESI+	24	2 52→126	18	250→90	30
Thiacloprid	$C_{10}H_9ClN_4S$	252.7	1.36	ESI+	18	2 50→169	18	252→90	30
Acethamprid	$C_{10}H_{11}ClN_4$	222.6	1.27	ESI+	20	223.2→207	20	223.3→126	20
Flonicamid	$C_9H_6F_3N_3O$	229.2	1.14	ESI-	10	228→188	10	228→81	10

MW: Molecular Weight; CV: Cone Voltage; CE: Collision Energy; Ion ratio: Area of qualitative ion/area of quantification ion

Table 1: Experimental parameters and UHPLC-MS/MS conditions of the six compounds in ESI+ and ESI- mode.

Compound	Matrix	Regression equation	R^2	LOQ (μg/kg⁻¹)	LOD (μg/kg⁻¹)
Thiamethoxam	Okra	y=8.8804x+180.91	0.9882	1.9	0.12
Imidacloprid	Okra	y=25855x+5974.7	0.9994	0.2	0.1
Clothianidin	Okra	y=18499x-3.6535	1	1.7	0.5
Acetamiprid	Okra	y=8.6611x-45.193	0.9996	2.2	0.5
Flonicamid	Okra	y=38033x-825.45	0.9984	0.13	0.09
Thiacloprid	Okra	y=4018.3x-5.9251	0.9961	5.9	2.5

Table 2: Comparison of matrix-matched calibration and solvent calibration at 10-1000 μg/L.

to the lowest point used in matrix matched calibration. Recoveries were determined for five replicates of the spikes samples at different three levels of each insecticide for Okra with standard working solutions. The spiked samples were allowed to equilibrate for 1 h prior to extraction.

Results and Discussion

Optimization of chromatography

The optional separation conditions including different mobile phase compositions (ACN/water, Acetonitrile, Water, ACN/1.0% formic acid aqueous solution) were established by injecting 3 μL of six insecticides mixture standard solution. The six insecticides were separated using Acetonitrile-water as shown in Figure 2. There were no interference peaks in region of chromatography and analysis time of the six insecticides was less than 2 min. UHPLC was performed using BEH C_{18} column (50 mm × 2.1 mm, 1.7 μm particle size) in this study allowed a considerable reduction of LC analysis (6.39 to 10.57 min) (Kamel, 2010), and (10 to 15 min) for acetamiprid, imidacloprid [15].

Optimization of MS/MS

Full-scan and MS/MS mass spectra were obtained from the infusion of 5 mg L^{-1} standard solution of these compounds in acetonitrile-water at a flow rate of 10 mL min^{-1}. The analysis of six insecticides was performed in MRM mode, to optimize MS/MS conditions of mass spectrometry and the target compounds presented and achieving a compromise between both positive and negative modes, for acetamiprid, Imidacloprid, Thiacloprid, Thiamethoxam, and clothianidin, in this study, ESI in positive mode was selected for subsequent experiments for the results demonstrated higher responses in positive mode than in negative mode. The chemical formulas, molecular weights, cone voltages, precursor ions, and collision voltages were also listed in Table 1.

As shown in Figure 2, six insecticides could be detected at the spiked level. When a non-spiked sample was also subjected to the entire procedure, no interfering peaks were observed in any of the samples. As shown in Table 2, the LODs for the six pesticides were estimated to be 0.09-2.5 μg kg^{-1}, and the LOQs for six insecticides were 0.13-5.9 μg kg^{-1}.

Precision and accuracy

Recovery studies were performed to validate the UPLC-MS/MS method by spiking the blank samples at three different concentration levels (0.01, 0.1, and 1 mg kg^{-1}) in Okra, and then analyzed in quintuplicate. The recoveries were calculated using the three-point matrix matched calibration curves presented in Table 3. Only one type of sorbent C_{18} is used in this work to investigate the influences on recovery rate in two matrices. The C_{18} is suitable to extract non-polar and medium-polar compounds from the polar samples, which are mainly used for antiphase extraction. The precision of the method was determined by the repeatability and reproducibility and expressed by the relative standard deviations (RSDs). The intra-day precision was measured by comparing the standard deviation of the recovery percentages of spiked samples run during the same day. The inter-day precision was determined by analyzing spiked samples for three distinct days.

In general, the mean recoveries ranged between 72.4% and 105.1% for the spiked levels (Table 3). And the RSD[b] (inter-day precision) for the proposed method were ranged from 9.1%-19.6%. These results showed that the developed residue method was satisfied for all six target compounds in Okra matrices.

Method application

In order to demonstrate the applicability of this method for

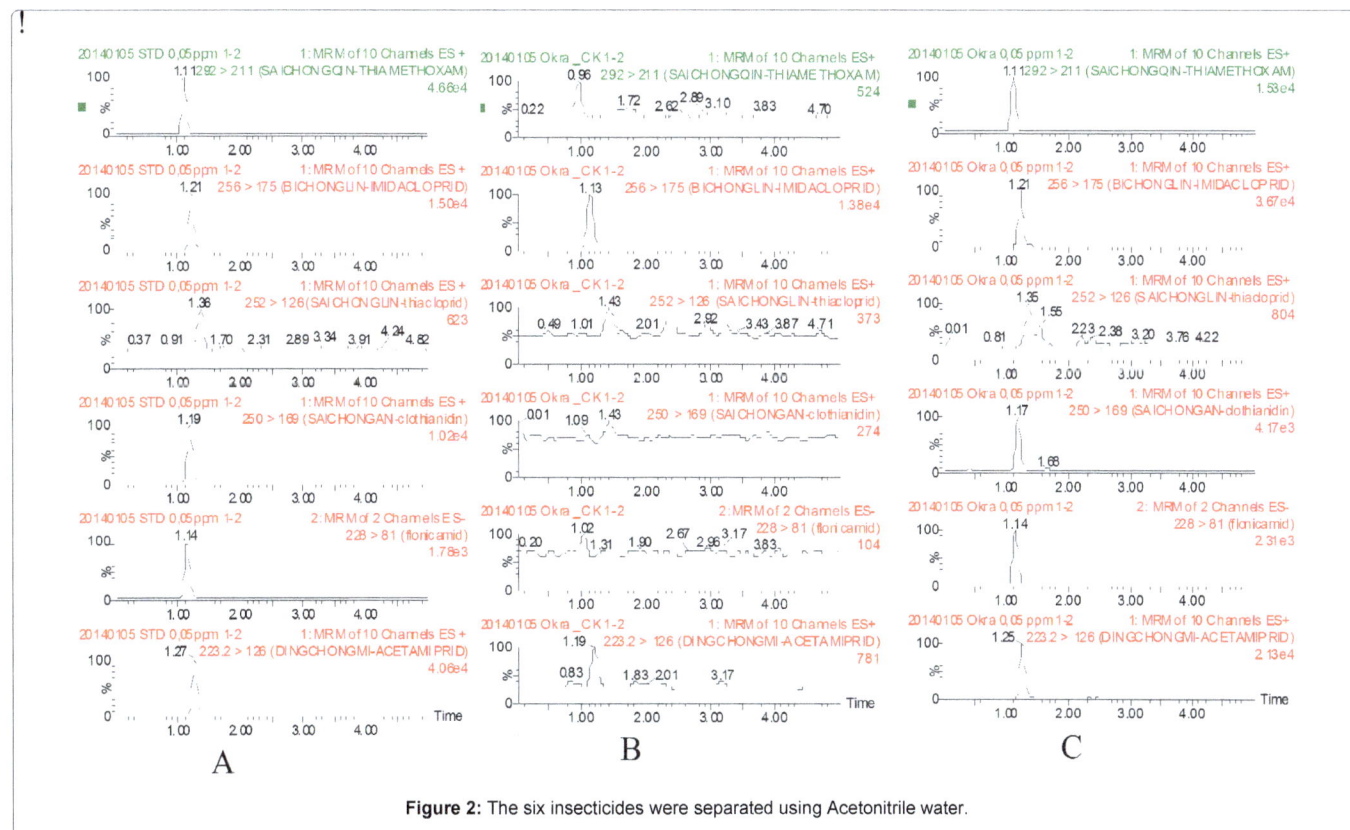

Figure 2: The six insecticides were separated using Acetonitrile water.

Matrix	Spiked level (mkg⁻¹)	Thiamethoxam			Imidacloprid			Thiacloprid		
		Recovery	RSDᵃ	RSDᵇ	Recovery	RSDᵃ	RSDᵇ	Recovery	RSDᵃ	RSDᵇ
Okra	0.01	79.6	4.7	10.0	75.7	12.8	13.0	79.5	11.5	10.8
	0.1	79.8	9.2	11.2	105.1	8.1	14.1	100.2	3.8	18.0
	1	76.1	2.4	10.7	81.0	3.2	9.1	78.3	2.2	9.4

Matrix	Spiked level (mkg⁻¹)	Clothianidin			Acetamiprid			Flonicamid		
		Recovery	RSDᵃ	RSDᵇ	Recovery	RSDᵃ	RSDᵇ	Recovery	RSDᵃ	RSDᵇ
Okra	0.01	78.1	20.0	19.6	104	14.8	18.4	88.8	11.5	10.2
	0.1	91.8	8.7	12.8	95.5	15.8	10.8	81.4	9.4	13.1
	1	73.7	4.8	12.2	91.7	7.2	17.1	72.4	2.4	15.5

ᵃIntra-day (n=5); ᵇInter-day (n=15)

Table 3: Recoveries (n=15, %) and RSD (%) for target compounds from different matrices in three spiked levels.

monitoring of this pesticides residue in Okra. Five okra samples were obtained from market in Beijing and randomly was analyzed. The samples were treated with the sample preparation method described in section 4.3.3. All samples and did not represent a threat for the consumer since they were not detected any residue and below the MRLs established by America (0.4-1 mg kg⁻¹) and Japan (0.7-5 mg kg⁻¹).

Conclusions

This work described the development and validation a simple LC-MS/MS method for the simultaneous determination of residues of six insecticides (Imidacloprid, acetamiprid, thiacloprid, thiamethoxam, clothianidin and flonicamid) in Okra. Extracts containing the target compounds were analyzed and validated by UHPLC–MS/MS. This method allowed separation of the six target pesticides in less than 2 min with good specificity. The recovery percentages were ranged from 72.4% to 105.1% with RSD in the range of 2.2% to 20.0% for all analytes in okra matrices. In addition, the specificity, calibration curves, precision, and reproducibility were tested successfully. As a conclusion, the proposed method is easy and useful in analysis these six insecticides. This method also can be recommended for monitoring studies in the okra to ensure food safety.

Acknowledgements

National Natural Science Foundation of China (31371970) supported this work.

References

1. Cserháti T, Forgács E, Deyl Z, Miksik I, Eckhardt A (2004) Chromatographic determination of herbicide residues in various matrices. Biomedical Chromatography 18: 350-359.

2. Ali IM, Khan AM, Abdul Rashid A, Ul-haq ME, Javed TM, et al. (2012) Epidemiology of Okra Yellow Vein Mosaic Virus (OYVMV) and Its Management through Tracer, Mycotal and Imidacloprid. American Journal of Plant Sciences 3: 1741-1745.

3. Zhang JJ, Wang Y, Xiang H, Xue M, Li WH, et al. (2010) Oxidative stress: Role in acetamiprid induced impairment of the male mice reproductive system. Agricultural Sciences in China 10: 786 -796.

4. Jeschke P, Moriya K, Lantzsch R, Seifert H, Lindner W, et al. (2001) Thiacloprid (Bay YRC 2894) - a new member of the chloronicotinyl insecticide (CNI) family. Pflanzenschutz- Nachrichten Bayer 54: 147-160.

5. Maiensfisch P, Huerlimann H, Rindlisbacher A, Gsell L, Dettwiler H, et al. (2001) The discovery of thiamethoxam: a second-generation neonicotinoid. Pest Management Science 57: 165-176.

6. Bailey JC, Scott-Dupree CD, Tolman JH, Harris CR, Harris BJ (2005) Alternative Agents for Control of European Corn Borer and Corn Flea Beetle on Sweet Corn. Journal of vegetable science 11: 27-46.

7. Morita M, Ueda T, Yoneda Y, Koyanagi T, Haga T (2007) Flonicamid, a novel insecticide with a rapid inhibitory effect on aphid feeding. Pesticide Science 63: 969-973.

8. Alberto N, Antonio GC, El-Khattabi R, Jose Luis V, Amadeo R, et al. (1997) Determination of Imidacloprid in Vegetable Samples by Gas Chromatography–Mass Spectrometry. Analyst 122: 579-581.

9. Pous X, Ruíz JR, Picó Y, Font G (2001) Determination of imidacloprid, metalaxyl, myclobutanil, propham, and thiabendazole in fruits and vegetables by liquid chromatography–atmospheric pressure chemical ionization–mass spectrometry. Fresenius Journal of Analytical Chemistry 371: 182-189.

10. Blasco C, Fernández M, Picó Y, Font G, Mañes J (2002) Simultaneous determination of imidacloprid, carbendazim, methiocarb and hexythiazox in peaches and nectarines by liquid chromatography–mass spectrometry. Analytica Chimica Acta 461: 109-116.

11. Fernández A, Amadeo R, Tejedor A, Agüera A, Contreras M, et al. (2000) Determination of Imidacloprid and Benzimidazole Residues in Fruits and Vegetables by Liquid Chromatography–Mass Spectrometry after Ethyl Acetate Multiresidue Extraction. Journal of AOAC International 83: 748-755.

12. Schöning R, Schmuck R (2003) Analytical determination of imidacloprid and relevant metabolite residues by LC MS/MS Bulletin of Insectology 56: 41-50.

13. Marin A, Vidal MJL, Egea GFJ, Frenich AG, Glass CR, et al. (2004) Assessment of potential (inhalation and dermal) and actual exposure to acetamiprid by greenhouse applicators using liquid chromatography–tandem mass spectrometry. Journal of Chromatography B 804: 269-275.

14. Obana H, Okihashi M, Akutsu K, Kitagawa Y, Hori S (2002) Determination of Acetamiprid, Imidacloprid, and Nitenpyram Residues in Vegetables and Fruits by High Performance Liquid Chromatography with Diode-Array Detection. Journal of Agricultural and Food Chemistry 50: 4464-4467.

15. Baigl AS, Akhter AN, Ashfaq M, Asi RM, Ashfaq U (2012) Imidacloprid residues in vegetables, soil and water in the southern Punjab, Pakistan. Journal of Agricultural Technology 8: 903-916.

16. Lehotay SJ (2006) Quick, Easy, Cheap, Effective, Rugged, and Safe Approach for Determining Pesticide Residues. Pesticide Protocols pp: 239-261.

17. Shams El Din AM, Azab MM, Abd El-Zaher TR, Zidan ZHA, Morsy RA (2012) Persistence of Acetamiprid and Dinotefuran in Cucumber and Tomato Fruits. American-Eurasian Journal of Toxicological Sciences 4: 103-107.

18. Lazić S, Šunjka D, Grahovac N, Guzsvány V, Bagi F, et al. (2012) Application of Liquid Chromatography with Diode-Array Detector for Determination of Acetamiprid and 6-chloronicotinic Acid Residues in Sweet Cherry Samples. Pesticides and Phytomedicine 27: 321-329.

19. Subhani Q, Huang PZ, Zhu YZ, Liu YL, Zhu Y (2014) Analysis of insecticide thiacloprid by ion chromatography combined with online photochemical derivatization and fluorescence detection in water samples. Chinese Chemical Letters 25: 415-418.

20. Rancan M, Rossi S, Sabatini GA (2006) Determination of Thiamethoxam residues in honeybees by high performance liquid chromatography with an electrochemical detector and post-column photochemical reactor. Journal of Chromatography A 1123: 60-65.

21. Hem L, Park HH, Shim HJ (2010) Residual Analysis of Insecticides (Lambda-cyhalothrin, Lufenuron, Thiamethoxam and Clothianidin) in Pomegranate

Using GC-µECD or HPLC-UVD. Korean Journal of Environmental Agriculture 29: 257-265.

22. Abd El-Zaher T, Nasr NI, Mahmoud AH (2011) Behavior of Some Pesticide residues in and on Tomato and Kidney Beans Fruits Grown in Open Field. American-Eurasian Journal of Toxicological Sciences 3: 213-218.

23. Hengel MJ, Miller M (2007) Analysis of flonicamid and its metabolites in dried hops by liquid chromatography-tandem mass spectrometry. Journal of agricultural and food chemistry 55: 8033-8039.

24. Xu Y, Shou LF, Wu YL (2011) Simultaneous determination of flonicamid and its metabolites in vegetables using QuEChERS and reverse-phase liquid chromatography-tandem mass spectrometry. Journal of Chromatography A 1218: 6663-6666.

25. Kamel (2010) A Refined Methodology for the Determination of Neonicotinoid Pesticides and Their Metabolites in Honey Bees and Bee Products by Liquid Chromatography–Tandem Mass Spectrometry (LC-MS/MS). Journal of Agriculture and Food Chemistry 58: 5926-5931.

26. Mellors SJ, Jorgenson WJ (2004) The Use of 1.5 micron Porous Ethyl-Bridged Hybrid Particles as a Stationary Phase Support for Reversed-Phase Ultra-High Pressure Liquid Chromatography, Analytical Chemistry 76: 5441-5450.

24

Quali-Quantitative Analysis by LC/DAD and GPC of the Polyphenols of "Uva Di Troia Canosina" Grape Seeds for the Development of an Industrial Nutraceutical Product

Martina Bava[1], Sebastiano Arnoldi[1], Lucia Dell'Acqua[1], Sergio Fontana[2], Flavia La Forgia[2], Giuseppe Mustich[1], Gabriella Roda[1]*, Chiara Rusconi[1], Giovanni Sorrenti[3], Giacomo Luca Visconti[1] and Veniero Gambaro[1]

[1]Dipartimento di Scienze Farmaceutiche, Università degli Studi di Milano, Via Mangiagalli 25, 20133 Milano, Italy
[2]Farmalabor Srl, Via Pozzillo ZI, 76012 Canosa di Puglia, Italy
[3]SIAN ASL BT, piazza Umberto I n°9, 76012 Canosa di Puglia (BT), Italy

Abstract

The quali-quantitative determination of the principal components of "uva di Troia canosina" seed extracts by LC/DAD analysis and the optimization of the extraction and purification processes for the development of an industrial nutraceutical product, are described. Two different fractions of seeds collected at different stages of fermentation were compared: "Tesi" 2 when there is a spontaneous stratification of the seeds at the bottom of the recipient and "Tesi" 4 at the end of fermentation. Percolation was applied and compared to maceration and the purification step carefully evaluated to obtain extracts free of contaminant species endowed with polyphenolic content comparable to commercial preparations such as Leucoselect® (Indena, SpA, Italy), Vitis Vinifera extract 95% (seeds), Vitis Vinifera dry extract 95%, Biovin grape seed and vinasse extract. (Farmalabor, Italy). In particular, "Tesi" 2 extract obtained by percolation and purified with a LLE extraction with ethyl acetate showed a polyphenolic content similar to Leucoselect®.

From the quantitative analyses it was evident, as expected, that "Tesi"2 has a higher polyphenolic content compared to "Tesi" 4, because during vinification the must extracts polyphenols from the seeds. On the other hand, "Tesi" 4 is particularly convenient since it is easily obtained and very economical, being a waste product.

The residual content of organic solvents (ethanol and ethyl acetate) and water was assessed in the grape extracts according to ICH rules by means respectively of HS/GC and Karl Fisher titration in order to meet the requirements for commercialization. Furthermore, the high molecular weight polyphenolic fraction of our extracts was investigated through gel permeation chromatography (GPC) and compared to that of Leucoselect®.

Keywords: Uva di Troia canosina; Polyphenols; Extraction; Purification; Quantitative analysis; LC/DAD

Introduction

In a previous work we carried out the analysis of the polyphenolic content of the seed extracts coming from four different fractions of "uva di Troia canosina" grape seeds sampled at different stages of the fermentation process [1]. "Uva di Troia canosina" grape is a *Vitis Vinifera* variety characterized by a small berry and cultivated around the city of Canosa in Apulia, a region of southern Italy. This kind of grape, although considered unproductive from the oenological point of view, shows high polyphenolic content and a great wine ageing potential [2].

Polyphenols, especially flavonols, flavan-3-ols (monomeric catechins, proanthocyanidins) and anthocyanidins, are responsible for many organoleptic characteristics of wine and grapes; [3] their concentration and composition are influenced by viticultural and environmental factors, such as climate conditions, maturity stage and production area [4-7]. Phenolic compounds, due to their strong free radical scavenging and high antioxidant properties, display many pharmacological benefits such as cardio protective, vasodilatatory, anticarcinogenic, anti-inflammatory, anti-allergic, antibacterial, immune-stimulating, anti-viral and estrogenic activities [8-12].

The determination of polyphenolic composition is very important to characterize different grape varieties, moreover, analytical methods [13] have been improved throughout the years [14] and to this end different extraction procedures [15,16] involving a variety of solvents [17] have been widely studied.

From the data obtained it was evident that the polyphenolic content is lower in the seeds belonging to Tesi 4 than in the Tesi 2 seeds because the polyphenols are extracted from the seeds to the must during fermentation. Moreover, extraction conditions and purification were studied in order to obtain suitable extracts for the preparation of a nutraceutical product based on the antioxidant activity of polyphenols [1].

In this paper we describe the quantitative determination of the principal components of "uva di Troia canosina" seed extracts respect to catechin and the optimization of the extraction and purification processes for the development of an industrial nutraceutical product. The residual content of organic solvents (ethanol and ethyl acetate) and water was measured the grape extracts according to ICH rules by means respectively of HS/GC and Karl Fisher titration in order to meet the

*Corresponding author: Gabriella Roda, Dipartimento di Scienze Farmaceutiche, Università degli Studi di Milano, Via Mangiagalli 25, 20133 Milano, Italy, E-mail: gabriella.roda@unimi.it

requirements for the commercialization of medicines and excipients. Moreover, the analysis of residual water and organic solvents was also useful for the determination of the title of the extracts in catechin + epicatechin, a parameter which allows the characterization and the comparison of products obtained with different methodologies, thus helping the optimization of extraction and purification procedures.

In particular, percolation was applied and compared to maceration and the purification step carefully evaluated to obtain extracts free of contaminant species such as high molecular weights polymers, fats, pectins and phlobaphenes. Seeds were not grounded, but extracted intact because the polyphenols are contained in the cuticle of the seed. Moreover intact seeds at the end of the extraction protocol can be used to obtain grape seed oil, thus exploiting in a more efficient way the waste products. Furthermore, the analytical method was optimized and validated and our extracts were compared to highly standardized grape seeds extracts such as Leucoselect (Indena, SpA, Italy), Vitis Vinifera extract 95% (seeds), Vitis Vinifera dry extract 95% (procyanidins), Biovin grape seed and vinasse extract. (Farmalabor, Italy).

Materials and Methods

Fruit sampling

"Uva di Troia canosina" grape was cultivated in Canosa (Apulia, South Italy) in 2010. The crop was immediately divided into 4 fractions, called "Tesi"1-4, which were subjected to 4 different stages of the fermentation processes [1].

In this work "Tesi" 2 and Tesi "4", from the grape harvest in 2010, were compared with "Tesi" 2 deriving from the experimental cultivation "Dr. Sergio Fontana" by Farmalabor, harvested in 2012. These fractions were chosen since from our previous work we noticed that "Tesi" 2 is the richest in polyphenols which allows the concomitant production of wine. On the other hand, "Tesi" 4 is less abundant in polyphenolic content but it is a waste product of wine industry and for this reason it is particularly convenient and easy to obtain.

Chemical and reagents [1]

Acetonitrile and Orthophosphoric acid 85% of HPLC grade quality were purchased from Sigma-Aldrich (St. Louis, USA) and Merck (Whitehouse station, USA), respectively. Absolute ethanol (Carlo Erba, Milano, Italy) and Acetone (VWR, Pennsylvania, USA) were of reagent grade. The resin SEPABEADS SP-207 was obtained by Resindion (Milan, Italy). Standards of Gallic acid, (+)-Catechin hydrate, (−)-Epicatechin were supplied by Sigma (Milano, Italy); Procyanidin B1 and Procyanidin B2 were obtained from Fluka (Milano, Italy). All standards were of purity >90%. Leucoselect was supplied by Indena SpA (Settala, Italy). Vitis vinifera 95% extract was obtained from Farmalabor Srl (Canosa, Italy). Milli-Q quality water was obtained with a Milli-Q (H2O) system by Millipore (Bedford, MA, USA).

Seed extract preparation

Loss on drying: Before starting the extraction procedure, the humidity percentage of each "Tesi" was evaluated, by drying about 10 g of seeds at 60°C for 5 days. The calculated humidity is reported in Table 1.

Maceration: Ethanol or acetone was added to intact frozen seeds (50 g) to cover them and to reach a final concentration of 70%(v/v), taking into account the amount of water contained in the vegetable material, in glass beakers protected from lights and air by aluminium foils; the maceration was carried out under magnetic stirring and the

first 3 h extraction started when the seeds reached room temperature; seeds were filtered under vacuum and were subjected to the second 3h extraction adding fresh solvent (50 mL); at the end of the second 3 h extraction, seeds were filtered and extracted again with fresh solvent (50 mL) during the whole night; the steps of extractions were repeated until the sixth and last extraction, that is, the 2nd overnight extraction; then, the solvent of the combined extracts was completely removed at 40°C and the extracts were dried in oven for at least 12 h at 60°C.

The extraction procedure was evaluated by comparison of the recovery percentages (%REC) of each extract, calculated as W_{fin} / W_{init} x 100, where W_{fin} is the final weight and the initial weight W_{init} considers the intrinsic water content of seeds. % recoveries are reported in Table 2. 100.0 mg of each dry extract were then dissolved in 10 mL of a 1:1 0.3% H_3PO_4/Acetonitrile mixture, thus obtaining a 10 mg/mL solution. These solutions were filtered on 0.45 μm nylon filters before HPLC analysis.

Percolation: A comparison between "Tesi" 2 and "Tesi" 4, from the grape harvest of 2010, was carried out. Percolation was performed for each "Tesi" on 400 g of frozen seeds with a 70/30 ethanol/water (v/v) mixture, taking into account the humidity percentage of each "Tesi". A multi-step extraction was carried out, consisting in two extractions per day with a contact time of 3 hours plus 2 overnight extractions, with a total number of 6 extractions in 48 hours. At the end of the protocol the solvent was evaporated and a gummy dark violet extract was obtained as in the case of maceration. Percentage recoveries were calculated as described in the previous paragraph and the values obtained are reported in Table 3.

Purification with ethyl acetate: 500 mg of each ethanol extract were dissolved in 2.5 mL of distilled water and extracted five times with 2.5 mL of ethyl acetate previously saturated with water. The organic solutions were combined and evaporated to dryness. Each residue (20.0 mg) was then dissolved with a 1:1 mixture of 0.3% H_3PO_4/Acetonitrile, obtaining a 4 mg/mL solution. These solutions were filtered on 0.45 μm nylon filters before HPLC analysis.

Purification with ethyl acetate after salting: Extracts (500 mg) were dissolved in 2.5 mL of distilled water and 420 mg of NaCl was added; the mixture was extracted five times with 2.5 mL of ethyl acetate previously saturated with water. The organic solutions were combined and evaporated to dryness. Each residue (20.0 mg) was then dissolved with a 1:1 mixture of 0.3% H_3PO_4/Acetonitrile, obtaining a 4 mg/mL solution. These solutions were filtered on 0.45 μm nylon filters before HPLC analysis.

Purification with adsorbent resins: The resin (SEPABEADS SP-207, Resindion, Milan, Italy) was suspended in absolute ethanol overnight. The extracts (10 g) were dissolved in 75 mL of water and filtered on cotton before loading into the column containing 100 mL of the activated adsorbent resin. The resin was abundantly washed with water in order to eliminate interfering substances. Potential loss of phenolic compounds was monitored by UV/VIS spectrophotometry between 200 and 700 nm and TLC; the washing water collected in four different fractions was analysed and no absorption was detected in the selected wavelengths. Finally, the analytes of interest, i.e. anthocyanins, were desorbed from the resin by dripping into the column about 500 mL of ethanol 95% with 0.01% of citric acid. The elution solvent was fractionated into two parts and analysed by spectrophotometry and TLC to control the complete elution of the analytes from the column.

Determination of water content by Karl Fisher titration

Analyses were performed by a volumetric titrator Karl-Fisher

Analisi	Gallic acid t_R (min)	Area (mAU)	(+)-Catechin t_R (min)	Area (mAU)	(-)-Epicatechin t_R (min)	Area (mAU)	Procyanidin B1 t_R (min)	Area (mAU)	Procyanidin B2 t_R (min)	Area (mAU)	(-)-Epicatechin gallate t_R (min)	Area (mAU)
1	4.27	6654.7	13.17	11364.9	19-25	11728.6	9.23	2939.3	16.32	4678.9	38.21	15276.9
2	4.24	5215.8	13.25	10810.2	19.31	11891.4	9.07	2520.3	16.53	4530.2	38.34	15273.7
3	4.24	5004.6	13.36	11192.5	19.41	11818.9	9.41	2315.7	16.51	4149.9	38.40	15061.1
4	4.24	4945.6	13.55	10796.4	19.65	11821.6	9.36	2934.3	16.24	4914.6	37.28	13915.2
5	4.24	4972.6	13.36	10755.8	19.81	11927.7	9.04	2615.9	16.72	3743	38.66	14200.7
Mean	4.246	5358.66	13.338	10983.96	19.486	11837.64	9.222	2665.1	16.46	4403.3	19.48	14745.5
St.dev	0.01	732.32	0.14	276.60	0.24	76.65	0.17	270.70	0.19	461.98	0.53	641.71
%CV	0.32	13.67	1.07	2.52	1.22	0.65	1.80	10.16	1.15	10.49	2.71	4.35

Table 1: Reproducibility of gallic acid, (+)-catechin, (-)-epicatechin, procyanidin B1, B2 and (-)-epicatechin gallate

Sample	% Water content	Maceration Total %Recovery Simple	LLE	Percolation Total %Recovery Simple	LLE	LLE salting	Resin
Tesi 2 (2010)	38.68	24.48	8.08	16.74	5.05	4.02	16.88
Tesi 4 (2010)	44.30	13.64	9.72	10.51	4.76	3.36	17.47
Tesi 2 (2012)	60.31	26.91	10.68				

Table 2: % Recovery of the extracts calculated on the dry weight before and after purification; LLE = liquid-liquid extraction with ethyl acetate.

Tesi	Purification	%H$_2$O	Ethanol Ppm Residual solvent	% Residual solvent		Ethyl acetate ppm Residual solvent	% Residual solvent
2	resin	13.1	12558,66	1.26			
	LLE	11.9				33238.43	3.32
	LE salting	14.9				39754.27	3.98
4	resin	15.1	308.97	0.03			
	LLE	13.5				4772.91	0.48
	LLE salting	12.8				677.63	0.07

Table 3: Amount of water and residual solvents in the purified grape extracts.

V20 (Mettler Toledo AG – Analytical; Switzerland) as described in the chapter 2.5.12 of the European Pharmacopoeia (method A). The extracts and Leucoselect® (5 mg) were dissolved in 1.5 mL of methanol and titrated with Hydranal˙ – Composite 5, methanol-free titrating agent (Fluka, Italy).

Determination of residual solvents by HS/GC

Apparatus and conditions: Analyses were carried out on an Ultra Thermo Electron Trace GC (Thermo Fisher, Waltham, MA, United States) with a split-splitless injection system and a HS 2000 Thermo Electron auto-sampler, coupled with a FID detector. The system was managed by a Thermo Electron Chrom Card 2.3 software. The GC was equipped with a "VF-624 ms" capillary column (30 m x 0,32 mm I.D., thickness 1.8 μm). The GC-FID system was operated under the following conditions: 50°C (10 min) – 250°C, 10°C/min; final isotherm 3 min. Temperatures: inlet 180°C; detector 250°C; split flow 30 mL/min; split ratio 15; carrier constant flow 1.3 mL/min (helium); detector gas flow: hydrogen 35 mL/min, air: 350 mL/min; make-up: 30 mL/min; signal range: 1. HS auto-sampler: syringe temperature: 100°C; incubation temperature: 100°C with alternate stirring; incubation time: 60 min; injection volume: 0.5 mL.

Preparation of standard solutions and samples: Standard solutions of propanol (IS), ethanol and ethyl acetate (200 μg/mL; 200 ppm) were prepared in DMSO and working standards samples were obtained mixing 1 mL of the standard solution of ethanol and 1 mL of the standard solution of propanol (IS). Blank samples were prepared mixing 1 mL DMSO with 1 mL IS standard solution. Samples were obtained dissolving 10.00 mg of purified extracts in 1 mL of DMSO, adding 1 mL of IS standard solution.

Calculation of the concentration of residual solvents: The calculation of the concentration of residual solvents was carried out determining the Response Ratio RR as follows.

Three standards solutions were analysed to determine RA$_{std}$ defined as:

$$RA_{std} = A_{std}/A_{IS}$$

Where A$_{std}$ is the area of peak of the standard solution; A$_{IS}$ is the area of the peak of the internal standard. Then for each solution the response ratio RR was calculated as:

$$RR = RA_{std} / C_{std}$$

Where C$_{std}$ is the concentration of the standard solution.

The RR$_{mean}$ was then determined as the mean of the three RR obtained for the single standard solutions.

The concentration of residual solvents was expressed in ppm. The calculation was performed applying the following formula:

$$ppm= RA_{sample}/RR_{mean}$$

where RA$_{sample}$ = ratio between the area of the peak corresponding to the solvent in the sample and the area of the peak corresponding to the IS; RR$_{mean}$ = mean response ratio calculated from the analysis of three working standards.

LC analyses

LC analyses were performed on a Varian™ Pro Star equipped with an auto-sampler mod. 410, two pumps mod. 210 and detector DAD mod. 335. The instrument was controlled by Software Galaxie.

Analyses of the grape extracts were carried out under conditions similar to those employed by Gabetta et al. [18] optimized for improving reproducibility.

Chromatographic column: Zorbax SB C18 250 x 4.6 mm i.d. particle size 5 μm (Agilent Technologies™); pre-column: Security Guard Cartridges C18 4 x 2.0 mm (Phenomenex™); column temperature: R.T.; detection wavelength: 278 nm; flow rate: 1.0 mL/min; Injection volume: 10 μL; solvent A: 0.3% H_3PO_4 in water; Solvent B: acetonitrile; mobile phase: solvents were filtered under vacuum on 0.45 μm membrane filters and degassed by immersion in ultrasonic bath for 15 minutes before column conditioning; linear gradients: 0-45 min, 10-20% B; 45-65 min, 20-60% B; 65-68 min 60% B; 68-69 min 60-10%B; 69-85 min, 10 % B.

All analyses were carried out in triplicate. The phenolic compounds in the samples were identified according to their elution order, comparing their retention times and spectroscopic spectra with those of the pure commercial standards and by means of sum tests.

Reproducibility of the LC analytical method: Due to the complexity of the vegetable matrix (Figure 1) reproducibility was assessed evaluating the areas and retention times (t_R) obtained analysing (+)-catechin, (-)-epicatechin, gallic acid, procyanidin B1, B2 and (-)-epicatechin gallate standard solutions five consecutive times in the same day (Figure 2). Reproducibility was expressed as %CV. The %CV calculated on the areas of the standards of the active principles and on the retention times show that the method is endowed with an adequate reproducibility (Table 1).

Linearity of the response of catechin: To evaluate the linearity of the analytical method (+)-catechin was taking into account in a range of concentrations from 0.1 mg/mL and 0.03 mg/mL. In this range six solutions of non-sequential concentrations (0.1, 0.07, 0.06, 0.05, 0.04, 0.03 mg/mL) were analysed (n=5). The linearity equation was y = 43258x + 214.99 with a good correlation coefficient (R^2 = 0.9962).

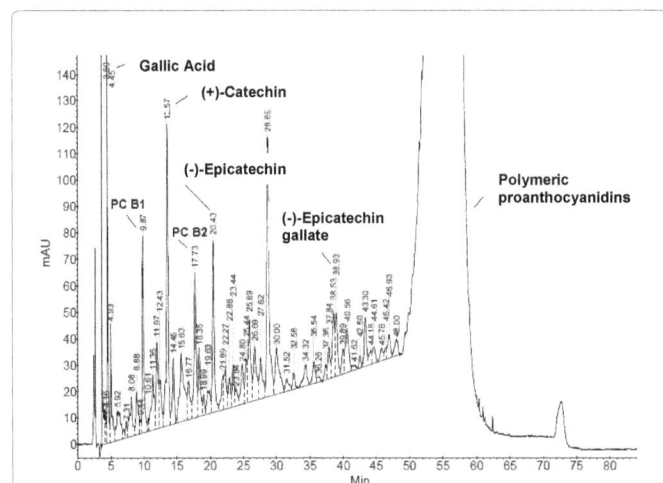

Figure 1: HPLC-DAD chromatogram of the crude extract obtained from Tesi 2 (2010) by maceration.

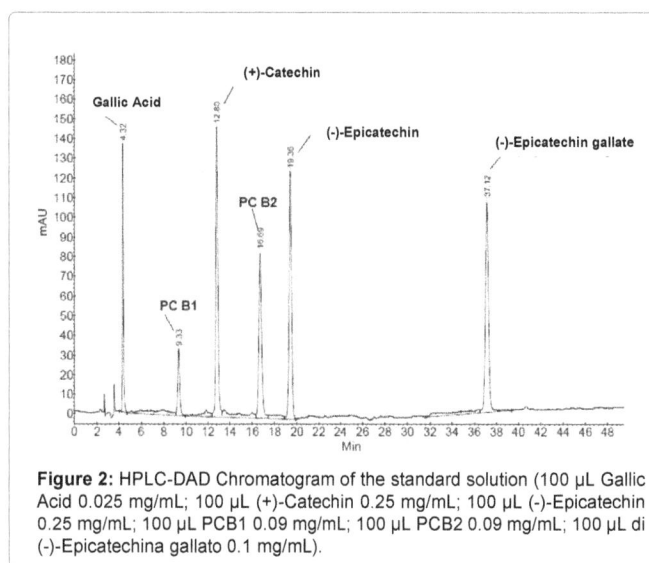

Figure 2: HPLC-DAD Chromatogram of the standard solution (100 μL Gallic Acid 0.025 mg/mL; 100 μL (+)-Catechin 0.25 mg/mL; 100 μL (-)-Epicatechin 0.25 mg/mL; 100 μL PCB1 0.09 mg/mL; 100 μL PCB2 0.09 mg/mL; 100 μL di (-)-Epicatechina gallato 0.1 mg/mL).

Quantification of the analytes

RR_{mean} (see 2.5.3) for catechin was calculated as the mean of three determinations of RR obtained analyzing three standard solutions of catechin (0.25 mg/mL) taking into account the purity (98%) of catechin.

The % title of the analytes was calculated respect to catechin according to the following formula:

$$\% \, T_{analyte} = [(A_{analyte} \times V_{sample}) / (RR_{mean} \times W_{sample})] \times 100$$

Where $A_{analyte}$ is the area of the analyte peak, V_{sample} is the volume of the sample solution (mL) and W_{sample} is the weight of the sample (mg). The title of the purified extracts was expressed as the sum of the title in catechin and epicatechin referred to the dry and solvent free preparation as follows:

$$\% \, T_{tot \, dry} = \% \, (\%T_{cat} + \%T_{epi}) \times [100 / (100 - \%H_2O - \%SR)]$$

Where % SR is the total amount of residual solvents determined by HS/GC and $\%H_2O$ is water content determined by Karl Fisher titration.

Gel Permeation Chromatography (GPC)

GPC analyses were carried out as described by Gabetta et al. [18] at room temperature on a PL Gel 5, particle size 5 μm, pore type 500 Å column (Agilent PL1110-6525) linked to a pre-column 0,5 μm (Supelco). The separation was performed with an isocratic elution mode using THF and an aqueous solution of LiBr 12 x 10^{-3} M (95:5 V/V) at a constant flow of 1,0 mL/min. The signal was detected at 280 nm and the injection volume was 10 μL. The samples were prepared dissolving 10.00 mg of extract in 10 mL of mobile phase (1 mg/mL).

Results and Discussion

The evaluation of the mass recovery and the analysis of the extracts were carried out. "Uva di Troia canosina" grape seeds underwent two different types of extraction procedures: maceration and percolation. The latter gave larger quantities of extracts in a shorter period of time and for this reason it can be applicable industrially, as it is cheaper and less time consuming compared to maceration. The vegetable material was maintained intact thus avoiding the extraction of oils, fats, pectins and mucilage contained inside the seeds. Furthermore, at the end of the extraction it is possible to recover the exhausted drug to obtain grape seed oil exploiting this waste material of the wine industry in the best possible way. The extraction protocol was optimized for maceration in our previous work [1], the same multi step procedure was applied to percolation, thus obtaining six cycles of extraction in two days. The vegetable material was not completely exhausted, but at the end of the protocol the amount of the active principles extracted is so low that it is not convenient to go on. Either maceration or percolation yielded semi-solid dark violet preparations.

The percentage recovery on the dry weight of the seeds is reported in Table 2.

The extraction yield decreases from "Tesi" 2 to "Tesi" 4 because the polyphenols are extracted by the must during the fermentation process. Comparing "Tesi" 2 obtained in 2010 with the harvest in 2012 we can see that the recovery is similar, although it is slightly higher in 2012. As far as percolation is concerned, the extraction yield results lower than maceration and this is due to the fact that at the end of the process a huge amount of solvent remains in the seeds and in the tubes of the system linked to the peristaltic pump. Moreover, as the extract is semi-solid, it is difficult to eliminate the solvent completely during evaporation. The sixth extraction has a low yield compared to the previous five, consequently we decided to stop the process at the fifth extraction.

In order to enrich and standardize the polyphenol composition of the "uva di Troia canosina " grape seed extracts and to reduce the high molecular weight polymers, different purification methodologies were applied. In this way gums, fats, oils and pectins which give to the extract a soft consistency, were eliminated, yielding a dry dark orange powder which was easier to handle and standardize. The crude extracts obtained by maceration were purified by means of a liquid-liquid extraction (LLE) with ethyl acetate (Figure 3), while in the case of percolation three different purification protocols were applied to the crude extracts: LLE with ethyl acetate (Figure4), LLE with ethyl acetate after salting (Figure 5) and purification with an adsorbent resin (Figure 6). In Table 2 the percentage recovery for the different procedures are reported.

It is evident that the yield after treatment with ethyl acetate, with or without salting, is considerably lower than the recovery obtained operating with SEPABEADS SP-207. The purification with ethyl acetate gave a higher yield in the case of maceration probably because with percolation the component of the extract soluble in water is eliminated. This fraction could be responsible for the higher weight of the extract obtained with maceration.

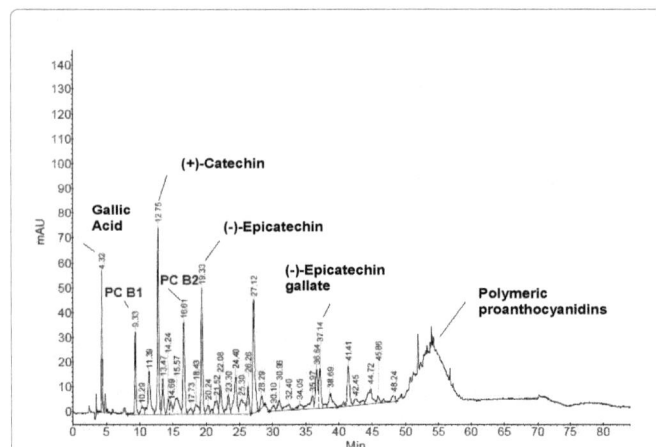

Figure 4: HPLC-DAD Chromatogram of the Tesi 2 (2010) extract purified with ethyl acetate obtained by percolation (0.5 mg/mL).

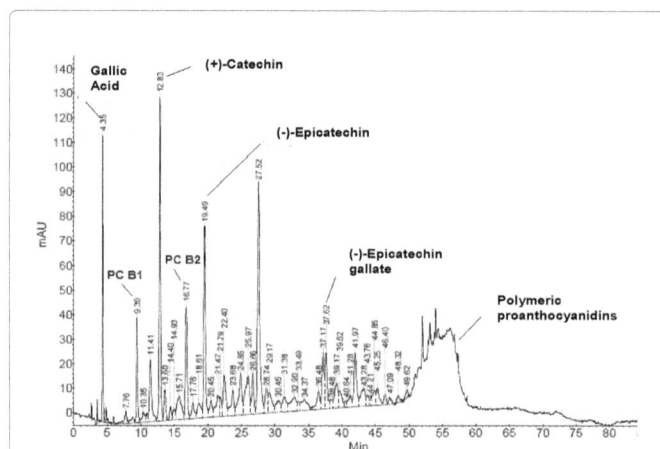

Figure 5: HPLC-DAD Chromatogram of the Tesi 2 (2010) extract purified with ethyl acetate and salting obtained by percolation (0.5 mg/mL).

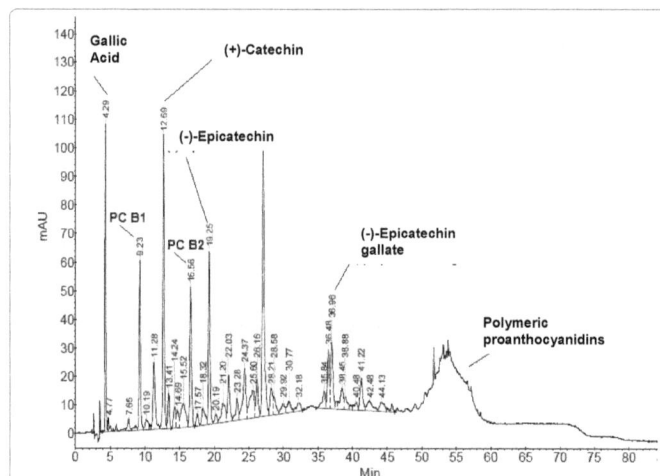

Figure 6: HPLC-DAD Chromatogram of the Tesi 2 (2010) extract purified with Sepabeads® SP-207 obtained by percolation (0.5 mg/mL).

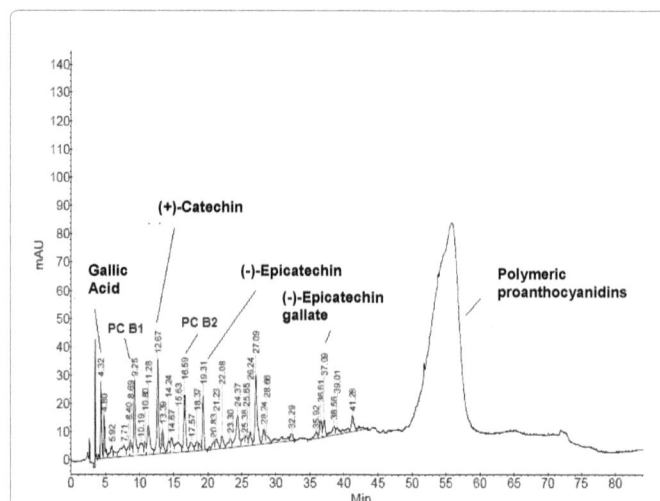

Figure 3: HPLC-DAD Chromatogram of the Tesi 2 (2010) extract purified with ethyl acetate obtained by maceration (0.5 mg/mL).

The search for residual solvents in the purified extracts was carried out for two reasons: first of all we wanted to verify if the residual content of solvents was below the limits imposed by ICH rules, thus allowing us to calculate the percentage title of the extracts in terms of catechin and epicatechin referred to the dry preparation. The determination was performed on the extracts obtained by percolation and purified with different methodologies in order to compare these different protocols. The solvents taken into consideration were ethanol and ethyl acetate whose permitted day exposure (PDE) is 5000 ppm (0.5%). From Table 3 it is possible to notice that "Tesi" 2 extracts have a residual content of organic solvents which overcomes ICH limits, while in the case of "Tesi" 4 this does not happen. Ethanol is not present in the extracts purified by means of a LLE with ethyl acetate, even though it is used for percolation.

The total amount of water was calculated as the mean of three determinations, taking into account the water contained in the methanol used for the preparation of the extracts, which is different from that used by the instrument. The results are reported in Table 3. It is evident that there are not significant differences among the various preparations.

"Uva di Troia canosina" grape seeds extracts were compared with commercial extracts such as Leucoselect® (Indena SpA), Vitis Vinifera Extract 95% (seeds), Vitis Vinifera dry extract 95% (procyanidins) and Biovin seeds and vinasse dry extracts (Farmalabor). The results are reported in Table 4. In particular, Leucoselect® was used as reference for the purified extracts because it is a patented highly standardized preparation and as control of the efficiency of the chromatographic separation. In fact, it contains about 15% of catechin and epicatechin, 80% of epicatechin gallate and its dimers, trimers and tetramers and 5% of pentamers, hexamers and heptamers.

The percentage title (%T) of the analytes was calculated in comparison to catechin as described in paragraph 2.6.2; three different solutions of the analytes were analysed and the results were expressed as the mean of these determinations. The chromatographic method resulted adequate, as the %T calculated for Leucoselect (15.14%) matched the result obtained by Gabetta et al. [18] (15-16%). This preparation has a significantly higher polyphenolic content than other commercial extracts.

From the data obtained in the analysis of the crude extracts, it is evident that the %T is higher in case of "Tesi" 2 respect to "Tesi" 4 in accordance with the fact that during fermentation polyphenols are extracted from the seeds to the must. "Tesi" 2 coming from the harvest in 2012 has higher polyphenolic content than the same "Tesi" of 2010. Moreover, the LC analysis confirmed the trends observed for the percentage recovery, indicating that this parameter is representative for polyphenolic content. From the comparison between the two extraction processes it is possible to notice that percolation gives better results, although comparable to those of maceration. The polyphenolic content of these crude extracts is dramatically lower compared with Leucoselect.

The purification step produces an increase and an enrichment of the polyphenolic content, in fact %T of the purified extracts is significantly higher than the corresponding crude extracts.

In Table 5 the total content of catechin and epicatechin for the commercial, crude and purified extracts is reported.

The three purification protocols led to different results. The purification with the adsorbent resin gave the best results in terms of recovery but the extracts had a lower quality which was evident in the chromatographic profile in which the broad peak, due to the polymeric substances, was higher than the chromatograms obtained with the other purification protocols (Figures 3-6) [1].

Purification with ethyl acetate, with or without salting, allowed us to obtain extracts with high polyphenolic content. In particular, "Tesi" 2 extract purified with ethyl acetate had a %T comparable to that of Leucoselect.

The purified extracts obtained by maceration gave results in terms of %T lower than those obtained by percolation, probably because in the latter process the component of the insoluble extract in water, which interferes with purification, is eliminated.

The salting of the extracts resulted the more problematic protocol from a practical point of view and it gave results comparable to the simple LLE extraction with ethyl acetate. For this reason salting was excluded. The purification with ethyl acetate led to the best quality, while the purification with the resin gave the highest recovery, thus these two purification methodologies yielded two different kind of products in terms of costs and commercial characteristics.

The profile of high molecular weight compounds in the purified extracts obtained by percolation, was carried out by means of gel permeation chromatography (GPC). The GPC analysis of Leucoselect (black trace) [18] compared with "Tesi" 2 and 4 extract obtained by percolation and purified with the resin (Bordeaux trace) and "Tesi" 2 and 4 extract obtained by percolation and purified with ethyl acetate (blue trace) gave the profiles reported in Figures 7 and 8. The region labelled A refers to epicatechin; region B corresponds to epicatechin gallate, dimers and their gallates, while region C is representative of trimers, tetramers, pentamers, hexamers, heptamers and their gallates.

"Tesi" 2 extract purified with ethyl acetate has a molecular weight distribution comparable to that of Leucoselect®, even if it has a reduced polymeric portion. The fine structure is therefore different, because different kinds of seeds were used (Figure 9). "Tesi" 4 extract purified with ethyl acetate shows a profile which is superimposable with "Tesi" 2 extract purified in the same way, but the signal intensity is lower, confirming the decrease in polyphenolic content passing from "Tesi" 2 to "Tesi" 4, observed in the LC analysis and in the percentage recovery. The profiles of the extracts purified with the resin are shifted towards the higher molecular weight region: the monomer content is lower as shown by the LC analysis, confirming the different features which the two purification methods give to grape extracts.

Conclusion

The aim of this work was the valorization of waste products of wine industries, in particular of the kind of grape called "Uva di Troia canosina" with a small berry, that is a variety considered not so convenient from an oenological point of view, because it has a higher skin/pulp ratio than other varieties. Previous studies [1] demonstrated that this grape biotype is particularly rich in polyphenols, compounds endowed with several beneficial properties. For this reason, we carried out a complete quali-quantitative characterization of the polyphenols contained in the seeds of this kind of grape, optimizing the extraction and purification protocol of the extracts. Two different fractions of seeds sampled at different stages of the fermentation were compared: "Tesi" 2 when there is a spontaneous stratification of the seeds at the bottom of the recipient and "Tesi" 4 at the end of fermentation. From the quantitative analyses it was evident that "Tesi"2 has a higher polyphenolic content compared to "Tesi" 4, as expected since during vinification the must extracts polyphenols from the seeds. On the

			%T					
			(1)	(2)	(3)	(4)	(5)	(6)
Commercial extracts		Leucoselect	8.91	6.23	3.39	4.50	4.57	3.86
		Vitis Vinifera seeds	2.96	4.40	0.37	0.36	1.03	0.43
		Vitis Vinifera procyanidins	2.18	2.98	0.71	0.31	0.67	0.70
		Biovin	2.37	2.54	0.42	2.46	3.06	0.47
Tesi 2 2010	Maceration	crude	0.44	0.30	0.27	0.26	0.29	0.13
		LLE	4.55	3.13	2.12	1.65	2.38	1.85
		enrichment	10.3	10.4	7.8	6.3	8.2	14.2
	Percolation	crude	0.49	0.28	0.23	0.33	0.34	0.16
		LLE	8.62	5.70	3.37	2.50	3.78	2.49
		enrichment	17.6	20.4	14.7	7.6	11.1	15.6
		LLE salting	6.49	4.31	3.04	3.53	3.78	2.72
		enrichment	13.2	15.4	13.2	10.7	11.1	17.0
		resin	2.47	1.45	1.14	1.65	1.80	0.76
		enrichment	5.0	5.2	5.0	5.0	5.3	4.8
Tesi 4 2010	Maceration	crude	0.41	0.24	0.28	0.24	0.32	0.09
		LLE	3.35	1.99	2.74	1.36	1.91	2.01
		enrichment	8.02	8.3	9.8	5.7	6.0	22.3
	Percolation	crude	0.44	0.25	0.29	0.27	0.29	0.12
		LLE	5.56	3.37	4.13	1.63	2.50	1.38
		enrichment	12.6	13.5	14.2	6.0	8.6	11.5
		LLE salting	4.94	2.96	4.38	2.26	2.96	1.19
		enrichment	11.2	11.8	15.1	8.4	10.2	9.9
		resin	1.97	1.10	1.40	1.33	1.32	0.53
		enrichment	4.5	4.4	4.8	4.9	4.6	4.4
Tesi 2 2012	Maceration	crude	0.69	0.35	0.27	0.41	0.44	0.21
		LLE	6.32	3.57	2.11	1.68	2.65	2.12
		enrichment	9.20	10.2	7.8	4.1	6.0	10.1

Table 4: Content (% Title) of the active principles contained in the commercial, crude and purified grape extracts. catechin=(1); epicatechin=(2); gallic acid=(3); Procyanidin B1=(4); Procyanidin B2=(5); epicatechin gallate=(6).

			Extract	Catechin + epicatechin (%T)
Commercial extracts			Leucoselect	15,14
			Vitis Vinifera seeds	7,36
			Vitis Vinifera procyanidins	5,16
			Biovin	4,91
Crude extracts	maceration		"Tesi" 2 2010	0.74
			"Tesi" 4 2010	0.65
			"Tesi" 2 2012	1.04
	percolation		"Tesi" 2 2010	0.77
			"Tesi" 4 2010	0.69
Purified extracts	maceration	LLE AcOEt	"Tesi" 2 2010	7.68
			"Tesi" 4 2010	5.34
			"Tesi" 2 2012	9.89
	percolation	Resin	"Tesi" 2 2010	3.92
			"Tesi" 4 2010	3.07
		LLE AcOEt	"Tesi" 2 2010	14.32
			"Tesi" 4 2010	8.92
		LLE	"Tesi" 2 2010	10.80
		AcOEt salting	"Tesi" 4 2010	7.91

Table 5: Content (% Title) catechin + epicatechin; of the commercial, crude and purified extracts

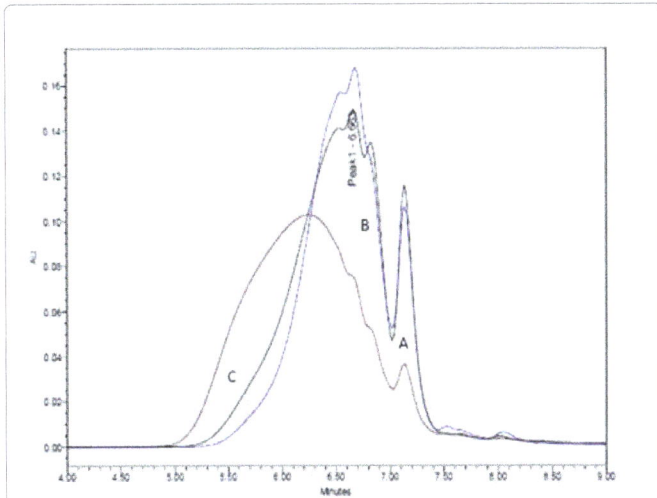

Figure 7: GPC profiles of Leucoselect® (black trace); "Tesi" 2 extract obtained by percolation and purified with the resin (Bordeaux trace) and "Tesi" 2 extract obtained by percolation and purified with ethyl acetate (blue trace).

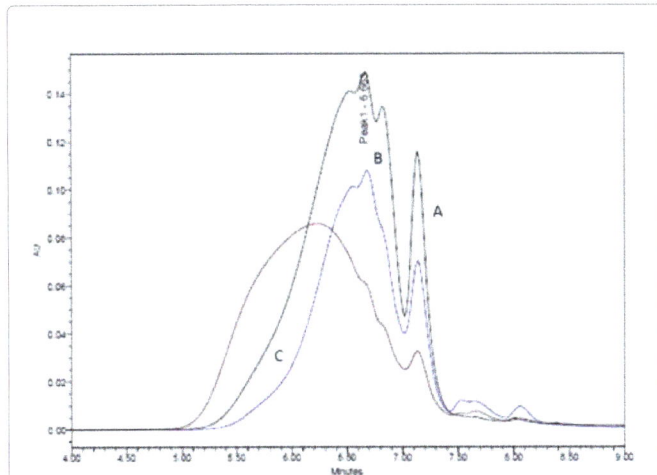

Figure 8: GPC profiles of Leucoselect® (black trace); "Tesi" 4 extract obtained by percolation and purified with the resin (Bordeaux trace) and "Tesi" 4 extract obtained by percolation and purified with ethyl acetate (blue trace).

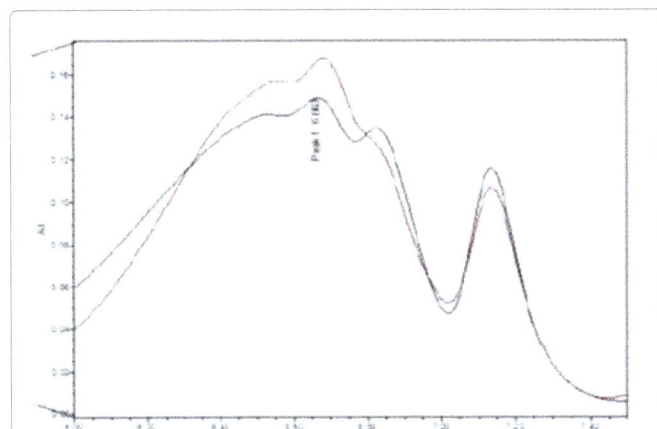

Figure 9: Comparison between Leucoselect® (Bordeaux trace) and "Tesi" 2 obtained by percolation and purified with ethyl acetate (black trace).

other hand, "Tesi" 4 is particularly convenient because it is very easy to obtain and very cheap, being a waste product. From an industrial point of view it could be more convenient to extract "Tesi" 4 seeds rather than "Tesi" 2 seeds. The purification process was carefully investigated in order to obtain the highest enrichment and a polyphenolic content comparable with commercial preparations sold on the market. In particular, "Tesi" 2 extract obtained by percolation and purified with a LLE extraction with ethyl acetate showed a polyphenolic content similar to Leucoselect*.

The residual content of organic solvents (ethanol and ethyl acetate) and water was determined on the grape extracts according to ICH rules by means respectively of HS/GC and Karl Fisher titration in order to meet the requirements for commercialization. Moreover, the high molecular weight polyphenolic fraction of our extracts was investigated by means of GPC and compared to Leucoselect*.

Acknowledgements

The authors would like to thank Dr. Giuseppe Ramaschi (Indena SpA) for performing GPC analyses.

References

1. Catalano D, Fontana S, Roda G, Dell'Acqua L, La Forgia F, et al. (2013) Methods for the Evaluation of Polyphenolic Content in "Uva Di Troia Canosina" Grape and Seeds at the Different Maceration Stages. ISRN Anal Chem: 1-9.

2. Suriano S, Tarricone L, Savino M, Rossi MR (2005) Caratterizzazione phenolic uve di di di Troia Aglianico grape coltivate e nel nord Barese. L'enologo 41: 71-80

3. Nicoletti I, Bello C, De Rossi A, Corradini D (2008) Identification and quantification of phenolic compounds in grapes by HPLC-PDA-ESI-MS on a semimicro separation scale. J Agric Food Chem 56: 8801-8808.

4. Muñoz S, Mestres M, Busto O, Guasch J (2008) Determination of some flavan-3-ols and anthocyanins in red grape seed and skin extracts by HPLC-DAD: Validation study and response comparison of different standards. Anal Chim Acta 628:104–110.

5. Cavaliere C, Foglia P, Gubbiotti R, Sacchetti P, Samperi R, et al. (2008) Rapid□ resolution liquid chromatography/mass spectrometry for determination and quantitation of polyphenols in grape berries. Rapid Commun Mass Spectrom 22: 3089-3099.

6. Mané C, Souquet JM, Olle D, Verries C, Veran F, et al. (2007) Optimization of simultaneous flavanol, phenolic acid, and anthocyanin extraction from grapes using an experimental design: application to the characterization of champagne grape varieties. J Agric Food Chem 55: 7224-7233.

7. Gómez-Alonso S, García-Romero E, Hermosín-Gutiérrez I (2007) HPLC analysis of diverse grape and wine phenolics using direct injection and multidetection by DAD and fluorescence. J Food Compos Anal 20: 618-626.

8. Maffei FR, Carini M, Aldini G, Bombardelli E, Morazzoni P, et al. (1994) Free radicals scavenging action and anti-enzyme activities of procyanidines from Vitis vinifera. A mechanism for their capillary protective action. Arzneim-Forsch/ Drug Res 44: 592-601.

9. Carini M, Stefani R, Aldini G, Ozioli M, Facino RM (2001) Procyanidins from Vitis vinifera seeds inhibit the respiratory burst of activated human neutrophils and lysosomal enzyme release. Planta med 67: 714-717.

10. Aldini G, Carini M, Piccoli A, Rossoni G, Facino RM (2003) Procyanidins from grape seeds protect endothelial cells from peroxynitrite damage and enhance endothelium-dependent relaxation in human artery: new evidences for cardio-protection. Life sciences 73: 2883-2898.

11. Waterhouse AL (2002) Wine Phenolics. Ann NY Acad Sci 957: 21-36.

12. Karvela E, Makrisb DP, Kalogeropoulosa N, Karathanosa VT (2009) The Effect of Ph on the Efficiency of Vinification By-Product Extracts to Inhibit Lipid Peroxidation in A Lecithin Liposome Model Matrix Talanta 79: 1311-1321.

13. Yang Y, Chien M (2000) Characterization of grape procyanidins using high-performance liquid chromatography/mass spectrometry and matrix-assisted laser desorption/ionization time-of-flight mass spectrometry. J Agric Food Chem 48: 3990-3996.

14. Lorrain B, Ky I, Pechamat L, Teissedre PL (2013) Evolution of analysis of polyphenols from grapes, wines, and extracts. Molecules 18: 1076-1100.

15. Ju ZY, Howard LR (2003) Effects of solvent and temperature on pressurized liquid extraction of anthocyanins and total phenolics from dried red grape skin. J Agric Food Chem 51: 5207-5213.

16. Flamini R (2003) Mass spectrometry in grape and wine chemistry. Part I: Polyphenols. Mass Spectrom Rev 22: 218-250.

17. Liang Z, Sang M, Fan P, Wu B, Wang L, et al. (2011) CIELAB coordinates in response to berry skin anthocyanins and their composition in Vitis. J of food science 76: C490-C497.

18. Gabetta B, Fuzzati N, Griffini A, Lolla E, Pace R (2000) Characterization of proanthocyanidins from grape seeds. Fitoterapia 71: 162-175.

25

Simultaneous Determination of Pharmaceutical and Pesticides Compounds by Reversed Phase High Pressure Liquid Chromatography

Gineys M, Kirner T, Cohaut N, Béguin F and Delpeux-Ouldriane S*

Centre National de la Recherche Scientifique, ICMN, Orleans cedex 2, France

Abstract

A reversed phase HPLC method has been developed for the simultaneous separation and quantification of a mixture, constituted by a total of thirteen selected pollutants: pharmaceutical products and others compounds including one hormone, a pesticide, a natural marker and solvents. For each pollutant, the calibration curve shows a good linearity ($r^2 \geq 0.998$) over the concentration range of $1.0 \times 10^{-5} - 5 \times 10^{-8}$ mol/L. The limit of detection at 230 nm is ranging from 5.0×10^{-6} to 4.0×10^{-5} mg/mL, whereas the limit of quantification is in the range $1.8 \times 10^{-5} - 1.2 \times 10^{-4}$ mg/mL depending on the selected pollutant. The separation of the complex mixture was performed on a Hypersil Gold column 100 mm × 2.1 mm (dp=3 µm) using a multi-step linear gradient at 210, 230 and 280 nm. This method is well adapted for the detection of pharmaceutical compounds or micropollutants at low concentrations as a routine analysis but also as an alternative method to sophisticated ones like HPLC-MS or HPLC-fluorescence.

Keywords: Pharmaceutical compounds; Hormones; Pesticides; HPLC-UV (DAD) method validation

Abbreviations: ACB: Clofibric acid; ASA: Acetylsalicylic acid; BPA: Bisphenol A; CAF: Caffeine; CBZ: Carbamazepine; DAD: Diode Array Detector; DFN: Diclofenac sodium salt; HPLC: High Pressure Liquid Chromatography; IBP: Ibuprofen; LOD: Limit of detection; LOQ: Limit of quantification; MCP: Mecoprop; 4NP: 4-Nonylphenol; OES: β-Estradiol; OFX: Ofloxacin; PCP: Pentachlorophenol; POL: Acetaminophen; SA: Salicylic acid

Introduction

Pharmaceuticals are widely used for health applications and ingested by humans and animals at high doses levels. Additionally, antibiotics are overused in breeding to minimize diseases distribution and epidemic threats. However, pharmaceuticals are not easily metabolized by the living organisms and therefore found in stools or urines. Others pollutions sources like industrial rejects, resulting from matter loss during fabrication processes or agricultural practices, are responsible for the dissemination of numerous compounds in the environment, like dyes, pesticides, solvents, chemical additives or organic pharmaceutical precursors and residues. Finally, the wastewater treatments technologies actually available are not able to remove efficiently all of these contaminants that are therefore detected at very low concentrations, in the range of several µg/L to several ng/L in water after treatment steps. These noxious compounds are found in many environmental compartments, as such as surface water, soil, sediment, ground water. Unlike hormones and pesticides, whose noxious effects on human health are no more to be proved, pharmaceutical products are actually still not considered by the REACH implementation. However, due to their regular occurrence in the environment, and potential risks on human health (antiobioresistance, carcinogenic properties) and ecosystems (fish feminization), these compounds are major targeted pollutants to be controlled and eliminated in a near future [1-3].

The recent developments in analytical technology allow setting up new normalized methods for the detection and the quantification of pharmaceuticals products, at very low concentrations and even trace levels in many water samples. For example, recently (February 2013) a normalized method XP T 90-223 for the assay of pharmaceutical products and their metabolites in water - dissolved fraction by liquid chromatography with tandem mass spectrometry coupled with a solid phase extraction, was developed. This method is efficient at a very low concentration level with a quantification limit ranging from 1 to 25 ng/L and shows a good relevancy on different water samples as ground water, surface water or water for human consumption [4]. Several studies have reported the separation and the quantification of a mixture of pharmaceuticals products (Carbamazepine, Diclofenac), hormones and pesticides with liquid chromatography coupled with fluorescence [5] or mass spectrometry for the detection phase [6]. Excellent results were obtained with quantification limits in the range of ng/L (10 to 1000 ng/L). This method is indeed rather sophisticated and technically heavy and therefore presents a high analysis cost. HPLC analytical methods using UV equipped with diode array detector (DAD) is well developed because of its easier accessibility, handling and lower cost as compared to mass or fluorescence detectors. DAD detection is less sensible but indeed permits to reach intermediate sensitivity levels, and to bring quick and cheap analyses. Many studies on pharmaceutical compounds (Diclofenac, Ofloxacin, Aspirine) using HPLC-UV method were achieved on different matrices (water, urine, plasma sample, tablet or drugs) with detection and quantification limits ranging from µg/L to mg/L [7-10]. If DAD detection is well adapted for known mixtures of water contaminants at intermediate concentration (to µg/L), for real matrices representative of treatment plants successive extraction, clean up and preconcentration steps were required. Recently, Zhou developed a hollow-fiber-supported ionic liquid microextraction method coupled with HPLC-UV in order to detect and quantify four endocrine disrupting compounds (bisphenol A, 17-β-estradiol, estrone and diethylstilbestrol) present in water surface samples. After extraction optimization, the proposed method allowed to reach a good

*****Corresponding author:** Sandrine Delpeux-Oudlriane, Centre National de la Recherche Scientifique, ICMN, Orleans cedex 2, 45071-40059, France E-mail: delpeux@cnrs-orleans.fr

linearity range (0.15-100 μg/L) and reproducibility but also to achieve very low detection limits: 0.03, 0.05, 0.10, 0.05 μg/L for bisphenol A, 17-β-estradiol, estrone and diethylstilbestrol respectively [11].

The aim of this study was therefore to develop an easy method able to perform the separation and the quantification of a wide diversity of targeted water pollutants as such as pharmaceuticals, pesticides and hormone molecules, in the range of μg/L using a reversed phase chromatography technique equipped with UV detection.

Materials and Methods

HPLC instrumentation

The HPLC system consists of a Dionex Ultimate 3000 equipped with a PDA detector (Accela 80 Hz, cell of 5 cm), coupled to a computer with Chromeleon 6.8 software package.

Reagents and chemicals

HPLC grade acetonitrile and orthophosphoric acid 85% were purchased from VWR international. Acetaminophen (POL), Caffeine (CAF), Ofloxacin (OFX), Acetylsalicylic acid (ASA), Carbamazepine (CBZ), Bisphenol A (BPA), β-Estradiol (OES), Clofibric acid 97% (ACB), Diclofenac sodium salt (DFN), Ibuprofen (IBP), Pentachlorophenol (PCP) were purchased from Sigma-Aldrich. Mecoprop (MCP) and 4-Nonylphenol (4NP) were purchased from Fluka.

Preparation of the mobile phase

100 μL of orthophosphoric acid 85% was mixed with 1000 mL of ultrapure water (σ_{water}=0,055 μS.cm^{-1}), stirred and was finally filtered using 0.2 μm disk (hydrophilic membrane). The pH of the solution is controlled and fixed to 2.9.

Preparation of standard solutions

The mixture was composed of different pharmaceuticals products as an analgesic (Acetaminophen), non sterodial anti-inflammatory (Diclofenac, Ibuprofen, and Acetylsalicylic acid), antibiotic (Ofloxacin), neuroleptic (Carbamazepine) and an anti-cholesterol (Clofibric acid). Additionally, the mixture contains other compounds as endocrine disruptor (Bisphenol A, 4-Nonylphénol), hormone (β-Estradiol), pesticide (Mecoprop), natural marker (Caffeine) and a solvent (Pentachlorophenol). The chemical formulas of these pollutants were drawn in Figure 1. All of these pollutants have been selected because of their nature and their regular occurrence in water treatment plants at increasing concentration [12,13]. A standard solution was directly prepared without intermediate steps by dissolving the appropriate mass of each pollutant in ultrapure water (σ_{water}=0.055 μS.cm^{-1}) to get a concentration of 1.0×10^{-5} mol/L per pollutant; the total pollutant concentration being equal to 1.3×10^{-4} mol/L. The weighted mass corresponding to the standard concentration (1.0×10^{-5} mol/L) was lower than the compounds water solubility whatever the molecules (Table 1). Appropriate dilutions of the 1.0×10^{-5} mol/L initial solution were realized to obtain eight calibration points in the concentration range of 1.0×10^{-5}-5.0×10^{-8} mol/L, i.e., 5.0×10^{-8}, 2.5×10^{-7}, 5.0×10^{-7}, 1.0×10^{-6}, 2.5×10^{-6}, 5.0×10^{-6}, 7.5×10^{-6} and 1.0×10^{-5} mol/L. A linear regression analysis was carried out at the concerned wavelength.

HPLC method

The chromatographic separation was performed in the reversed phase mode using a Hypersil Gold C$_{18}$ column at 25°C (100 × 2.1 mm with a particle size of 3 μm). This column permitted to obtain during the separation thin and symmetric peaks and consequently an optimized

resolution and sensibility. Furthermore, it can be used in a wide pH range, particularly at very low pH. The low activity of the silanol groups indeed reduced the peak trailing of the molecules, particularly in the case of basic compounds. The eluents were water (A) at pH 2.9 through acidification by orthophosphoric acid 0.01% (v/v) and acetonitrile (B). The following multi-step linear gradient was applied: from 10% B to 80% B in 25.45 min (slope of 2.75 mL/min), followed by a plateau for 2 min then a decrease from 80% B to 10% B in one minute and a final plateau of 3 min at the initial conditions. The flow rate was set to 0.25 ml/min and the volume of injection to 50 μL. The Accela PDA detector allowing to follow three simultaneous wavelengths, a compromise was established to obtain the highest sensitivity for a maximum of pollutant and so optimized answers. The overlap of the pollutant UV spectra, extracted from PDA data, allowed selecting three wavelengths: 210, 230 and 280 nm (Figure 2). In the mixture, peaks identification was accomplished comparing the retention time and UV spectra of each peak and the ones obtained for each selected pollutant injected separately with the same analytical method.

Results and Discussion

Analysis of pharmaceutical, hormone and pesticide products

The typical chromatogram of the mixture obtained with the reversed phase method described here, is illustrated in Figure 3. The thirteen compounds constituting the mixture were separated in thirty minutes with a good resolution and return to baseline for each peak. All of the pollutants were fully recovered except Acetylsalicylic acid (ASA), almost completely hydrolyzed in acidic medium in Salicylic acid (SA) and Acetic acid. System suitability tests were performed and chromatographic parameters such as retention time, resolution (R$_s$), selectivity (α), capacity factor (k') and asymmetry for each pollutant are reported in Table 2. The pollutants elution depends on the molecule affinity for the mobile and the stationary phase. Within the few first minutes of elution, the more hydrophilic compounds like Acetaminophen, Caffeine or Ofloxacine (Log D=1.1, -0.8 and -2.0 respectively) which present a better affinity for the polar mobile phase (water - acetonitrile, 80/20), are separated and eluted more quickly than the more hydrophobic compounds like Pentachlorophenol and 4-Nonylphenol (Log D=4.4 and 5.4 respectively). The increase of the organic solvent ratio along the analysis permits to improve the affinity of the hydrophobic compounds for the less polar mobile phase and therefore to promote their elution. Presenting a better affinity for the C$_{18}$ stationary phase, the hydrophobic compounds are more strongly retained and as a consequence eluted more slowly from the column, after 23.0 and 28.7 min respectively. The resolution R$_s$, represents the column ability to separate two components. For a resolution lower than 1.5, the components are considered as not totally separated, whereas for a resolution higher than 1.5, the separation is complete. The resolution parameters are ranging from 2 to 20, therefore showing a good separation for almost all the pollutants studied here. Nevertheless, in the chromatogram, the trio Bisphenol A, β-Estradiol and Clofibric acid seems to be co eluted. At a higher magnification, it appears finally that the return to baseline is suitable (Figure 4). In order to demonstrate that the separation was complete, a purity test using PDA data has been performed. The PPI (Peak Purity Index) were drawn on each peak of the chromatogram. The PPI representation has a rectangular shape in the case of a pure eluted molecule and becomes curved in the case of an impure peak. In this study, the PPI has a rectangular shape whatever the compound of the mixture, proving that the chromatographic separation is fully accomplished. Furthermore, it demonstrated that no co elution with others compounds takes place, particularly in the

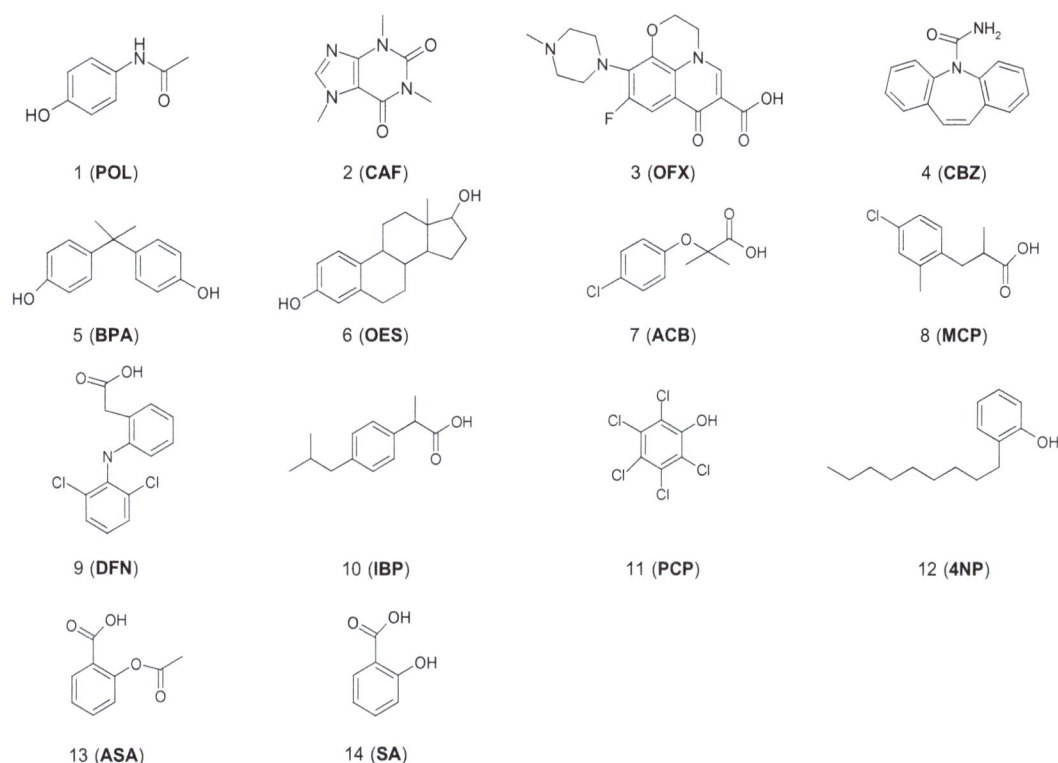

Figure 1: Structure of compounds constituting the selected mixture - 1: Acetaminophen, 2: Caffeine, 3: Ofloxacin, 4: Carbamazepin, 5: Bisphenol A, 6: β-Estradiol, 7: Clofibric acid, 8: Mecoprop, 9: Diclofenac, 10: Ibuprofen, 11: Pentachlorophenol, 12: 4-Nonylphenol, 13: Acetylsalicylic acid, 14: Salicylic acid.

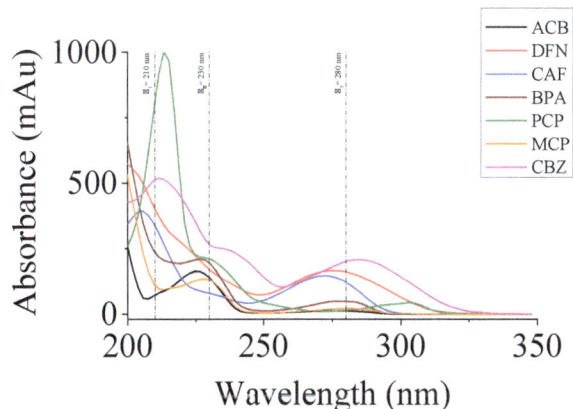

Figure 2: UV spectra using PDA detector of some compounds constituting the selected mixture.

case of the trio Bisphenol A, β-Estradiol and Clofibric acid. The 1.5 R_s value indeed demonstrates that the separation is efficient and sufficient enough to perform the quantitative analysis of the three components of the peak. The capacity factor k' is usually used to describe the migration rate of an analyte on the column. For high retention factor value, greater than 20, it is established that the elution time is too high and that the separation conditions needs to be optimized. In our case, the last eluted compound, 4-Nonylphenol, shows a capacity factor lower than 20. The developed method therefore permits to achieve a complete and fast separation of the thirteen pollutants within thirty minutes. Furthermore the selectivity factor α represents, like the resolution, the

column capacity to separate two compounds. It allows to calculate the separation power of the column for each pair of pollutant (α>1). In our case, the selectivity (α) and the number of theoretical plates (N) which were always higher than 1 and 2000 respectively proved that the Hypersyl Gold column was adequate and efficient to perform the pollutants mixture separation. However, with a selectivity factor α near to one, the separation of the massive peak constituted by BPA, OES and ACB appears to be at the limit of acceptance. The peak asymmetry permitted to evaluate the column quality. Ideally, the asymmetry of a perfectly gaussian peak is equal to one. Experimentally, the asymmetry was ranging from 1.2 to 5, where quantification is considered as less and less precise. In our case, the peak asymmetry was ranging from 1.1 to 1.5 (except Ofloxacin with an asymmetry of 5.6), reflecting the good quality of the column. Furthermore, the tailoring factor values were less than 2 and demonstrated consequently that the system suitability requirement was reached (Table 2). In the case of OFX, the high asymmetry value is explained by the interaction between the molecule and the residual silanol groups of the chromatographic column. Such interactions between the cationic adsorbate and dissociated silanol groups appear specifically in the chromatogram through an increasing peak trail. In order to reduce the peak trailing of this molecule, addition of triethylamine in the mobile phase could be considered.

Linearity

The linearity of the method was evaluated using eight calibration points, in the concentration range from 1.0×10^{-5} to 5.0×10^{-8} mol/L. The calibration curve was obtained by plotting peak area *versus* the concentration of standard solutions. The linear regression equation $y=(Slope)x+Offset$, is presented in Table 3 for each pollutant. The obtained standard calibration curves possess an acceptable degree of

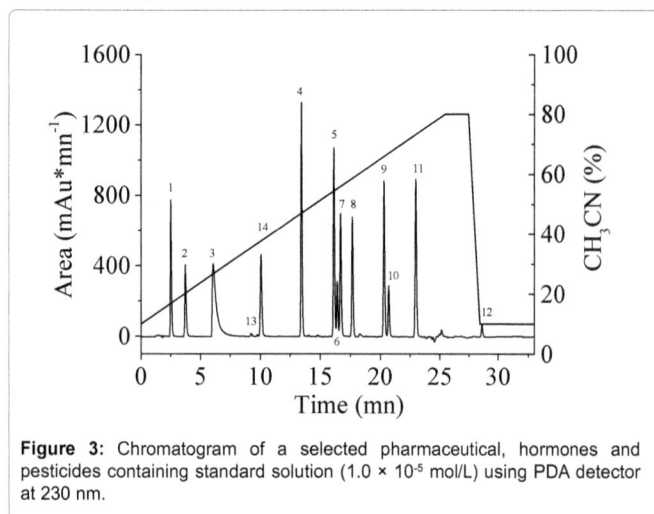

Figure 3: Chromatogram of a selected pharmaceutical, hormones and pesticides containing standard solution (1.0 × 10⁻⁵ mol/L) using PDA detector at 230 nm.

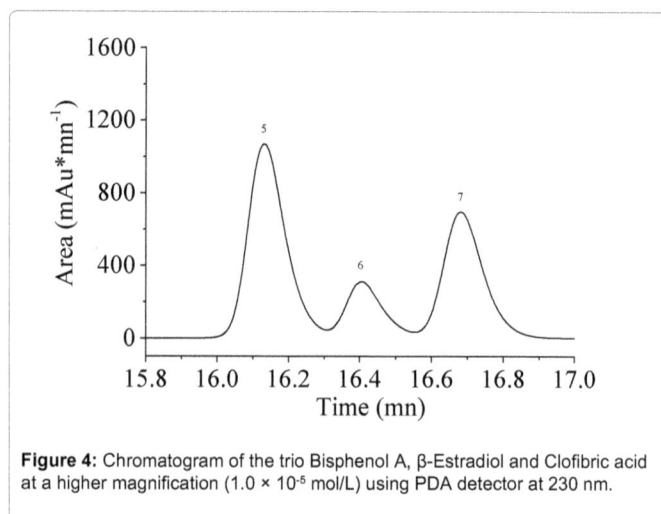

Figure 4: Chromatogram of the trio Bisphenol A, β-Estradiol and Clofibric acid at a higher magnification (1.0 × 10⁻⁵ mol/L) using PDA detector at 230 nm.

Pollutants	pK_A	Speciation*	Log D*	solubility* (mg/mL)	solubility** (mg/mL)	[Pollutant]$_{mixture}$ (mg/mL)
POL	9.5	Neutral	1.1	11.10	11.10	1.5×10^{-3}
CAF	0.6	Neutral	-0.8	43.80	43.70	1.9×10^{-3}
OFX	5.5 / 8.2	Cationic	-2.0	347.30	4.40	3.6×10^{-3}
SA	3.5	Neutral	0.9	3.7	280.30	1.8×10^{-3}
CBZ	2.3 / 13.9	Neutral	3.2	0.05	0.05	2.4×10^{-3}
BPA	9.8	Neutral	4.3	0.16	0.16	2.3×10^{-3}
OES	10.3	Neutral	3.7	0.03	0.03	2.7×10^{-3}
ACB	3.4	Neutral	2.4	0.32	81.10	2.1×10^{-3}
MCP	3.5	Neutral	3.6	0.18	7.10	2.1×10^{-3}
DFN	4.0	Neutral	3.9	0.02	1.16	3.2×10^{-3}
IBP	3.8	Neutral	3.8	0.06	0.70	2.1×10^{-3}
PCP	5.0	Neutral	4.4	0.01	0.04	2.7×10^{-3}
4NP	10.3	Neutral	5.4	0.01	0.01	2.2×10^{-3}

Table 1: Adsorbate characteristics. *: at pH 2.9; **: at pH 5.9.

Pollutants	Retention time (mn)	Retention factor (k')	Asymmetry	Tailoring factor	Theorical plates (N)	Pollutants pair	Resolution (R_s)	Selectivity (α)
POL	2.5	0.8	1.5	1.7	3 690	POL - CAF	6.5	2.2
CAF	3.7	1.6	1.4	1.5	5 150	CAF - OFX	5.2	2.1
OFX	6.2	3.3	5.6	4.5	1 930	OFX - SA	8.3	1.8
SA	10.1	6.1	1.3	1.3	28 100	SA - CBZ	14.9	1.4
CBZ	13.5	8.4	1.2	1.3	93 340	CBZ - BPA	14.0	1.2
BPA	16.2	10.3	1.1	1.2	115 200	BPA - OES	1.5	1.0
OES	16.4	10.5	1.2	1.3	147 800	OES - ACB	1.5	1.0
ACB	16.7	10.7	1.3	1.3	110 580	ACB - MCP	4.9	1.1
MCP	17.7	11.4	1.2	1.2	119 500	MCP - DFN	12.5	1.2
DFN	20.4	13.2	1.2	1.2	169 150	DFN - IBP	1.9	1.0
IBP	20.7	13.5	1.2	1.2	169 730	IBP - 4NP	8.9	1.1
PCP	23.0	15.1	1.2	1.2	171 330	PCP - 4NP	21.4	1.3
4NP	28.7	19.0	1.1	1.1	288 740			

Table 2: Chromatographic parameters for each compound of the mixture.

linearity in the selected concentration range, for each pollutant. More particularly, the higher correlation coefficient r^2 of 0.999 is observed at 230 nm, except for 4-Nonylphenol whose extinction coefficient is low whatever the wavelength [14,15].

Precision and accuracy

The intra-day precision of the method was determined for each compound by measuring standard repeatability characteristics, at three concentration levels (1.0×10^{-5}, 7.5×10^{-6} and 5.0×10^{-6} mol/L respectively) by making five repeated analysis performed on the same day [14,15]. The inter-day precision of the analytical method was determined at the same three concentration levels previously selected for the intra-day precision but repeated day by day over a period of five days. Precision is evaluated by the estimation of the relative standard deviation values (RSD) and the results are given in Table 4. The method was found to be precise with RSD value within 0.04-0.5% for intra-day experiment and RSD value within 0.07-3.6% for inter-day experiment. In both cases, % RSD values were found within 5% limit, indicating that the current method is repeatable for each pollutant. The RSD lower values correspond to the analytical method precision. Furthermore in the case of Ofloxacin, the quantification is less precise as compared to others pollutants with a RSD% value between 1 and 5%. Taking into account the OFX peak asymmetry of 5.6, the RSD value can be therefore linked to the peak asymmetry value, for which values higher than 5 lead also to a decrease of the quantification precision. The method accuracy was determined by comparing the experimental amount obtained from the calibration curve with the theoretical amount fixed in the standard solution prepared by weighting. The accuracy is ranging from 97.83 to 104.46% depending on the selected pollutant and the concentration level. The calculated values are much closed to the nominal values, suggesting that the analytical method possesses a good accuracy. The extremely high values can be directly correlated to the rigorous way for preparing the standards, especially the use of an accurate weighting scale (absolute accuracy of 10^{-5} g).

Limits of detection and quantification

The detection and quantification limits of an individual compound were determined at the three selected detection wavelengths by calculating signal/noise ratio (S/N=3) and (S/N=10), respectively for each compound (Table 5) [14,15]. The limits of detection differ depending on the selected pollutant and the wavelength used. For example, the limit of detection was ranging from 2.0×10^{-6} to 5.0×10^{-5} mg/mL at 210 nm, from 5.0×10^{-6} to 4.0×10^{-5} mg/mL at 230 nm and much higher at 280 nm. Owning a high extinction coefficient at 280 nm, Ofloxacin (2.0×10^{-6} mg/mL) and Caffeine (1.0×10^{-6} mg/mL) behave differently. In fact, for these two molecules sensitivities are higher and therefore it becomes possible to perform a quantitative analysis to lower concentration limits. The same behavior was observed for the limit of quantification. It was ranging from 5.0×10^{-6} to 1.8×10^{-4} mg/mL at 210 nm, from 2.0×10^{-5} to 1.2×10^{-4} mg/mL at 230 nm and much higher at 280 nm except for OFX (1.0×10^{-5} mg/mL) and CAF (3.0×10^{-6} mg/mL) whose quantification limits are much lower at that wavelength.

Pollutants	Slope	Offset	r²
POL	53.9	0.06	99.99
CAF	28.2	0.05	99.99
OFX	43.8	-1.64	99.89
SA	48.7	0.03	99.99
CBZ	64.0	2.14	99.94
BPA	53.0	0.75	99.90
OES	10.6	-0.13	99.97
ACB	39.5	0.59	99.97
MCP	42.0	0.50	99.98
DFN	37.5	0.16	99.97
IBP	17.3	0.11	99.99
PCP	47.4	0.27	99.96
4NP	5.5	-0.45	94.75

Table 3: Regression analysis of the calibration data for each compound of the mixture at 230 nm.

Pollutants	n	Intra-day				Inter-day				Accuracy (%)		
		RSD₁ (%)	RSD₂ (%)	RSD₃ (%)	RSD (%)	RSD₁ (%)	RSD₂ (%)	RSD₃ (%)	RSD (%)	E₁	E₂	E₃
POL	5	0.04	0.07	0.03	0.05	0.07	0.29	0.09	0.15	99.7	100.3	101.3
CAF	5	0.05	0.04	0.03	0.04	0.34	1.16	0.14	0.55	99.4	101.2	100.5
OFX	5	0.38	0.12	0.26	0.25	1.16	3.25	4.63	3.01	98.4	104.5	96.8
SA	5	0.08	0.03	0.04	0.05	1.02	2.08	0.34	1.15	100.8	98.4	101.1
CBZ	5	0.09	0.09	0.07	0.08	0.43	0.44	0.21	0.36	98.9	101.0	102.0
BPA	5	0.14	0.08	0.08	0.10	0.30	0.13	0.58	0.34	98.0	101.4	103.3
OES	5	0.51	0.34	0.40	0.42	1.42	0.89	1.88	1.40	97.8	99.6	98.6
ACB	5	0.16	0.11	0.10	0.12	0.39	0.44	0.51	0.45	99.0	100.5	101.8
MCP	5	0.05	0.05	0.03	0.04	0.07	0.32	0.16	0.18	99.5	100.7	101.7
DFN	5	0.05	0.05	0.08	0.06	0.75	0.26	1.12	0.71	99.0	100.8	100.9
IBP	5	0.16	0.12	0.13	0.14	2.22	3.61	0.32	2.05	100.9	97.0	101.0
PCP	5	0.08	0.05	0.04	0.06	0.82	0.48	1.63	0.98	99.0	101.0	101.6

Table 4: Precision and accuracy of the HPLC method.

Pollutants	LOD × 10⁵ (mg/mL)			LOQ × 10⁵ (mg/mL)		
	210 nm	230 nm	280 nm	210 nm	230 nm	280 nm
POL	1.0	1.3	4.1	3.4	4.2	14
CAF	0.5	1.1	0.1	1.6	3.6	0.3
OFX	5.3	3.7	0.3	18	12	1.0
SA	0.2	0.9	1.8	0.5	3.0	6.1
CBZ	0.2	0.5	0.6	0.5	1.8	1.9
BPA	1.0	0.7	2.0	3.4	2.2	6.7
OES	1.9	3.0	4.8	6.5	9.9	16
ACB	3.0	1.0	8.0	10	3.2	27
MCP	2.1	1.0	2.6	7.0	3.2	8.5
DFN	0.8	0.7	0.7	2.6	2.5	2.2
IBP	1.5	1.6	220	5.0	5.3	740
PCP	0.6	2.5	18	2.0	8.3	60
4NP	3.0	1.7	4.8	9.8	5.7	16

Table 5: Analytical parameters of the proposed method (limit of detection (LOD) and limit of quantification (LOQ)) for each compound of the mixture.

Conclusion

In this study, a wide variety of pollutants were investigated, pharmaceutical products, pesticides, hormone, and solvents; these contaminants being selected in relation with their occurrence in tertiary treatment plants. The wastewater treatments technologies actually available, as such as ozonation, membrane or adsorption processes, are indeed not able to remove efficiently all of these contaminants that are therefore detected at very low concentrations. Suitable analytical methods, allowing the detection and the precise quantification of these pollutants in water samples, need therefore to be developed. The HPLC-UV method described in this work appears as simple, precise, reproducible and sensitive in the low range of concentration i.e., μg/L. Furthermore, this method remains as an alternative to the more sophisticated methods like HPLC-MS or HPLC-fluorescence, in particular as far as intermediate concentrations are concerned (to μg/L). It can be used for routine analysis and is well adapted for many research studies, particularly concerning water treatment remediation. In our case, the developed method has been employed to study the adsorption of emerging pollutants on activated carbon cloths but also in order to evaluate the regeneration potentialities of the carbon material after loading using electrochemical techniques. This technique is well adapted for synthetic mixtures of water contaminants or samples containing an identified pollution. The limits of the method appear when considering real matrices and samples coming from treatment plants; the co-elution risks being more important as far as an increasing number of molecules, especially non-identified pollutants constituting the water matrix, are concerned. Considering its relatively weak sensibility, this method is however easy to handle and could find applications for the determination of composition and concentration of industrial or hospital effluents, particularly at high concentration levels (mg/L - μg/L).

Acknowledgements

The authors thank ANR, the French National Research Agency, in particular the ECOTECH program for the financial support of the PARME project.

References

1. Jones OA, Voulvoulis N, Lester JN (2001) Human pharmaceuticals in the aquatic environment a review. Environ Technol 22: 1383-1394.

2. Bottoni P, Caroli S, Caracciolo AB (2010) Pharmaceuticals as priority water contaminants. Toxicological and Environmental Chemistry 92: 3549-3565.

3. Fent K, Weston AA, Caminada D (2006) Ecotoxicology of human pharmaceuticals. Aquat Toxicol 76: 122-159.

4. XP T90-223 (Février 2013), Qualité de l'eau - Dosage de certains résidus médicamenteux dans la fraction dissoute des eaux - Méthode par extraction en phase solide et analyse par chromatographie en phase liquide couplée à la spectrométrie de masse en tandem (LC-MS/MS).

5. Patrolecco L, Ademollo N, Grenni P, Tolomei A (2013) Simultaneous determination of human pharmaceuticals in water samples by solid phase extraction and HPLC with UV-fluorescence detection. Microchemical Journal 107: 165-171.

6. Wille K, Claessens M, Rappé K, Monteyne E, Janssen CR, et al. (2011) Rapid quantification of pharmaceuticals and pesticides in passive samplers using ultra high performance liquid chromatography coupled to high resolution mass spectrometry. J Chromatogr A 1218: 9162-9173.

7. Shervington LA, Abba M, Hussain B, Donelly J (2005) The simultaneous separation and quantification of five quinolone antibiotics reversed-phase HPLC: Application to stability studies on an ofloxacin tablet formulation. J Pharm Biomed Anal 39: 769-775.

8. Rezaee M, Yamini Y, Shariati S, Esrafili A, Shamsipur M (2009) Dispersive liquid-liquid microextraction combined with high-performance liquid chromatography-UV detection as a very simple, rapid and sensitive method for the determination of bisphenol A in water samples. J Chromatogr A 1216: 1511-1514.

9. Mowafy HA, Alanazi FK, El Maghraby GM (2012) Development and validation of an HPLC-UV method for the quantification of carbamazepine in rabbit plasma. Saudi Pharm J 20: 29-34.

10. Franeta JT, Agbaba D, Eric S, Pavkov S, Aleksic M, et al. (2002) HPLC assay of acetylsalicylic acid, paracetamol, caffeine and phenobarbital in tablets. Farmaco 57: 709-713.

11. Zhou Y, Zhang Z, Shao X, Chen Y, Wu X, et al. (2014) Hollow-fiber-supported liquid-phase microextraction using an ionic liquid as the extractant for the pre-concentration of bisphenol A, 17-ß-estradiol, estrone and diethylstilbestrol from water samples with HPLC detection. Water Sci Technol 69: 1028-1035.

12. Miège C, Choubert JM, Ribeiro L, Eusèbe M, Coquery M (2009) Fate of pharmaceuticals and personal care products in wastewater treatment plants - Conception of a database and first results. Environ Pollut 157: 1721-1726.

13. Luo Y, Guo W, Ngo HH, Nghiem LD, Hai FI, et al. (2014) A review on the occurrence of micropollutants in the aquatic environment and their fate and removal during wastewater treatment. Sci Total Environ 473-474: 619-641.

14. ICH Guideline Q2 (R1) (2005) Validation of Analytical Procedures: Texts and Methodology.

15. Huber L (2010) Validation of Analytical Methods. Agilent Technologies.

Outlining a Multidimensional Approach for the Analysis of Coffee using HPLC

Sercan Pravadali-Cekic[1,2], Danijela Kocic[1,2], Paul Stevenson[1,3] and Andrew Shalliker R[1,2]*

[1]*Australian Centre for Research on Separation Sciences (ACROSS), Tasmania, Australia*
[2]*School of Science and Health, University of Western Sydney (Parramatta), NSW, Australia*
[3]*School of Life and Environmental Sciences, Deakin University (Waurn Ponds), VIC, Australia*

Abstract

This study investigated approaches for the profiling of coffee using two multidimensional approaches: (1) a multi-detection process and (2) a multi-separation process employing HPLC. The first approach compared multi-detection techniques of conventional High Performance Liquid Chromatography (HPLC) hyphenated with a detector (DPPH•, UV-Vis and MS), and multiplexed mode via HPLC with an Active Flow Technology (AFT) column in Parallel Segmented Flow (PSF) format with DPPH• detection, UV-Vis and MS running simultaneously. Multiplexed HPLC-PSF enabled the determination of key chemical entities by reducing the data complexity of the sample whilst obtaining a greater degree of molecule-specific information within a fraction of the time it takes using conventional multi-detection processes. DPPH•, UV-Vis and MS (TIC) were multiplexed for the analysis of espresso coffee and decaffeinated espresso coffee. Up to 20 DPPH• peaks were detected for each sample, and with direct retention time peak matching, 70% of DPPH• peaks gave a UV-Vis response for the espresso coffee and 95% for the decaffeinated espresso coffee. The second approach involved the use of a two-dimensional (2D) HPLC system to expand the separation space and separation power for the analysis of coffee, focusing on the resolution and detection of co-eluting and overlapping peaks, which was beyond the limits of conventional HPLC in resolving complex samples. The 2DHPLC analysis resulted with the detection of 176 peaks and a closer observation showed the presence of an additional 17 peaks in a cut section where in 1D mode only one peak was observed.

Introduction

Coffee, like all natural products, is complex in the nature of, and the number of compounds. It is popular for its flavour, aroma, caffeine content and antioxidant properties [1,2]. It has been reported that coffee contains hundreds of constituents [3,4] with a variety of compounds including, carbohydrates/fibre, nitrogenous compounds, lipids, minerals, acids and esters and melanoidins [3]. The analysis and characterisation of such complex samples can be rather challenging, in particular when screening for a number of compounds and/or isolating compounds of interest, such as, antioxidants. In industries, including but not limited to food and pharma, efficient and straightforward analytical techniques are sought in order to provide significant information for sample characterisation and chemical profiling. For example, screening for antioxidant compounds in coffee is achieved with various methods of detection some of which could be classified as hyphenated methods with HPLC [5-11]. The aim of this study is the comprehensive characterisation of complex samples, especially coffee to provide a greater understanding of their nature in an efficient way. It also highlights the importance of comprehensive analyses in the identification of peaks of interest, where misidentification can occur due to degrees of uncertainty.

This study focuses on multidimensional approaches to chemical profiling and characterisation of coffee in a time efficient yet comprehensive manner. From a multi-detection perspective, two detection processes are compared: The first utilises conventional HPLC coupled with a single detector, where for each detection method a sample injection is made, so that sample information is gathered through the use of a variety of detectors, including mass spectrometry (MS), 2, 2-diphenyl-1-picrylhydrazyl free radical (DPPH•) for the detection of reactive oxygen species (ROS), potentially antioxidants, and ultraviolet-visible detector (UV-Vis) as a general chromophore detection process. In the second process the detectors were multiplexed using Active Flow Technology (AFT) columns operated in Parallel Segmented Flow (PSF) mode. These AFT columns are a new column format that have been designed to provide more efficient separation

utilising radial flow stream splitting to separate the radial central flow through the chromatography column from the flow near the wall [12]. The AFT column therefore corrects for the column bed heterogeneity, a major determinant in poor separation performance [13]. The outlet port on the AFT column contains multiple exit ports; a single radial central exit port that manages the radial central flow through the column and peripheral port(s) that manage the flow from the column wall region [12-15]. Although primarily designed to improve separation performance and the interface to MS detection [16], an added benefit from the design is that more than one detector can be used simultaneously, even if each detector is destructive. In fact, since the flow stream is effectively split on the column, the multiple detection process is actually multiplexed [17-19].

These AFT columns have been used in several analyses that have involved DPPH, chemiluminescence, MS, UV and fluorescence detections [17]. The current study extends the application of multiplexed detection further, showing how DPPH•, UV-Vis and MS detectors can be used in a multiplexed arrangement with AFT (PSF) columns for chemical profiling of coffee, and in particular the search for antioxidant activity.

***Corresponding author:** Shalliker RA, Australian Centre for Research on Separation Sciences (ACROSS), School of Science and Health, University of Western Sydney (Parramatta), NSW 1797, Room LZ.1.48, Corner of Pemberton St and Victoria Rd, Parramatta, 2151, NSW, Australia
E-mail: R.Shalliker@uws.edu.au

Multidimensional HPLC is another technique that can be employed to overcome the limitations of uni-dimensional HPLC [20]. When dealing with complex samples, the peak capacity of a uni-dimensional separation (separations employing just one separation step) is often exceeded because of the vast number of components in the sample. Columns simply cannot be made long enough to yield the required peak capacity and then enable the separation within a suitable timeframe [20]. Thus, for the analysis of complex samples a move to multidimensional separation strategies is required, since the peak capacity of a two-dimensional system is equal to the product of the peak capacity of each single dimension, minus the amount of separation space lost due to correlation. A combination of multiplexed detection, and multi-dimensional HPLC would therefore form the basis of a very powerful analytical profiling tool.

Experimental

Chemicals and reagents

Mobile phase solvents were HPLC grade. Methanol was purchased from Merck (Kilsyth, VIC, Australia) and Ultrapure Milli-Q water (18.2 MΩ) was prepared in-house and filtered through a 0.2 μm filter. 2, 2-Diphenyl-1-picrylhydrazyl free radical (DPPH$\cdot$) was purchased from Merck. In conventional HPLC, the DPPH$\cdot$ reagent (1 mgmL^{-1}) was prepared in methanol. In multiplexed mode, 0.1% formic acid was added to the mobile phases and DPPH$\cdot$ reagent to cater for the mass spectral component of the multiplexed analysis. Caffeine was purchased from Sigma-Aldrich (Castle Hill, NSW, Australia).

Sample preparation

The espresso coffee (Nestlé Nespresso – Ristretto and Decaffeinated, Sydney, NSW Australia) samples were prepared fresh prior to analysis by extraction using an espresso coffee machine (Dè Longhi - Model EN 95.S), diluted fourfold with Milli-Q water, then filtered through 0.45 μm pore filter. A 0.05 mgmL^{-1} caffeine standard was prepared in 60:40 (water:methanol).

Instrumentation

Multiplexed chromatographic experiments were carried out using a Thermo UHPLC system coupled with TSQ Vantage mass spectrometer equipped with HESI II source (Thermo Scientific, San Jose, USA). The LC component was a Dionex Ultimate 3000 equipped with a quaternary pump, auto injector with an in-line degassing unit and RS diode array detector. The Thermo Vantage TSQ was operated as supplied from the manufacturer.

Two-dimensional chromatographic experiments were undertaken on a Waters 600E Multi Solvent Delivery LC System, equipped with a Waters 717 plus auto injector, two - Waters 600E pumps, two - Waters 2487 series UV-Vis detectors (set at 280 nm) and two - Waters 600E system controllers. Separation dimensions were interfaced *via* a 10-port, two position switching valve that was electrically actuated.

Chromatographic analysis

Conventional mode (Uni-dimensional mode): Separations were conducted using a 250 × 4.6 mm C18 reversed phase column (Hypersil Gold, 5 μm P$_d$, 175Å) from Thermo Scientific (Runcorn, Cheshire, United Kingdom). All chromatographic analyses were undertaken using gradient elution with the initial mobile phase of 100% water running to a final mobile phase of 100% methanol, at a rate of 1%min^{-1} then held for 20 minutes at 100% methanol. Flow-rate was set at 0.8 mLmin^{-1} and injection volumes were 80 μL.

Mass spectrometry detection: Mass spectrometry (MS) detection was conducted using electrospray ionisation in positive ion mode. Using Full Scan detection method, Total Ion Count (TIC) analysis was carried out on the TSQ Vantage mass spectrometer equipped with HESI II source under the following conditions: Vaporiser temperature 500°C, capillary temperature, 350°C; sheath gas set at a rate of 60 unites; auxiliary gas flow 40; sweep gas flow at 5 units, and spray voltage, 1.5 kV [21].

UV-Vis detection: UV-Vis detection was undertaken using the Ultimate 3000 RS diode array detector nominally set 280 nm.

DPPH$\cdot$ detection for ROS: The conventional chromatographic DPPH$\cdot$ experiments were conducted on the Thermo UPLC system. The post-column eluent stream was combined with the DPPH$\cdot$ reagent at a T-piece, with a flow rate of 0.8 mLmin^{-1} (1:1, eluent stream:DPPH$\cdot$ reagent), using an additional stand-alone Shimadzu Prominence LC-20AD pump and a Degassex model DG-440 inline degasser unit to deliver the DPPH$\cdot$ reagent. The combined eluent stream then entered a reaction coil (100 μL), which was maintained at a temperature of 60°C within a column heater. The eluent then entered the UV RS diode array detector set at a wavelength of 520 nm to detect the radical scavenging compounds within the eluent. A flow diagram illustrating the setup of the conventional HPLC system with each of the detectors is illustrated in Figure 1.

Multiplexed detection mode using AFT columns: An AFT column in parallel segmented (PSF) mode, of the same dimensions and column packing material as the conventional column was used for multiplexed detection separations. This PSF column was equipped with four exit ports, enabling up to four separate modes of detection to be employed (Figure 2). In this study, the flow ratio through of the four detectors was set at: 18.5% of the total flow rate from the central port, which was connected to the MS detector, 22.5% of the flow from peripheral port 1 was connected to a UV-Vis detector and 59% of the flow from peripheral port 2 was connected to a DPPH$\cdot$ detector via a T-piece. Peripheral port 4 was not used and blocked. The chromatographic conditions of the PSF column are the same as the conditions used in 'Conventional Mode', and the detection mode of the MS and UV-Vis detectors was also the same as that used for the conventional mode of analysis. The DPPH$\cdot$ detection process was, however, slightly different, largely to compensate for the reduced flow rate eluting to that detector.

DPPH$\cdot$ detection: A Shimadzu analytical HPLC system, comprising a Shimadzu LC-20ADvp quaternary pump, Shimadzu SIL-10ADvp auto injector, Shimadzu SPD-M10Avp PDA detector and a Degassex model DG-440 inline degasser unit (Phenomenex, Lane Cove NSW Australia) was used for multiplexed DPPH$\cdot$ detection. Since the flow rate through the DPPH$\cdot$ reagent detector (peripheral port 3) was reduced, the volumetric flow of the DPPH$\cdot$ reagent was reduced accordingly. Specifically, the eluent stream from peripheral port 2 had a flow rate of 0.47 mLmin^{-1}, so the DPPH$\cdot$ reagent was set at 0.47 mLmin^{-1} to maintain the 1:1, eluent stream:DPPH$\cdot$ reagent ratio. The combined eluent stream then entered a reaction coil (100 μL), which was maintained at a temperature of 60°C within a column heater, and then entered the Shimadzu PDA detector set at a wavelength of 520 nm to detect the radical scavenging compounds within the eluent. Figure 2 illustrates this setup.

Two-dimensional HPLC: The first dimension consisted of a 100 × 4.6 mm Hypersil GOLD Cyano (CN) (5 μm P$_d$, 175Å) column and the second dimensional, a 100 × 4.6 mm Hypersil GOLD C18 (5 μm P$_d$, 175Å) column. Both columns were supplied by ThermoFisher Scientific (Runcorn, Cheshire, United Kingdom). Gradient elution was

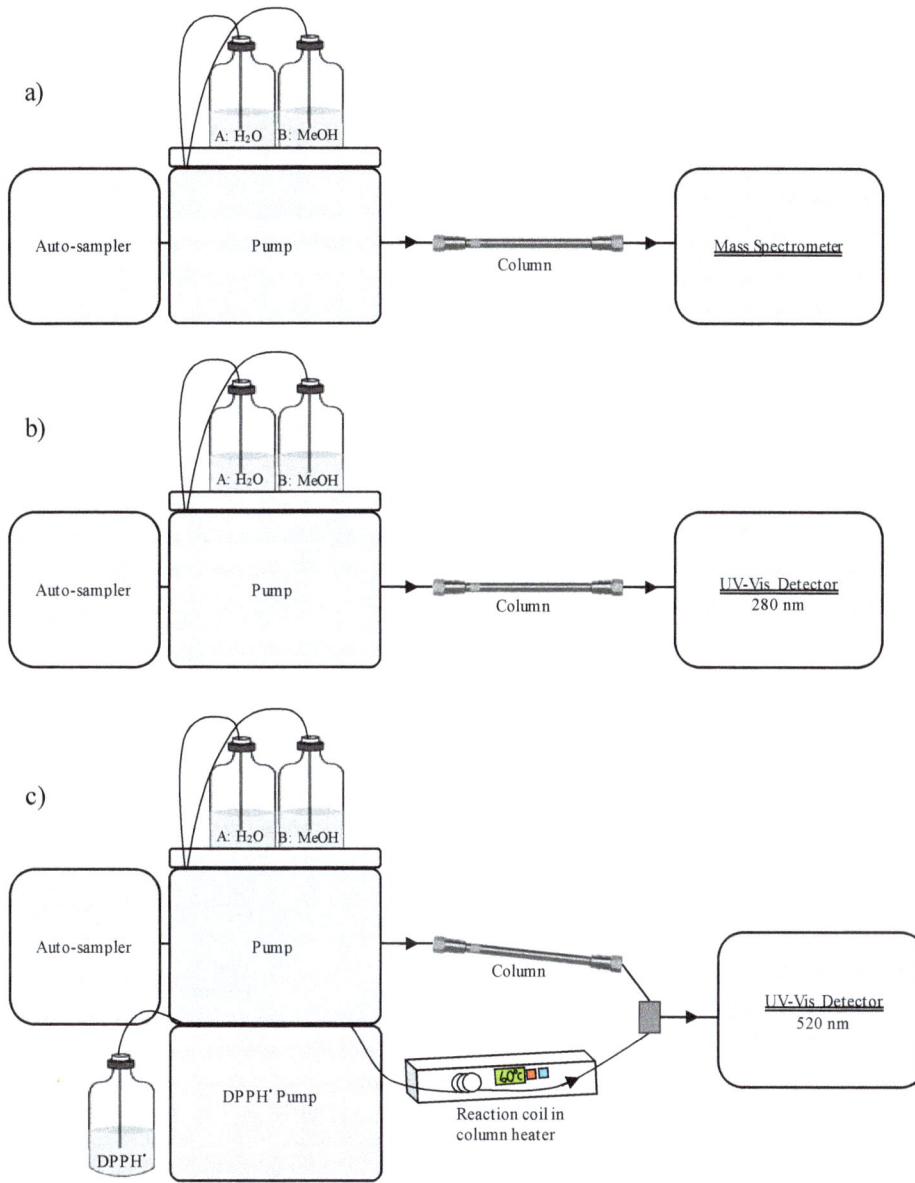

Figure 1: Illustration of conventional HPLC setup with: (a) MS detector, (b) UV-Vis detector and (c) DPPH˙ detector.

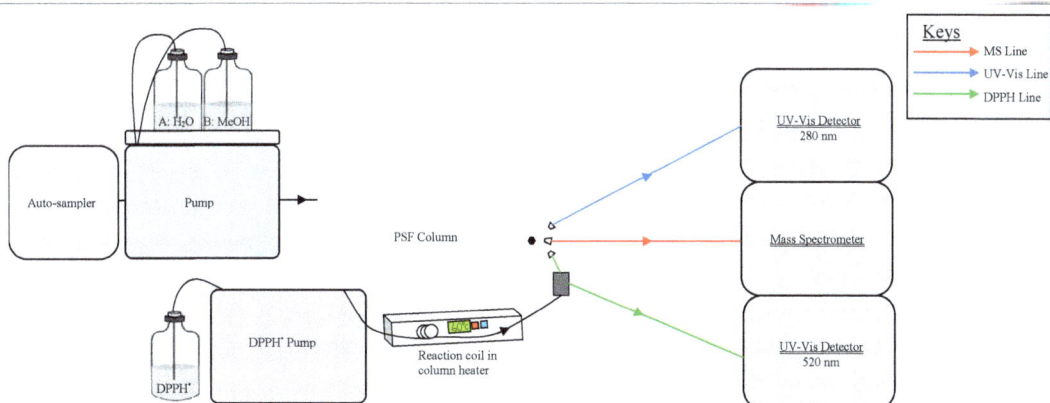

Figure 2: Illustration of multiplexed HPLC setup with DPPH˙, UV-Vis and MS detectors.

undertaken in both dimensions. In the first dimension, gradient elution began with initial mobile phase of 100% water running to a final mobile phase of 100% organic solvent (methanol), at a rate of 20%min⁻¹ then held for 4 minutes and returned to initial conditions in 1 minute. Flow-rate was set at 2.0 mL/min and injection volume was 100 µL into the first dimension. The 2D analyses of the sample was undertaken using a comprehensive incremental heart-cutting approach [8,10] involving the transfer of a 200 µL aliquot sampled from the first dimension, into the second dimension, followed by elution in the second dimension. The second dimension consisted of a C18 column, also undertaking gradient elution from 100% water to 100% organic solvent (methanol) at a rate of 20%min⁻¹, held for 4 minutes and returned to initial conditions in 1 minute, with a flow rate of 2.0 mL min⁻¹. This process was repeated (following reinjection of the sample) at every 0.4 minute interval across the first dimension, starting at 0.8 minutes in the first dimension, being completed after a total of 16 cuts. Both dimensions were temperature regulated at 45°C using a column heater.

Data analysis

The multiplexed data was analysed using Origin and Microsoft Excel. The MS data was processed using LC Quan software (Thermo Fisher Scientific, San Jose, USA). The two-dimensional chromatographic data was processed using Mathematica (with home-built programs) [22]. Two-dimensional chromatographic surface plots were produced using this software.

Results and Discussion

Conventional HPLC multi-detection

The Ristretto coffee sample was analysed using conventional HPLC with each of the following detectors; UV-Vis, DPPH· and MS – TIC. The chromatograms from each detector are shown in Figure 3. A new sample was prepared for each analysis mode. The coffee sample showed a positive response to the DPPH· reagent, but determining which compounds having the DPPH· signal corresponded to the respective components in UV-Vis and MS chromatograms required that each chromatographic response be manually aligned– a source of potential uncertainty.

Multiplexed HPLC

A multiplexed HPLC analysis was carried out using an AFT column in PSF mode allowing the compounds that responded to DPPH· to be easily matched up to the UV-Vis response and MS based on retention time and where a positive response was seen from the MS detector, the molecular mass of the peak was recorded. Table 1 lists the retention time of the DPPH· peaks, and the response of such peaks in the UV-Vis and/or the MS detector, which thus provided the molecular mass.

Coffee is known to have a high antioxidant content, with numerous components exhibiting response to the DPPH· reagent [5,6,17]. In this study, the coffee analysis with the DPPH· reagent resulted in 20 well resolved peaks, 13 of which also showed a response in the UV-Vis and MS detectors as indicated by red dotted lines across the multiplexed chromatograms in Figure 4. There were four regions that had the same chromatographic profile within each chromatogram, indicated by the red boxes. In these boxes, where UV-Vis and MS showed little response to these peaks, there was a strong response from the DPPH· detector. Four components that responded to the DPPH· reagent were not detected either by the UV-Vis or MS detectors (represented by the blue dotted lines) and three of the components that responded to the DPPH· reagent gave no UV-Vis or MS response at all (represented by the green dotted line). The molecular mass of the components that

responded to the MS are recorded in Table 1. The component that eluted at the 10 minute mark according to the DPPH· assay resulted in a strong response, with no response from the UV-Vis and with a very small response from the MS detector.

An interesting and reoccurring facet of the analysis of coffee with DPPH· antioxidant detection is that peak B in Figure 4 (a), which corresponds almost perfectly with the elution of caffeine shows a very strong antioxidant response. Peak A is labelled for comparison purposes. Caffeine, however, is not a reactive oxygen scavenging compound, or an antioxidant and shows no response to the DPPH· reagent when tested solely. A casual analysis of this assay may result in an analyst mistakenly assigning this antioxidant response to caffeine as the retention time is almost perfectly coincident with caffeine; rather, this antioxidant is a minor component (according to UV and MS detections) that co-elutes with caffeine. The analysis of decaffeinated coffee, for example, illustrates this interesting facet of the separation. The decaffeinated coffee sample was analysed using exactly the same methodology that was employed for the caffeinated coffee, with multiplexed detection. The resulting chromatographic responses for each of the three detectors are shown in Figure 5. In total 18 DPPH· peaks were recorded, 16 of which were also detected by the UV-Vis and MS detectors. One of the DPPH· peaks also only gave a UV-Vis response and one DPPH· peak gave no UV-Vis or MS response. Aside from a decrease in the signal intensities between the caffeinated and decaffeinated samples there was a very similar retention profile across all three detectors, except, however, the absence of the caffeine band in the UV-Vis and MS detection responses. The antioxidant responses at around the 10 minute mark in the DPPH· detector, however, still showed the presence of the two components that co-eluted with the caffeine in caffeinated coffee sample, showing clearly that caffeine was not responsible for the antioxidant behaviour, rather a co-eluting component at much lower concentration than caffeine.

The chromatographic response for caffeine using UV and DPPH· detection, shown in Figure 6 further verifies this phenomenon. This nice example of sample and detection selectivity, and the confusion that co-eluting species can produce in real complex samples shows the importance of sample analysis with multiple detectors and indicates that for complete sample analysis more powerful separations are required in order to resolve more components, some of which may have important biological function. In this case, a minor component

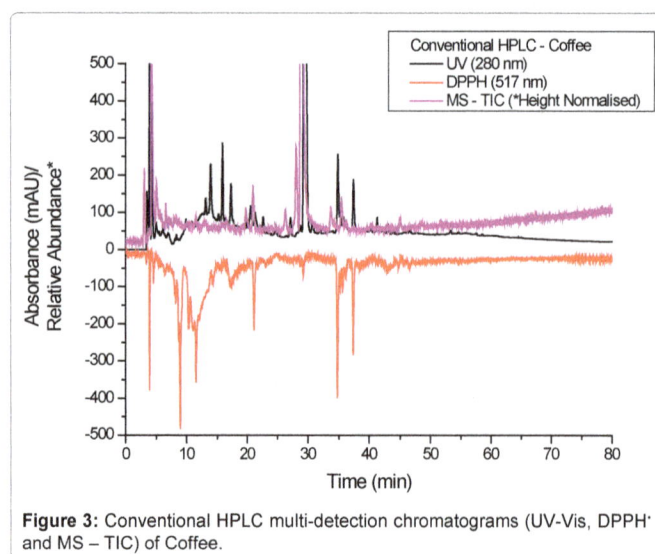

Figure 3: Conventional HPLC multi-detection chromatograms (UV-Vis, DPPH· and MS – TIC) of Coffee.

Coffee DPPH· Peak Response and Mass					
DPPH· Peak	Retention Time (min)	DPPH· Response	UV-Vis Response	MS Response	Mass
1	6.74	Yes	Yes	Yes	123.84
2	7.54	Yes	Yes	Yes	125.83
3	8.94	Yes	No	No	-
4	10.05	Yes	No	Yes	135.83
5	13.15	Yes	No	No	-
6	16.22	Yes	No	No	-
7	18.14	Yes	Yes	Yes	126.82
8	19.40	Yes	Yes	Yes	162.81
9	20.46	Yes	Yes	Yes	187.89
10	24.71	Yes	Yes	Yes	162.81
11	26.27	Yes	Yes	Yes	162.80
12	26.97	Yes	Yes	Yes	194.87
13	31.84	Yes	Yes	Yes	162.80
14	32.02	Yes	Yes	Yes	162.78
15	32.56	Yes	Yes	Yes	176.82
16	33.94	Yes	Yes	Yes	176.83
17	41.26	Yes	Yes	No	-
18	42.72	Yes	No	Yes	284.93
19	46.07	Yes	Yes	Yes	190.83
20	49.21	Yes	Yes	Yes	162.75
Decaffeinated Coffee DPPH· Peak Response and Mass					
DPPH· Peak	Retention Time (min)	DPPH· Response	UV-Vis Response	MS Response	Mass
1	6.86	Yes	Yes	Yes	123.81
2	7.63	Yes	Yes	Yes	125.78
3	9.01	Yes	Yes	No	-
4	10.28	Yes	Yes	Yes	184.80
5	16.45	Yes	Yes	Yes	162.87
6	17.44	Yes	Yes	Yes	162.78
7	18.59	Yes	Yes	Yes	126.80
8	19.62	Yes	Yes	Yes	162.79
9	24.86	Yes	Yes	Yes	162.79
10	26.57	Yes	Yes	Yes	162.80
11	27.22	Yes	Yes	Yes	162.80
12	31.73	Yes	Yes	Yes	162.79
13	32.29	Yes	Yes	Yes	162.75
14	32.93	Yes	Yes	Yes	336.80
15	34.08	Yes	Yes	Yes	336.79
16	39.08	Yes	No	No	-
17	41.41	Yes	Yes	Yes	204.82
18	46.17	Yes	Yes	Yes	100.84

Table 1: Detected DPPH· peaks response to UV-Vis and MS.

in the coffee sample with strong antioxidant activity co-eluted with the major component (caffeine). Extraction of this low concentration component from the coffee sample would therefore require better separation, perhaps multidimensional.

Two-dimensional HPLC

In the multiplexed section of this study it was apparent that although a long separation in gradient mode was carried out on the coffee samples in uni-dimensional HPLC, there were many peaks overlapping and co-eluting due to the complex nature of the sample. The limitations of separation power and resolution of uni-dimensional HPLC for complex samples are well known [23]. The separation power and resolution of a uni-dimensional system is impractical for complex samples, essentially due to the insufficient peak capacity [20,24-27] and

the need for more time efficient techniques. Thus, for the analysis of complex samples a move to multidimensional separation strategies is required, ergo a two-dimensional (2D) HPLC separation was carried out on the Ristretto Espresso coffee sample. Based on a previous study by Mnatsakanyan et al. [7] a 2DHPLC system containing a Cyano (CN) phase in the first dimension and a octadecyl (C18) phase in the second dimension with a water:methanol solvent system, under gradient conditions, gave highly orthogonal retention behaviour for the analysis of the same coffee sample. For this reason, this system of selectivity was chosen. Initially, a uni-dimensional (1D) HPLC separation using the CN column was carried out for the coffee sample and is illustrated in Figure 7. The chromatogram shows bimodal distribution of coffee peaks, however, many peaks co-elute as a result the saturation of the separation space.

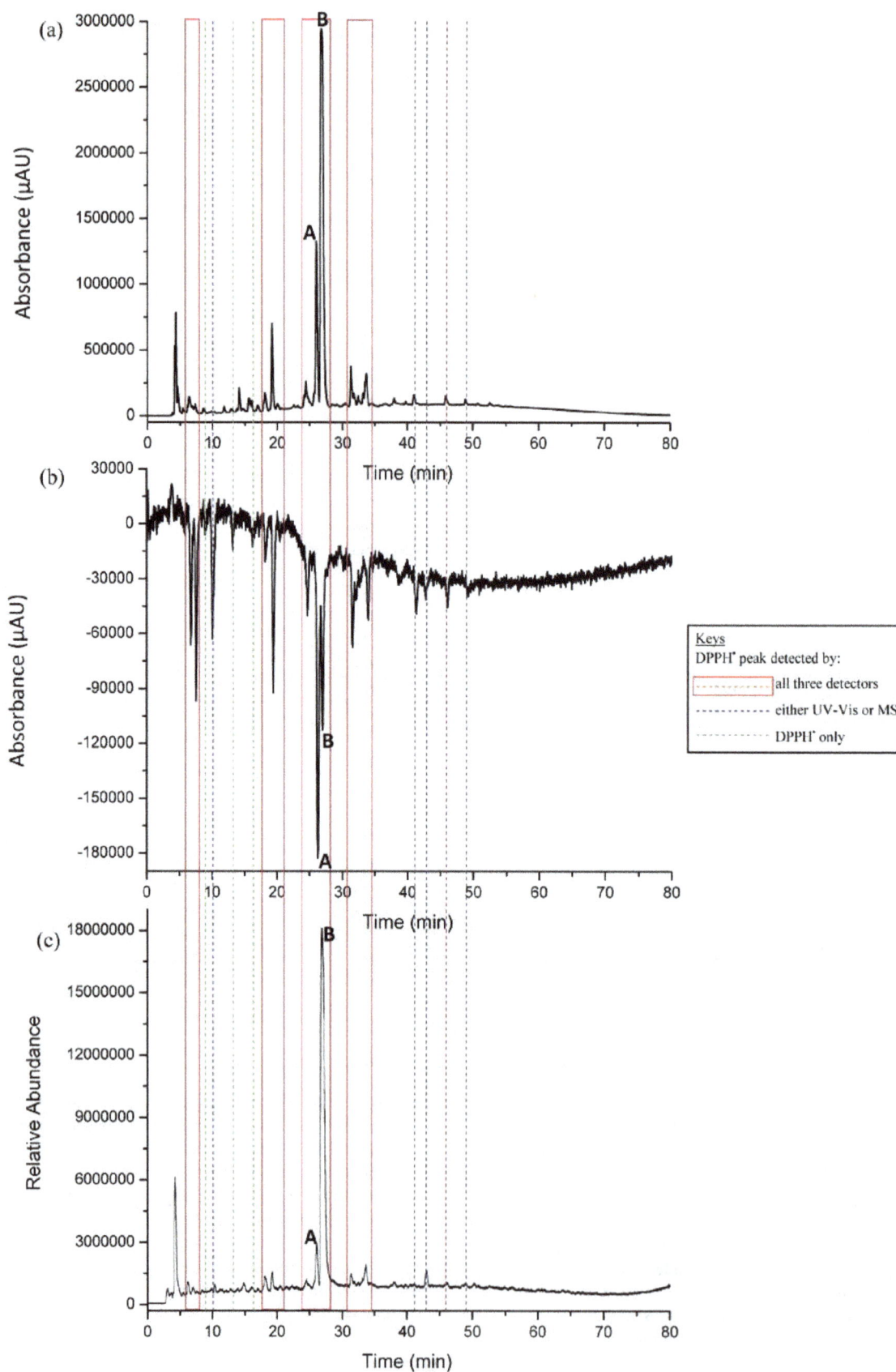

Figure 4: Coffee sample multiplex chromatograms (a) UV-Vis 280 nm, (b) DPPH˙ 520 nm, (c) MS - TIC.

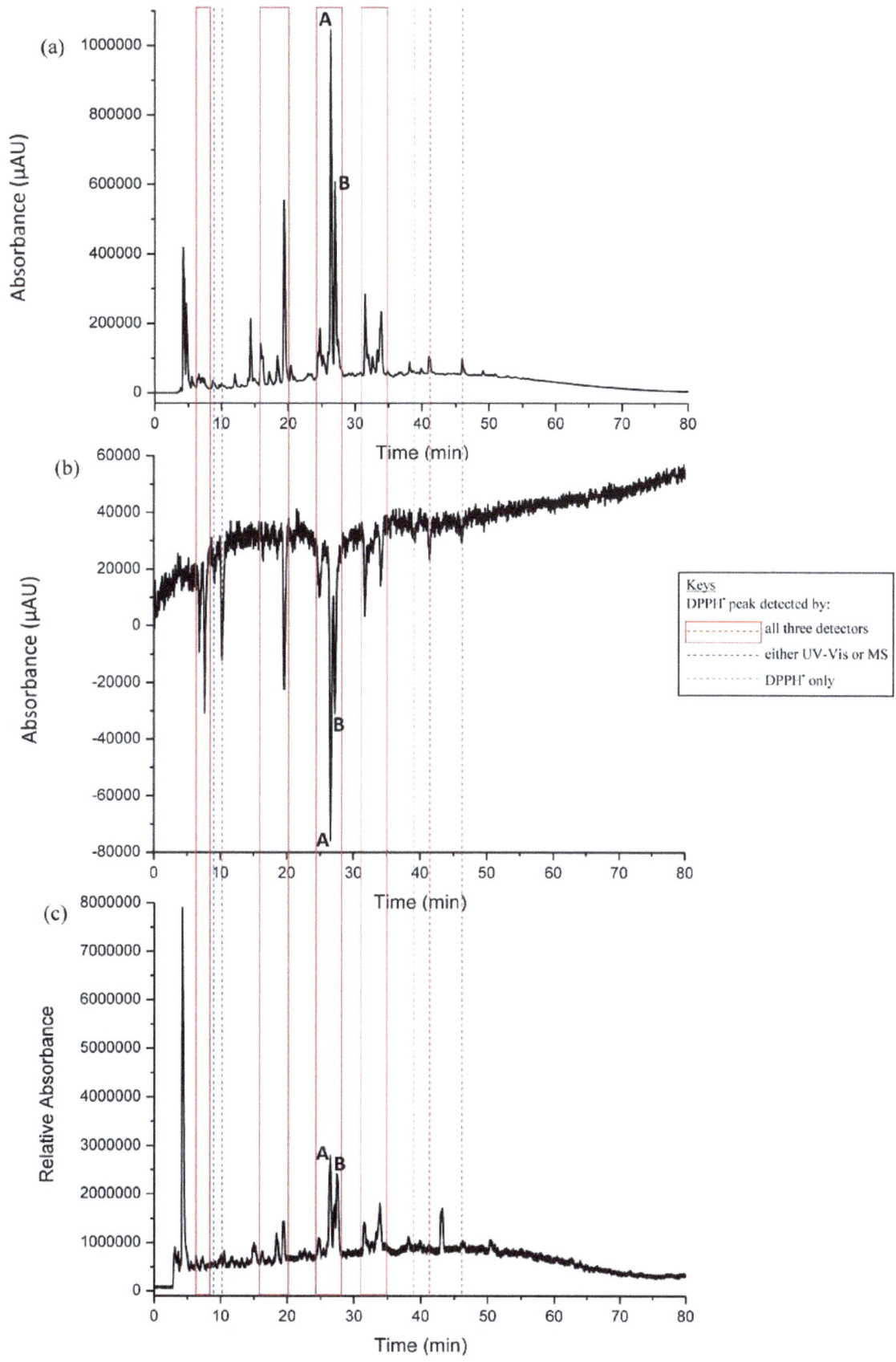

Figure 5: Decaffeinated sample multiplexed chromatograms (a) UV-Vis 280 nm, (b) DPPH˙ 520 nm, (c) MS - TIC.

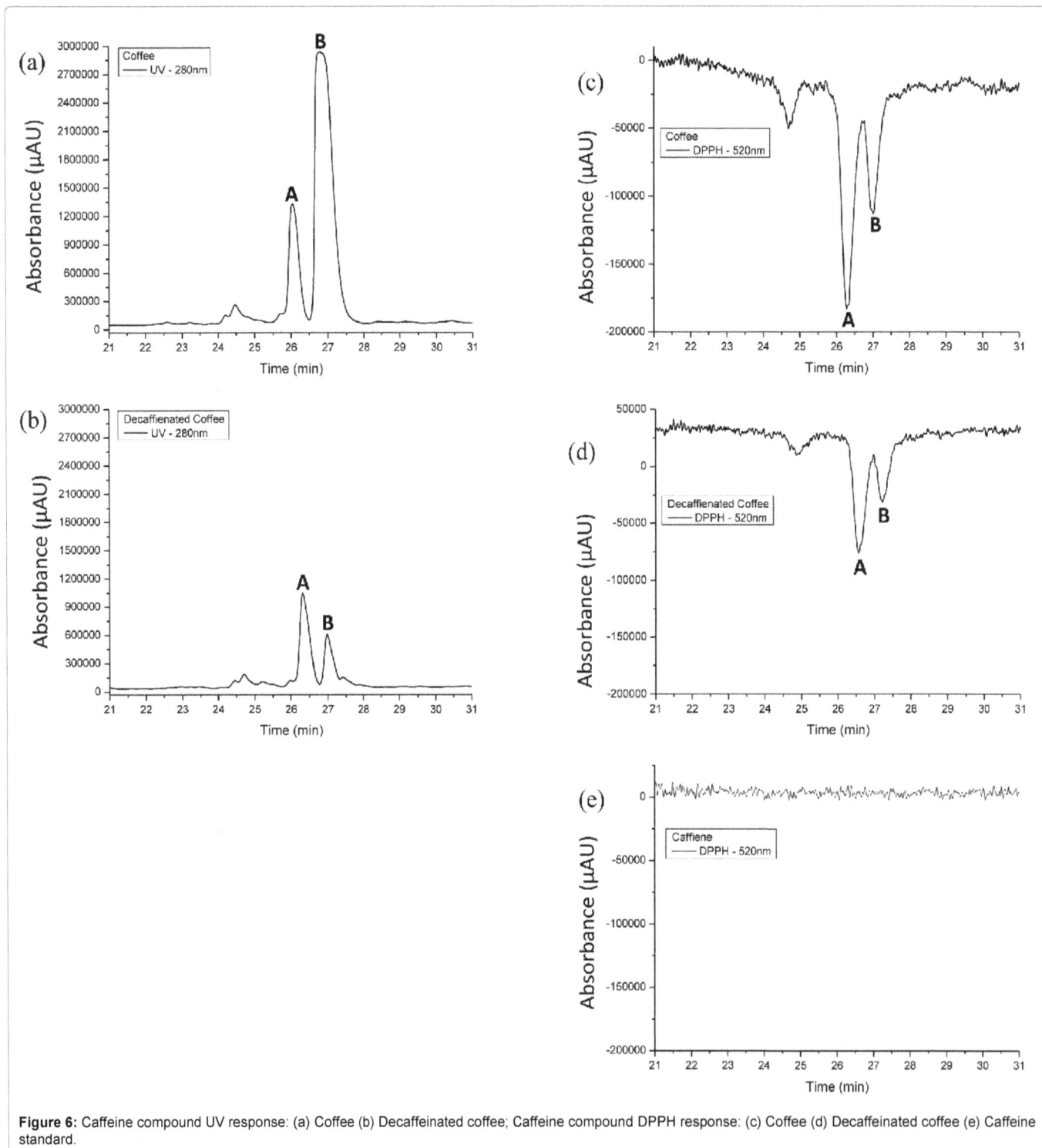

Figure 6: Caffeine compound UV response: (a) Coffee (b) Decaffeinated coffee; Caffeine compound DPPH response: (c) Coffee (d) Decaffeinated coffee (e) Caffeine standard.

The coffee sample was then analysed using 2DHPLC in a comprehensive heart-cutting approach where at every 0.4 minutes, a 200 μL aliquot from the first dimension was transferred into the second dimension for further separation. In total, 176 peaks were identified (compounds that responded to the UV set at 280 nm). Figure 8 illustrates the two-dimensional surface plot of the coffee sample. The identified peaks are represented by the white dots and the second dimension separation chromatograms of 3 cut times (1.6, 2.0 and 2.4 minutes) are also shown. It is evident that under 1DHPLC many peaks

were co-eluting and by expanding the separation space through the use of a 2DHPLC many underlying peaks were resolved. For example, in the 1D chromatogram (Figure 7) only a single peak - caffeine (indicated by the yellow circle) can be seen at around the 2.0-2.1 minute mark, however, when a 200 μL aliquot cut from this region was transferred to the second dimension for further separation an additional 17 peaks were resolved from under this band. This example illustrates the power of 2DHPLC compared to 1DHPLC for the analysis of complex samples, such as coffee. Through the use of different selectivity conditions,

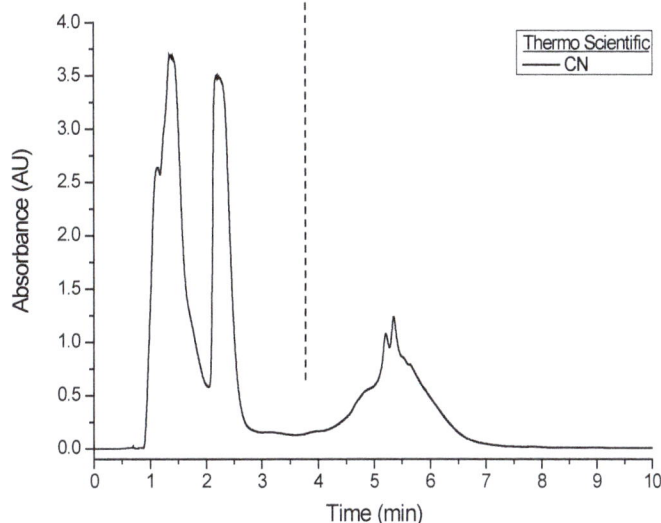

Figure 7: Uni-dimensional separation of Coffee on Cyano phase under gradient conditions of 20%min⁻¹ gradient, from 100% water to 100% methanol at 2 mLmin⁻¹, held for 4 minutes and returned to initial conditions in 1 minute (yellow circle indicates caffeine peak).

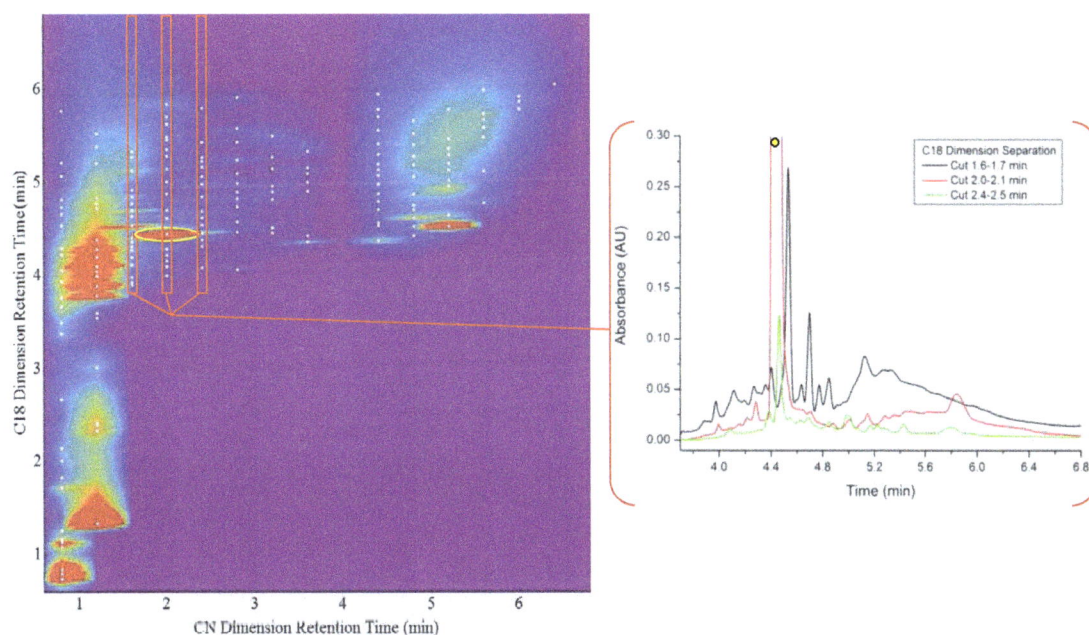

Figure 8: Two-dimensional separation of Coffee - 1DCN/2DC18 with heart-cut segments from CN phase on C18 separation (cut times 1.6-2.4 minutes). Both dimensions ran at a 20%min⁻¹ gradient, from 100% water to 100% methanol at 2 mLmin⁻¹ the held for 4 minutes and returned to initial conditions in 1 minute (yellow circle indicates caffeine peak).

the separation power and efficiency can be increased to enable more detailed sample characterisation. Clues to the presence of multiple co-eluting species were, in the first instance, provided by the multiplexed detection process. Hence demonstrating the synergistic benefits of the multidimensional/multi-detection process for comprehensive sample analysis.

Conclusions

The characterisation and profiling of coffee was achieved using two multidimensional approaches: (1) using multi-detection processes, in particular using a new column technology that enable multiplexed

detection, (2) multidimensional HPLC, which utilised two separation steps coupled in time. The first approach involved the detection selectivity for antioxidants *via* DPPH• detection, chromophore functionality through UV-Vis and information for molecular mass by MS. The coffee characterisation through detection selectivity was initially carried out in conventional mode, where for each detector a sample injection was made followed by the HPLC separation then detection. A significant amount of information was derived about compound functionalities of the sample. However, this process was time consuming and required greater amount of sample. Multiple detectors could be used in a series to reduce lengthy run-times whilst only requiring one sample injection, however, this would lead to band broadening, which can be substantial and diminish the HPLC efficiency. Furthermore, only one destructive detector can be used in any single detector series, in this study two of the detectors were destructive. The coupling of an AFT column in PSF mode gave the opportunity to run multiple detectors in multiplexed mode, where all detectors regardless of HPLC-detector requirements were run simultaneously in a single injection and separation of sample. Multiplexed detection with AFT columns provided a large amount of sample information, three detectors operated simultaneously within the run time of one injection and allowed for the exact match of retention time of peaks within each detection mode. Also, two of these detectors were destructive detectors. A follow up study would involve the coupling of 2DHPLC with AFT columns in multiplexed mode that encompasses the separation power of a multidimensional HPLC system as well as the versatility and efficiency of multiplexed detection, resulting in a comprehensive characterisation and profiling of a sample with 2DHPLC efficiency.

References

1. Robards K (2003) Strategies for the determination of bioactive phenols in plants, fruit and vegetables. J Chromatogr A 1000: 657-691.

2. Leo ML Nollet, Toldra F (2012) Food Analysis by HPLC, CRC Press, Taylor and Francis, Boca Raton, Florida.

3. Farah A (2012) Coffee, Wiley-Blackwell, pp: 21.

4. Clarke RJ, Macrae R (1985) Coffee: Chemistry, Springer, Netherlands.

5. Camenzuli M, Ritchie HJ, Dennis GR, Shalliker RA (2013) Reaction flow chromatography for rapid post column derivatisations: the analysis of antioxidants in natural products. J Chromatogr A 1303: 62-65.

6. Mnatsakanyan M, Goodie TA, Conlan XA, Francis PS, McDermott GP, et al. (2010) High performance liquid chromatography with two simultaneous on-line antioxidant assays: Evaluation and comparison of espresso coffees. Talanta 81: 837-842.

7. Mnatsakanyan M, Stevenson PG, Shock D, Conlan XA, Goodie TA (2010) The assessment of pi-pi selective stationary phases for two-dimensional HPLC analysis of foods: application to the analysis of coffee. Talanta 82: 1349-1357.

8. Mnatsakanyan M, Stevenson PG, Conlan XA, Francis PS, Goodie TA (2010) The analysis of café espresso using two-dimensional reversed phase–reversed phase high performance liquid chromatography with UV-absorbance and chemiluminescence detection. Talanta 82: 1358-1363.

9. Richelle M, Tavazzi I, Offord E (2001) Comparison of the antioxidant activity of commonly consumed polyphenolic beverages (coffee, cocoa, and tea) prepared per cup serving. J Agric Food Chem 49: 3438-3442.

10. Stevenson PG, Mnatsakanyan M, Francis AR, Shalliker RA (2010) A discussion on the process of defining 2-D separation selectivity. J Sep Sci 33: 1405-1413.

11. McDermott GP, Conlan XA, Noonan LK, Costin JW, Mnatsakanyan M (2011) Screening for antioxidants in complex matrices using high performance liquid chromatography with acidic potassium permanganate chemiluminescence detection. Anal Chim Acta 684: 134-141.

12. Camenzuli M, Ritchie HJ, Ladine JR, Shalliker RA (2011) The design of a new concept chromatography column. Analyst 136: 5127-5130.

13. Shalliker RA, Ritchie H (2014) Segmented flow and curtain flow chromatography: overcoming the wall effect and heterogeneous bed structures. J Chromatogr A 1335: 122-135.

14. Camenzuli M, Ritchie HJ, Shalliker RA (2013) Improving HPLC separation performance using parallel segmented flow chromatography. Microchem J 111: 3-7.

15. Shalliker RA, Camenzuli M, Pereira L, Ritchie HJ (2012) Parallel segmented flow chromatography columns: conventional analytical scale column formats presenting as a 'virtual' narrow bore column. J Chromatogr 1262: 64-69.

16. Camenzuli M, Goodie TA, Bassanese DN, Francis PS, Barnett NW, et al. (2012) The use of parallel segmented outlet flow columns for enhanced mass spectral sensitivity at high chromatographic flow rates. Rapid Commun Mass Spectrom 26: 943-949.

17. Camenzuli M, Ritchie HJ, Dennis GR, Shalliker RA (2013) Parallel segmented flow chromatography columns with multiplexed detection: An illustration using antioxidant screening of natural products. Microchem J 110: 726-730.

18. Camenzuli M, Ritchie HJ, Shalliker RA (2013) Evaluating active flow technology HPLC columns as a platform for multiplexed detection. Microchem J 110: 473-479.

19. Camenzuli M, Terry JM, Shalliker RA, Conlan XA, Barnett NW (2013) Parallel segmented outlet flow high performance liquid chromatography with multiplexed detection. Anal Chim Acta 803: 154-159.

20. Guiochon G (2006) The limits of the separation power of unidimensional column liquid chromatography. J Chromatogr A 1126: 6-49.

21. Kocic D, Pereira L, Foley D, Edge T, Mosely JA, et al. (2013) High through-put and highly sensitive liquid chromatography-tandem mass spectrometry separations of essential amino acids using active flow technology chromatography columns. J Chromatogr A 1305: 102-108.

22. Stevenson PG, Mnatsakanyan M, Guiochon G, Shalliker RA (2010) Peak picking and the assessment of separation performance in two-dimensional high performance liquid chromatography. Analyst 135: 1541-1550.

23. Felinger A (1998) Data Anlaysis and Signal Processing in Chromatography, Elsevier Science B.V, Netherlands, pp: 331-413.

24. Dugo P, Cacciola F, Kumm T, Dugo G, Mondello L (2008) Comprehensive multidimensional liquid chromatography: theory and applications. J Chromatogr A 1184: 353-368.

25. Gray MJ, Dennis GR, Slonecker PJ, Shalliker RA (2004) J Chromatogr A 1041: 101-110.

26. Stoll DR, Li X, Wang X, Carr PW, Porter SE, et al. (2007) Fast, comprehensive two-dimensional liquid chromatography. J Chromatogr A 1168: 3-43.

27. Toups EP, Gray MJ, Dennis GR, Reddy N, Wilson MA, et al. (2006) Multidimensional liquid chromatography for sample characterisation. J Sep Sci 29: 481-491.

27

ICP-OES Determination of Titanium (IV) in Marine and Wastewater Samples after Preconcentration onto Unloaded and Reagent Immobilized Polyurethane Foams Packed Columns

El-Shahawi MS *, Bashammakh AS, Alwael H, Al-Sibaai AA and Al-Ariqe HK

Department of Chemistry, Faculty of Science, King Abdulaziz University, P.O. Box 80203, Jeddah 21589, Saudi Arabia

Abstract

A novel method that utilizes untreated and polyurethane foam (PUF) physically immobilized with the reagent 4-(2-pyridylazo) resorcinol (PAR) or 2, 3, 5-triphenyl-2H-tetrazolium chloride (TZ$^+$Cl$^-$) as solid phase extractor packed column has been developed for preconcentration and subsequent determination of titanium(IV) ions in marine and wastewater samples. The method is based on retention of the titanium(IV) traces present in aqueous media at pH 3-4 onto the reagent treated PUF packed column, followed by recovery with HNO$_3$ (2.0 mol dm^{-3}) and subsequent ICP-OES determination. The uptake of titanium species onto the unloaded- and reagent impregnated PUF was fast and followed a first-order rate equation. Titanium sorption onto PUF followed Langmuir, Freundlich and Dubinin-Radushkevich (D-R) type isotherm models. Thus, a dual-mode sorption mechanism involving absorption related to "weak-base anion exchange" and an added component for "surface adsorption" seems a more likely retention model. PAR-immobilized PUF packed column has been applied successfully for complete collection of titanium(IV) species in fresh and wastewater samples at low level of titanium (<0.5 ng Ti mL^{-1}) at pH 3-4. The retained titanium species were then recovered (97-101%) from the packed column with HNO$_3$ and determined by ICP-OES. The proposed PUF packed column method was further applied for analysis of picomolar concentrations of dissolved Ti species in marine water.

Keywords: ICP-OES; Titanium (IV); 4-(2-Pyridylazo) resorcinol (PAR), Polyurethane foam packed column; Preconcentration; Marine and wastewater

Introduction

Titanium is well known for its excellent corrosion resistance, having ability to withstand attack by dilute H$_2$SO$_4$ and HCl or even moist chlorine. These properties make titanium highly resistant to the usual kinds of metal fatigue. Titanium alloys are principally used for aircrafts and missiles, where light weight strength and ability to withstand extremes of temperature are important. In nature, titanium exists in its most stable and common oxidation state (IV). Many organic compounds of titanium such as phthalates, oxalates, tetraethylate and butyltitanate are widely synthesized and used extensively. Titanium is naturally present in sea and ocean water (picomolar) and in food at only trace levels (μg Kg^{-1}) [1]. The presence of TiO$_2$ as an excipient in most pharmaceutical preparations, pigment, as a particulate food additive and in human intestinal tissue has been proposed that an abnormal response in the pathogenesis of Cohn's disease additive [2-4]. Hence, determination of titanium (IV) at trace levels in various samples is of paramount importance.

Few methods for the separation and subsequent determination of titanium in food and industrial wastewater samples are known [4]. A reversed-phase liquid chromatographic method for the determination of titanium with 5,5'-methylenedisalicylohydroxamic acid (MEDSHA) has been described by Bagur et al. [5]. Separation and determination of titanium (IV) at trace levels in different matrices including industrial wastewater have been reported employing solid phase extraction (SPE) [6-11]. SPE has several advantages e.g. simple operation, low cost, no time consuming, good selectivity, higher preconcentration factor, rapid phase separation and the ability to be combined with different modern analytical techniques [6-11].

Polyurethane foam (PUF) sorbent in the last four decades UF has been tested as an excellent support in reversed phase extraction chromatography, gas-solid and gas-liquid partition chromatography

[12-23]. The cellular structures and the available surface area of the PUF in both foamed and micro spherical forms make it suitable as an excellent extractor and as a column filing material with good capacity for firmly retaining various loading and extracting agents [21]. Preconcentration, separation and subsequent sensitive determination of titanium (IV) in various matrices at trace levels is of prime importance and necessitates. Thus, the present article is focused on: i. studying the retention profile of titanium (VI) onto PAR or TZ$^+$Cl$^-$ treated and untreated PUF; ii. Developing a convenient and low cost extraction procedure for separation and subsequent ICP-OES determination of titanium (IV) species in water samples employing PAR immobilized PUF in packed column and finally iii. Assigning the most probable sorption mechanism of Ti retention.

Experimental

Reagents and materials

All chemicals and solvents used were of analytical reagent grade and were used without further purification. Doubly deionized water was used throughout. Stock solution (0.1 % w/v) of BDH 4-(2-pyridylazo) 4-resorcinol (BDH, Poole, England) and 2, 3, 5-triphenyl-2H-

*Corresponding author: El-Shahawi MS, Department of Chemistry, Faculty of Science, Damietta University, Damietta, Egypt

E-mail: malsaeed@kau.edu.sa, mohammad_el_shahawi@yahoo.co.uk*

tetrazolium chloride (TZ$^+$Cl$^-$) were prepared by dissolving the required weight in few drops of ethanol and the solution was then completed with water. Stock solution (1 mg mL^{-1}) of titanium (IV) nitrate (BDH) was used for the preparation of diluted solutions (0.05-150 µg Ti mL^{-1}) in water. Stock solutions (1.0 % w/v) of BDH sodium dodecyl sulphate (SDS), tetrabutylammonium bromide, TBA$^+$Br, (BDH) and Triton X-100 (Analar) were prepared in water. Foam cubes (10-15 mm edge) of commercial white sheets of polyether type based PUF were cut from the foam sheets, purified and finally dried at 80°C [20]. The immobilized reagent PUF cubes were prepared by mixing the dried foam cubes with an aqueous solution (50 mL g^{-1} dry foam) containing PAR or TZ$^+$Cl$^-$ (0.1 % w/v) with efficient stirring for 30 min, squeezed and finally dried as reported [21]. The reagent PAR-immobilized PUF was packed in the glass columns (2 cm, 10 mm ID) by applying vacuum method of foam packing [13,18]. All containers used were pre-cleaned by soaking in HNO$_3$ (20 % w/v) and rinsed with de-ionized water before use.

Apparatus

A Perkin - Elmer (Lambda 25, Shelton, CT,USA) spectrophotometer (190-1100 nm) with 10 mm (path width) quartz cell was used for recording the electronic spectra and measuring the absorbance of the complex species of titanium. A Perkin Elmer Inductively Coupled Plasma-Optical Emission Spectrometry (ICP-OES, Optima 4100 DC Shelton, CT, USA) was operated at the optimum instrumental parameters for titanium determination before and after extraction with the reagent treated PUFs under the optimum operational parameters (Table 1). A Soxhlet extractor and a lab-line Orbital mechanical shaker (Corporation Precision Scientific, Chicago, USA) with a shaking rate in the range of 10-250 rpm were used were used for the foam purification and for shaking in batch experiments, respectively. De-ionized water was obtained from Milli-Q Plus system (Millipore, Bedford, MA, USA). A thermo Orion model 720 pH Meter (Thermo Fisher Scientific, MA, USA) was employed for pH measurements with absolute accuracy limits being defined by NIST buffers. Self-made columns (16, 10 and 2 cm h and 10 mm id) were used in flow experiments.

Recommended procedures

Batch experiments: In a dry 100 mL polyethylene bottle, an accurate weight (0.05 ± 0.001 g) of the unloaded or PAR immobilized foam cubes was shaken for 1 hr in a mechanical shaker with 50 mL of an aqueous solution containing titanium(IV) ions at 100 µg mL^{-1} concentration at 25 ± 0.1°C at the required pH employing Britton-Robinson (B-R) buffer (pH 2.5-11.5). After phase separation, the aliquot solution was analyzed for titanium by ICP-OES under the optimal operational parameters of the instrument (Table 1). The amount of titanium (IV) retained at equilibrium, q$_e$ on the foam cubes was then determined from the difference between the concentration of titanium (IV) measured in solution before (C_i) and after (C_f) shaking with the unloaded or reagent loaded foam cubes employing the equation:

Rf power, kW	1050
Plasma gas (Ar) flow rate, L min^{-1}	15
Auxiliary gas (Ar) flow rate, L min^{-1}	0.2
Nebulizer gas (Ar) flow rate, L min^{-1}	0.80
Pump rate, mL min^{-1}	1.5
Observation height, mm	15
Integration time	auto
Plasma view	radial
Wavelength, nm	Ti 334.940

Table 1: ICP-OES Operational parameters of titanium determination by ICP-OES.

$$q_e = \left(C_i - C_f\right) v / w \qquad (1)$$

where v and w are the volume (mL) of the aqueous solution and the weight (g) of the foam cubes, respectively. The extraction percentage (%E) and the distribution ratio (D) of the titanium sorption onto the unloaded and the reagent loaded foam were then calculated as reported [20]. Following these procedures, the influence of different parameters was critically investigated. The values of E and D are the average of three independent measurements and the precision in most cases was ± 2%.

Column experiments: An accurate weight (0.50 ± 0.01 g) of the unloaded or PAR-immobilized PUF was packed in a column using the vacuum method of foam packing [23]. The aqueous solutions (0.1-10 L) containing titanium (IV) at various concentrations (0.01-10 µg mL^{-1}) adjusted with Britton-Robinson buffer of pH = 3-4 were percolated through the foam column at 15-20 mL min^{-1} flow rate. The sample and blank foam packed columns were then washed with 100 mL of B-R buffer solution at the same pH. Complete retention of titanium (IV) took place on the unloaded and PAR or TZ$^+$Cl$^-$ -immobilized PUF as indicated from the ICP-OES determination of titanium in the effluent solutions. The sorbed titanium (IV) species were then recovered quantitatively from sorbent packed column with HNO$_3$ (50 mL, 2M) at 5 mL min^{-1} flow rate. Equal fractions (10 mL) of the eluate were then collected and the titanium species were then determined with ICP-OES.

Analysis of titanium (IV) in fresh, sea and wastewater samples: A 10 mL of concentrated HNO$_3$ was added to 0.1-0.5 L of tap, seawater or industrial wastewater samples. The mixture was boiled until the volume of the sample solution is reduced to two-third and the solution was allowed to cool down and filtered through a Whatman No 1 filter-paper. The pH of the solution was adjusted to 3-4 with B-R buffer. The water samples were then spiked with (or without) titanium (IV) at a total concentration 0.1-10 µg mL^{-1} and diluted to the original volume with water in a volumetric flask. The water samples were then percolated through the unloaded or PAR-loaded PUF packed column at 20-25 mL min^{-1} flow rate. The retained titanium (IV) species on the PUF column were then recovered with HNO$_3$ solution (25 mL, 2M) at 5 mL min^{-1} flow rate. Titanium (IV) concentration before extraction and after recovery in the eluate was finally determined by ICP-OES.

Results and Discussion

The concentrations of heavy metals in natural water and wastewater samples are frequently lower than their limit of detection (LOD). Therefore, recent years have seen an upsurge of interest in developing solid sorbents [22, 23] and exploring them for the separation and chemical speciation of metal ions [20]. PUF sorbent represents an inexpensive and efficient separation and preconcentration media with steadily versatile applications in inorganic and organic complex species [22].

Retention profile of titanium (IV) onto the PUF

The retention behavior of titanium (IV) ions from the aqueous solutions by the untreated and reagent-immobilized PUF cubes after 1 hr shaking at different pH employing B-R buffer (pH 2.5-11.5) were investigated. The uptake of titanium (IV) onto the unloaded and PAR loaded PUF increases on raising solution upto pH 3.1-4.0 and decreased markedly on increasing the solution pH (Figure 1). The observed behavior of titanium(IV) species sorption onto the unloaded and TZ$^+$Cl$^-$ loaded PUF at pH 3.1-4.0 is most likely attributed to the formation of binary and ternary complex ion associates of titanium (IV) with PUFs via protonated ether (-CH$_2$-O$^+$H-CH$_2$-) oxygen linkage of the PUF and TZ$^+$Cl$^-$ loaded PUF, respectively. The reagent

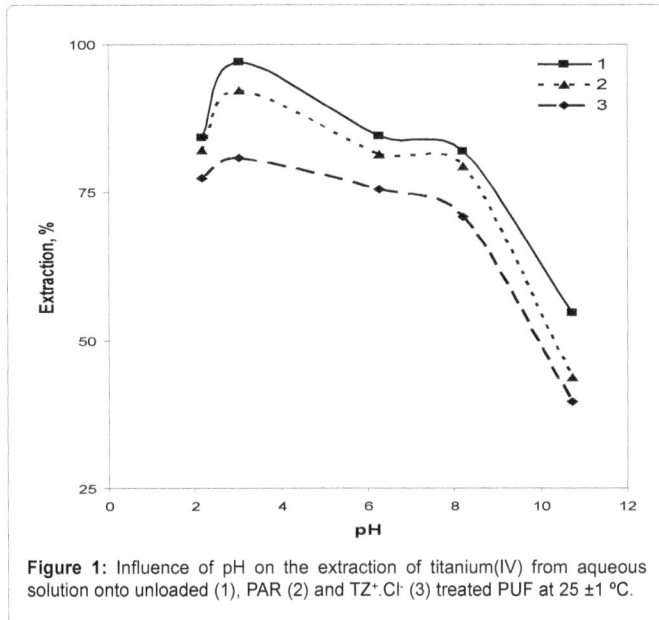

Figure 1: Influence of pH on the extraction of titanium(IV) from aqueous solution onto unloaded (1), PAR (2) and TZ$^+$.Cl$^-$ (3) treated PUF at 25 ±1 °C.

PAR-immobilized PUF showed also similar trend of sorption with better extraction performance (Figure 1). The low titanium sorption onto PUF at pH higher than pH 4.0 is attributed to the instability, hydrolysis or incomplete extraction of the produced ion associates of titanium (IV) with the unloaded PUF or TZ$^+$Cl$^-$. On the other hand, PAR molecules immobilized onto PUF are most likely complexed with titanium (IV) species in the solution via ligand exchange or ligand addition mechanism [2]. Thus, in the subsequent work, the aqueous solution was adjusted at pH 3-4.

A possible explanation of the observed trend involves a "weak-base anion exchanger" mechanism for the unloaded and TZ$^+$ Cl$^-$ treated PUF and "cation chelation or ligand addition extraction" mechanism for PAR-immobilized may be preceded [18,20] as follows:

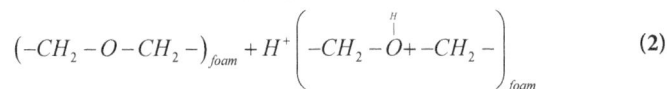

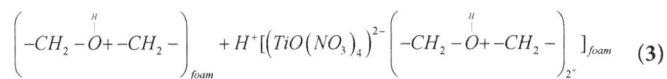

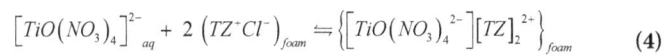

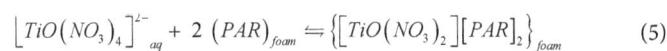

$$\left(-CH_2-O-CH_2-\right)_{foam} + H^+ \left(-CH_2-\overset{H}{O}+-CH_2-\right)_{foam} \quad (2)$$

$$\left(-CH_2-\overset{H}{O}+-CH_2-\right)_{foam} + H^+[\left(TiO\left(NO_3\right)_4\right)^{2-}\left(-CH_2-\overset{H}{O}+-CH_2-\right)]_{foam} \quad (3)$$

$$\left[TiO\left(NO_3\right)_4\right]^{2-}_{aq} + 2\left(TZ^-Cl^-\right)_{foam} \rightleftharpoons \left\{\left[TiO\left(NO_3\right)_4^{2-}\right]\left[TZ\right]_2^{2+}\right\}_{foam} \quad (4)$$

$$\left\lfloor TiO\left(NO_3\right)_4\right\rfloor^{2-}_{aq} + 2\left(PAR\right)_{foam} \rightleftharpoons \left\{\left[TiO\left(NO_3\right)_2\right]\left[PAR\right]_2\right\}_{foam} \quad (5)$$

The influence of shaking time (1-60 min) on the sorption of titanium (IV) from the aqueous solution at pH 3-4 onto the unloaded and TZ$^+$Cl$^-$ or PAR- immobilized PUF was investigated. The extraction was found fast, followed a first-order rate equation and the equilibrium was attained in ~10 min (Figure 2). Thus, a shaking of 20 min time was adopted in the subsequent work. The calculated half-life time ($t_{1/2}$) of the equilibrium sorption to reach 50% saturation of the sorption capacity of PUF loaded PAR and TZ$^+$Cl$^-$ and untreated PUF (Figure 2) was in the range 1.5-2 min. The values of E and D of titanium sorption onto unloaded and PAR-immobilized PUF were found better in comparison with TZ$^+$Cl$^-$ immobilized PUF. Thus, in the subsequent experiments, the unloaded and PAR immobilized foams were used.

The effect of cation size (Na$^+$, K$^+$, NH$_4^+$ and Ca^{2+}) as chloride

salts at concentration 0.05% w/v on titanium sorption by treated and untreated PUF was studied. In the unloaded foam, the uptake followed the sequence:

Na$^+$ > K$^+$ > NH$_4^+$ > Ca^{2+}

Different trend was observed for PAR or TZ$^+$Cl$^-$ immobilized, with reasonable increase (5-10 %) of titanium (IV) sorption in the presence of K$^+$ and the retention percentage followed the following order:

K$^+$ > NH$_4^+$ > Na$^+$ ≈ Ca^{2+}

The reduction of the repulsive forces between adjacent sorbed titanium (IV) complex ion associates in the unloaded PUF membrane may account for the trend observed [22,23]. Thus, the ion-dipole interaction of NH$_4^+$ with the oxygen sites of the PUF is not the predominating factor in the extraction step. The added K$^+$ ions is most likely reduce the number of water molecules available to solvate the titanium ions which would therefore, be forced out of the solvent phase onto the PUF. Thus, "weak-base anion exchange and "cation chelation or "ligand addition extraction" are the most probable sorption mechanism for the sorbent.

The influence of the surfactants SDS, TBA$^+$Br and Triton-X 100 on titanium (IV) sorption from aqueous solution onto the unloaded- and loaded PUF was investigated. Titanium sorption onto PUF sorbent increased in the presence of SDS (0.1 % w/v) and leveled off on raising the surfactant concentration. This behavior is most likely attributed to the increase of the solution viscosity leading to progressive change in the physical properties of the micro environment of the produced complex ion associates of unloaded or TZ$^+$Cl$^-$ treated PUF and the chelate formed with PAR-treated PUF [18,20], respectively. The increase in the solution viscosity enhances the dissociation and/or the formation of aggregate complexes with low diffusion constants [24,25]. The competition between the surfactant and the anionic complex of titanium (IV) may also predominate in the observed trend. Also, the surfactant may reacts directly with the anionic complex of titanium and this may retard the extraction process [25].

Sorption isotherms of titanium (IV) by PUF

The sorption profile of titanium (IV) from the bulk aqueous solution onto the untreated and PAR treated PUF was determined over

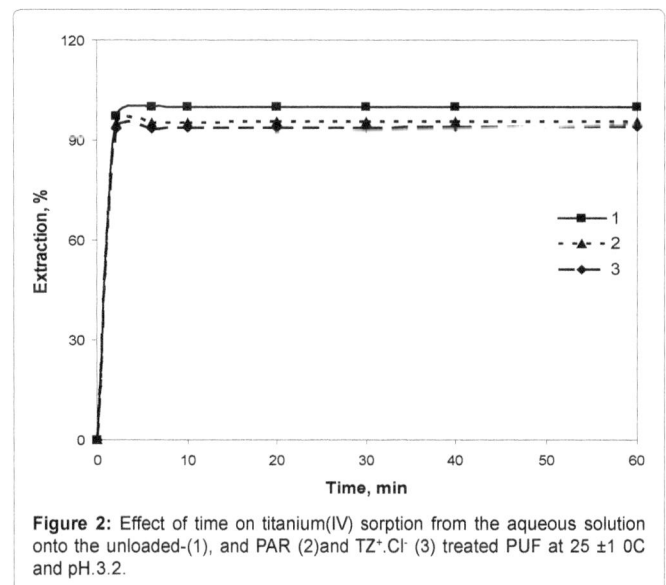

Figure 2: Effect of time on titanium(IV) sorption from the aqueous solution onto the unloaded-(1), and PAR (2)and TZ$^+$.Cl$^-$ (3) treated PUF at 25 ±1 0C and pH.3.2.

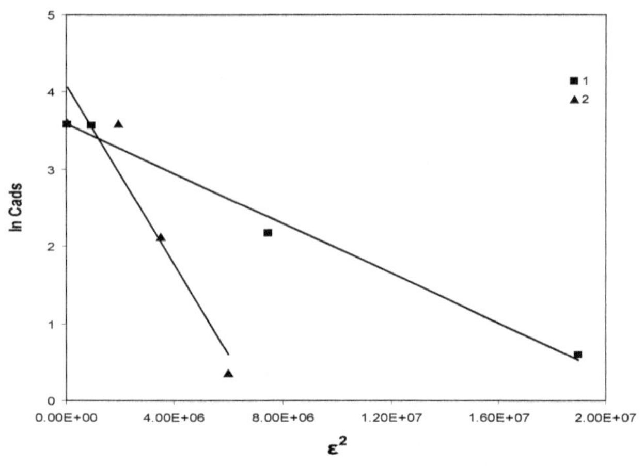

Figure 3: Dubinin-Radushkevich (D-R) plots for titanium sorption (0.05-150 µg mL^{-1}) onto unloaded (1), PAR (2) loaded PUF at 25 ±1 0C and pH.3.2.

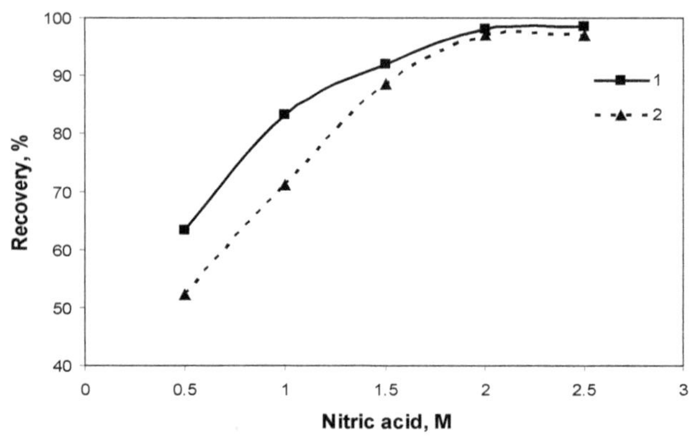

Figure 4: Influence of HNO$_3$ concentration on the recovery of titanium(IV) from the unloaded (1) and PAR (2) immobilized PUF packed (0.3 ±0.01 g) columns at room temperature and 5 mL min^{-1} flow rate.

a wide range of concentrations. In the aqueous, the amount of titanium (IV) retained onto unloaded and reagent treated PUF varied linearly with the corresponding amount of titanium (IV) at low or moderate Ti concentration. Thus, the titanium (IV) sorption onto the PUF were subjected to Freundlich [26] and Dubinin-Radushkevich [27] isotherms over a wide range of equilibrium concentrations. The Freundlich model [26] is expressed as follows:

$$\log C_{ads} = \log A + \frac{1}{n}\log C_e \qquad (6)$$

where C_e is the equilibrium concentration (M) of titanium(IV) in solution, C_{ads} is the sorbed titanium(IV) ions concentration (mmol g^{-1}) and A and $1/n$ are the Freundlich parameters related to the maximum sorption capacity of solute (mol g^{-1}). The values of A and $1/n$, computed from the intercepts and slopes of linear plots of log C$_{ads}$ *versus* log C$_e$ over the entire range of titanium(IV) concentrations (0.05-150 µg mL^{-1}) were 0.0156 ± 0.004, 0.0173 ± 0.003 mol.g^{-1} and 0.571 ± 0.07, 0.642 ± 0.18 onto the PUF, respectively. The value of $1/n < 1$ indicated that, the isotherms do not predict any saturation of the surface of the solid sorbent by the adsorbate and the sorption capacity is slightly reduced at lower concentration.

The linear form of Dubinin-Radushkevich (D-R) model [27]

postulated within the adsorption space close to the PAR-Treated PUF adsorbent surface is expressed as follows:

$$\ln C_{ads} = \ln K_{DR}\beta\varepsilon^2 \qquad (7)$$

where K_{DR} is the maximum amount of titanium (IV) retained onto PAR treated PUF, β is a constant related to the energy of the transfer of the solute from the bulk solution to the solid sorbent and ε is Polanyi potential which is given by the equation:

$$\varepsilon = RT\ln(1+\frac{1}{C_e}) \qquad (8)$$

where R is the gas constant (kJ mol^{-1} K^{-1}) and T is the absolute temperature (298 K) in Kelvin. The plot of ln C$_{ads}$ *vs.* ε^2 was linear (Figure 3) indicating that the D-R isotherm is obeyed for titanium (IV) sorption onto the sorbent over the entire concentration range. The computed values of β and K_{DR} from the slope and intercept of Figure 3 were found in the range 0.0027-0.0032 mmol^2 kJ^{-2} and 105-127 µmol g^{-1}, respectively. These results and the data reported earlier [18-20] suggested a dual sorption mechanism involving absorption related to "weak-base anion exchange" or "cation chelation" and an added component for "surface adsorption" mechanism for the uptake of titanium (IV) ions by unloaded and PAR immobilized PUF. This model can be expressed as follows [20]:

$$C_r = C_{abs} + C_{ads} = DC_{aq} + SK_LC_{aq} / 1 + K_LC_{aq} \qquad (9)$$

where C_r and C_{aq} are the equilibrium concentrations of titanium (IV) ions onto the PUF and in solution, respectively. C_{abs} and C_{ads} are the equilibrium titanium (IV) ions onto the PUF as an absorbed and adsorbed species, respectively, S and K_L are the saturation value for the Langmuir adsorption, the distribution coefficient and the Langmuir constant. This equation can be solved for D as reported earlier [20,21] as follows:

$$D = K_L + SK_L / 1 + K_LC_{aq} \qquad (10)$$

The D values are dependent on the titanium ions concentration confirming the proposed mechanism. These results suggested the use of PAR loaded PUF in flow mode for complete collection, recovery and subsequent ICP-OES Ti determination in water.

Chromatographic separation of titanium (IV)

Preliminary investigation on the use of PAR -PUF packed column for the collection of titanium (IV) ions from aqueous media has indicated that, the column performance towards titanium ions is good. Thus, aqueous solutions of deionized and tap water samples (1.0 L) containing various concentrations (0.01-10 µg mL^{-1}) of titanium(IV) at pH 3-4 were percolated through unloaded- and PAR treated PUF packed columns at 20-25 mL min^{-1} flow rate. Analysis of titanium in the effluent solution versus a reagent blank indicated complete (98 ± 3.1%) retention of titanium. Hence, a series of various eluating agents e.g. HNO$_3$, EDTA, NAF and HCl has been tested for recovery of titanium (IV) from the PUF packed column. Nitric acid (50 mL, 2.0M) was found suitable for complete recovery of titanium (IV) from the packed column at 5.0 ml/min flow rate. The obtainable results are demonstrated in (Figure 4).

The performance of the developed unloaded and PAR-immobilized PUF columns was determined by passing 0.5 L (10 µg mL^{-1}) of titanium (IV) solution at pH 3-4 through the PUF packed column at 20 mL min^{-1} flow rate. Complete sorption of titanium (IV) onto PAR loaded foam column took place at 20 mL min^{-1}. The retained titanium (IV) species were recovered with 50 mL HNO$_3$ (2.0M). The results are demonstrated in Figure 5. The height equivalent to theoretical plates (HETP) and

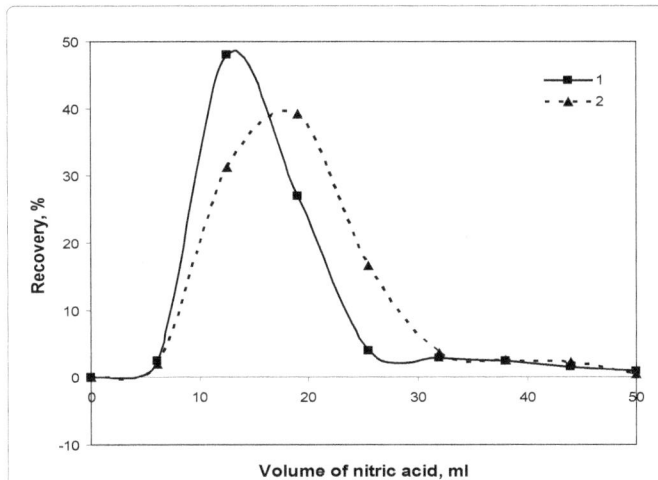

Figure 5: Elution curves of titanium(IV) from the unloaded (1) and PAR (2) immobilized PUF packed column (0.3 ± 0.01 g) using HNO₃ (2M) at 5 mL min⁻¹ flow rate.

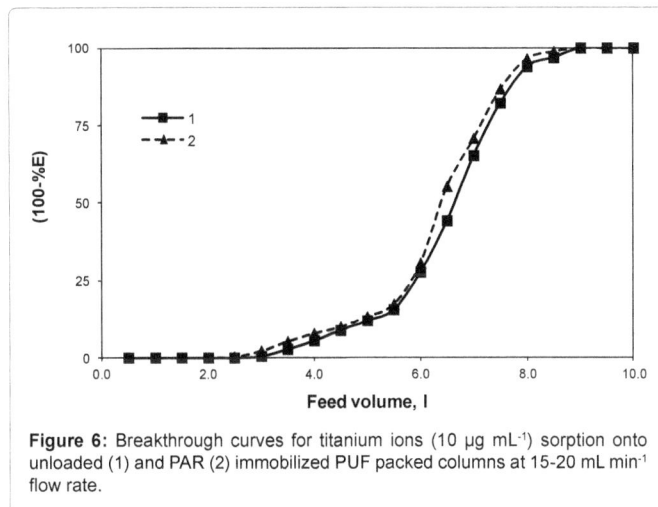

Figure 6: Breakthrough curves for titanium ions (10 μg mL⁻¹) sorption onto unloaded (1) and PAR (2) immobilized PUF packed columns at 15-20 mL min⁻¹ flow rate.

Titanium, μg mL⁻¹	Recovery, %*			
	Unloaded		PAR-immobilized	
	A	B	A	B
0.05	102 ± 2	100 ± 2	101 ± 2.8	101 ±2.6
0.5	99 ± 1.5	99 ± 1.5	98.2 ± 1.6	98.2 ±1.8
5	98 ± 1	98 ± 1	98 ± 2.4	98.5 +1 7
10	97.5 ± 1	97 ± 1	96.5 ± 2.9	97 ±2.1

*Average of five measurements ±relative standard deviation

Table 2: Results of analysis of tap (A) and wastewater (B) samples using the developed unloaded and PAR-immobilized PUF packed columns.

the number of plates (N) were then calculated from the elution curves (Figure 5) employing Gluenkauf equation [22]:

$$N = \frac{8V_{max}^2}{w_e} = \frac{L}{HETP} \tag{11}$$

where V_{max} is the volume of eluting at maximum elution of solute, w_e is the width of the chromatogram peak at (1/e) times maximum recovery of solute and L is the length of the PUf bed in the packed column. HETP and N values were found in the range 0.5-0.75 ± 0.04 mm and 80 ± 4, respectively. The HETP and N values evaluated from the breakthrough capacity curves (Figure 6) were found also in the range

0.74 ± 0.01 mm; 78 ± 3 (n=5) for titanium retention onto unloaded PUF and 0.51 ± 0.02 mm and 83 ± 2 for PAR-loaded PUF packed column. The critical capacities of titanium (IV) ions sorption onto the unloaded and loaded foam packed column calculated from Figure 6 were 56.3 ± 2.2 and 60.4 ± 1 mg titanium per gram PUF, respectively at 20-25 mL min⁻¹ flow rate. The calculated capacity from Par-PUF packed column was higher than capacity calculated from batch mode (40.4 ± 1 mg/g PUF).

Interference Study

The analytical utility of the PAR-immobilized PUF packed column for the retention and recovery of titanium ions (10 μg mL⁻¹) from aqueous solutions (100 mL) was tested in the presence of a relatively high excess (100-1000 times) of the diverse ions (Fe^{3+}, Al^{3+}, Ca^{2+}, Mg^{2+}, Cr^{3+}, V^{4+}, Ni^{2+}, Mn^{2+}, Co^{2+}, Cu^{2+}, Zn^{2+}, Hg^{2+}, and Cd^{2+}) relevant to waste water. The tolerance less than ± 2% change in the recovery of titanium ions is considered free from interference. Good extraction efficiency (>97 ± 3%) for titanium (IV) ions sorption and recovery were achieved successfully in the presence of the investigated divers ions except Al and Fe. Addition of NaF (100 μg mL⁻¹) prevented the effect of both Al and Fe.

Analytical applications

Analysis of titanium (IV) ions in tap- and wastewater samples: The validity of the proposed unloaded and PAR-loaded PUF packed columns for the collection, recovery and ICP-AES determination of titanium (IV) ions in tap and wastewater samples was tested as described in the experimental procedures. Low concentrations (0.01-1.0 μg mL⁻¹) of the spiked titanium ions in tap and/or wastewater samples were retained quantitatively as indicated from ICP-OES analysis of Ti in the effluent solutions. The retained species of titanium on the unloaded and PAR-PUF columns were then recovered by HNO₃ (25 mL, 2M) at 3-5 mL min⁻¹ flow rate and subsequently determined by ICP-OES. The results are summarized in Table 2. Satisfactory results (96.5-102 ± 2.9%) for the recovery of titanium (IV) ions in tap and wastewater samples were achieved by the proposed PUF packed columns and the standard ICP-OES. The results revealed absence of Ti in the tested samples in good agreement with the data obtained by ICP-OES method. On these bases, titanium ions are not detectable in tap and the wastewater samples.

Analysis of titanium (IV) ions in seawater samples: Satisfactory results (97 ± 2.7 %) for the preconcentration, recovery and subsequent ICP-AES determination of very low concentration of titanium (≤ 0.5μgL⁻¹) spiked in Red seawater samples (Jeddah, Saudi Arabia) by the proposed method was attempted. The titanium (IV) concentration (0.05 μgL⁻¹) obtained by the proposed packed column was in acceptable agreement with the data achieved by ICP-mass spectrometry (ICP-MS) and cathodic voltamertric [28] methods.

Conclusion

The present paper demonstrates application of PAR-immobilized PUF solid sorbent packed column for complete removal of titanium (IV) from wastewater samples and subsequent ICP-OES determination. The method is simple to operate and low cost then the conventional method. The PAR-PUF packed column was found stable and it can be reused for many times, without decrease in the extraction and recovery percentage of titanium (over 95%). Work is still continuing for online chemical speciation of inorganic titanium (III) & (IV) and organo-titanium (IV) compounds using PAR-immobilized PUF packed column and ICP-OES.

References

1. National Toxicology Program (1979) Bioassay of Titanium Dioxide for Possible Carcinogenicity. Natl Cancer Inst Carcinog Tech Rep Ser 97: 1-123.

2. Smith HM (2002) High Performance Pigment. Wiley-VCH Verlag Gmbh & Co KGaA.

3. Powell JJ, Harvey RSJ, Thompson RPH, Ashwood P, Wolstencroft R, et al. (2000) Immune potentiation of ultrafine dietary particles in normal subjects and patients with inflammatory bowel disease. Journal of Autoimmun 14: 99-105.

4. Satake M, Nagahiro T, Puri BK (1992) Column preconcentration of titanium in aluminium and zinc alloys with oxalic acid-ascorbic acid and the ion-pair of sodium 1,2-dihydroxybenzene-3,5-disulphonic acid and tetradecyldimethylbenzylammonium chloride supported on naphthalene using spectrometry. Talanta 39: 1349-1354.

5. Bagur G, Sánchez-Viñas M, Gázquez D (1997) Determination of Titanium (IV) as an Additive in Organic Matrices by Reversed-Phase High-Performance Liquid Chromatography with 5,5â€²-Methylenedisalicylohydroxamic Acid. Journal of Chromatographic Sciences 35: 131-134.

6. Orians KJ, Boyle EA (1993) Determination of picomolar concentrations of titanium, gallium and indium in sea-water by inductively-coupled plasma-mass spectrometry following an 8-hydroxyquinoline chelating resin preconcentration. Anal Chim Acta 282: 63-74.

7. Gong B, Li X, Wang F, Chang X (2000) Synthesis of spherical macroporous epoxy-dicyandiamide chelating resin and properties of concentration and separation of trace metal ions from samples. Talanta 52: 217-223.

8. Chang X, Yang X, Wei X, Wu K (2001) Efficiency and mechanism of new poly(acryl-phenylamidrazone phenylhydrazide) chelating fiber for adsorbing trace Ga, In, Bi, V and Ti from solution. Anal Chim Acta 450: 231-238.

9. Anthemidis AN, Zachariadis GA, Stratis JA (2003) Gallium trace on-line preconcentration/separation and determination using a polyurethane foam mini-column and flame atomic absorption spectrometry. Application in aluminum alloys, natural waters and urine. Talanta 60: 929-936.

10. Suvarapu LN, SEO YK, Baek SO (2012) Spectrophotometric Determination of Titanium(IV) by Using 3,4-Dihydroxybenzaldehydeisonicotinoyl-hydrazone(3,4-DHBINH) as a Chromogenic Agent Chem Sci Trans.1: 171-179.

11. Lobinski R, Marczenko,Z (1997) Spectrochemical Trace Analysis for Metals and Metalloids", Wilson & Wilson's Comprehensive Analytical Chemistry, Elsevier, Amsterdam 236.

12. Bashammakh AS, Bahaffi SO, Al-Shareef FM, El-Shahawi MS (2009) Development of an Analytical Method for Trace Gold in Aqueous Solution using Polyurethane Form Sorbents: Kinetic and Thermodynamic Characteristic of Gold (III) Sorption. Anal Sci (Japan) 25: 413-418.

13. El-Shahawi MS, Othman MA, Abdel-Fadeel MA (2005) Kinetics, thermodynamic and chromatographic behaviour of the uranyl ions sorption from aqueous thiocyanate media onto polyurethane foams. Anal Chim Acta 546: 221-228.

14. Saeed MM, Ahmed M (2004) Retention, kinetics and thermodynamics pro-file of cadmium adsorption from iodide medium onto polyurethane foam and its separation from zinc bulk. Anal Chem Acta 525: 289–297.

15. Marchisio PF, Sales A, Cerutti S, Marchevski E, Martinez LD (2005) On-line preconcentration/determination of lead in Ilex paraguariensis samples (mate tea) using polyurethane foam as filter and USN-ICP-OES. J Hazardous Materials 124: 113-118.

16. Hasany SM, Saeed MM, Ahmed M (2001) Sorption of traces of silver ions onto polyurethane foam from acidic solution. Talanta 54: 89-98.

17. Bhaskar M, Aruna P, Jeevan RG, Radhakrishnan G (2004) β-Cyclodextrin-polyurethane polymer as solid-phase extraction material for the analysis of carcinogenic aromatic amines. Anal Chim Acta 509: 39–45.

18. El-Shahawi MS, Nassif HA (2003) Retention and thermodynamic characteristics of mercury(II) complexes onto polyurethane foams Anal. Chim. Acta 481: 29-39.

19. Abou-Mesalam MM, El-Naggar IM, Abdel-Hai MS, El-Shahawi MS (2003) Use of open-cell resilient polyurethane foam loaded with crown ether for the preconcentration of uranium from aqueous solutions. J Radioanal Nucl Chem 258: 619-625.

20. El-Shahawi MS, El-Sonbati MA (2005) Retention profile, kinetics and sequential determination of selenium(IV) and (VI) employing 4,4'-dichlorodithizone immobilized-polyurethane foams. Talanta 67: 806-815.

21. Palágyi S, Braun T (1992) Unloaded polyether type polyurethane foams as solid extractants for trace elements. J Radioanal Nucl Chem 163: 69-79.

22. Braun T, Navratil JD, Farag AB (1985) Polyurethane Foam Sorbents in Separation Science, CRC Press Inc., Boca Raton, Florida, USA.

23. Cassella RJ, Bitencourt DT, Branco AG, Ferreira SLC, de Carvalho MS (1999) On-line preconcentration system for flame atomic absorption spectrometry using unloaded polyurethane foam: determination of zinc in waters and biological materials. J Anal At Spectrom 14: 1749-1753.

24. Condamines N, Musikas C (1992) The Extraction by N.N-Dialkylamides. II. Extraction of Actinide Cations. Sol Ext Ion Exchange 10: 69-100.

25. Gupta VK, Srivastava SK, Tyagi R (2000) Design parameters for the treatment of phenolic wastes by carbon columns (obtained from fertilizer waste material). Water Res 34: 1534-1550.

26. Freundlich H (1926) Capillary and colloid chemistry, Methuen, London.

27. Dubinin M, Radushkevich LV (1974) Procc Acad Sci USSR. Phys Chem Soc 55: 331.

28. Yokoi K and van den Berg CMG (1992) Simultaneous determination of titanium and molybdenum in natural waters by catalytic cathodic stripping voltammetry. Anal Chim Acta 257: 293-299.

28

Study on Quantitative Structure-Retention Relationships (QSRR) for Oxygen-Containing Organic Compounds Based on Gene Expression Programming (GEP)

Zhang X[1,2]*, Shi L[2], Ding L[2], Sun Z[2], Song L[2], Qu H[3] and Sun T[1]

[1]College of Sciences, Northeastern University, 110004 Shenyang, Liaoning, People's Republic of China
[2]Liaoning Key Laboratory of Petrochemical Engineering, Liaoning Shihua University, 113001 Fushun, Liaoning, People's Republic of China
[3]Research Institute of PetroChina, Fushun Petrochemical Company, 113004 Fushun, Liaoning, China

Abstract

Gene Expression Programming (GEP) is a novel genetic algorithm, a highly effective, stable random searching method. We take GEP to make models of Quantitative Structure-Retention Relationship (QSRR) for a series of oxygen-containing organic compounds of GC retention index, and compare the predictive results with Artificial Neural Network (ANN) and Multiple Linear Regression (MLR). The correlation coefficient on OV-1 column is 0.9919, 0.9891 and 0.9911 for GEP, ANN and MLR respectively, on SE-54 column is 0.9955, 0.9892, and 0.9917. It is shown that the predicted results by GEP are in good agreement with experimental ones, better than those of ANN and MLR.

Keywords: Gene Expression Programming (GEP); Oxygen-containing organic compounds; Artificial Neural Network (ANN); Quantitative Structure-Retention Relationship (QSRR)

Introduction

Chromatography in itself is not an accurate analytical technique, but rather a separation one. The identification of oxygen-containing organic compounds can be made with the method of gas chromatographic peak in comparison with that of a standard sample of each compound. Because samples of pure compounds are not always available, it is important to develop QSRR that can efficiently predict retention parameters by using theoretical descriptors computed from chemical structure.

Quantitative Structure-Retention Relationships (QSRR) [1] establish the relationship between a chemical structure and its chromatographic retention value, which has been demonstrated to be a powerful tool for the investigation of chromatographic parameters. The main advantage of QSRR is the ability to distinguish in quantitative theoretical terms, packing materials of different chemical nature of the organic ligand and/or organic or inorganic support [2], Furthermore, it can be of valuable assistance in the prognosis of the behavior of new molecules, even before they are actually synthesized [3].

An important property that has been extensively studied in QSRR is the chromatographic retention index. The retention index is a generally accepted type of data used for the identification of chemical compounds by gas chromatography. A retention index is a continuous quantitative variable that relates the retention of a solute to the retention of a set of standard compounds. Retention indices are much less dependent on experimental factors (e.g., Temperature, flow, column, length etc.) than retention times. While Kovats retention indice [4] have linear collerations with column temuprature. And they were obtained by the logarithmic interpolation method.

QSRR on the Kovats retention indices have been reported for different types of organic compounds. The Kovats retention index is the most popular dependent variable in QSRR studies because of its reproducibility and accuracy. In many cases, the precision and accuracy of the QSRR models are not sufficient for identification purposes; still the models are useful to elucidate retention mechanisms, to optimize the separation of complex mixtures or to prepare experimental designs.

Topological descriptors computed on the basis of molecular graph are easy to be calculated with present computing facilities. Due to the simplicity and efficiency of graph-theoretical approaches, we take novel polarizability effect index (PEI), odd-even index (OEI), the sum eight values X_{1CH} of every C-H bond adjacency matrix S_{x1CH}.

An interesting and increasing application of QSRR is to test various chemometric methods from multiple linear regression (MLR) methods to Artifical neural network (ANN) methods. Multiple linear regression (MLR) is without doubt the most frequently applied technique in building QSRR models.

Gene Expression Programming (GEP) is a new evolutionary algorithm that evolves from computer programs (they can take many forms: mathematical expressions, neural networks, decision trees, polynomial constructs, logical expressions, and so on). The computer programs of GEP, irrespective of their complexity, are all encoded in liner chromosomes. Then the liner chromosomes are expressed or translated into expression trees (the branched structures). Thus, in GEP, the genotype (the liner chromosomes) and the phenotype (the expression trees) are different entities (both structurally and functionally), and because of this apparently trivial fact, this new evolutionary system can finally make a difference, successfully assisting researchers in the design of robust and accurate computer models [5].

The aim of the present research is to develop a general model capable of predicting the gas chromatographic retention index of oxygen-containing organic compounds based on GEP, compared with the result predicted by traditional linear MLR and another powerful non-linear ANN method.

*Corresponding author: Zhang Xiaotong, College of Sciences, Northeastern University, 110004 Shenyang, Liaoning, People's Republic of China
E-mail: xt_zhang2015@163.com

Materials and Methods

Data set

The Kovats retention index of 91 molecules (include esters, ketones, and alcohols) taken from reference [6] were presented in Table 1. Kovat's retention index of all compounds was obtained under the same conditions on two stationary phases: OV-1 (dimethylpolysiloxane) and SE-54(5% phenyl -95% dimethylpolysiloxane) 74 molecules were used as training set for model generation and 17 molecules were used as test set for model prediction. The corresponding experimental and predicted values of the RI for all the molecules studied in this work are shown in Table 1.

GEP theory

Gene Expression Programming (GEP) was first proposed formally by Candida Ferreira in 2001. It was an elegant and efficient solution to expression-mutation problems. GEP, which is an extraordinarily powerful tool, is a subset of Genetic Algorithms, except it uses genomes whose strings of numbers represent symbols. GEP-an evolutionary algorithm inherits both the evolutionary simplicity of Genetic Algorithms (GA) and the expressional power in Genetic Programming (GP) by utilizing a genotype/phenotype representation system. The string of symbols can further represent equations, grammars, or logical mappings.

Ferreira [5] proposes the use of a set of genetic operators: Replication, Mutation, IS Transposition, RIS Transposition, Gene Transposition, 1-Point Recombination, 2-Point Recombination, Gene Recombination. As Ferreira comments, the advantages of a Genetic Representation like the one in GEP are simple entities: linear, compact, relatively small, easy to manipulate genetically. The genetic operators applied to them are less restricted than those used in GP [5].

Fortunately for us, in GEP, thanks to the simple rules that determine the structure of expression trees and their interactions, it is possible to infer immediately the phenotype given the sequence of a gene. It is easy for a computer program to follow these three rules while performing mutations, and it never has to check whether the resulting expression has valid syntax. By allowing a broad range of mutations, the process can efficiently explore a high dimensional space, and the expressions can change in size as functions are replaced by terminals and terminals by functions.

GEP are evolutionary tools inspired in the Darwinian principle of natural selection and survival of the fittest individual and uses populations of candidate solutions to a given problem in order to evolve new ones. These methods use an initial random population and apply genetic operators to this population until the algorithm finds an individual that satisfies some termination criteria. The evolving populations undergo selective pressure and their individuals are submitted to genetic operators.

Gene representation: GEP genes are composed of a head and a tail. The head contains symbols that represent both functions (elements from the function set F) and terminals (elements from the terminal set T), whereas the tail contains only terminals. Therefore, two different alphabets occur at different regions within a gene. For each problem, the length of the head h is chosen, whereas the length of the tail t is a function of h and the number of arguments of the function with the most arguments n, and is evaluated by the equation:

$$t=h(n-1)+1 \quad (1)$$

Consider a gene composed of {Q, *, /, -, +, a, b}. In this case $n=2$.

For instance, for $h=15$ and $t=16$, the length of the gene is $10+11=21$. One such gene is shown below (the tail is shown in bold):

012345678901234567890123456789

*b+a-aQab+//+b+**babbabbbababbaaa** (2)

It codes for the following ET:

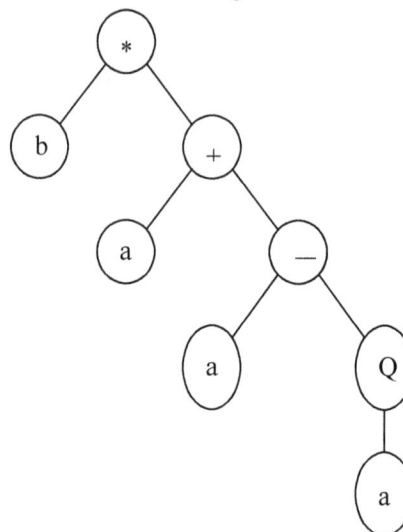

A K-expression can be mapped into an expression tree (ET) following a first-order procedure. A branch of the ET stops growing when the last node in this branch is a terminal. For example, the ET shown above corresponds to chromosome (2). In this case, the open reading frames end at position 7, whereas the gene ends at position 30.

Chromosomes in GEP are usually composed of more than one gene of equal length. For each problem or run, the number of genes, as well as the length of the head, is chosen. Each gene codes for a sub-ET and the sub-ETs can be linked by pre-defined rules forming a more complex multi-subunit ET.

Selection method and genetic operators: References [3] suggests there is no difference between different selection methods. It is strongly advised to use a simple elitism in any GEP implementation. The elitism means copying the best (or few best) individual to the offspring population without modifying them. GEP uses the well-known roulette-wheel method for selecting individuals.

In GEP, individuals were selected according to fitness by roulette wheel sampling coupled with the cloning of the best individual. The fitter the individual is, the higher the probability of leaving more offspring. Thus, during replication the genomes of the selected individuals are copied as many times as the outcome of the roulette. The roulette is spun as many times as there are individuals in the population, always maintaining the same population size.

GEP uses simple elitism of the best individual of a generation, preserving it for the next one. Replication is an operation that aims to preserve several good individuals of the current generation for the next one. In fact, this is a do-nothing probabilistic operation that takes place during selection (using the roulette-wheel method), and replicated individuals will be subjected to the action of the genetic operators. The mutation operator aims to introduce random modifications into a given chromosome. A particularity of this operator is that some integrity rules must be obeyed so as to avoid syntactically invalid individuals. In the head of a gene, both terminals and functions are permitted (except

No.	Compounds	Rlov-1$_{exp}$	Rlse-54$_{exp}$	Rlov-1$_{pre}$	Rlse-54$_{pre}$	Rlov-1$_{pre}$	Rlse-54$_{pre}$	Rlov-1$_{pre}$	Rlse-54$_{pre}$
				MLR		ANN		GEP	
Training									
1	3,3-Dimethyl-1-butanol	778.77	763.63	776.89	788.80	782.22	801.60	760.36	761.76
2	3-Methyl-3-hexanol	826.62	841.11	814.26	836.41	831.95	835.48	815.86	836.80
3	2,2,4-Trimethyl-3-pentanol	881.49	894.00	847.18	861.53	885.53	899.84	835.29	853.68
4	4-Methyl-1-pentanol	821.19	836.97	809.72	823.22	822.74	844.74	804.57	811.72
5	2-Pentanone	666.34	687.79	671.52	690.74	673.76	678.33	677.08	703.33
	Isopropyl acetate	646.54	661.78	660.33	679.07	667.98	673.74	661.57	682.94
6	Propyl formate	605.79	623.60	618.75	632.11	615.20	626.31	625.90	634.54
7	Isobutyl formate	673.40	689.84	699.24	712.76	681.99	690.37	707.24	711.61
8	4-Ethyl-3-hexanol	953.25	967.63	944.88	962.88	947.61	963.90	944.37	947.38
9	Butyl formate	707.64	725.53	711.12	725.78	699.74	724.78	724.97	736.01
10	2,4-Dimethyl-2-pentanol	775.91	789.03	783.98	803.39	794.79	805.14	776.97	813.11
11	2-Hexanone	767.93	790.03	769.55	788.32	761.54	781.05	774.87	806.38
12	1-Heptanol	955.05	971.73	932.05	945.08	953.74	967.40	948.86	955.97
13	Isobutyl propionate	852.83	869.02	855.00	872.73	847.41	863.52	851.15	865.93
14	2-Ethyl hexanal	934.65	954.71	936.68	953.13	942.20	942.37	944.85	952.54
15	2,2,4,4-Tetramethyl-3-pentanone	900.00	914.09	873.03	882.66	903.85	909.81	859.99	882.02
16	3,3-Dimethyl-2-butanone	693.05	711.58	698.23	715.29	682.14	702.27	690.34	714.12
17	Butyl butyrate	979.36	997.07	974.40	991.25	971.71	987.65	972.98	988.62
18	2-Methyl-3-hexanone	819.95	838.42	830.93	849.58	821.44	832.20	832.23	853.74
19	2,2-Dimethyl-1-propanol	657.34	670.46	643.91	656.76	661.46	673.92	614.17	619.31
20	2-Methyl-2-heptanol	916.43	930.38	915.41	933.85	926.28	939.32	912.96	926.32
21	Methyl propionate	615.21	630.43	590.22	608.00	614.74	627.78	594.77	616.39
22	Methyl isobutyrate	670.97	686.58	671.39	690.34	666.38	682.36	665.89	686.90
23	3,6-Dimethyl-3-heptanol	986.60	1000.00	991.41	1008.20	973.88	995.18	977.26	988.84
24	3-Methyl-2-butanone	640.92	661.44	648.96	668.02	652.44	665.53	649.82	671.46
25	3-Methyl-1-butanol	719.03	734.39	719.66	734.47	711.24	732.08	703.23	711.51
26	2,2-Dimethyl-3-pentanol	805.63	818.97	792.82	809.97	796.39	806.42	776.29	803.69
27	2-Ethylbutyl acetate	956.99	974.66	976.12	991.47	970.80	987.58	979.65	969.91
28	Isobutyl isobutyrate	900.00	915.56	938.84	954.97	915.96	919.04	928.49	932.01
29	Methyl hexanoate	907.01	925.46	895.42	912.68	901.95	917.79	895.05	914.34
30	6-Methyl-2-heptanol	951.10	965.00	956.36	970.21	956.90	967.592	963.69	964.78
31	3-Heptanone	865.79	886.89	853.08	871.55	853.70	867.21	862.98	887.18
32	2-Methyl pentanal	742.38	762.95	782.17	796.35	748.25	768.33	784.17	793.87
33	3-Pentanone	676.41	700.00	663.17	683.25	668.70	674.72	670.08	698.03
34	Butyl isobutyrate	938.55	954.26	937.89	954.15	938.62	946.85	937.12	950.04
35	Ethyl hexanate	982.90	1000.00	982.29	998.68	973.80	989.91	980.57	994.26
36	2-Methyl-2-pentanol	717.57	731.39	717.23	739.35	695.81	696.02	711.67	741.92
37	Pentyl acetate1	896.36	914.88	884.62	901.68	901.11	912.56	890.22	910.48
38	2-Ethyl-1-butanol	825.94	841.00	818.16	834.20	814.70	831.34	806.85	828.43
39	Propyl butyrate	881.53	898.88	874.68	893.58	883.32	892.54	873.51	898.84
40	2-Octanone	968.77	991.27	966.20	980.92	970.00	983.02	982.88	997.12
41	2,4-Dimethyl-3-pentanone	779.01	795.28	789.16	806.77	779.97	788.80	787.96	819.42
42	Ethyl isovalerate	838.35	854.28	854.48	872.20	847.38	863.18	850.94	865.74
43									
44	Butyl acetate	796.18	814.16	785.22	803.81	796.99	823.92	787.28	813.77
45	Methyl butyrate	705.61	722.96	690.13	709.13	702.13	721.05	692.14	714.29
46	2-Methyl butanal	636.32	657.70	623.15	643.91	632.64	659.32	633.06	659.06
47	5-Methyl-3-heptanol	943.58	957.88	950.68	967.13	952.21	966.46	952.35	960.19
48	2-Pentanol	602.00	700.00	687.43	707.18	680.20	702.12	680.42	707.20
49	4-Heptanol	875.42	890.00	860.20	879.34	866.68	878.27	862.49	886.06
50	3-Methyl-2-butanol	666.02	680.26	665.02	684.60	656.48	687.14	653.13	674.30
51	3-Methyl-2-pentanone	734.76	754.92	752.44	771.85	745.13	755.42	754.10	786.74
52	4-Heptanone	853.35	873.44	849.97	868.78	850.96	861.75	859.57	884.48
53	2-Methyl-2-hexanol	817.33	831.38	820.98	841.95	831.84	838.71	816.22	843.30
54	2-Methyl-2-butanol	626.20	640.33	632.83	655.60	616.68	658.45	629.46	657.68
55	Ethyl butyrate	784.04	784.04	779.76	799.69	790.90	802.39	777.71	807.41
56	2-Ethyl-4-methyl-1-pentanol	972.00	972.00	988.87	999.82	967.29	979.45	996.57	978.56
57	3-Hexanone	764.84	764.84	757.80	777.76	757.62	773.78	763.51	797.35
58	2,2,4-Trimethyl-1-pentanol	930.00	930.00	929.19	937.52	928.37	942.34	925.75	909.71
59	Isobutyl acetate	757.65	757.65	757.87	775.33	757.92	771.30	755.04	769.79
60	1-Hexanol	852.96	852.96	871.65	887.15	856.71	869.69	854.25	870.01
61	3-Ethyl-3-pentanol	843.09	843.09	814.63	838.50	832.49	846.81	814.02	847.98
62	2,2-Dimethyl-pentanol	867.57	867.57	852.42	864.52	859.41	882.94	840.44	846.05
63	4-Methyl-2-pentanol	744.14	744.14	756.88	774.96	753.54	757.67	746.66	767.60
64	Methyl isovalerate	761.30	761.30	775.82	793.62	758.72	766.78	762.58	776.21
65	5-Methyl-3-hexanol	838.15	838.15	843.71	861.74	848.59	862.99	841.32	861.91
66	2,2-Dimethyl-3-heptanone	964.66	964.66	981.29	993.86	958.27	980.05	966.94	984.12
67	2,3-Dimethyl-2-pentanol	823.66	823.66	803.38	826.32	813.98	837.36	798.54	823.59
68	2-Methyl-1-pentanol	818.35	818.35	802.63	817.40	805.03	823.58	791.29	801.34
69	Isobutyl alcohol	611.31	626.00	722.87	745.99	674.91	682.21	710.77	742.57
70	2,4-Dimethyl-3-heptanol	821.18	821.18	809.41	827.53	811.85	821.81	794.67	826.13
71	1-Butanol	646.48	646.48	644.18	659.18	640.48	663.39	619.70	635.72
72	5-Methyl-2-hexanone	836.53	836.53	847.63	863.78	835.65	868.52	854.67	870.33
73	5-Methyl-3-hexanone	816.74	816.74	832.51	850.18	824.34	842.85	838.03	857.75
74	2-Heptanol	885.57	885.57	879.65	897.16	888.20	902.17	882.22	901.00
Test set									
1	Propyl acetate	693.34	713.63	683.42	702.30	617.44	637.53	689.44	711.89
2	Ethyl propionate	694.19	711.16	684.77	704.71	618.47	639.56	685.73	710.38
3	Butyl propionate	891.40	909.12	875.79	894.30	886.56	899.39	876.32	900.71
4	4-Methyl-2-pentanone	721.24	741.61	742.54	760.17	647.07	678.29	742.86	762.88
5	2-Heptanone	868.70	891.01	866.39	883.51	820.26	869.43	877.86	899.02
6	3-Pentanone	684.21	700.00	676.76	697.32	625.59	636.85	672.45	700.92
7	3-Methyl-1-pentanol	828.82	845.00	831.83	846.72	769.50	787.37	823.22	841.33
8	4-Octanol	975.50	990.22	956.58	973.54	960.82	979.67	962.16	980.55
9	Propyl propionate	792.58	809.79	764.89	784.53	713.60	724.93	771.67	802.09
10	1-Pentanol	750.40	766.59	736.52	752.01	656.55	668.02	726.71	744.17
11	Isobutyl butyrate	940.26	956.57	935.85	951.45	942.12	962.02	940.19	951.73
12	2-Methyl-3-pentanone 7	733.02	752.40	730.85	750.29	642.27	675.20	729.12	753.16
13	2,6-Dimethyl-4-heptanone	954.66	970.95	987.70	999.46	983.67	991.68	990.96	981.93
14	2,3-Dimethyl-2-butanol	715.26	729.44	714.70	737.64	635.23	672.76	706.62	734.66
15	3-Hexanol	780.36	795.07	776.55	797.04	711.05	732.12	769.10	803.06
16	3,5-Dimethyl-3-hexanol	883.13	896.48	881.05	900.30	890.45	910.45	880.08	893.57
17	3-Octanol	981.97	996.71	962.43	978.90	962.95	981.74	968.49	985.66

Table 1: Data set and corresponding experimental (exp.) and predicted (cal.) values of RI.

for the first position, where only functions are allowed). However, in the tail of a gene only terminal is allowed.

Mutation, Inversion, Transposition and Recombination

Mutation: Mutations can occur anywhere in the chromosome. Simple mutation just replaces symbols in genes with replacement symbols. However, the structural organization of chromosomes must remain intact. Symbols in the heads of genes can be replaced by functions or terminals (variables and constants). Symbols in the tail sections can be replaced only by terminals. Randomly change symbols in a chromosome. Symbols in the tail of a gene may not operate on any arguments. Typically two-point mutation per chromosome is used. It is worth noticing that in GEP there are no constraints neither in the kind of mutation nor the number of mutations in a chromosome: in all cases the newly created individuals are syntactically correct programs.

Inversion: Inversion reverses the order of symbols in a section of a gene. A portion of a chromosome is chosen to be inserted in the head of a gene. The tail of the gene is unchanged. Thus symbols are removed from the end of the head to make room for the inserted string. Typically a probability of 0.1 of insertion is used.

Transposition: Transposition selects a group of symbols and moves the symbols to a different position within the same gene. Gene transposition moves entire genes around in the chromosome. One gene in a chromosome is randomly chosen to be the first gene. All other genes in the chromosome are shifted downwards in the chromosome to make place for the first gene.

An IS element is a variable-size sequence of elements extracted from a random starting point within the genome (even if the genome was composed of several chromosomes). Another position within the genome is chosen as the insertion point.

This target site must be within the head part of a gene and cannot be the first element (gene root). The IS element is sequentially inserted in the target site, shifting all elements from this point onwards and a sequence with the same number of elements is deleted from the end of the head, so that the structural organization is maintained. This operator simulates the transposition found in the evolution of biological genomes. RIS is similar to the IS transposition, except that the insertion sequence must have a function as the first element and the target point must be also the first element of a gene (root).

The transposable elements of GEP are fragments of the genome that can be activated and jump to another place in the chromosome: (1) Short fragments with a function or terminal in the first position that transpose to the head of genes, except to the root (insertion sequence elements or IS elements); (2) Short fragments with a function in the first position that transpose to the root of genes (root IS elements or RIS elements); (3) Entire genes that transpose to the beginning of chromosomes.

Recombination: During recombination, two chromosomes are randomly selected, and genetic material is exchanged between them to produce two new chromosomes.

The cross over operation this can be one point (the chromosomes are split in two and corresponding sections are swapped), two point (chromosomes are split in three and the middle portion is swapped) or gene (one entire gene is swapped between chromosomes) recombination. Typically the sum of the probabilities of recombination is 0.7.

In GEP there are three kinds of recombination: one-point, two-point, and gene recombination. (1) One-point: During one-point recombination, the chromosomes crossover a randomly chosen point to form two daughter chromosomes; (2) Two-point: In two-point recombination the chromosomes are paired and the two points of recombination are randomly chosen. The material between the recombination points is afterwards exchanged between the two chromosomes, forming two new daughter chromosomes; (3) Gene recombination: recombines entire genes. This operator randomly chooses genes in the same position in two parent chromosomes to form two new off springs. In gene recombination an entire gene is exchanged during crossover. The exchanged genes are randomly chosen and occupy the same position in the parent chromosomes. It is worth noting that this operator is unable to create new genes: the individuals created are different arrangements of existing genes.

Fitness function: A fitness function is the most important part of any EA application. Fitness function given with above equations allows for fulfilling all of the set conditions. In GEP, fitness is based on how well an individual model the data. If the target variable has continuous values, the fitness can be based on the difference between predicted values and actual values. Evolution stops when the fitness of the best individual in the population reaches some limit that is specified for the analysis or when a specified number of generations have been created or a maximum execution time limit is reached.

All of the fitness functions produce fitness scores in the range 0.0 to 1.0 with 1.0 being ideal fitness – that is, the individual exactly fits the data. If a function is unviable – for example, it takes the square root of a negative number or divides by zero – then its fitness score is 0.0.

GEP evolution process: The GEP evolution begins with the random generation of linear fixed-length chromosomes for individuals of the initial population. The chromosomes are translated into ETs and subsequently into mathematical expressions, and the fitness of each individual is evaluated based on a pre-defined fitness function. The individuals are then selected by fitness to reproduce with modification. The individuals of this new generation are, in their run, subject to the same developmental process. The selection and reproduction is accomplished by roulette-wheel sampling with elitism, which guarantees the survival and cloning of the best individual to the next generation. Variation in the population is introduced by applying one or more genetic operators to selected chromosomes, including crossover, mutation and insertion.

Models

GEP model: The GEP program was coded by the combination of MATLAB and VC++. The MATLAB software has the advantage of computing matrix conveniently and programming efficiently, but its operating efficiency is relatively low. So VC++ was combined for its powerful function and the characteristics of higher operating efficiency with MATLAB. In this paper, MATLAB engine was used to achieve the combination with VC++ programming. There are two steps: (1) Add MATLAB engine library header files and library functions of the path. (2) Add libmx.lib libeng.lib libmex.lib to complete the import of the corresponding MATLAB engine static link library.

From the data in Table 1, GEP method was used that 6 topological index as input, output for its retention index. During the run, parameter values were needed to adjust constantly in order to achieve the optimal results. The set of optimal parameter values were listed in Table 2 and the predicting results of test set on OV-1 and SE-54 were listed in Figures 1 and 2. It can be seen from the figures that the predictive values of gas chromatography retention index of oxygen-organic compounds

Parameters	Values
Generation	2000
Population Size	100
Function Set	"+" "-" "*" "/" "sin" "cos" "sqrt" "exp" "ln"
Head Size	8
Number Of Genes	3
Linking Function	+
Mutation Rate	0.044
1-Point recombination rate	0.3
Gene recombination	0.3
Gene	0.1
IS transposition rate	0.1
RIS transposition rate	0.1
Gene transposition rate	0.1
Selection range	100

Table 2: Parameters of GEP models.

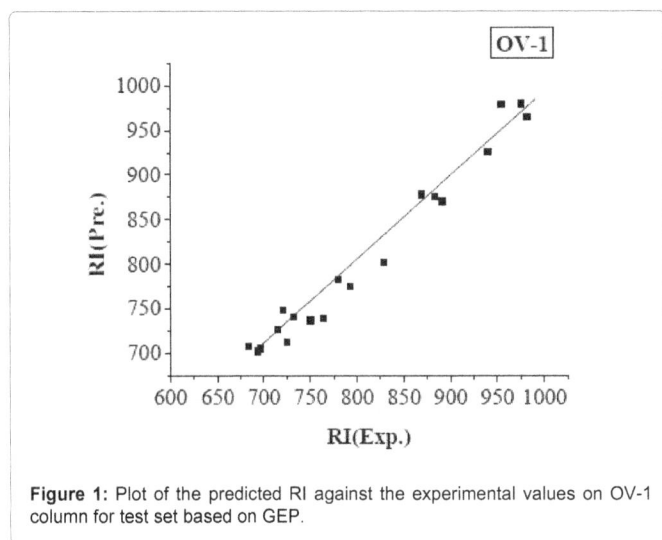

Figure 1: Plot of the predicted RI against the experimental values on OV-1 column for test set based on GEP.

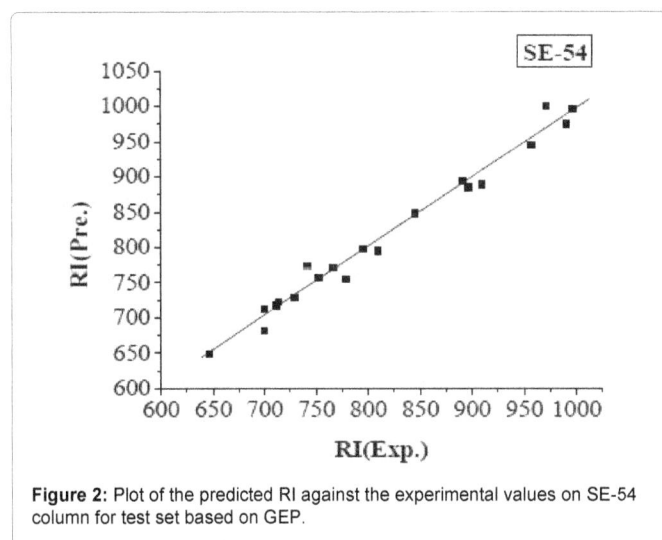

Figure 2: Plot of the predicted RI against the experimental values on SE-54 column for test set based on GEP.

were in good agreement with the experimental data.

ANN model: Non-linear statistical treatment of QSRR data is expected to provide models with better predictive quality as compared with related MLR models. In this perspective, functioning and

applications of ANN have been adequately described elsewhere [7-10]. Extensive use of ANN, which has inherent ability to incorporate nonlinear and cross-product terms into the model and does not require prior knowledge of the mathematical function as well, largely rests on its flexibility and less sensitivity to collinearity among variables. The theory behind ANN and their use in chromatography have been reported elsewhere [11-13].

Multi-layer feed forward networks, with good self-learing ability and adaptability is widely used in the field of QSRR modeling [14]. Commonly, they consist of three layers: one input layer formed by a number of neurons that equal to the number of descriptors, one out neuron (providing the model response) and a number of hidden neurons fully connected to both input and out neurons. Among the available learning algorithms, back-propagation of errors is one of the most widely used [8,15].

Usually, there are four steps involved in ANN modeling: (1) assembling the training data of input (independent variables) and output (dependent variables), (2) deciding the network architecture, (3) training the network, and (4) simulating the network response to new inputs. The training process is simply an optimization process which aims at finding the set of weight and biases associated with each layer that will minimize the error objective function related to the deviations of the network predictions from the true response output data of the training set.

Before data set was used for the training of ANN, it was normalized separately. Its minimum value was set to zero and maximum to one. The proper number of nodes in the hidden layer was determined by training the network with different number of nodes in the hidden layer. The root-mean-square error (RMSE) value measures how good the outputs are in comparison with the target values. In this paper, following a troubleshooting study to investigate the effects of the number of hidden layers and the number of neurons involved in these hidden layers, a 2-3-1 network, with tansig-logsig transfer functions, was found to be the most optimum in terms of the root mean squared errors (RMSE) obtained.

ANN with basic back-propagation of errors learning algorithm was used in this study to predict oxygen-containing retention index. A three-layer network with a sigmoid transfer function was designed for ANN. The ANN program was coded in MATLAB 7 for windows [15].

The MLR: For regression analysis, data set was randomly divided into two groups: training and test sets. The training set, composed of 74 molecules, was used for the model generation. The test set, composed of 17 molecules, was used to evaluate the generated model. The program used for MLR analysis was compiled in Statistical Product and Service Solutions (SPSS version 19.0 IBM) software. In MLR analysis, in order to minimize the information overlap in descriptors and to reduce the number of descriptors required in regression equation, the concept of non-redundant descriptors was used in this study. The best equation was selected on the basis of the highest multiple correlation coefficients (R) and the lowest root mean squared error (RMS). The linear equation between these descriptors and the retention parameters of fluid catalytic cracking (FCC) gasoline was:

$$y = b_0 + b_1 x_1 + b_2 x_2 + ... + b_p x_p \qquad (1)$$

Where b_0 is the intercept and b_j is the regression coefficient for descriptor j. The statistical results obtained by using the two molecular descriptors based on MLR are listed in Table 3 and plotted against the experimental values in Figures 3 and 4.

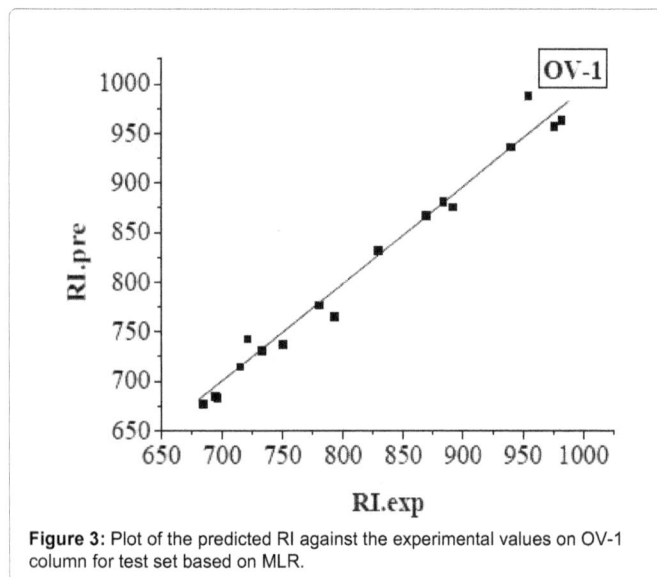

Figure 3: Plot of the predicted RI against the experimental values on OV-1 column for test set based on MLR.

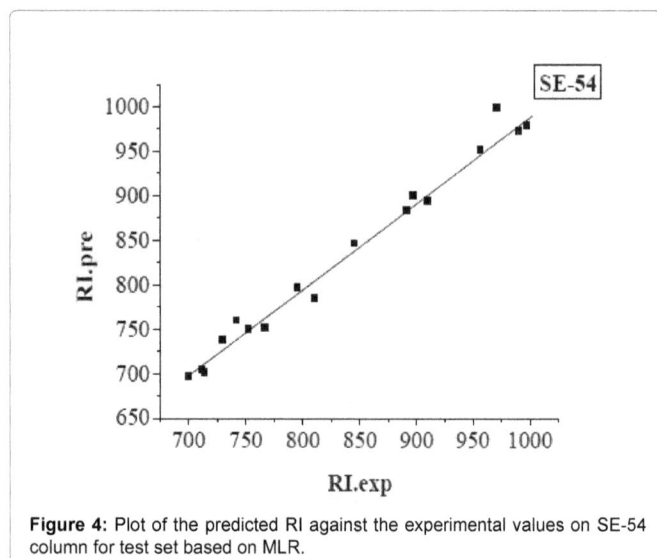

Figure 4: Plot of the predicted RI against the experimental values on SE-54 column for test set based on MLR.

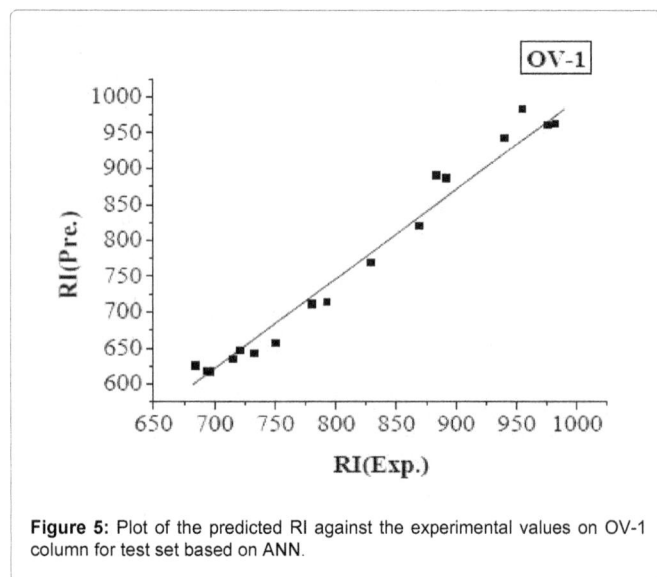

Figure 5: Plot of the predicted RI against the experimental values on OV-1 column for test set based on ANN.

It is common to consider four statistical parameters for regression equation. These parameters are the number of descriptors, correlation coefficient (R) for training and test sets, root mean squared error (RMS) for training and test sets, and F statistic. A reliable MLR model is one that has high R and F values, low RMS and least number of descriptors. In addition to these, the model should have a high predictive ability. Consequently, among different models, the best model was chosen, whose specifications are presented in Table 3. Here the corresponding descriptors used in MLR were applied as inputs for ANN in order to compare the performance of the two models.

Results

The main aim of the present work was developing a QSRR model to predict the retention parameter (RI) of oxygen-containing compounds appeared in Table 1. A linear model of MLR was developed, whose specifications are given in Table 3. All statistic tests were performed at a significance level of 5%. MLR model performance was measured by three metrics: (1) R, which gives the fraction explained variance for the analyzed set, was used to measure the model's fit performance. (2) Root Mean Squared error (RMS), which can give the bias in the prediction, was used to evaluate the model's predictive precision: the lower the RMS, the better the prediction precision. It can be calculated as below:

$$RMS = \sqrt{\frac{\sum_{i=1}^{n}(d_i - o_i)^2}{n}} \tag{3}$$

where d_i is the target value, o_i is the experimental value and n is the number of compounds in analyzed set. (3) The variance ratio of calculated and observed activities F.

After the linear model was gained, non-linear characteristics of the retention parameter were also performed using ANN. Here a feed-forward neural network with basic error back-propagation algorithm was constructed to model the nonlinear QSRR models. Therefore, a 2-3-1 BP-ANN, with tansig-logsig transfer functions, was developed. Figure 5 demonstrates the plot of the ANN predicted versus the experimental values of the RI for the data set. A correlation coefficient of this plot indicates the reliability of the model. As can been seen in Table 4, the correlation coefficient R on OV-1 and SE-54 for the ANN models are larger than that of MLR models respectively, which indicates that the ANN models are slightly improved to MLR models. The residuals of calculated values of RI by ANN are plotted against the experimental values in Figure 6. The propagation of the residuals on both sides of zero line indicates that no symmetric error exists in the development of ANN model.

Descriptors	OV-1 coefficient	Std. Error	SE-54 coefficient	Std. Error	OV-1 test values	SE-54 test value
Constant	1951.898	450.621	2028.14	345.24	4.332	5.875
OEI	48.089	4.351	50.562	3.333	11.053	15.169
SX1CH	-43.038	151.408	3.36	116	-0.284	0.029
N2/3	3.331	51.507	26.672	39.462	0.065	0.676
χeq × PEI	-459.326	162.605	-504.236	124.579	-2.825	-4.048
χeq	-173.948	12.183	-159.685	9.334	-14.278	-17.108
MPEIm × IMPEIm	5.589	1.039	5.694	0.796	5.382	7.156

Table 3: Model parameters value and coefficients for MLR model.

OV-1	Test sets	SE-54	Test sets
Method	R	Method	R
MLR	0.9911	MLR	0.9917
ANN	0.9891	ANN	0.9892
GEP	0.9909	GEP	0.9955

Table 4: Result of correlation coefficient (R) with MLR, ANN and GEP for the test set.

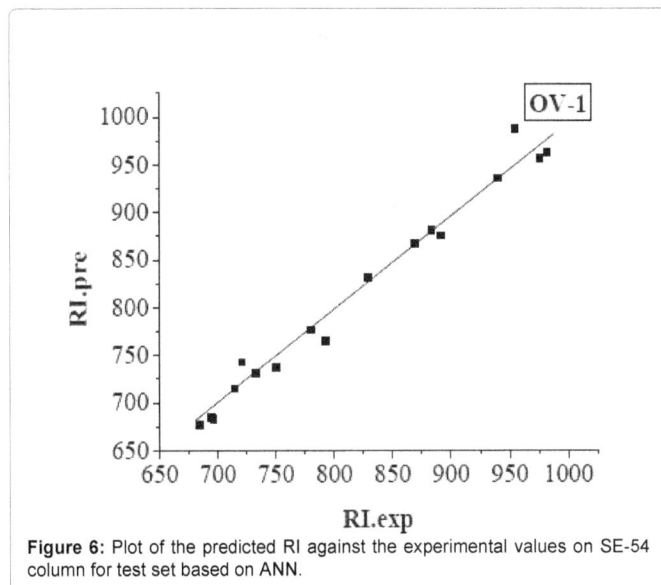

Figure 6: Plot of the predicted RI against the experimental values on SE-54 column for test set based on ANN.

Acknowledgements

Correspondence author is grateful to the financial support from National Natural Science Foundation of China, Lanzhou University of China and Liaoning Province Education Department of China, which made this work possible.

References

1. Lu C, Abrahan F, Adamowiciz L (2007) QSRR study for gas and liquid chromatographic retention indices of polyhalogenated biphenyls using two 2D decriptors. Chromatographia 66: 717-724.

2. Daghir-Wojtkowiak E, Studzi'nska S, Buszewski B, Kaliszanc R, JanMarkuszewski M (2014) Quantitative structure–retention relationships of ionic liquid cations in characterization of stationary phases for HPLC. Royal Society of Chemistry 6: 1189-1196.

3. Ghasemi J, Saaidpour S, Brown SD (2007) QSPR study for estimation of acidity constants of some aromatic acids derivatives using multiple linear regression (MLR) analysis. J Mol Struct Theochem 805: 27-32.

4. Liu F, Liang Y, Cao C, Zhou N (2007) Theoretical prediction of the Kovat's retention index for oxygen-containing organic compounds using novel topological indices. Anal Chim Acta 594: 279-289.

5. Ferreira C (2006) Gene Expression Programming: Mathematical Modelling by an Artificial Intelligence, Angra do Heroismo, Portugal.

6. Zhang XM, Lu PC (1996) Unified equation between Kováts indices on different stationary phases for select types of compounds. J Chromatogr A 731: 187-199.

7. Noorizaden H, Farmany A (2010) QSRR models to predict rentention indices of cyclic compounds of essential oils. J Chemometrics Laboratory 72: 563-569.

8. Gorynski K, Bojko B, Nowaczyk A, Bucinski A, Pawliszyn J, et al. (2013) Quantitative structure-retention relationships models for prediction of high performance liquid chromatography retention time of small molecules: Endogenous metabolites and banned compounds. Analytical Chimica Acta 797: 13-19.

9. Mior R, Morés S, Welz B, Carasek E, de Andrade JB (2013) Determination of sulfur in coal using direct solid sampling and high-resolution continuum source molecular absorption spectrometry of the CS molecule in a graphite furnace. Talanta 106: 368-374.

10. Gupta VK, Khani H, Ahmadi-Roudi B, Mirakhorli S, Fereyduni E, et al. (2011) Prediction of capillary gas chromatographic retention times of fatty acid methyl esters in human blood using MLR, PLS and back-propagation artificial neural networks. Talanta 83: 1014-1022.

11. D'Archivio AA, Incani A, Ruggieri F (2011) Retention modeling of polychlorinated biphenyls in comprehensive two-dimensional gas chromatography. Anal Bioanal Chem 399: 903-913.

12. Fatemi MH (2002) Simultaneous modeling of the Kovats retention indices on OV-1 and SE-54 stationary phases using artificial neural networks. J Chromatogr A 955: 273-280.

13. Zhang X, Zhang X, Li Q, Sun Z, Song L, et al. (2014) Support Vector Machine Applied to Studyon Quantitative Structure–Retention Relationships of Polybrominated Diphenyl Ether Congeners. Chromatographia 77: 1387-1398.

14. Xiaotong Z, Xingming L (2004) GBP network application in the modified paraffin wax quality prediction. Journal of Petrochemical Universities 21: 1-5.

15. MATLAB, The Language of Technical Computing.

29

Phytochemical Characterization of Active Constituents from Extracts of *Ixora Javanica* D.C Flowers

Sunitha Dontha*, Hemalatha Kamurthy, Bhagavan Raju Mantripragada

Malla Reddy College of Pharmacy, Hyderabad, Telangana, India

Abstract

The investigation of chemical compounds from natural products is fundamentally important for the development of new drugs, especially in view of the vast worldwide flora. The present work is extraction, isolation and characterization of active constituents from *Ixora javanica* D.C flowers. Freshly collected flowers were shade dried, powdered and subjected to continuous soxhilation with various solvents like petroleum ether (60-80°), ethyl acetate and ethanol (70%) of increasing polarity. All the three obtained extracts were concentrated and after preliminary phytochemical investigation, column chromatographed using silica gel.

Three triterpenoidal compounds (Maslinic acid, Ursolic acid and Lupeol), one flavonoidal compound (Formononetin) and two flavonoidal glycoside compounds (Quercetin-3-rutinoside, Quercetin-3-glucoside) have been isolated and further these isolated structures were established by spectral analysis (IR, ^{1}H NMR spectra, ^{13}C NMR and Mass) and direct comparison with authentic samples. This is the first report of occurrence of these compounds from *Ixora javanica* D.C flowers.

Kewords: *Ixora javanica* D.C flowers; Maslinic Acid; Ursolic Acid; Formononetin; Quercetin-3-rutinoside; Quercetin-3-glucoside; Lupeol

Introduction

Natural products once served humankind as the source of all drugs, and higher plants provided most of these therapeutic agents. Today, natural products and their derivatives still represent over 50% of all drugs in clinical use, with higher plant-derived natural products representing 25% of the total [1]. Traditional medicine has served as a source of alternative medicine, new pharmaceuticals, and healthcare products. Medicinal plants are important for pharmacological research and drug development, not only when plant constituents are used directly as therapeutic agents, but also as starting materials for the synthesis of drugs or as models for pharmacologically active compounds [2]. The World Health Organization estimates that 80% of the people in developing countries of the world rely on traditional medicine for their primary health care, and about 85% of traditional medicine involved the use of plant extracts. This means that about 3.5 to 4 billion people in the world rely on plants as sources of drugs [3].

Ixora is a genus of flowering plants in the Rubiaceae family. It consists of tropical evergreen trees and shrubs and holds around 500 species with its centre of diversity in Tropical Asia. Most of the species are grown as ornamental plants. Phytochemical studies of some species like *Ixora coccinea* [4], *I. finalaysonia* [5,6], *I. arborea* [7], etc., indicated the presence of important phytochemicals [8-11] such as lupeol, ursolic acid, oleanolic acid, sitosterol, rutin, lecocyanadin, anthocyanins, proanthocyanidins, glycosides of kaempferol and quercetin. Hence an effort was made to investigate the chemical constituents from another species of Ixora i.e., *Ixora javanica* (*I. j*).

From the literature survey, it was revealed that no substantial work was carried out on *I. javanica* D.C flowers (Figure 1) both in the chemical investigation and pharmacological activities. Hence an effort was made to investigate the chemical constituents of the different extracts of *I. javanica* flowers. Some current drugs are not beneficial in all cases, due to their side effects and potency. Hence search for other alternatives seems necessary and beneficial. This study was aimed to phytochemical investigation of the pure compounds to be used for treating various ailments (like anti-oxidant, hepatoprotective, anti-cancer activities, etc.).

Experimental Procedure

Materials and methods

Solvents: Petroleum ether (60-80°), ethyl acetate, ethanol (70%), chloroform, methanol, ethanol (70%), benzene, etc.

Chemical and reagents: 1. Laboratory grade (L. R) chemicals were used for isolation. 2. Analytical grade (A. R) reagents were used for analytical work.

Absorbents: 1. Silica Gel (ACME Chemical works, Mumbai) was used for TLC. 2. Silica Gel of mesh size 60-120 (Merck, Bombay) was used for column chromatography.

Equipments: All the Melting points were recorded in a Toshiwal electrically heated melting point apparatus and were uncorrected. I.R spectra of the compounds were recorded using Thermo Nicolet Nexus 670 I.R spectrophotometer. ^{1}H NMR spectra were taken on Varian EM-360 (300 MHZ) NMR spectrometer using CDCl$_3$ as solvent. ^{13}C NMR was recorded on Bruker instrument with CDCl$_3$ as solvent at 300 MHz. Mass spectra were recorded on a GC-MS, data on E:/ISO/21184-1.QGD.

General experimental procedures

Collection and authentication *I. javanica* flowers: *I. javanica* D.C flowers were collected from local areas of Hyderabad and authenticated by Dr. B. Bhadraiah, HOD, Department of Botany, Osmania University, Hyderabad. Voucher specimens ((IJ/2012/23) are kept at Malla Reddy College of Pharmacy, Dhulapally, Hyderabad, Telangana, India.

*Corresponding author: Dontha Sunitha, Associate Professor, Malla Reddy College of Pharmacy, Hyderabad, Telangana-500 014, India
E-mail: basasunitha@gmail.com

Figure 1: *Ixora javanica* D.C flowers and leaves.

Preparation of different extracts from flowers of *I. javanica* D.C: The shade dried flowers of *I. javanica* were reduced to fine powder and around 5 kg were subjected to continuous successive extractions with different solvents of increasing polarity like petroleum ether (60-80°), ethyl acetate and ethanol (70%) into 15 batches of each 250-300 gm each in an ethane extractor. These extracts were filtered and air dried. The air dried extracts were weighed and percentage yields of extraction were calculated. The colour and consistency of the extracts were recorded. The organic extracts were dried in vacuum desiccators and the solvents were evaporated in vacuum and extracts were stored in desiccators.

After drying the respective extracts were weighed and percentage yield were recorded (Table 1).

Preliminary phytochemical tests: The preliminary phytochemical studies were performed for testing the different chemical groups present in different extracts. The chemical group tests were performed using different reagents for the identification of different groups. Petroleum ether (60-80°) extract indicated the presence of steroids and triterpenoids whereas ethyl acetate extract gave positive results for the presence of flavonoids, carbohydrates and tri terpenoids and ethanolic extract (70%) showed the presence of steroids, triterpenoids, flavonoids, alkaloids, ethane, carbohydrates, tannins and glycosides (Table 2).

Isolation and characterization of various compounds

Isolation of phytoconstituents from petroleum ether extract: 150 g of silica gel for column chromatography was activated in hot air oven at 110°C for one hour. The petroleum ether was used to build the silica gel in the glass column. The glass wool was fixed at the bottom. The activated silica gel was charged into column in small proportion with gentle tapping after each addition in order to ensure uniform packing. The small quantity of solvent (petroleum ether 60-80°) system was allowed to remain on the top of the column. The air bubbles present in the column were removed by gentle tapping to get a uniform bed of adsorbent.

The concentrated petroleum ether extract (35 g) of *I. javanica* was dissolved in a small quantity of the n-hexane and adsorbed on silica gel and then charged into column. The chromatogram was allowed to develop overnight, taking care to prevent the drying of the column by plugging the open end with adsorbent cotton. The elution of solvent system was started after complete saturation of the column. Column being successively eluted with increasing polarities of n-hexane, n-hexane: ethyl acetate, ethyl acetate. The elution was carried out with n-hexane:ethyl acetate in graded mixture i.e., 98: 2, 96: 4, 94: 6, 92: 8,.........up to 50:50 and finally carried out with 100% ethyl acetate. From above elution's, two different fractions were collected (i.e., fractions A and B).

Fraction A was eluted from n-hexane: ethyl acetate (98:2) resulted in a single compound which was confirmed by TLC (n-hexane:ethyl acetate, 8.5:1.5). This elute was collected and concentrated to about 10 ml crude residue substance. It was recrystallized with acetone. The product was designated as compound I-1

Fraction B was eluted from n-hexane:ethyl acetate (84:16), gave a single compound which was confirmed by TLC (n-hexane:ethyl acetate, 8:2). This elute collected was concentrated to crude residue and further recrystalized with acetone. The product was designated as **compound I-2.**

Isolation of phytoconstituents from ethyl acetate extract: The concentrated ethyl acetate extract (33 g) of *I. javanica* was dissolved in a small quantity of the chloroform (20 ml) and adsorbed on silica gel and then charged into column (same as 3.3.1 section). The elution were carried out with chloroform: methanol in graded mixture i.e., 95:05, 90:10, 85:15, and 80:20 ...up to 100% methanol. From above elution's three different fractions were collected (fraction C, D and E).

Fraction C was eluted from chloroform: methanol (90:10) and it gave a single spot by using TLC plate in chloroform: methanol (9:1) as mobile phase and its resolution factor was found to be 0.82, the compound was designated as **compound I-3.**

Fraction D was eluted from chloroform: methanol (80:20) and it gave a single spot by using TLC plate in toluene: methanol (4:1) as mobile phase and its resolution factor was found to be 0.61. The product was designated as **compound I-4.**

Fraction E was eluted from chloroform: methanol (80:20) and it gave a single spot by using TLC plate in chloroform: methanol (9:1) as mobile phase and its resolution factor was found to be 0.55. The product was designated as **compound I-5.**

Isolation of phytoconstituents from ethanolic extract: The concentrated ethanolic extract (30 g) was dissolved in a small quantity of the same solvent (chloroform, 20 ml) and adsorbed on silica gel and then charged into column. The column being successively eluted with chloroform: methanol in graded mixture i.e., 95:05, 90:10, 85:15, 80:20,.........up to 100% methanol. From above elution's, only one fraction was collected (i.e., fraction F). The concentration of other elutes gave only brown resinous masses which were not processed further.

Fraction F was eluted from chloroform: methanol (90:10) and it gave single spot by using TLC plate in toluene: methanol (4:1) as mobile phase and its resolution factor was found to be 0.81. The product was designated as **compound I-6.**

S No	Extracts	Nature of extract	Colour	Yield (%)
1.	Petroleum ether extract	Fine powder	Yellow orange	8.9
2.	Ethyl acetate extract	Sticky mass	Light yellow	6.7
3.	Ethanol (70%) extract	Sticky mass	Dark green	12.3

Table 1: Percentage yield of different extracts of flowers of *Ixora javanica* D.C.

Extracts	Chemical constituents
Petroleum ether (60-80°) extract	Steroids, triterpenoids
Ethyl acetate extract	Flavonoids, carbohydrates and tri terpenoids.
Ethanolic extract (70%)	Steroids, triterpenoids, flavonoids, alkaloids, saponins, carbohydrates, tannins and glycosides.

Table 2: Results of preliminary phytochemical analysis of different extracts of *I. javanica* flowers.

Results and Discussion

Extraction and preliminary qualitative analysis

The shade dried of flowers of *I. javanica* was subjected to successive solvent extraction method by using petroleum ether, ethyl acetate and ethanol (70%). All these extracts were concentrated and calculated for their percentage yield. The yields were found to be 8.9%, 6.7% and 12.3% respectively.

The results of phytochemical investigation of petroleum ether extract shown the presence of sterols, triterpenoids, whereas, ethyl acetate extract shown the presence of flavonoids, triterpenoids and carbohydrates and finally ethanolic extract (70%) indicated the presence of steroids, triterpenoids, flavonoids, alkaloids and glycosides.

Phytochemical characterization of isolated compounds from different extracts of *I. Javanica* flowers

From petroleum ether extract: The two different fractions were obtained from petroleum ether extract.

From fraction A

A yellow orange fine powder was resulted as a single compound from the fraction A. It was designated as **compound I-1,** which was further recrystalized from acetone, with M.P 162-166°C range. IR (KBr) showed absorption band at 3446.24 cm^{-1} of –OH group, a band at 1700.00 cm^{-1} indicated C=O stretching, a peak at 1460.77 cm^{-1} indicated –CH$_2$ bending, a peak at 1376.21 cm^{-1} indicated CH$_3$ bending and a peak at 1169.43 cm^{-1} indicated C-O stretching.

^{1}H NMR spectral data, one vinylic portion peak at 5.4 δ ppm, methylene proton peaks appeared as multiplet in between 2.4 to 1.2 δ ppm. Free four -CH$_3$ protons observed as multiplet at 0.99 δ ppm.

^{13}C NMR spectral data exhibited the presence of 30 carbon signals of both aliphatic and cyclic carbons are in their respective ppm. Spectral data matched exactly with that of maslinic acid (Table 3).

Further the structure was confirmed as **"maslinic acid"** (Figure 2) with its molecular ion peak 472 [M$^+$] (18%) and confirmed with authentic sample.

From fraction B

Whereas fraction B obtained a single triterpenoidal compound, the elution was with n-hexane: ethyl acetate (84:16) which gave a single compound and was confirmed by TLC (n-hexane: ethyl acetate, 8:2). It was designated as **compound I-2,** and was further crystallized from acetone, M.P 284-286°C.

IR (KBr) showed absorption bands at 3430 cm^{-1} which indicated the presence of –OH stretching, 2920.59 cm^{-1} and 2850.90 cm^{-1} defined both CH$_2$ and CH$_3$ stretchings, 1742.76 cm^{-1} indicated C=C stretching, 1462.02 cm^{-1} defined –CH$_3$ deformation and a peak at 1163.98 cm^{-1} indicated C-O stretching of secondary alcohol.

In ^{1}H NMR spectral data, one vinylic proton peak was observed at 5.6 δ ppm, 25 methylene proton peaks appeared as multiplet in between 2.75 to 1.0 δ ppm. Free four CH$_3$ observed as multiplet at 0.88 δ ppm, methene group of proton peaks appeared as multiplet in between 1.6 δ ppm, proton of carboxylic group was appeared as singlet at 10.3 δ ppm.

^{13}C NMR spectral data exhibited the presence of 30 carbon signals of both aliphatic and cyclic carbons in their respective ppm. The ^{13}C NMR spectral data matched with that of ursolic acid (Table 3). Its

identity as **"ursolic acid"** (Figure 3) was further confirmed by mass spectra with its molecular ion peak at 457 [M+H]$^+$ (54%).

From ethyl acetate extract: Three different fractions (C, D, E) were obtained from ethyl acetate extract.

From fraction C

All the collected elutes were combined and subjected for concentration, which was designated as **compound I-3.** This was further recrystallized from methanol, resulted as yellow crystalline needles with M.P 241-243°C.

The IR spectra of compound I-3 showed absorption band for -OH stretching at 3399.14 cm^{-1}, aromatic CH stretching observed at 2925.03 cm^{-1}, C=O stretching absorption peak at 1732.11 cm^{-1}, C=C stretching indicated the absorption peak at 1515.43 and 1459.37 cm^{-1} and finally C-O stretching at 1169 and 1074.54 cm^{-1}.

In ^{1}H NMR, at 5.42 δ ppm indicated the doublet of four –OH groups, 4.0 to 3.1 δ ppm denoted the –OH groups of sugar moiety.

^{13}C NMR spectral data exhibited the presence of 21 carbon signals of both aromatic and aliphatic carbons in their respective ppm. ^{13}C NMR spectral data matched exactly with that of quercetin-3-glucoside (Table 4). Further the structure was confirmed as **"quercetin-3-glucoside"** (Figure 4) with its molecular ion peak 464 [M+H]$^+$ (20%) and confirmed with authentic sample.

From fraction D

Elutes with chloroform: methanol (80:20) showed the presence of a single compound, which was confirmed by TLC (Toluene: Methanol, 4:1). All the collected elutes were combined and subjected for concentration, which was designated as **compound I-4.**

The IR spectra of compound **I-4** showed absorption band for OH-stretching at 3389.15 cm^{-1}, aromatic C-H stretchings were observed at 2925.03 cm^{-1}, C=O stretching absorption at 1742.20 cm^{-1}, C=C stretching indicated the absorption at 1459.37 and 1374.59 cm^{-1} and finally C-O stretching at 1239.59 cm^{-1}, 1169.00 cm^{-1} and 1074.54 cm^{-1}.

In ^{1}H NMR, the singlet of aromatic protons was observed at 7.25,

Carbon No.	Compound		
	I.1	I.2	I.6
1	48.2	37.7	22.1
2	76.5	27.1	31.8
3	80.5	74.4	76.5
4	38.3	39.5	44.3
5	55.3	56.5	49.3
6	18.1	19.9	23.9
7	34	33.2	19.9
8	39.9	39.9	62.1
9	50.3	45.7	37.9
10	37.8	37.1	50.5
11	23.7	22.8	41.1
12	129.7	121.9	27.7
13	150.8	140.8	50.9
14	42.9	42.2	38.5
15	29.1	29.1	41.9
16	24.9	25.1	32
17	47.9	50.1	51.9
18	42.8	51.1	52.1

Table 3: ^{13}C-NMR Spectral data of isolated compounds.

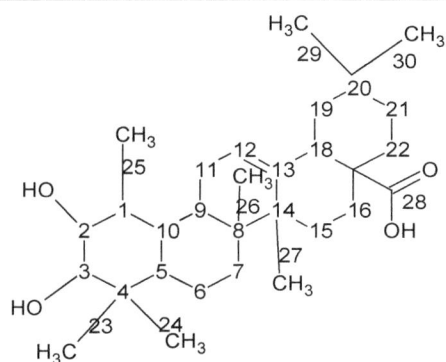

Figure 2: Maslinic acid.

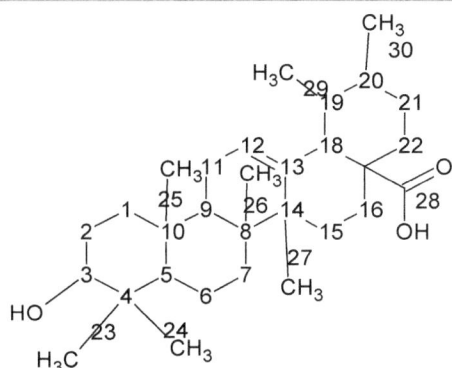

Figure 3: Ursolic acid.

Table 4: ^{13}C-NMR Spectral data of isolated compounds.

Carbon No.	Compound I.3	Compound I.4	Compound I.5
	C-O	C-O	C-O
1	155.2	153.4	149.5
2	136.1	130.1	119.3
3	179.9	174.5	179.5
4	160.7	161.8	132.2
5	99.5	98.3	115.3
6	173.2	166.5	167.7
7	105.2	94	100.5
8	156.1	156.8	160.3
9	105.3	102.5	121.2
10	124.5	122.8	127.5
1'	142.7	112.3	127.1
2'	143.2	45.9	95.1
3'	126.2	144.5	160.2
4'	125.7	116.2	32.2
5'	125.3	121.8	34.2
6'	-	107.5	-
1"	-	70.4	-
2"	-	70.1	-

6.9 and 6.25 δ ppm. Aromatic –OH groups was observed at 3.52 δ ppm. And ^{1}H of CH group in rhamnoglycosyl was observed at 3.48 to 3.97 δ ppm.

^{13}C NMR spectral data exhibited the presence of 27 carbon signals of both aromatic and aliphatic carbons in their respective ppm and the spectral data matched with exactly with that of Quercetin-3-rutinoside

(Table 4). Further the structure was confirmed as "**Quercetin-3-rutinoside**" (Figure 5) with its molecular ion peak 610 [M+H] + (41%).

From fraction E

A light brown coloured compound was obtained by recrystalization with ethanol, M.P 291-293°C. Spectral characterization of IR (KBr) indicated the presence of phenolic –OH absorption at 3405.27 cm^{-1}, 1441 and 1373 cm^{-1} indicated both CH$_2$ and CH$_3$ bending, C=C stretching absorption at 1741.23 cm^{-1}, absorption at 1164.18 and 1050 cm^{-1} indicated C-O stretching of secondary alcohol, and absorption at 759 and722 cm^{-1} indicated C-H deformation.

In ^{1}H NMR spectral data, a peak of –OCH$_3$ group appeared as a singlet at 3.63 δ ppm, aromatic –OH group was observed at 5.4 δ ppm, 7H aromatic protons of ring A and B appeared as multiplets at 7.9 to 6.0 δ ppm.

^{13}C NMR spectral data exhibited the presence of 16 carbon signals of both aromatic and aliphatic carbons in their respective ppm. ^{13}C NMR spectral data matched exactly with that of formononetin (Table 4). Hence identified as "**formononetin**" (Figure 6) and was further confirmed by mass spectra with its molecular ion peak at 268 [M+H]$^+$ (16%) and also by Co. chromatography with authentic sample.

From ethanolic extract

A single fraction (F) was obtained from ethanolic extract.

From fraction F

Elutes with chloroform: methanol (90:10) showed the presence of single compound, which was confirmed by TLC (toluene: methanol 4:1). All the collected elutes were combined and subjected for concentration, which was designated as **compound I-6**. This was further recrystalized from methanol, resulted as light brown colour crystalline needles with M.P 218-220°C.

The IR spectra of compound I-6 showed absorption band for –OH stretching at 3421.4 cm^{-1}, aromatic –CH stretching observed at 2919.58 and 2848.80 cm^{-1}, C=C stretching absorption at 1714.15 cm^{-1}, C-O stretching of 2°alcohol indicates the absorption at 1062.98 cm^{-1}.

In ^{1}H NMR, doublet of protons of –C=CH, H-29 1a, 1b were observed at 5.35, 4.83 δ ppm, multiplet of –OH group observed at 3.75 δ ppm and multiplets of 6 –CH$_3$ of 18H were observed at 0.99 δ ppm.

^{13}C NMR spectral data exhibited the presence of 30 carbon signals of both aromatic and aliphatic carbons in their respective ppm. Spectral

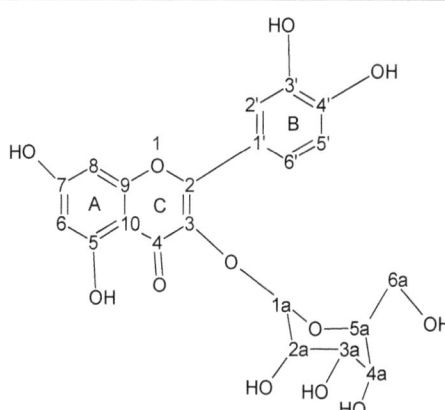

Figure 4: Quercetin-3-glucoside.

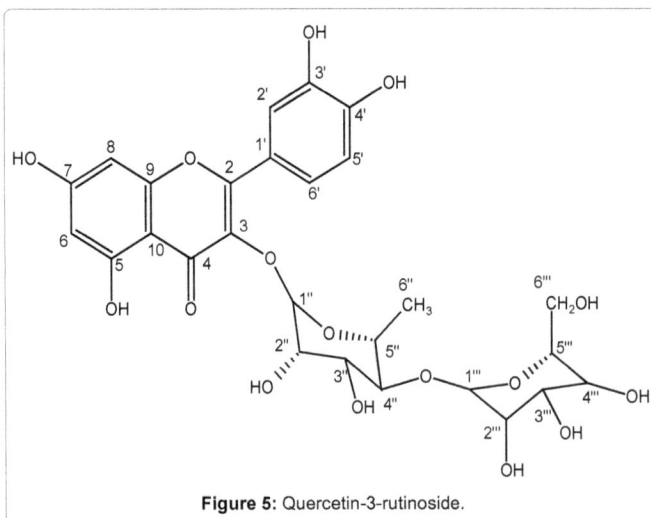

Figure 5: Quercetin-3-rutinoside.

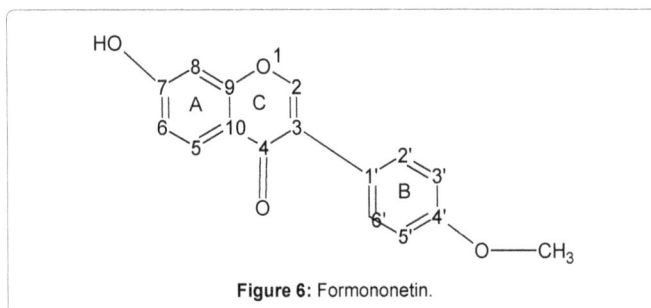

Figure 6: Formononetin.

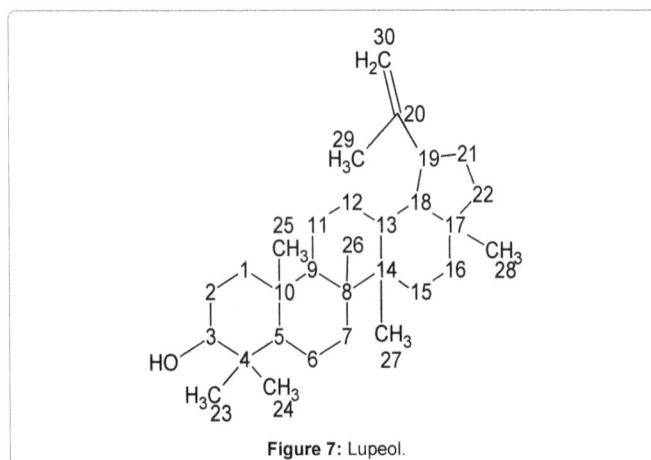

Figure 7: Lupeol.

as a better alternative to chemical based pharmaceuticals. Further these compounds have to be screened for different activities based on literature available and were to be used for treating various ailments.

Acknowledgements

The authors acknowledge Dr. B. Bhadraiah, HOD, Department of Botany, Osmania University, Hyderabad, for authentication of the plant. The authors express their sincere thanks to M. Sudhakar, Principal, Malla Reddy College of Pharmacy, for providing necessary facilities for phytochemical investigation. We are thankful to IICT (Indian Institute of Chemical Technology) for providing the analytical data.

References

1. Balandrin NF, Kinghorn AD, Farnsworth NR (1993) Human Medicinal Agents from Plants Kinghorn, ACS Symposium Series 2-12.

2. Mukherjee PK (2003) GMP for Indian Systems of Medicine, Business Horizons, New Delhi. pp: 99-112.

3. Verma S, Singh SP (2008) Current and future status of herbal medicines. Veterinary World 1: 347-350.

4. Elumalai A, Chinna E, Yetcharla V, Burle SK, Chava N (2012) Phytochemical and pharmacological profile of Ixora coccinea Linn. International Journal of Pharmacy and Life Sciences 3: 1563-1567.

5. Faten Darwish MM, Zedan Ibraheim Z (2003) Phytochemical Study of *Ixora finlaysoniana* Wall. Ex. G. Don growing in Egypt. Bull Pharm Sci 26: 91-96.

6. Metcalfe CR, Chalk L (1970) Anatomy of Dicotyledones. The Clarendon Press, Oxford. pp: 1500-1950.

7. Aktar F, Kaisar A, Hamidul Kabir ANM, Hasan CM, Rashid MA (2009) Phytochemical and biological investigations of Ixora arborea Roxb. Dhaka Univ J Pharm Sci 8: 161-166.

8. Baliga MS, Kurian PJ (2012) Ixora coccinea Linn.: traditional uses, phytochemistry and pharmacology. Chin J Integr Med 18: 72-79.

9. Gloria UO, Gibson UN (2011) Chemical Composition of Essential Oil of Ixora coccinea flower from Port Harcourt, Nigeria. Int J of Academic Research 3: 381-384.

10. Ikram A, Versiani MA, Shamshad S, Ahmed SK, Ali ST, et al. (2013) Ixorene, a new Dammarane triterpene from the leaves of Ixora coccinea Linn. Rec Nat Prod 7: 302-306.

11. Ragasa CY, Tiu F, Rideout JA (2004) New cycloartenol esters from Ixora coccinea. Nat Prod Res 18: 319-323.

data was matched exactly with that of authentic sample lupeol (Table 3). Further the structure was confirmed as **"lupeol"** (Figure 7) with its molecular ion peak 426 [M⁺] (55%).

Conclusions

Maslinic acid and ursolic acid were isolated from petroleum ether extract by column chromatography using n-hexane: ethyl acetate as mobile phase. Quercetin-3-glucoside, quercetin-3-rutinoside and formononetin were eluted from ethyl acetate extract by column chromatography using chloroform: methanol as mobile phase. Lupeol was isolated from ethanol extract by using chloroform: methanol as mobile phase.

Based on the results, the obtained compounds of *I. javanica* D.C flowers are effective pharmaceutical compounds which will serve

Permissions

List of Contributors

Singh A, Tandon S and Sand NK
Department of Chemistry (Division of Agricultural Chemicals), College of Basic Sciences and Humanities, G. B. Pant University of Agriculture & Technology, Pantnagar 263145, U.S. Nagar, Uttarakhand, India

Valentina Bongiovanni and Andrea Cavallero and Daniela Talarico
Kerry Ingredients and Flavors, Via Capitani di Mozzo 12/16, 24030, Mozzo, BG, Italy

Maria Laura Colombo
Department of Science and Pharmaceutical Technology, University of Turin, Via P. Jury 9, 10125 Turin, Italy

Samadrita Saha
Calcutta Institute of Pharmaceutical Technology and Allied Health Science, Howrah, West Bengal, India

Nasim Sepay
Department of Chemistry, Presidency University, West Bengal, India

Pranjal Das, Sarmisttha Kundu, Milan Kumar Maiti and Goutam Mukhopadhyay
BCDA College of Pharmacy and Technology, Kolkata, West Bengal, India

Chitte RR and Nagare SL
Biotechnology, Vidya Pratishthan's School of Biotechnology, Vidyanagari, Baramati, Maharashtra, India
Agriculture Biotechnology, Vidya Pratishthan's School of Biotechnology, Vidyanagari, Baramati, Maharashtra, India

Date PK and Shinde BP
Agriculture Biotechnology, Vidya Pratishthan's School of Biotechnology, Vidyanagari, Baramati, Maharashtra, India

Mark Bokhart
College of Natural Science, Chemistry Department, Michigan State University, East Lansing, MI, USA
Diagnostic Center for Population and Animal Health, Toxicology Section, Michigan State University, Lansing, MI, USA
Department of Pathobiology and Diagnostic Investigation, College of Veterinary Medicine, Michigan State University, East Lansing, MI, USA

Andreas Lehner and Margaret Johnson
Diagnostic Center for Population and Animal Health, Toxicology Section, Michigan State University, Lansing, MI, USA

John Buchweitz
Diagnostic Center for Population and Animal Health, Toxicology Section, Michigan State University, Lansing, MI, USA
Department of Pathobiology and Diagnostic Investigation, College of Veterinary Medicine, Michigan State University, East Lansing, MI, USA

Sharma DK, Anil Kumar and Mahender
Department of Chemistry, Himachal Pradesh University, Shimla, Himachal Pradesh, India

Binsalom A and Campbell K
Institute for Global Food Security, School of Biological Sciences, Queen's University, UK

Chianella I
Cranfield Health, Cranfield University, Cranfield, Bedfordshire, MK43 0AL, UK

Zourob M
Department of Chemistry, Alfaisal University, Riyadh, KSA

Atul S Rathore, Sathiyanarayanan L and Kakasaheb R Mahadik
Centre for Advanced Research in Pharmaceutical Sciences, Poona College of Pharmacy, Bharati Vidyapeeth Deemed University, Pune, Maharashtra, India

Ali A, Uddin J and Musharraf SG
HEJ Research Institute of Chemistry, International Center for Chemical and Biological Sciences, University of Karachi, Karachi, Pakistan

Ansari HN and Firdous S
Department of Chemistry, University of Karachi, Karachi, Pakistan

Rajamanickam V, Herwig C and Spadiut O
Research Division Biochemical Engineering, Institute of Chemical Engineering, Vienna University of Technology, Vienna, Austria

Christian Doppler Laboratory for Mechanistic and Physiological Methods for Improved Bioprocesses, Institute of Chemical Engineering, Vienna University of Technology, Vienna, Austria

Winkler M, Meyer L and Flotz P
Christian Doppler Laboratory for Mechanistic and Physiological Methods for Improved Bioprocesses, Institute of Chemical Engineering, Vienna University of Technology, Vienna, Austria

Ramesh Kumar K, Sneha Xiavour, Swarna Latha, Vijay Kumar and Sukumaran
Bhavans Vivekananda College, Department of Biochemistry, Osmania University, Hyderabad, India

Pei-Shan Wu, Yu-Ting Kuo and Bih-Show Lou
Chemistry Division, Center for General Education, Chang Gung University, 259 Wen-Hwa 1st Road, Kwei-Shan, Tao-Yuan 333, Taiwan, ROC

Shen-Ming Chen and Ying Li
Department of Chemical Engineering and Biotechnology, National Taipei University of Technology, No. 1, Section 3, Chung-Hsiao East Road, Taipei 106,Taiwan, ROC

M Luísa S Silva
Centre of Chemical Research, Autonomous University of Hidalgo State, Carr. Pachuca-Tulancingo km 4.5, 42184 Pachuca, Hidalgo, México

Atul S Rathore, Sathiyanarayanan L and Kakasaheb R Mahadik
Centre for Advanced Research in Pharmaceutical Sciences, Poona College of Pharmacy, Bharati Vidyapeeth (Deemed University), Pune, Maharashtra, India

Arti Gupta and Sonia Pandey
Maliba Pharmacy College, UkaTarsadia University, Bardoli, Gujarat, India

Navin R Sheth
Department of Pharmacognosy, Saurashtra University, Rajkot, Gujarat, India

Jitendra Singh Yadav
Vidyabharati Trust College of Pharmacy, Umrakh, Gujarat, India

Sebastiano Arnoldi, Gabriella Roda, Eleonora Casagni, Lucia Dell'Acqua, Michele Dei Cas, Fiorenza Fare, Chiara Rusconi, Giacomo Luca Visconti and Veniero Gambaro
Department of Pharmaceutical Sciences, University of Milan, Via Mangiagalli 25, 20133, Milan, Italy

Reddithota J Krupdam, Darshana Gour and Govind Patel
National Environmental Engineering Research Institute, Nehru Marg, Nagpur 440020, India

Arpan Chakraborty
Department of Pharmaceutical Technology, Jadavpur University, Kolkata, West Bengal, India

Arka Bhattacharjee
Department of Pharmaceutical Sciences and Technology, Birla Institute of Technology, Mesra, Ranchi, India

Pallab Dasgupta and Goutam Mukhopadhyay
BCDA College of Pharmacy and Technology, Kolkata, West Bengal, India

Debasmita Manna
NSHM College of Pharmaceutical Sciences, Kolkata, West Bengal, India

Won Chun OH
Department of Material Science and Engineering, Hanseo University, South Korea

Nor Azreen Mohd Jamil, Noraswati Mohd Nor Rashid and Norasfaliza Rahmad
Agro-Biotechnology Institute, Malaysia, National Institute of Biotechnology, C/O MARDI Headquarters, 43400 Serdang, Selangor, Malaysia

Attarde DL and Bhambar RS
Department of Pharmacognosy, Mahatma Gandhi Vidyamandir's Pharmacy College, Panchavati, Nashik, Maharashtra, India

Pal SC
Department of Pharmacognosy, RG Sapakal College of pharmacy, Kalyani Hills, Trimbakeshwar, Nashik, Maharashtra, India

Alfredo Aires
Centre for the Research and Technology for Agro-Environment and Biological Sciences, CITAB, Universidade de Trás-os-Montes e Alto Douro, UTAD, Quinta de Prados, 5000-801 Vila Real, Portugal

Rosa Carvalho
Agronomy Department, Universidade de Trás-os-Montes e Alto Douro, UTAD, Quinta de Prados, 5000-801 Vila Real, Portugal

Marco Malferrari and Francesco Francia
Laboratorio di Biochimica e Biofisica, Dipartimento di Farmacia e Biotecnologie, FaBiT, Università di Bologna, 40126 Bologna, Italy

Husham NM Hussan
Pesticides Residue Analysis Laboratory, Crop Protection Research Centre, Agricultural Research Cooperation, Wad Medani, Sudan

Xingang Liu, Fengshou Dong, Jun Xu and Yongquan Zheng
State Key Laboratory for Biology of Plant Diseases and Insect Pests, Institute of Plant Protection, Chinese Academy of Agricultural Sciences, Beijing 100193, P.R. China

Martina Bava, Sebastiano Arnoldi, Lucia Dell'Acqua, Giuseppe Mustich, Gabriella Roda, Chiara Rusconi, Giacomo Luca Visconti and Veniero Gambaro
Dipartimento di Scienze Farmaceutiche, Università degli Studi di Milano, Via Mangiagalli 25, 20133 Milano, Italy

Sergio Fontana and Flavia La Forgia
Farmalabor Srl, Via Pozzillo ZI, 76012 Canosa di Puglia, Italy

Giovanni Sorrenti
SIAN ASL BT, piazza Umberto I n°9, 76012 Canosa di Puglia (BT), Italy

Gineys M, Kirner T, Cohaut N, Béguin F and Delpeux-Ouldriane S
Centre National de la Recherche Scientifique, ICMN, Orleans cedex 2, France

Sercan Pravadali-Cekic, Danijela Kocic and Andrew Shalliker R
Australian Centre for Research on Separation Sciences (ACROSS), Tasmania, Australia
School of Science and Health, University of Western Sydney (Parramatta), NSW, Australia

Paul Stevenson
Australian Centre for Research on Separation Sciences (ACROSS), Tasmania, Australia
School of Life and Environmental Sciences, Deakin University (Waurn Ponds), VIC, Australia

El-Shahawi MS, Bashammakh AS, Alwael H, Al-Sibaai AA and Al-Ariqe HK
Department of Chemistry, Faculty of Science, King Abdulaziz University, P.O. Box 80203, Jeddah 21589, Saudi Arabia

Sun T
College of Sciences, Northeastern University, 110004 Shenyang, Liaoning, People's Republic of China

Zhang X
College of Sciences, Northeastern University, 110004 Shenyang, Liaoning, People's Republic of China
Liaoning Key Laboratory of Petrochemical Engineering, Liaoning Shihua University, 113001 Fushun, Liaoning, People's Republic of China

Shi L, Ding L, Sun Z and Song L
Liaoning Key Laboratory of Petrochemical Engineering, Liaoning Shihua University, 113001 Fushun, Liaoning, People's Republic of China

Qu H
Research Institute of PetroChina, Fushun Petrochemical Company, 113004 Fushun, Liaoning, China

Sunitha Dontha, Hemalatha Kamurthy and Bhagavan Raju Mantripragada
Malla Reddy College of Pharmacy, Hyderabad, Telangana, India

Index